W9-AGF-517

CE 406

second edition

Applied
Numerical
Analysis

second edition

Applied Numerical Analysis

CURTIS F. GERALD

California Polytechnic State University
San Luis Obispo

Addison-Wesley Publishing Company

Reading, Massachusetts
Menlo Park, California · London
Amsterdam · Don Mills, Ontario
Sydney

Copyright © 1978, 1970, by Addison-Wesley Publishing Company, Inc. Philippines copyright 1978, 1970, by Addison-Wesley Publishing Company, Inc.

All rights reserved. No part of this publication may be reproduced, stored in a retrieval system, or transmitted, in any form or by any means, electronic, mechanical, photocopying, recording, or otherwise, without the prior written permission of the publisher. Printed in the United States of America. Published simultaneously in Canada. Library of Congress Catalog Card No. 77–79469.

ISBN 0-201-02696-1
DEFGHIJKL-MA-79

Preface

This second edition of *Applied Numerical Analysis* continues the emphasis that the first edition placed on applications. The range of topics, much broader than in most elementary texts, has been retained, so the student is introduced to a wide selection of numerical procedures, and several newer methods have been added. The level of the material is still directed to the sophomore and junior student in engineering, science, mathematics, and computer science. An informal tone and dependence more on demonstrations than on rigorous proofs makes it accessible to a student who has had only the usual calculus background and an introduction to ordinary differential equations. This also makes the book a valuable guide and reference to the practicing engineer or applied mathematician.

The book has been strengthened in many ways. The already extensive problem sets have been expanded, and each chapter now includes applied problems and projects, in addition to standard exercises for practice in using the methods. These applied problems are placed in a separate section at the end of each chapter, while the exercises are keyed to the chapter sections, in order to ease the instructor's burden in making assignments. Answers to a few selected exercises are included in the text; in the exercise sections for each chapter we have identified such exercises by a small pointer (triangle) in the margin.

Significant new material has been added and many of the chapters have been extensively rewritten to improve the clarity of exposition and to provide additional illustrative examples. The order of chapters has been modified; most significantly, the chapter on solving linear systems has been moved up to become Chapter Two. At the same time, topics from linear algebra are more extensively covered so that the treatment is now relatively independent of any prior exposure in this area. The discussion of eigenvalues and eigenvectors is greatly expanded; however, for motivational reasons it continues to be placed in a later chapter with boundary-value problems, and so it is closer to the chapters on partial-differential equations, where the concepts are used in discussing stability. Other topics have been regrouped to improve the logical flow of ideas. Back reference to earlier topics frequently occurs when interrelated subjects are considered.

The strong interdependence between computers and numerical analysis is recognized by completely rewritten computer programs. FORTRAN is still the language of choice. The programs are grouped at the conclusion of each chapter for ease of reference and also because the student should normally gain some insight and understanding of the methods by personally performing the computations before he calls on the computer to grind out numbers for him. Most of the computer programs are in the form of subroutines, which the instructor may wish to incorporate in the FORTRAN library for ready access by student-written programs. The programs are written so as to be readily understandable by the student, and some concessions of efficiency are made in the interests of clarity. In the first chapter, an example of the program-development process is presented in order to help the students learn how they should write programs. An informal structured language is used to describe many of the algorithms. Since FORTRAN is not a structured language, the programs may serve as examples of how one makes the transition from algorithm to computer program.

A section on errors, with emphasis on computer arithmetic and the round-off problem, is included. Rather than place this separately as an introduction, it is incorporated within Chapter One at a time when the student is ready to appreciate the importance of the topic. Throughout the book the subject of errors is continually stressed in connection with the effectiveness and efficiency of the methods. Alternative methods are contrasted and compared from the standpoint of the computational effort required for a desired accuracy of result.

There is more than enough material for a full year's course but, by a judicious selection of topics, the book can serve shorter courses. Because the book is readable by itself, many students will find it a helpful reference to topics and procedures during their professional careers, even though they have not specifically covered the material in a class.

I wish to thank several colleagues who have suggested improvements in the book: Paul W. Davis (Worcester Polytechnic Institute), Richard Franke (Naval Postgraduate School), Stanley Preiser (Polytechnic Institute of New York), and Stanley L. Spiegel (University of Lowell).

San Luis Obispo   C. F. G.
November 1977

Contents

one

Solution of Nonlinear Equations

two

Solving Sets of Equations

three

Interpolating Polynomials

four

Numerical Differentiation and Numerical Integration

five

Numerical Solution of Ordinary Differential Equations

six

Boundary-Value Problems and Characteristic-Value Problems

seven

Numerical Solution of Elliptic Partial-Differential Equations

eight

Parabolic Partial-Differential Equations

nine

Hyperbolic Partial-Differential Equations

ten

Curve-Fitting, Splines, and Approximation of Functions

Bibliography and References

Appendixes

Answers to Selected Exercises

Index

Chapter One

Solution
of Nonlinear Equations

1.1 THE LADDER IN THE MINE

It is not uncommon, in applied mathematics, to have to solve a nonlinear equation. If you worked for a mining company the following might be a typical problem:

> *There are two intersecting mine shafts that meet at an angle of 123°, as shown in Fig. 1.1. The straight shaft has a width of 7 feet, while the entrance shaft is 9 feet wide. What is the longest ladder that can negotiate the turn? You can neglect the thickness of the ladder members, and assume it is not tipped as it is maneuvered around the corner. Your solution should provide for the general case in which the angle, A, is a variable, as well as the widths of the shafts.*

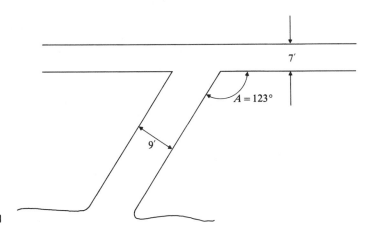

Figure 1.1

As the analysis below shows, to solve this problem we must solve a transcendental equation for the value of C:

$$\frac{9 \cos (\pi - 123° - C)}{\sin^2 (\pi - 123° - C)} - \frac{7 \cos C}{\sin^2 C} = 0,$$

and then substitute C into

$$\ell = \frac{9}{\sin (\pi - 123° - C)} + \frac{7}{\sin C}.$$

Finding the solution to an algebraic or transcendental equation, as we must do here, is the topic of this first chapter.

Here is one way to analyze our ladder problem. Visualize the ladder in successive locations as we carry it around the corner; there will be a critical position in which the two ends of the ladder touch the walls while a point along the ladder touches the corner where the two shafts intersect. (See Fig. 1.2.) Let C be the angle between the ladder and the wall when in this critical position.

Consider a series of lines drawn in this critical position—their lengths vary with the angle C, and the following relations hold (angles are expressed in radian measure):

$$\ell_1 = \frac{w_2}{\sin B}; \qquad \ell_2 = \frac{w_1}{\sin C};$$

$$B = \pi - A - C;$$

$$\ell = \ell_1 + \ell_2 = \frac{w_2}{\sin (\pi - A - C)} + \frac{w_1}{\sin C}.$$

The maximum length of ladder that can negotiate the turn is the minimum of ℓ as a function of angle C. We hence set $d\ell/dC = 0$.

$$\frac{d\ell}{dC} = \frac{w_2 \cos (\pi - A - C)}{\sin^2 (\pi - A - C)} - \frac{w_1 \cos C}{\sin^2 C} = 0.$$

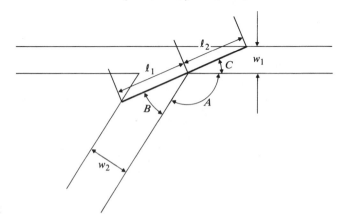

Figure 1.2

We can solve the problem if we can find the value of C that satisfies this equation. With the critical angle determined, the ladder length is given by:

$$\ell = \frac{w_2}{\sin{(\pi - A - C)}} + \frac{w_1}{\sin{C}}.$$

In this chapter we study methods to find the roots of an equation such as in our ladder-in-the-mine example. Much of algebra is devoted to the "solution of equations." In simple situations, this consists of a rearrangement to exhibit the value of the unknown variable as a simple arithmetic combination of the constants of the equation. For second-degree polynomials, this can be expressed by the familiar *quadratic formula*. For third- and fourth-degree polynomials, formulas exist but are so complex as to be rarely used; for higher-degree equations it has been proved that finding the solution through a formula is impossible. Most transcendental equations (involving trigonometric or exponential functions) are likewise intractable.

Even though it is difficult if not impossible to exhibit the solution of such equations in explicit form, numerical analysis provides a means where a solution may be found, or at least approximated as closely as desired. Many of these numerical procedures follow a scheme that may be thought of as providing a series of successive approximations, each more precise than the previous one, so that enough repetitions of the procedure eventually give an approximation which differs from the true value by less than some arbitrary error tolerance. Numerical procedures are thus seen to resemble the *limit concept* of mathematical analysis.

1.2 METHOD OF HALVING THE INTERVAL

The first numerical procedure that we shall study is that of *interval-halving.*[*] Consider the cubic

$$f(x) = x^3 + x^2 - 3x - 3 = 0.$$

At $x = 1$, f has the value -4. At $x = 2$, f has the value $+3$. Since the function is continuous, it is obvious that the change in sign of the function between $x = 1$ and $x = 2$ guarantees at least one root on the interval $(1, 2)$. See Fig. 1.3.

Suppose we now evaluate the function at $x = 1.5$ and compare the result to the function values at $x = 1$ and $x = 2$. Since the function changes sign between $x = 1.5$ and $x = 2$, a root lies between these values. We can obviously continue this interval-halving to determine a smaller and smaller interval within which a root must lie. For this example, continuing the process leads eventually to an approximation to the root at $x = \sqrt{3} = 1.7320508075\ldots$. The process is illustrated in Fig. 1.4.

---

[*] The method, also known as the *Bolzano method*, is of ancient origin. Some authors call it the *bisection method.*

Figure 1.3

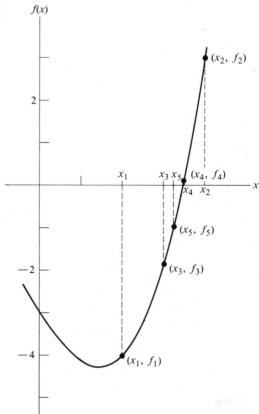

Figure 1.4

While a graphic method, as illustrated in Fig. 1.4, may be suitable if we want only an approximate answer, to obtain more accuracy we need to write a rule to do it mathematically. We should also express our algorithm (the technical name for a systematic procedure) in a way that makes it easy to implement the method with a computer program. We shall adopt a style of expressing algorithms that emphasizes the orderly structure.

Method of halving the interval (Bisection method)

To determine a root of $f(x) = 0$, accurate within a specified tolerance value, given values of x_1 and x_2 such that $f(x_1)$ and $f(x_2)$ are of opposite sign,

DO   WHILE $\frac{1}{2}|x_1 - x_2| \geq$ tolerance value,
      Set $x_3 = (x_1 + x_2)/2$.
      IF $f(x_3)$ of opposite sign to $f(x_1)$:
            Set $x_2 = x_3$.
      ELSE Set $x_1 = x_3$.
      ENDIF.
ENDDO.

The final value of x_3 approximates the root.

Note. The method may give a false root if $f(x)$ is discontinuous on $[x_1, x_2]$.

Applying the method to $f(x) = x^3 + x^2 - 3x - 3 = 0$, we get the results of Table 1.1. The repetition of our algorithm is called *iteration* and the successive approximations are termed the *iterates*.

The entries in Table 1.1 indicate the necessity of representing values of the argument, x, as well as of the function, $f(x)$, only approximately when we carry a limited number of decimal figures. In floating-point operations on digital computers, there is a similar inaccuracy in our work because computers retain only a limited number of significant digits. Note that this is true in all computations, not just in numerical methods. We shall give attention to such "round-off errors" later. The distinction between *numerical methods* and *numerical analysis* is that the latter term implies the consideration of errors in the procedure used. Certainly the blind use of any calculation method without concern for its accuracy is foolish.

Whether one rounds to the nearest fractional value or chops off the extra digits will make a difference in the effect of the round-off errors. In Table 1.1, the figures have been chopped after five places, which is similar to the action of most digital computers.

In addition to the limitation on accuracy because we retain only a limited number of figures in our work, there is an obvious limitation if we terminate the procedure itself too soon. One important advantage of the interval-halving

Table 1.1 Method of halving the interval for $f(x) = x^3 + x^2 - 3x - 3 = 0$.

Iteration number	x_1	x_2	x_3	$f(x_1)$	$f(x_2)$	$f(x_3)$	Maximum error in x_3
1	1	2	1.5	-4.0	3.0	-1.875	0.5
2	1.5	2	1.75	-1.875	3.0	0.17187	0.25
3	1.5	1.75	1.625	-1.875	0.17187....	-0.94335....	0.125
4	1.625	1.75	1.6875	-0.94335....	0.17187	—0.40942	0.0625
5	1.6875	1.75	1.71875	-0.40942....	0.17187	-0.12478	0.03125
6	1.71875	1.75	1.73437....	-0.12478....	0.17187....	+0.02198	0.015625*
7	1.71875	1.73437....	1.72656....		0.17187....		0.0078125
∞			1.73205....			-0.00000....	

* Actual error in x_3 after 5 iterations is 0.01330.

Table 1.2 Halving the interval for $f(x) = e^x - 3x = 0$.

Iteration number	x_1	x_2	x_3	$f(x_1)$	$f(x_2)$	$f(x_3)$	Maximum error in x_3
1	1.0	2.0	1.5	−0.028172	1.38906	−0.01831	0.5
2	1.5	2.0	1.75	−0.01831	1.38906	0.50460	0.25
3	1.5	1.75	1.625	−0.01831	0.50460	0.20342	0.125
4	1.5	1.625	1.5625	−0.01831	0.20342	0.08323	0.0625
5	1.5	1.5625	1.53125	−0.01831	0.08323	0.03020	0.03125
6	1.5	1.53125	1.51562	−0.01831	0.03020	0.00539	0.015625*
.	.	.	.				
.	.	.	.				
.	.	.	.				
∞			1.51213 . . .				

* Actual error in x_3 after 5 iterations is −0.01912.

method, beyond its simplicity, is our knowledge of the accuracy of the current approximation to the root. Since a root must lie between the x-values where the function changes sign,† the error in the last approximation can be no more than one-half the last interval of which it is the midpoint, and this interval is known exactly since the original difference, $|x_1 - x_2|$, is halved at each iteration. For other methods, the accuracy determination is much more difficult.

The accuracy of a computed value is usually expressed either as the *absolute error* (true value − approximate value) or as the *relative error* (absolute error divided by true value). The relative error is often the better measure of accuracy for very large or very small values. Sometimes the accuracy is expressed as the number of digits that are correct; in other cases, the number of correct digits after the decimal point is used. When the true value is unknown, it is impossible to express the accuracy with exactness, and approximate accuracy must be specified. Frequently we will put bounds on the size of the error.

The method of halving the interval applies equally well to transcendental equations, as do the other methods of this chapter. Table 1.2 shows the results when we apply the method to $f(x) = e^x - 3x = 0$, which has a root between $x = 1$ and $x = 2$.

The method of interval halving requires that starting values be obtained before the method can begin. This is true of most methods for root finding. Getting these starting values can be done by making a rough graph, by trial calculations, or by writing a search program on a computer or programmable

† Observe that if the function is discontinuous, $f(x)$ may change sign without having a root in the interval. Unknown functions should be examined for continuity before attempting to evaluate their roots.

calculator. Perhaps the best way is through interactive graphics, letting the computer draw curves at the direction of the user, and varying the parameters at the console to find approximate values of roots.

1.3 METHOD OF LINEAR INTERPOLATION

While the interval-halving method is easy and has simple error analysis, it is not very efficient. For most functions, we can improve the rate at which we converge to the root. One such method is the method interpolation.* Suppose we assume that the function is linear over the interval (x_1, x_2), where $f(x_1)$ and $f(x_2)$ are of opposite sign. From the obvious similar triangles in Fig. 1.5 we can write†

$$\frac{x_2 - x_3}{x_2 - x_1} = \frac{f(x_2)}{f(x_2) - f(x_1)},$$

$$x_3 = x_2 - \frac{f(x_2)}{f(x_2) - f(x_1)}(x_2 - x_1).$$

Figure 1.5

---

* This is also known as the method of false position, and by the Latinized version *regula falsi*. It is also a very old method.

† Note that, since $[f(x_2) - f(x_1)]/(x_2 - x_1)$ is the slope of the secant line, which approximates the slope of the function in the neighborhood of the root, the equation can be considered to be $x_3 = x_2 - f(x_2)/(\text{slope of function})$. Compare to Newton's method, in the next section.

Table 1.3   Method of linear interpolation for $f(x) = x^3 + x^2 - 3x - 3 = 0$.

Iteration number	x_1	x_2	x_3	$f(x_1)$	$f(x_2)$	$f(x_3)$
1	1.0	2.0	1.57142	−4.0	3.0	−1.36449
2	1.57142	2.0	1.70540	−1.36449	3.0	−0.24784
3	1.70540	2.0	1.72788	−0.24784	3.0	−0.03936
4	1.72788	2.0	1.73140	−0.03936	3.0	−0.00615
5	1.73140	2.0	1.73194*			

* Error in x_3 after 5 iterations is 0.00011.

We then compute $f(x_3)$ and again interpolate linearly between the values at which the function changes sign, giving a new value for x_3. Repetition of this will give improving estimates of the root. Table 1.3 shows the results of this method for the same polynomial discussed in Section 1.2. The method appears to be somewhat faster than the method of halving the interval, giving about the same accuracy after three steps as was obtained there in seven. It is intuitively obvious that the speed with which the successive approximations approach the zero of the function will depend on the degree to which the function departs from a straight line in the interval of consideration. In other words, the rate of convergence will be related to the rate of change of the slope of the curve, which is measured by the magnitude of the second derivative.

An algorithmic statement of this method is shown below.

Method of linear interpolation (*regula falsi* method)

To determine a root of $f(x) = 0$, given values of x_1 and x_2 such that $f(x_1)$ and $f(x_2)$ are of opposite sign,

> DO   WHILE $|x_2 - x_1| \geqslant$ tolerance value 1, or
> $|f(x_3)| \geqslant$ tolerance value 2,
> Set $x_3 = x_2 - f(x_2) \dfrac{x_2 - x_1}{f(x_2) - f(x_1)}$.
> IF $f(x_3)$ of opposite sign to $f(x_1)$:
> Set $x_2 = x_3$.
> ELSE   Set $x_1 = x_3$.
> ENDIF.
> ENDDO.

Note. The method may give a false root if $f(x)$ is discontinuous on $[x_1, x_2]$.

Table 1.3 discloses a serious fault of the interpolation method: The approach to the root is one-sided. If $f(x)$ has significant curvature between x_1 and x_2, this can be most damaging to the speed with which we approach the root, as shown in Fig. 1.6.

A remedy for this is the Modified Linear Interpolation method. We replace the value of $f(x)$ at the stagnant end position with $f(x)/2$.* This helps, as Fig. 1.7 shows.

An algorithm for this modification to the method of linear interpolation is shown below.

Modified linear interpolation method

To determine a root of $f(x) = 0$, given values of x_1 and x_2 such that $f(x_1)$ and $f(x_2)$ are of opposite sign,

> Set SAVE = $f(x_1)$; set F1 = $f(x_1)$; set F2 = $f(x_2)$.
> DO WHILE $|x_1 - x_2| \geq$ tolerance value 1, or
> $\qquad\qquad |f(x_3)| \geq$ tolerance value 2,
>
> $\qquad$ Set $x_3 = x_2 - \text{F2}\,\dfrac{x_2 - x_1}{\text{F2} - \text{F1}}$.
>
> $\qquad$ IF $\qquad f(x_3)$ of opposite sign to F1:
> $\qquad\qquad$ Set $x_2 = x_3$.
> $\qquad\qquad$ Set F2 = $f(x_3)$.
> $\qquad\qquad$ IF $\qquad f(x_3)$ of same sign as SAVE:
> $\qquad\qquad\qquad$ Set F1 = F1/2.
> $\qquad\qquad$ ENDIF.
> $\qquad$ ELSE $\:$ Set $x_1 = x_3$.
> $\qquad\qquad$ Set F1 = $f(x_3)$.
> $\qquad\qquad$ IF $\qquad f(x_3)$ of same sign as SAVE:
> $\qquad\qquad\qquad$ Set F2 = F2/2.
> $\qquad\qquad$ ENDIF.
> $\qquad$ ENDIF.
> $\qquad$ Set SAVE = $f(x_3)$.
> ENDDO.

Table 1.4 shows the modified method applied to our example polynomial problem. While the modified method normally converges faster, in this example there is no difference. The changing values of x_1 and x_2 do better confine the root, however.

---

* This halving of the ordinate at the other end of the interval is omitted when we step "beyond" the root, as occurs on the third iteration in Fig. 1.7.

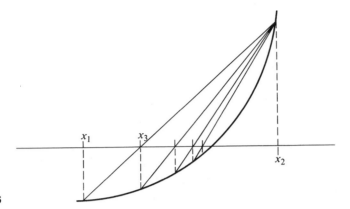

Figure 1.6

There is one other way that we can improve the method of linear interpolation. Instead of requiring that the function have opposite signs at the two values used for interpolation, we can choose the two values nearest the root (as indicated by the magnitude of the function at the various points) and interpolate or extrapolate from these. Usually the nearest values to the root will be the last two values calculated. This makes the interval under consideration shorter and hence improves the assumption that the function can be represented by the line through the two points.

Table 1.5 shows the calculations according to this scheme, which is known as the *secant method.** The example illustrates a more rapid convergence: x_6 is more accurate than was x_7 by linear interpolation.

We leave the development of the algorithm for the secant method as an exercise for the student.

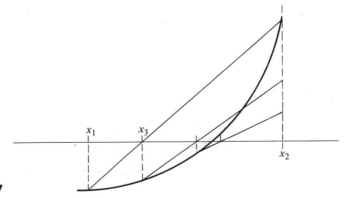

Figure 1.7

---

* So called because the line through two points on a curve is the secant line.

Table 1.4 Modified linear interpolation for $f(x) = x^3 + x^2 - 3x - 3 = 0$.

Iteration number	x_1	x_2	x_3	F1	F2	$f(x_3)$	SAVE
1	1.0	2.0	1.57142	-4.0	3.0	-1.3449	-4.0
2	1.57142	2.0	1.77557	-1.36449	1.5*	0.42369	-1.36449
3	1.57142	1.77557	1.72720	-1.36449	0.42369	-0.04576	0.42369
4	1.72720	1.77557	1.73191	-0.04576	0.42369	-1.332×10^{-3}	-0.04576
5	1.73191	1.77557	1.732183†	-1.332×10^{-3}	0.21184*		

* These function values are old F2/2.
† Error in x_3 after 5 iterations is -0.00013.

Table 1.5 Secant method for $f(x) = x^3 + x^2 - 3x - 3 = 0$.

Iteration number	x_1	x_2	x_3	$f(x_1)$	$f(x_2)$	$f(x_3)$
1	1.0	2.0	1.57142	−4.0	3.0	−1.36449
2	2.0	1.57142	1.70540	3.0	−1.36449	−0.24784
3	1.57142	1.70540	1.73513	−1.36449	−0.24784	0.02920
4	1.70540	1.73513	1.73199	−0.24784	0.02920	−0.0005755
5	1.73513	1.73199	1.73205*			

* Error in x_3 after 5 iterations is $<10^{-6}$.

It is important not to extrapolate to a root from two points whose functional values are of the same sign when knowledge is lacking that a real root is nearby. Figure 1.8 illustrates the futility of searching for a root that is not there. This is especially important in a computer program, since the successive calculated values are usually not apparent as soon as they are computed, as they are in a hand computation. In addition, it will be observed that the secant method can lead to a division by zero when $f(x_2) = f(x_1)$. It may also shoot off to find a root different from the expected one.

The methods based on linear interpolation are not limited to polynomials, of course. Table 1.6 compares the methods so far discussed when each is used to find the root of the equation

$$3x + \sin x - e^x = 0.$$

The trigonometric term is, of course, evaluated with the x-value in radians. Each method began with $x_1 = 0$ and $x_2 = 1$.

Note that all the methods we have been using require an initial estimate of the root we are computing. It often requires as much thought and effort to get a good starting value as it does to refine it to acceptable accuracy. Sometimes one's knowledge of the physical problem will suggest a starting value. When this is not available, one normally finds starting values by initial trial-and-error computations, or by making a rough graph of the function. We later discuss some methods that are self-starting for polynomials.

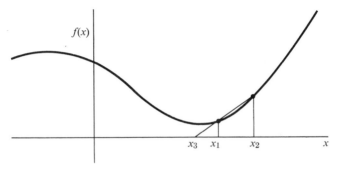

Figure 1.8

Table 1.6 Comparison of methods for finding the roots of $f(x) = 3x + \sin x - e^x = 0$.

Iteration number	Interval halving		Linear interp.		Mod. lin. interp.		Secant method	
	x_3	$f(x_3)$	x_3	$f(x_3)$	x_3	$f(x_3)$	x_3	$f(x_3)$
1	0.5	0.330704	0.470990	0.265160	0.470990	0.265160	0.470990	0.265160
2	0.25	−0.286621	0.372277	0.029533	0.372277	0.029533	0.372277	0.029533
3	0.375	0.036281	0.361598	2.94×10^{-3}	0.351514	−0.022356	0.359904	-1.29×10^{-3}
4	0.3125	−0.121899	0.360538	2.90×10^{-4}	0.360727	7.64×10^{-4}	0.360424	5.53×10^{-6}
5	0.34375	−0.041956	0.3604334	2.93×10^{-5}	0.3604224	1.80×10^{-6}	0.3604218	2.13×10^{-7}
Error after five iterations	0.01667		-1.17×10^{-5}		-7×10^{-7}		-1×10^{-7}	

1.4 NEWTON'S METHOD

One of the most widely used methods of solving equations is *Newton's method.**
Figure 1.9 gives a graphical description. Starting from an initial estimate which is
not too far from a root, x_1, we extrapolate along the tangent to its intersection
with the x-axis, and take that as the next approximation. This is continued until
either the successive x-values are sufficiently close, or the value of the function is
sufficiently near zero.†

The calculation scheme follows immediately from the right triangle shown in
Fig. 1.9, which has the angle of inclination of the tangent line to the curve at
$x = x_1$ as one of its acute angles:

$$\tan \theta = f'(x_1) = \frac{f(x_1)}{x_1 - x_2}, \qquad x_2 = x_1 - \frac{f(x_1)}{f'(x_1)}.$$

We continue the calculation scheme by computing

$$x_3 = x_2 - \frac{f(x_2)}{f'(x_2)},$$

or, in more general terms,

$$x_{n+1} = x_n - \frac{f(x_n)}{f'(x_n)}, \qquad n = 1, 2, 3, \ldots$$

Newton's algorithm is widely used because, at least in the near neighborhood
of a root, it is more rapidly convergent than any of the methods so far discussed.

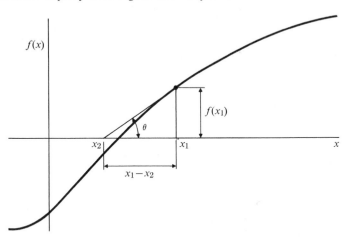

Figure 1.9

---

* Newton did not publish an extensive discussion of this method, but solved a cubic
polynomial in *Principia* (1686). The version given here is considerably improved over his
original example.
† Which criterion should be used often depends on the particular physical problem for
which the equation applies. Customarily, agreement of successive x-values to a specified
tolerance is required.

We show in a later section that the method is quadratically convergent, by which we mean that the error of each step approaches proportionality to the square of the error of the previous step. The net result of this is that the number of decimal places of accuracy nearly doubles at each iteration. However, offsetting this is the need for two function evaluations at each step, $f(x_n)$ and $f'(x_n)$.

When Newton's method is applied to $f(x) = 3x + \sin x - e^x = 0$, we have the following calculations:

$$f(x) = 3x + \sin x - e^x,$$
$$f'(x) = 3 + \cos x - e^x.$$

If we begin with $x_1 = 0.0$, we have:

$$x_2 = x_1 - \frac{f(x_1)}{f'(x_1)} = 0.0 - \frac{-1.0}{3.0} = 0.33333;$$

$$x_3 = x_2 - \frac{f(x_2)}{f'(x_2)} = 0.33333 - \frac{-0.068418}{2.54934} = 0.36017;$$

$$x_4 = x_3 - \frac{f(x_3)}{f'(x_3)} = 0.36017 - \frac{-6.279 \times 10^{-4}}{2.50226} = 0.3604217.$$

After three iterations, the root is correct to seven significant digits. The rule for doubling the number of correct digits would indicate six correct digits in the last result because three are repeated from the previous one. Comparing this with the results in Table 1.6, we see that Newton's method converges considerably more rapidly than the previous methods. In comparing numerical methods, however, one usually counts the number of times functions must be evaluated. Because Newton's method requires two function evaluations per step, the comparison is not as one-sided in favor of Newton's method as at first appears; the three iterations with Newton's method required six function evaluations. Five iterations with the previous methods required seven evaluations.

A more formal statement of the algorithm for Newton's method, suitable for implementation in a computer program, is shown here.

Newton's method

To determine a root of $f(x) = 0$, given a value x_1, reasonably close to the root,

> DO   WHILE $|x_2 - x_1| \geq$ tolerance value 1, or
> $\qquad\qquad |f(x_2)| \geq$ tolerance value 2, or
> $\qquad\qquad f'(x_1) \neq 0$,
> $\quad$ Set $x_2 = x_1 - f(x_1)/f'(x_1)$.
> $\quad$ Set $x_1 = x_2$.
> ENDDO.

Note. The method may converge to a root different from the expected one or diverge if the starting value is not close enough to the root.

In applying Newton's method to polynomials, it is most efficient to evaluate $f(x_n)$ and $f'(x_n)$ by use of synthetic division.* We illustrate this by the same cubic polynomial as has been used before, $x^3 + x^2 - 3x - 3 = 0$, which has a root at $x = \sqrt{3}$. We begin with the value $x = 2$. We utilize the remainder theorem to evaluate $f(2)$, and evaluate $f'(2)$ as the remainder when the reduced polynomial (of degree two here) is divided by $(x - 2)$:

$$
\begin{array}{r|rrrr}
x_1 = 2 & 1 & 1 & -3 & -3 \\
 & & 2 & 6 & 6 \\
\hline
 & 1 & 3 & 3 & 3 \leftarrow \text{remainder} = f(2) \\
 & & 2 & 10 & \\
\hline
 & 1 & 5 & 13 & \leftarrow \text{second remainder} = f'(2)
\end{array}
$$

$$x_2 = 2 - \frac{3}{13} = 1.76923\ldots$$

$$
\begin{array}{r|rrrr}
x_2 = 1.76923 & 1 & 1 & -3 & -3 \\
 & & 1.76923 & 4.89940 & 3.36048 \\
\hline
 & 1 & 2.76923 & 1.89940 & 0.36048 \\
 & & 1.76923 & 8.02957 & \\
\hline
 & 1 & 4.53846 & 9.92897 &
\end{array}
$$

$$x_3 = 1.76923 - \frac{0.36048}{9.92897} = 1.73292.$$

Similarly,

$$x_4 = 1.73292 - \frac{0.00823}{9.47487} = 1.73205.$$

The value of x_4 is correct to five decimals. To observe the improvement in accuracy, consider the successive errors:

	Error	Number of correct figures
$x_1 = 2$	0.26895	1
$x_2 = 1.76923$	0.03718	2
$x_3 = 1.73292$	0.00087	4
$x_4 = 1.73205$	0.00000	6+

In order to compute with five decimal places, as in this example, we used a desk calculator. (If the student has access to an electronic calculator with storage for two or more values, he will find it especially well adapted to this method.)

---

* The mechanics of synthetic division, whereby we divide a polynomial by the factor $x - x_1$, are explained in most algebra books.

The initial value at which Newton's method is begun can make a considerable difference. For example, if this problem is started with $x_1 = 1$, the following values result:

x	$f(x)$	$f'(x)$
$x_1 = 1$	-4	2
$x_2 = 3$	24	30
$x_3 = 2.2$	5.888	15.92
$x_4 = 1.83015$		

From here on, the convergence is rapid, for we are using iterates just about as near the root as in the previous example.

After a first root is found (as shown by a remainder that is very small), one normally proceeds to determine additional roots from the reduced polynomial (whose coefficients are in the third row of the synthetic-division tableau). This makes the computations somewhat shorter. In the example, the reduced equation is a quadratic, so the quadratic formula would be used, but if a higher-degree polynomial were being solved, Newton's method employing synthetic division would be employed to improve an initial estimate of a second root. The process is then repeated until the reduced equation is of second degree.

This technique of working with the reduced function can be used even if the function is not a polynomial. After a root r of $f(x) = 0$ has been found, the new function $F(x) = f(x)/(x - r)$ will have all the roots of $f(x)$ except the root. r. This procedure is called *deflating* the function. One must remember that a discontinuity has been introduced at $x = r$, however. It is suggested that the student explore how deflation works on nonpolynomial functions by graphing $f(x) = (x - 1)(e^x - \cos x)$, then $g(x) = f(x)/x$, $h(x) = f(x)/(x - 1)$, and comparing the graphs. You may also wish to compare to $h(x)/x$ and to functions derived by deflating $f(x)$ with other roots than $x = 0$ and $x = 1$.

It should be observed that using deflated functions can be subject to unexpected errors. If the first root is determined only approximately, the coefficients of the reduced equation are themselves not exact and the succeeding roots are subject to not only round-off errors and the errors that occur when iterations are terminated too soon, but also to inherited errors residing in the nonexact coefficients. Some functions are extremely sensitive in that small changes in the value of the coefficients cause large differences in the roots. One can avoid this difficulty by returning to the original function, perhaps after using the above scheme to locate approximate values for all the roots. Removing roots in order of increasing magnitude is said to minimize the difficulty, and the use of double-precision arithmetic will further help preserve accuracy.

It is of interest to develop the synthetic-division algorithm and to establish the remainder theorems. The scheme is also the most efficient way to evaluate polynomials and their derivatives in a computer program.

Write the nth-degree polynomial* as

$$P_n(x) = a_1 x^n + a_2 x^{n-1} + \ldots + a_n x + a_{n+1}.$$

We wish to divide this by the factor $(x - x_1)$, giving a reduced polynomial $Q_{n-1}(x)$ of degree $n - 1$, and a remainder, b_{n+1}, which is a constant:

$$\frac{P_n(x)}{x - x_1} = Q_{n-1}(x) + \frac{b_{n+1}}{x - x_1}.$$

Rearranging yields

$$P_n(x) = (x - x_1)Q_{n-1}(x) + b_{n+1}.$$

Note that at $x = x_1$,

$$P_n(x_1) = (0)[Q_{n-1}(x_1)] + b_{n+1},$$

which is the remainder theorem: The remainder on division by $(x - x_1)$ is the value of the polynomial at $x = x_1$, $P_n(x_1)$.

If we differentiate the last equation, we get

$$P_n'(x) = (x - x_1)Q_{n-1}'(x) + (1)Q_{n-1}(x) + 0.$$

Letting $x = x_1$, we have

$$P_n'(x_1) = Q_{n-1}(x_1).$$

We evaluate the Q-polynomial at x_1 by a second division whose remainder equals $Q_{n-1}(x_1)$.

We now develop the synthetic-division algorithm, writing $Q_{n-1}(x)$ in form similar to $P_n(x)$:

$$P_n(x) = a_1 x^n + a_2 x^{n-1} + \ldots + a_n x + a_{n+1}$$

$$= (x - x_1)Q_{n-1}(x) + b_{n+1}$$

$$= (x - x_1)(b_1 x^{n-1} + b_2 x^{n-2} + \ldots + b_{n-1}x + b_n) + b_{n+1}.$$

Multiplying out and equating coefficients of like terms in x, we get

Coef. of x^n:	$a_1 = b_1$		$b_1 = a_1$	
x^{n-1}:	$a_2 = b_2 - b_1 x_1$		$b_2 = a_2 + b_1 x_1$	
x^{n-2}:	$a_3 = b_3 - b_2 x_1$	or	$b_3 = a_3 + b_2 x_1$	
x:	$a_n = b_n - b_{n-1} x_1$		$b_n = a_n + b_{n-1} x_1$	
Const.:	$a_{n+1} = b_{n+1} - b_n x_1$		$b_{n+1} = a_{n+1} + b_n x_1$.	

---

* It is well to avoid zero subscripts for the coefficients so that when a computer program is written we do not have to resubscript to avoid the *forbidden zero subscript* in Fortran.

The general form is $b_i = a_i + b_{i-1}x_1$ by which all the b's except b_1 may be calculated. If this is compared to the synthetic divisions above, it is seen to be identical except that we now have a vertical array. The horizontal layout is easier for hand computation. For evaluation of the derivative, a set of c values is computed from the b's in the same way in which the b's are computed from the a's.

Synthetic division is also known as the *nested multiplication* method of evaluating polynomials. Consider the fifth-degree polynomial, evaluated at $x = x_1$:

$$a_1 x_1^5 + a_2 x_1^4 + a_3 x_1^3 + a_4 x_1^2 + a_5 x_1 + a_6.$$

We can rewrite this as

$$(\{[(a_1 x_1 + a_2)x_1 + a_3]x_1 + a_4\}x_1 + a_5)x_1 + a_6.$$

In the original form, $5 + 4 + 3 + 2 + 1 = 15$ multiplications are required, plus five additions. In the nested form, only five multiplications are required, plus five additions; it is obviously the more efficient method.

Comparing with the equations $b_2 = a_2 + b_1 x_1 = a_1 x_1 + a_2$ and $b_i = a_i + b_{i-1}x_1$ for synthetic division, we see that the successive terms are formed in exactly the same way so that synthetic division and nested multiplication are two names for the same thing.

In some cases Newton's method will not converge. Figure 1.10 illustrates this situation. Starting with x_1, one never reaches the root r. We shall develop the analytical condition for this in a later section and show that Newton's method is quadratically convergent in most cases.

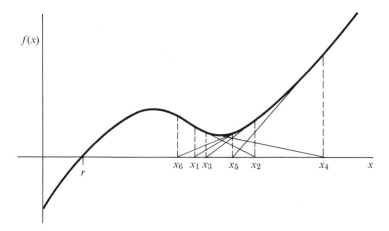

Figure 1.10

1.5   USE OF $x = g(x)$ FORM* $\left(\text{Iteration}\right)$

We now discuss another method that is of general applicability, and which also lets us develop some necessary theory. We begin with the equation $f(x) = 0$, and rearrange it into an equivalent expression of the form

$$x = g(x), \quad \text{such that if } f(r) = 0, \quad r = g(r).$$

Under suitable conditions, which we develop below, the algorithm

$$x_{n+1} = g(x_n), \quad n = 1, 2, 3, \ldots$$

will converge to a zero of $f(x)$. Consider a simple example:

$$f(x) = x^2 - 2x - 3 = 0,$$

which has obvious roots at $x = 3$, $x = -1$.

Rearranging yields:

$$x = \sqrt{2x + 3},$$

so $g(x) = \sqrt{2x + 3}$. Starting with $x_1 = 4$, we get

$$x_2 = \sqrt{11} = 3.316,$$
$$x_3 = \sqrt{9.632} = 3.104,$$
$$x_4 = \sqrt{9.208} = 3.034,$$
$$x_5 = \sqrt{9.068} = 3.011,$$
$$x_6 = \sqrt{9.022} = 3.004.$$

(This is only one of many possible rearrangements.) The various iterates appear to converge to $x_\infty = 3$.

The equation $f(x) = x^2 - 2x - 3 = 0$ can be rearranged in other ways also. For example, $x = 3/(x - 2)$ is an alternative rearrangement of form $x = g(x)$. If $x_1 = 4$.

$$x_2 = 1.5$$
$$x_3 = -6,$$
$$x_4 = -0.375,$$
$$x_5 = -1.263,$$
$$x_6 = -0.919,$$
$$x_7 = -1.028,$$
$$x_8 = -0.991,$$
$$x_9 = -1.003.$$

Note that this converges, but to the root at $x = -1$, and that the iterates oscillate rather than converge monotonically.

---

* The method is called by some authors simply "the method of iteration."

Consider a third rearrangement:

$$x = \frac{x^2 - 3}{2}.$$

From $x_1 = 4$ we get

$$x_2 = 6.5,$$
$$x_3 = 19.635,$$
$$x_4 = 191.0,$$

which obviously is diverging.

Figure 1.11 illustrates the several cases; (a) shows monotonic convergence, (b) shows oscillatory convergence, and (c) shows divergence. For a function $x = g(x)$, the solution is at the intersection of the line $y_1 = x$ with the curve $y_2 = g(x)$. In every case, we move vertically to the curve and then horizontally to the line, and repeat.

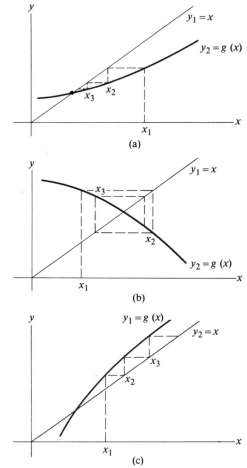

Figure 1.11

An algorithm for this method is shown below.

Iteration with the form x = g(x)
To determine a root of $f(x) = 0$, given a value x_1, reasonably close to the root,

> Rearrange the equation to an equivalent form $x = g(x)$.
> DO WHILE $|x_2 - x_1| \geq$ tolerance value,
> > Set $x_2 = g(x_1)$.
> > Set $x_1 = x_2$.
> ENDDO.

Note. The method may converge to a root different from the expected one, or it may diverge. Different rearrangements will converge at different rates.

We now study the conditions that are needed for convergence. We iterate with

$$x_{n+1} = g(x_n).$$

Let $x = r$ be a solution to $f(x) = 0$, so $f(r) = 0$ and $r = g(r)$. Subtracting, and then multiplying and dividing by $(x_n - r)$, we have

$$x_{n+1} - r = g(x_n) - g(r) = \frac{g(x_n) - g(r)}{(x_n - r)}(x_n - r).$$

If $g(x)$ and $g'(x)$ are continuous on the interval from r to x_n, the mean-value theorem* lets us write

$$x_{n+1} - r = g'(\xi_n)(x_n - r),$$

where ξ_n lies between x_n and r.

If we define the error of the ith iterate as $e_i = x_i - r$, and $e_{i+1} = x_{i+1} - r$, we can write

$$e_{i+1} = g'(\xi_i)e_i.$$

Taking absolute values, we get

$$|e_{i+1}| = |g'(\xi_i)| \cdot |e_i|.$$

Now suppose that $|g'(x)| \leq K < 1$ for all values of x in an interval of radius h about r. If x_1 is chosen in this interval, x_2 will also be in the interval and the algorithm will converge, since

$$|e_{n+1}| \leq K|e_n| \leq K^2|e_{n-1}| \leq \ldots \leq K^n|e_1|.$$

---

* Appendix A reviews certain calculus principles, including this theorem.

In summary, if $g(x)$ and $g'(x)$ are continuous on an interval about a root r of the equation $x = g(x)$, and if $|g'(x)| \leq 1$ for all x in the interval, then $x_{n+1} = g(x_n)$, $n = 1, 2, 3, \ldots$ will converge to the root $x = r$ provided that x_1 is chosen in the interval. Note that this is a sufficient condition only, since for some equations convergence is secured even though not all the conditions hold.[*]

Since the above demonstration shows that $e_{i+1} = g'(\xi_i) \cdot e_i$, it is obvious that the rate of convergence is rapid if $|g'(x)|$ is small in the interval. If the derivative is negative, the errors alternate in sign, giving oscillatory convergence. As the iterates get closer to the root r, the values of $g'(\xi_i)$ approach the constant value $g'(r)$ because the ξ's are squeezed into smaller intervals about r. In the limit, each error becomes proportional to the previous error. For this reason, the method is sometimes called *linear iteration*.

Even though the proportionality between successive errors is true only in the limiting situation, if we assume that each of the errors is proportional to the previous one, we can develop an acceleration technique called *Aitken acceleration*, that is often useful:

Assume that
$$|e_n| = |x_n - r| = K^{n-1} e_1$$
or
$$x_n = r + K^{n-1} e_1.$$
Similarly, if
$$|e_{n+1}| = |x_{n+1} - r| = K^n e_1,$$
and
$$|e_{n+2}| = |x_{n+2} - r| = K^{n+1} e_1,$$
then
$$x_{n+1} = r + K^n e_1,$$
and
$$x_{n+2} = r + K^{n+1} e_1.$$

Substitute these expressions into
$$x_{n+3} = \frac{x_n x_{n+2} - x_{n+1}^2}{x_{n+2} - 2x_{n+1} + x_n}.$$

It is found that
$$\frac{x_n x_{n+2} - x_{n+1}^2}{x_{n+2} - 2x_{n+1} + x_n} = \frac{r + K^{n-1} e_1)(r + K^{n+1} e_1) - (r + K^n e_1)^2}{(r + K^{n+1} e_1) - 2(r + K e_1) + (r + K^{n-1} e_1)}$$
$$= \frac{r(K^{n+1} - 2K^n + K^{n-1}) e_1}{(K^{n+1} - 2K^n + K^{n-1}) e_1} = r.$$

---

[*]The analytical test that $|g'(x)| < 1$ is often awkward to apply. A constructive test is to merely observe whether the successive x_i values converge. In a computer program it is worthwhile to determine whether $|x_3 - x_2| < |x_2 - x_1|$.

From three successive estimates of the root, x_1, x_2, and x_3, we extrapolate to an improved estimate. Since the assumption of constant ratio between successive errors is not normally true, our extrapolated value is not exact, but it is usually improved. One uses this technique by calculating two values beginning with x_1, extrapolating, calculating two new values, extrapolating again, etc.

A different form is useful to avoid the round-off problem that occurs in subtracting large numbers of nearly the same magnitude. Define*

$$\Delta x_i = x_{i+1} - x_i,$$
$$\Delta^2 x_i = \Delta(\Delta x_i) = \Delta(x_{i+1} - x_i) = x_{i+2} - 2x_{i+1} + x_i.$$

Our acceleration scheme becomes

$$r = x_n - \frac{(\Delta x_n)^2}{\Delta^2 x_n} = \frac{x_n x_{n+2} - x_{n+1}^2}{x_{n+2} - 2x_{n+1} + x_n}.$$

The differences are most readily computed in a table. We illustrate with the iterates from the first example in this section:

$$f(x) = x^2 - 2x - 3 = 0,$$
$$x_{n+1} = \sqrt{2x_n + 3}, \qquad x_1 = 4,$$

x	Δx	$\Delta^2 x$
$x_1 = 4.000$		
	0.684	
$x_2 = 3.316$		0.472
	0.212	
$x_3 = 3.104$		

The accelerated estimate is

$$r = 4.000 - \frac{(0.684)^2}{0.472} = 3.009.$$

We have jumped ahead about two iterations. If $g(x)$ is expensive to compute, we have gained. Often Aitken acceleration gives bigger jumps than this.

1.6 CONVERGENCE OF NEWTON'S METHOD

We now use the result of the previous section to show a criterion for convergence of Newton's method. The algorithm

$$x_{n+1} = x_n - \frac{f(x_n)}{f'(x_n)}, \qquad n = 1, 2, 3, \ldots$$

---

* We shall deal quite extensively with such differences in a later chapter.

is of the form $x_{n+1} = g(x_n)$. Successive iterations converge if $|g'(x)| < 1$. Since

$$g(x) = x - \frac{f(x)}{f'(x)},$$

$$g'(x) = 1 - \frac{f'(x)f'(x) - f(x)f''(x)}{[f'(x)]^2} = \frac{f(x)f''(x)}{[f'(x)]^2}.$$

Hence if

$$\left| \frac{f(x)f''(x)}{[f'(x)]^2} \right| < 1$$

on an interval about the root r, the method will converge for any initial value x_1 in the interval. The condition is sufficient only, and requires the usual continuity and existence of $f(x)$ and its derivatives. Note that $f'(x)$ must not be zero.

Now we show that Newton's method is quadratically convergent. Since r is a root of $f(x) = 0$, $r = g(r)$. Since $x_{n+1} = g(x_n)$, we can write

$$x_{n+1} - r = g(x_n) - g(r).$$

Let us expand $g(x_n)$ as a Taylor series* in terms of $(x_n - r)$, with the second-derivative term as the remainder:

$$g(x_n) = g(r) + g'(r)(x_n - r) + \frac{g''(\xi)}{2}(x_n - r)^2,$$

where ξ lies in the interval from x_n to r.

Since

$$g'(r) = \frac{f(r)f''(r)}{[f'(r)]^2} = 0$$

because $f(r) = 0$ (r is a root), we have

$$g(x_n) = g(r) + \frac{g''(\xi)}{2}(x_n - r)^2.$$

Letting $x_n - r = e_n$, we have

$$e_{n+1} = x_{n+1} - r = g(x_n) - g(r) = \frac{g''(\xi)}{2} e_n^2.$$

Each error is (in the limit) proportional to the square of the previous error, i.e., Newton's method is quadratically convergent.†

---

* See Appendix A for a review of Taylor series.
† If $f'(x) = 0$ at $x = r$ (hence a multiple root), the rate of convergence will not be quadratic. For a double root, it can be shown that convergence is linear.

1.7 BAIRSTOW'S METHOD FOR QUADRATIC FACTORS

The methods so far considered are difficult to use to find a complex root of a polynomial. It is true that Newton's method works satisfactorily provided that one begins with an initial estimate which is complex-valued but, in a hand computation, performing the multiplications and divisions of complex numbers is awkward. There is no problem in a computer program if complex arithmetic capabilities exist, but the execution is slower.

For polynomials, the complex roots occur in conjugate pairs if the coefficients are all real-valued. For this case, if we extract the quadratic factors that are the products of the pairs of complex roots, we can avoid complex arithmetic because such quadratic factors have real coefficients. We first develop the algorithm for synthetic division by a trial quadratic, $x^2 - rx - s$, which is hopefully near to the desired factor of the polynomial:

$$P_n(x) = a_1 x^n + a_2 x^{n-1} + \ldots + a_{n+1}$$
$$= (x^2 - rx - s)Q_{n-2}(x) + \text{remainder}$$
$$= (x^2 - rx - s)(b_1 x^{n-2} + b_2 x^{n-3} + \ldots + b_{n-2}x + b_{n-1}) + b_n(x - r) + b_{n+1}.$$

(The remainder is the linear term, $b_n(x - r) + b_{n+1}$, written in this form to provide later simplicity. If $x^2 - rx - s$ is an exact divisor of $P_n(x)$, then b_n and b_{n+1} will both be zero.) The negative signs in the factor are also for later simplification.

On multiplying out and equating coefficients of like powers of x, we get

$$\begin{aligned}
a_1 &= b_1 & b_1 &= a_1 \\
a_2 &= b_2 - rb_1 & b_2 &= a_2 + rb_1 \\
a_3 &= b_3 - rb_2 - sb_1 & b_3 &= a_3 + rb_2 + sb_1 \\
a_4 &= b_4 - rb_3 - sb_2 \quad\text{or}\quad & b_4 &= a_4 + rb_3 + sb_2 \qquad (1.1) \\
&\vdots & &\vdots \\
a_n &= b_n - rb_{n-1} - sb_{n-2} & b_n &= a_n + rb_{n-1} + sb_{n-2} \\
a_{n+1} &= b_{n+1} - rb_n - sb_{n-1} & b_{n+1} &= a_{n+1} + rb_n + sb_{n-1}.
\end{aligned}$$

We would like b_n and b_{n+1} to both be zero, for that would show $x^2 - rx - s$ to be a quadratic factor of the polynomial. This will normally not be so; if we properly change the values of r and s, we can make the remainder zero, or at least make its coefficients smaller. Obviously b_n and b_{n+1} are both functions of the two parameters r and s. Expanding these as a Taylor series for a function of two variables[†] in terms of $(r^* - r)$ and $(s^* - s)$, where $(r^* - r)$ and $(s^* - s)$ are presumed small so that terms of higher order than the first are negligible, we

---

† The Appendix reviews this.

obtain

$$b_n(r^*, s^*) = b_n(r, s) + \frac{\partial b_n}{\partial r}(r^* - r) + \frac{\partial b_n}{\partial s}(s^* - s) + \ldots,$$

$$b_{n+1}(r^*, s^*) = b_{n+1}(r, s) + \frac{\partial b_{n+1}}{\partial r}(r^* - r) + \frac{\partial b_{n+1}}{\partial s}(s^* - s) + \ldots$$

Let us take (r^*, s^*) as the point at which the remainder is zero, and

$$r^* - r = \Delta r, \qquad s^* - s = \Delta s.$$

(Δr and Δs are increments to add to the original r and s to get the new values r^* and s^* for which the remainder is zero.) Then

$$b_n(r^*, s^*) = 0 \doteq b_n + \frac{\partial b_n}{\partial r}\Delta r + \frac{\partial b_n}{\partial s}\Delta s,$$

$$b_{n+1}(r^*, s^*) = 0 \doteq b_{n+1} + \frac{\partial b_{n+1}}{\partial r}\Delta r + \frac{\partial b_{n+1}}{\partial s}\Delta s.$$

All the terms on the right are to be evaluated at (r, s). We wish to solve these two equations simultaneously for the unknown Δr and Δs, so we need to evaluate the partial derivatives.

Bairstow showed that the required partial derivatives can be obtained from the b's by a second synthetic division by the factor $x^2 - rs - s$ in just the same way that the b's are obtained from the a's. Define a set of c's by the relations shown below at the left, and compare these to the partial derivatives in the right columns:

$$c_1 = b_1 \qquad\qquad \frac{\partial b_1}{\partial r} = \frac{\partial a_1}{\partial r} = 0 \qquad\qquad \frac{\partial b_1}{\partial s} = \frac{\partial a_1}{\partial s} = 0$$

$$c_2 = b_2 + rc_1 \qquad\qquad \frac{\partial b_2}{\partial r} = r\frac{\partial b_1}{\partial r} + b_1 = b_1 = c_1 \qquad\qquad \frac{\partial b_2}{\partial s} = \frac{\partial a_2}{\partial s} + r\frac{\partial b_1}{\partial s} = 0$$

$$c_3 = b_3 + rc_2 + sc_1 \qquad\qquad \frac{\partial b_3}{\partial r} = r\frac{\partial b_2}{\partial r} + b_2 = c_2 \qquad\qquad \frac{\partial b_3}{\partial s} = r\frac{\partial b_2}{\partial s} + s\frac{\partial b_1}{\partial s} + b_1 = b_1 = c_1$$

$$c_4 = b_4 + rc_3 + sc_2 \qquad\qquad \frac{\partial b_4}{\partial r} = r\frac{\partial b_3}{\partial r} + b_3 + s\frac{\partial b_2}{\partial r}$$
$$= b_3 + rc_2 + sc_1 = c_3 \qquad\qquad \frac{\partial b_4}{\partial s} = r\frac{\partial b_3}{\partial s} + s\frac{\partial b_2}{\partial s} + b_2$$
$$= b_2 + rc_1 = c_2$$

$$\vdots \qquad\qquad\qquad \vdots \qquad\qquad\qquad \vdots$$

$$c_n = b_n + rc_{n-1} + sc_{n-2} \qquad\qquad \frac{\partial b_n}{\partial r} = r\frac{\partial b_{n-1}}{\partial r} + b_{n-1} + s\frac{\partial b_{n-2}}{\partial r} \qquad\qquad \frac{\partial b_n}{\partial s} = r\frac{\partial b_{n-1}}{\partial s} + s\frac{\partial b_{n-2}}{\partial s} + b_{n-2}$$
$$= b_{n-1} + rc_{n-2} + sc_{n-3} = c_{n-1} \qquad\qquad = b_{n-2} + rc_{n-3} + sc_{n-4} = c_{n-2}$$

$$c_{n+1} = b_{n+1} + \qquad\qquad\qquad c_n \qquad\qquad\qquad\qquad c_{n-1}$$

Hence the partial derivatives that we need are equal to the properly corresponding c's. Our simultaneous equations become

$$-b_n = c_{n-1}\Delta r + c_{n-2}\Delta s,$$

$$-b_{n+1} = c_n\Delta r + c_{n-1}\Delta s.$$

By Cramer's rule

$$\Delta r = \frac{\begin{vmatrix} -b_n & c_{n-2} \\ -b_{n+1} & c_{n-1} \end{vmatrix}}{\begin{vmatrix} c_{n-1} & c_{n-2} \\ c_n & c_{n-1} \end{vmatrix}}$$

$$\Delta s = \frac{\begin{vmatrix} c_{n-1} & -b_n \\ c_n & -b_{n+1} \end{vmatrix}}{\begin{vmatrix} c_{n-1} & c_{n-2} \\ c_n & c_{n-1} \end{vmatrix}}$$

A statement of the algorithm for Bairstow's method is given in Section 10 of this chapter.

Example. Find the quadratic factors of

$$x^4 - 1.1x^3 + 2.3x^2 + 0.5x + 3.3 = 0.$$

Use $x^2 + x + 1$ as starting factor ($r = -1, s = -1$). (Frequently $r = s = 0$ are used as starting values if no information as to an approximate factor is known.) Equations (1.1) lead to a double synthetic-division scheme as follows:

	a_1	a_2	a_3	a_4	a_5
	1	−1.1	2.3	0.5	3.3
$r = -1$		−1.0	2.1	−3.4	0.8
$s = -1$		—	−1.0	2.1	−3.4
	1	−2.1	3.4	−0.8	0.7
		−1.0	3.1	−5.5	
		—	−1.0	3.1	b_{n+1}
	1	−3.1	5.5	−3.2	b_n
		c_{n-2}	c_{n-1}	c_n	

Note that the equations for b_2 and c_2 have no term involving s. The dashes in the above tableau represent these missing factors. Then

$$\Delta r = \frac{\begin{vmatrix} 0.8 & -3.1 \\ -0.7 & 5.5 \end{vmatrix}}{\begin{vmatrix} 5.5 & -3.1 \\ -3.2 & 5.5 \end{vmatrix}} = \frac{2.23}{20.33} = 0.11, \qquad r^* = -1 + 0.11 = -0.89,$$

$$\Delta s = \frac{\begin{vmatrix} 5.5 & 0.8 \\ -3.2 & -0.7 \end{vmatrix}}{20.33} = \frac{-1.29}{20.33} = -0.06, \qquad s^* = -1 - 0.06 = -1.06.$$

The second trial yields

	1	−1.1	2.3	0.5	3.3
−0.89		−0.89	1.77	−2.68	0.06
−1.06		—	−1.06	2.11	−3.17
	1	−1.99	3.01	−0.07	0.17
		−0.89	2.56	−4.01	
		—	−1.06	3.05	
	1	−2.88	4.51	−1.03	

$$\Delta r = \frac{\begin{vmatrix} 0.07 & -2.88 \\ -0.17 & 4.51 \end{vmatrix}}{\begin{vmatrix} 4.51 & -2.88 \\ -1.03 & 4.51 \end{vmatrix}} = \frac{-0.175}{17.374} = -0.010, \qquad r^* = -0.89 - 0.010 = -0.900,$$

$$\Delta s = \frac{\begin{vmatrix} 4.51 & 0.07 \\ -1.03 & -0.17 \end{vmatrix}}{17.374} = \frac{-0.694}{17.374} = -0.040, \qquad s^* = -1.06 - 0.040 = -1.100.$$

The exact factors are $(x^2 + 0.9x + 1.1)(x^2 - 2x + 3)$.

1.8 QUOTIENT–DIFFERENCE ALGORITHM

All the methods so far discussed require a starting value sufficiently near to the root or to the quadratic factor being sought. A relatively efficient method to determine all the roots of a polynomial without starting values is the Q–D or quotient–difference algorithm. We present the method without elaboration.*

---

* Henrici (1964) discusses the method in some detail.

For the nth-degree polynomial

$$P_n(x) = a_1 x^n + a_2 x^{n-1} + \ldots + a_n x + a_{n+1},$$

we form an array of q and e terms, starting the tableau by calculating a first row of q's and a second row of e's:

$$q^{(1)} = -a_2/a_1, \qquad \text{all other } q\text{'s are zero,}$$
$$e^{(i)} = a_{i+2}/a_{i+1}, \qquad i = 1, 2, \ldots, n - 1,$$
$$e^{(0)} = e^{(n)} = 0.$$

The start of the array is

$e^{(0)}$	$q^{(1)}$	$e^{(1)}$	$q^{(2)}$	$e^{(2)}$	$q^{(3)}$	$\ldots$	$e^{(n-1)}$	$q^{(n)}$	$e^{(n)}$
	$\dfrac{-a_2}{a_1}$		0		0	$\ldots$		0	
0		$\dfrac{a_3}{a_2}$		$\dfrac{a_4}{a_3}$		$\ldots$	$\dfrac{a_{n+1}}{a_n}$		0

A new row of q's is computed by the equation

$$\text{new } q^{(i)} = e^{(i)} - e^{(i-1)} + q^{(i)},$$

using terms from the e and q rows just above. Note that this algorithm is "e to right minus e to left plus q above." A new row of e's is now computed by the equation

$$\text{new } e^{(i)} = \left(\frac{q^{(i+1)}}{q^{(i)}} \right) e^{(i)};$$

"q to right over q to left times e above." The example in Table 1.7 isolates the roots of the quartic

$$P_4(x) = 128x^4 - 256x^3 + 160x^2 - 32x + 1$$

by continuing to compute rows of q's and then e's until all the e-values approach zero. When this occurs, the q-values assume the values of the roots. Since the method is slow to converge, it is generally used only to get approximate values, which are then improved by Newton's method.

If the polynomial has a pair of conjugate complex roots, one of the e's will not approach zero but will fluctuate in value. The sum of the two q-values on either side of this e will approach r and the product of the q above and to the left times the q below and to the right approaches $-s$ in the factor $x^2 - rx - s$. Two equal roots behave similarly.

Table 1.8 shows the result of the method for the polynomial

$$(x - 1)(x - 4)(x^2 - x + 3) = x^4 - 6x^3 + 12x^2 - 19x + 12.$$

Table 1.7 Example of Q–D method for $P(x) = 128x^4 - 256x^3 + 160x^2 - 32x + 1$.

$e^{(0)}$	$q^{(1)}$	$e^{(1)}$	$q^{(2)}$	$e^{(2)}$	$q^{(3)}$	$e^{(3)}$	$q^{(4)}$	$e^{(4)}$
	2.000		0		0		0	
0		−0.625		−0.200		−0.031		0
	1.375		0.425		0.169		0.031	
0		−0.193		−0.079		−0.006		0
	1.182		0.539		0.242		0.037	
0		−0.088		−0.036		−0.001		0
	1.094		0.591		0.277		0.038	
0		−0.048		−0.017		−0.000		0
	1.046		0.622		0.294		0.038	
0		−0.028		−0.008		−0.000		0
	1.018		0.642		0.302		0.038	
0		−0.018		−0.004		−0.000		0
	1.000		0.656		0.304		0.038	
0		−0.012		−0.002		−0.000		0
	0.988		0.666		0.306		0.038	
0		−0.008		−0.001		−0.000		0
	0.980		0.673		0.307		0.038	
0		−0.005		−0.001		−0.000		0
	0.975		0.677		0.308		0.038*	

* The true values of the roots are 0.96194, 0.69134, 0.30866, and 0.03806.

Table 1.8 Q–D method with complex roots, for $P(x) = x^4 - 6x^3 + 12x^2 - 19x + 12$.

$e^{(0)}$	$q^{(1)}$	$e^{(1)}$	$q^{(2)}$	$e^{(2)}$	$q^{(3)}$	$e^{(3)}$	$q^{(4)}$	$e^{(4)}$
	6.000		0		0		0	
0		−2.000		−1.583		−0.632		0
	4.000		0.417		0.951		0.632	
0		−0.208		−3.610		−0.420		0
	3.792		−2.985		4.141		1.052	
0		0.164		5.008		−0.107		0
	3.956		1.859		−0.974		1.159	
0		0.077		−2.624		0.127		0
	4.033		−0.842		1.777		1.032	
0		−0.016		5.538		0.074		0
	4.017		4.712		−3.687		0.958	
0		−0.019		−4.333		−0.019		0
	3.998		0.398		0.627		0.977	
0		−0.002		−6.826		−0.030		0
	4.000		−6.426		7.423		1.007	
0		0.003		7.885		−0.004		0
	4.003		1.456		−0.466		1.010	

Factors are $(x - 4)(x - 1)(x^2 - x + 3)$.

$$q^{(1)} \text{ converging to } 4. \qquad q^{(4)} \text{ converging to } 1.$$

Since $e^{(2)}$ does not approach zero, $q^{(2)}$ and $q^{(3)}$ represent a quadratic factor:

$$r \doteq q^{(2)} + q^{(3)} = 1.456 - 0.466 = 0.990;$$
$$s \doteq -(-6.426)(-0.466) = -2.995.$$

This quadratic factor is $x^2 - rx - s = x^2 - 0.990x - (-2.995)$.

Note that one cannot compute the first q and e rows if one of the coefficients in the polynomial is zero, for division by zero is undefined. In such a case, we change the variable to $y = x - 1$. (Subtracting 1 from the roots of the equation is an arbitrary choice, but this facilitates the reverse change of variable to get the roots of the original equation after the roots of the new equation in y have been found.)

For example, if $f(x) = x^4 - 2x^2 + x - 1 = 0$, we let $y = x - 1$ and use repeated synthetic division to determine the coefficients of $f(y) = 0$. The successive remainders on dividing by $x - 1$ are the coefficients of $f(y)$:

1	0	−2	1	−1	$\lfloor 1$
	1	1	−1	0	
1	1	−1	0	−①	
	1	2	1		
1	2	1	①		
	1	3			
1	3	④			
	1				
1	④				
①					

Therefore,

$$f(y) = y^4 + 4y^3 + 4y^2 + y - 1.$$

We proceed to find the roots of $f(y) = 0$, and then get the roots of $f(x) = 0$ by adding 1.

1.9 OTHER METHODS

We mention two other methods that are sometimes used. *Graeffe's method* finds values for all the roots of a polynomial directly from its coefficients without requiring starting values. It is based on the fact that if the roots are all different and widely separated, then for the polynomial

$$P_n(x) = a_1 x^n + a_2 x^{n-1} + \ldots + a_n x + a_{n+1},$$

the roots are given by

$$r_1 \doteq -\frac{a_2}{a_1}, \qquad r_2 \doteq -\frac{a_3}{a_2}, \qquad \ldots, \qquad r_n \doteq -\frac{a_{n+1}}{a_n}. \qquad (1.2)^*$$

In order to separate the roots of the given polynomial, it is converted to another polynomial whose roots are the negative squares of the original roots. After enough repetitions of the root-squaring operation the relations of Eq. (1.2) give the values of the roots, provided no multiple roots occur.† The method is laborious and is used less often than the Q–D algorithm.

Muller's method is an interpolation method that uses quadratic interpolation rather than linear as in Section 1.3. A second-degree polynomial is made to fit three points near a root, $[x_0, f(x_0)]$, $[x_1, f(x_1)]$, $[x_2, f(x_2)]$, and the proper zero of this quadratic, using the quadratic formula, is used as the improved estimate of the root. The process is then repeated using the set of three points nearest the root being evaluated.

The procedure for Muller's method is developed by writing a quadratic equation that fits through three points in the vicinity of a root, in the form $av^2 + bv + c$. (See Fig. 1.12.) The development is simplified if we transform axes to pass through the middle point, by letting $v = x - x_0$.

Let $h_1 = x_1 - x_0$ and $h_2 = x_0 - x_2$. We evaluate the coefficients by evaluating $p_2(v)$ at the three points:

$$v = 0; \qquad a(0)^2 + b(0) + c = f_0;$$
$$v = h_1: \qquad ah_1^2 + bh_1 + c = f_1;$$
$$v = -h_2: \qquad ah_2^2 - bh_2 + c = f_2.$$

Figure 1.12

---

* We use the notation $r_1 \doteq -a_2/a_1$ to indicate approximate equality of the two quantities.
† Scarborough (1950) discusses the method, including its application in cases of multiple and complex roots.

From the first equation, $c = f_0$. Letting $h_2/h_1 = \gamma$, we can solve the other two equations for a and b:

$$a = \frac{\gamma f_1 - f_0(1 + \gamma) + f_2}{\gamma h_1^2(1 + \gamma)}, \qquad b = \frac{f_1 - f_0 - ah_1^2}{h_1}.$$

After computing a, b, and c, we solve for the root of $av^2 + bv + c = 0$ by the quadratic formula, choosing the root nearest to the middle point, x_0. This value is:

$$\text{Root} = x_0 - \frac{2c}{b \pm \sqrt{b^2 - 4ac}},$$

with the sign in the denominator taken to give the largest absolute value of the denominator (that is, if $b > 0$, choose plus; if $b < 0$, choose minus; if $b = 0$, choose either).

We take the root of the polynomial as one of a set of three points for the next approximation, taking the three points that are most closely spaced (i.e., if the root is to the right of x_0, take x_0, x_1, and the root; if to the left, take x_0, x_2, and the root). We always reset the subscripts to make x_0 be the middle of the three values.

Example Find a root of $f(x) = \sin x - x/2$ near $x = 2.0$.

Let

$$x_0 = 2.0, \qquad f(x_0) = -0.09070, \qquad h_1 = 0.2.$$
$$x_1 = 2.2, \qquad f(x_1) = -0.29150, \qquad h_2 = 0.2;$$
$$x_2 = 1.8, \qquad f(x_2) = 0.07385, \qquad \gamma = 1.0.$$

Then

$$a = \frac{(1.0)(-0.29150) - (-0.09070)(2.0) + 0.07385}{(1.0)(0.2)^2(2.0)} = -0.45312,$$

$$b = \frac{-0.29150 - (-0.09070) - (-0.45312)(0.2)^2}{0.2} = -0.91338,$$

$$c = -0.09070;$$

and

$$\text{Root} = 2.0 - \frac{2(-0.09070)}{-0.91338 - \sqrt{(0.91338)^2 - 4(-0.45312)(-0.09070)}} = 1.89526.$$

For the next iteration, we have:

$$x_0 = 1.89526, \qquad \overset{middle.}{f(x_0)} = 1.9184 \times 10^{-4}, \qquad h_1 = 0.10474;$$
$$x_1 = 2.0, \qquad \overset{top}{f(x_1)} = -0.09070, \qquad h_2 = 0.09526;$$
$$x_2 = 1.8, \qquad \underset{low}{f(x_2)} = 0.07385, \qquad \gamma = 0.9095.$$

Then

$$a = \frac{(0.9095)(-0.09070) - (1.9184 \times 10^{-4})(1.9095) + 0.07385}{(0.9095)(0.10474)^2(1.9095)} = -0.47280,$$

$$b = \frac{-0.09070 - 1.9184 \times 10^{-4} - (-0.47280)(0.10474)^2}{0.10474} = -0.81826,$$

$$c = 1.9184 \times 10^{-4},$$

and

$$\text{Root} = 1.89526 - \frac{(2)(1.9184 \times 10^{-4})}{-0.81826 - \sqrt{(0.81826)^2 - 4(-0.47280)(1.9184 \times 10^{-4})}}$$

$$= 1.895494.$$

After this second iteration, the root is accurate to seven significant digits. Experience shows that Muller's method converges nearly quadratically, in a way similar to Newton's method. It does not require the evaluation of derivatives, however, and (after we have obtained the starting values) needs only one function evaluation per iteration.

1.10 ERRORS AND COMPUTER ARITHMETIC

We have previously observed that it is essential not to neglect the study of the errors of our numerical techniques. For several of our methods, we have analyzed how the errors of successive iterations decrease (or increase if the method diverges). It is time that we look more deeply into the several sources of error that interact to affect the accuracy of our result. We can list these several sources as:

Truncation Error   This term is given to the errors caused by the method itself (the term originates from the fact that numerical methods can usually be compared to a truncated Taylor series) and is the error to which we have so far paid most attention. For iterative methods, this error usually can be reduced by repeated iterations but, since life is finite and computer time is expensive, we must be satisfied with approximations to the exact analytical answer.

Round-off Error   All computing devices represent numbers with some imprecision. Digital computers, which are the normal devices for implementing numerical methods, will nearly always use floating-point numbers with a fixed word length. The true values are not exactly expressed by such representations. We call this error *round-off*, whether the decimal fraction is rounded or chopped after the final digit. We discuss this in more detail below.

Error in Original Data   Real-world problems, in which an existing or proposed physical situation is modeled by a mathematical equation, frequently have coefficients that are imperfectly known. The model itself may not perfectly reflect the behavior of the situation either. The numerical analyst can do nothing to overcome such errors by his choice of method, but he needs to be aware of such uncertainties; in particular, he may need to perform tests to see how sensitive his results are to changes in the input information. Since the reason for performing the computation is to permit some decision with validity in the real world, sensitivity analysis is of extreme importance. As Hamming says, "the purpose of computing is insight, not numbers."

Blunders   It is anticipated that the student will use a digital computer (or at least an electronic calculator) in his professional use of numerical analysis. He will probably use such computing tools extensively while he is learning the topics covered in this text. Such machines make mistakes only most infrequently but, since humans are involved in programming, operation, preparing the input, and interpreting the output, blunders or gross errors do occur, rather more frequently than we like to admit. The solution here is *care*, coupled with a careful examination of the results for reasonableness. Sometimes a test run with known results is worthwhile, but this is no guarantee of freedom from foolish error. When hand computation was more common, *check sums* were frequently computed—these would ordinarily reveal the mistake and permit its correction.

Propagated Error   This is more subtle than the other errors. By propagated error we mean the error in the succeeding steps of the process due to the occurrence of an earlier error. This is in addition to the local error made at that step; it is somewhat analogous to errors in the initial conditions. The methods discussed in this chapter do not reflect this type of error, except in the case of finding additional zeros of a function using the reduced or deflated equation. Here the reduced equations reflect errors in the previous stages. In other examples of numerical methods treated in later chapters, propagated error is of critical importance. If errors are magnified continuously as the method continues, eventually they will completely overshadow the true value, destroying its validity; we call such a method *unstable*. (For a stable method (the desirable kind), errors made at early points die out as the method continues. This will be covered more thoroughly in later chapters.)

Each of these types of error, while interacting to a degree, may occur even in the absence of the other kinds. For example, round-off error will occur even if truncation error is absent, as in an analytical method. Likewise, truncation errors would cause inaccuracies even if one could attain perfect precision in the calculations. The usual error analysis of a numerical method treats the truncation error as if such perfect precision did exist.

Floating-Point Numbers

To examine round-off error in detail, we need to understand how numeric quantities are represented in computers. In nearly all cases, they are stored as floating-point quantities, which are very much like scientific notation. Different computers use slightly different techniques, but the general procedure is similar. We discuss that used in IBM 360 and 370 computers as representative.

A single-precision floating-point number is stored as a 32-bit quantity, with 24 bits representing a fraction having 6 hexadecimal digits. One bit is used to indicate the sign of the fraction and seven bits are a binary number representing the exponent and its sign. While the fraction is stored as a signed magnitude, the exponent is "biased 64"; that is, the value of the exponent plus 64 is stored as a binary value. The exponent is applied to 16 as a base. The sign bit and the seven exponent bits form the first eight bits of the number. The fraction is normalized; that is, the exponent is adjusted so that the first (hexadecimal) digit of the fraction is nonzero. The usual way of writing the bits of a floating-point number is as its hexadecimal equivalent. A few examples will illustrate.

Value (base ten)	Hexadecimal equivalnet	Stored as
a) 3.1415926	3.243F69A25 . . .	413243F6
		Sign & exponent Fraction
b) −107.3251	−6B.5339COEB . . .	C26B5339
		Sign & exponent Fraction
c) $0.2023965 * 10^{-4}$	$0.15390A282A \ldots * 16^{-3}$	3D15390A
		Sign & exponent Fraction

In example (a) the sign bit is 0 (for a positive fraction); the exponent of 16 is 1 plus 64 (the bias), equalling 65 (binary 1000001). Combining, we have 01000001 or hexadecimal 41 for the first two hexadecimal digits. The fraction is chopped after six hexadecimal digits.

In example (b) the sign bit is 1; the exponent is $2 + 64 = 66 = (1000010)_2$. Combining, $(11000010)_2 = (C2)_{16}$. In example (c), the sign bit is 0; the exponent is $-3 + 64 = 61 = (0111101)_2$. Combining, $(00111101)_2 = (3D)_{16}$. Nearly all decimal fractions cannot be expressed exactly in base 16.

It is hard for us to think in number bases other than ten, so we will discuss the arithmetic accuracy of floating-point operations using normalized base-ten numbers. The other number bases that are actually used in computers behave analogously. To simplify our treatment, assume only three digits in the fraction part, and one decimal digit for the exponent. We supply signs for the fraction and the exponent as separate symbols.

When two floating-point numbers are added or subtracted, the digits in the fraction of the number with the smaller exponent must be shifted to align the decimal points; the sum may need to be shifted and the exponent adjusted to normalize the result. For example,

d)
$$.137 \times 10^1 + .269 \times 10^{-1}$$

$.137 \quad \times 10^1$
$+.00269 \times 10^1$ Align decimal points
$\overline{.13969 \times 10^1} \rightarrow$ Chopped $= .139 \times 10^1$
$+.0005$
$\overline{.14019 \times 10^1} \rightarrow$ Rounded* $= .140 \times 10^1$

e)
$$.485 \times 10^4 - .482 \times 10^4$$

$.485 \times 10^4$
$-.482 \times 10^4$
$\overline{.003 \times 10^4}$
$.300 \times 10^2$ Normalized $\rightarrow$ Chopped $= .300 \times 10^2$
$+.0005$
$\overline{.3005 \times 10^2} \rightarrow$ Rounded $= .300 \times 10^2$

f)
$$.378 \times 10^4 + .727 \times 10^4$$

$.378 \times 10^4$
$+.727 \times 10^4$
$\overline{1.105 \quad \times 10^4}$
$.1105 \times 10^5$ Normalized $\rightarrow$ Chopped $= .110 \times 10^5$
$+.0005$
$\overline{.1110 \times 10^5} \rightarrow$ Rounded $= .111 \times 10^5$

Observe that rounding requires an extra operation, which can be done with either hardware or software. Either way adds cost, so most computers chop rather than round. Pay particular attention to the loss of precision in example (e). There is only one digit of accuracy in the result, although the difference is represented as if the trailing digits were significant. This loss of significance when two nearly equal numbers are subtracted is the major reason for the loss of accuracy in floating-point operations.†

Multiplication of two n-digit numbers gives a result of $2n$ digits. The floating-point registers of a computer normally allow for this, converting the $2n$-digit product to an n-digit result. Such double-length registers are also used in

---

* To round to 3 digits after the decimal point, we add 0.0005 and then chop the digits beyond three.

† One way to detect such insignificant zeros is to provide for changing the entering digit on a left shift of the fraction during normalizing. Two runs with different fill digits will then give different results. This is pretty costly. Another way (also costly) is to repeat the computer run using double-precision, and compare the results.

division. Some examples:

g) $.403 \times 10^6 * .197 \times 10^{-1}$

Multiply fractions	Add exponents
.403	6
$* .197$	-1
.079391	5

$.079391 \times 10^5$

$.79391 \ \times 10^4$ Normalize → Chopped $= .793 \times 10^4$

$+ .0005$

$.79441 \ \times 10^4 \to$ Rounded $= .794 \times 10^4$

h) $.356 \times 10^{-2} \div .156 \times 10^4$

Divide fractions	Subtract exponents
.356	-2
$\div \ .156$	-4
2.28205	-6

$2.2805 \ \times 10^{-6}$

$.22820 \times 10^{-7}$ Normalize → Chopped $= .228 \times 10^{-7}$

$+ \ .0005$

$.22870 \times 10^{-7} \to$ Rounded $= .228 \times 10^{-7}$

In multiplication and division, the initial shifting to align decimal points can be omitted. The multiplication step is usually considerably slower than addition; the overall time for floating-point multiplication is typically from 2.5 to 5 times that for floating-point addition or subtraction. Floating-point division is usually the slowest of all (4 to 9 times that for addition).

Many computers provide for two (or even three) levels of precision by using a double-length word for double precision. Usually the number of bits used for the exponent remains the same, so the number of digits in the fraction more than. doubles.* Since decimal numbers are usually converted to a different number base when stored as a floating-point value, the number of significant decimal digits equivalent to the accuracy of the fraction is not an integer. Different computer manufacturers have approached the problem of floating-point representation with wide variations of word lengths, so the accuracy provided is greatly different (from 6 to 16 equivalent decimal digits in single-precision, for example).

Converting numbers to the computer's internal number base will often introduce some error. Terminating decimal fractions may be nonterminating in

---

* Double precision therefore requires twice the memory space to store the numbers. The execution time is also increased, sometimes several fold.

binary or hexadecimal [$(0.6)_{10} = (0.100110011001\ldots)_2$, for example]. In addition, a finite word length has many "gaps" in its number set, in that the number of different values representable is finite. In effect, we have to map the infinite set of mathematical numbers into a finite set of computer numbers. For our simplified example of three-digit fractions, there are only 900 different fraction values—all the mathematical numbers between 0.1 and 1.0 must be translated to one of these 900 values. In each decade, as represented by a constant value of the exponent, there also are only 900 different values. The spacing between values in the different decades is therefore different.

Zero is a special situation among the floating-point numbers. It cannot be normalized, of course, so special conventions are adopted. In most systems, the fraction digits are all zeros. The exponent must also be set to the most negative value; if this is not done, alignment of decimal points when adding will shift out significant digits from the addend. Zero is relatively isolated from the other values. In our simplified example system, the nearest neighbors to zero are $\pm 0.1 * 10^{-9}$. Trying to represent any magnitudes smaller than this causes a program error called *exponent underflow*. In some FORTRAN systems, on underflow the number is replaced by floating-point zero and execution continues. This may produce other errors if subsequent division is performed. Similarly, an attempt to represent numbers of magnitude larger than $0.999 * 10^9$ in our system gives rise to *exponent overflow*. Normally, this terminates execution, but some systems replace the number with the largest possible floating-point quantity and then continue.

Peculiar things happen in floating-point arithmetic. For example, adding 0.001 one thousand times usually doesn't equal 1.0. In some instances, multiplying a number by unity doesn't reproduce the number. In many calculations, changing the order in which operations are performed will produce different results.

Absolute versus Relative Error The absolute error of a given result is frequently used as a measure of accuracy; the conventional definition is

$$\text{Absolute error} = \text{True value} - \text{Approximate value},$$

so we get the true value by adding the absolute error to the approximation. However, a given error is usually much more serious when the magnitude of the true value is small. For example, 1036.52 ± 0.010 is accurate to five significant digits and is frequently of more than adequate precision, while 0.005 ± 0.010 is a clear disaster. The relative error,

$$\text{Relative error} = \frac{\text{True value} - \text{Approximate value}}{\text{True value}},$$

is often a better indicator of the accuracy. It is more nearly scale-independent, a most desirable property. When the true value is zero, the relative error is undefined. It follows that the round-off error due to finite-fraction length in floating-point numbers is more nearly constant when expressed as relative error than when expressed as absolute error. Observe that the loss of significant digits when nearly equal floating-point numbers are subtracted produces a particularly severe relative error.

1.11 PROGRAMMING OF METHODS TO SOLVE EQUATIONS

Numerical procedures for obtaining the roots of an equation are slow and tedious when done by hand, although electronic calculators ease the task. Digital computers are the usual tool used by mathematicians, scientists, and engineers, and are ideally adapted to implementing the algorithms of this chapter. We illustrate the program-development process with two examples, the first one applied to the "Ladder in the Mine" problem that served to introduce this chapter. Recall that that problem requires us to solve the transcendental equation for C:

$$\frac{w_2 \cos (\pi - A - C)}{\sin^2 (\pi - A - C)} - \frac{w_1 \cos C}{\sin^2 C} = 0,$$

with w_1, w_2, and A supplied as constants. After finding the angle C, we get the maximum length from

$$\ell_{max} = \frac{w_2}{\sin (\pi - A - C)} + \frac{w_1}{\sin C}.$$

For our first example, we will find a root, $0° < C < (180° - A)$, using the Modified Linear Interpolation algorithm. Our problem has $w_1 = 7$, $w_2 = 9$, and $A = 123°$. The algorithm was prevously given in Section 1.3. We repeat it here.

Modified linear interpolation method

To determine a root of $f(x) = 0$, given values of x_1 and x_2 such that $f(x_1)$ and $f(x_2)$ are of opposite sign,
      Set SAVE $= f(x_1)$; set F1 $= f(x_1)$; set F2 $= f(x_2)$.
      DO WHILE $|x_1 - x_2| \geq$ tolerance value 1, or
                  $|f(x_3)| \geq$ tolerance value 2,

      Set $x_3 = x_2 - F2 \dfrac{x_2 - x_1}{F2 - F1}$.

        IF $f(x_3)$ of opposite sign to F1:
                Set $x_2 = x_3$.
                Set F2 $= f(x_3)$.
                IF $f(x_3)$. of same sign as SAVE:
                        Set F1 $=$ F1/2.
                ENDIF.

ELSE   Set $x_1 = x_3$.
          Set F1 $= f(x_3)$.
          IF $f(x_3)$ of same sign as SAVE:
                Set F2 $=$ F2/2.
          ENDIF.
     ENDIF.
     Set SAVE $= f(x_3)$.
ENDDO.

Before we can write our program, we must establish the design of our program modules. We must also refine the algorithm so that decisions are unambiguous.

a)   A main program will read in the starting values of x_1 and x_2 and call the subroutine MDLNIN to execute the algorithm. Reflection reveals that the function is continuous in C and has only one zero. Suitable values for x_1 and x_2 are 5° and 118°; these certainly bracket the value that makes the function zero.

b)   A function subprogram will compute values of $f(x)$. It then is easy to find roots of a different function by replacing the function subprogram. To provide flexibility, we will pass the name of the function subprogram as a parameter to the MDLNIN subroutine. This requires us to declare the name of the function subprogram as EXTERNAL. We pass the constants of the problem, w_1, w_2, and A through COMMON. The values of x_1 and x_2 are passed as parameters; the value of the root is returned in XR.

c)   In the subroutine, it will be appropriate to test initially whether $f(x_1)$ and $f(x_2)$ are indeed of opposite sign, printing a message if they are not.

d)   The phrase "while $|x_1 - x_2| \geq$ tolerance value 1" must be quantified. We shall compare this value to an input parameter, XTOL, to provide for terminating the calculation. In some situations, it may be preferable to terminate when the relative error $|x_1 - x_2|/x_1$ is less than some predefined tolerance.

e)   A second termination criterion is whether the function value is extremely small; it is possible that this indicates an x-value very near a root. The subroutine compares values of $f(x_3)$ to the parameter FTOL, to stop when the function is "sufficiently small."

f)   It is always good practice to limit the maximum number of iterations. The parameter NLIM does this.

g)   The utility of the subroutine is increased if the calling program can easily determine on which condition the subroutine terminated. The parameter I signals which of the various conditions was satisfied on termination. It is also desirable to sometimes print out the successive iterates, but at other times this is not needed. The parameter I is used (at the time the subroutine is called) to indicate whether or not to print out each iterate.

In our programs, we try to provide adequate documentation by comment cards. Some documentation systems also require a flow chart;* we provide one in Fig. 1.13.

Program 1 (see Fig. 1.14) was run with input data as follows:

$$
\begin{aligned}
X1 &= \quad 5.0 \\
X2 &= 118.0 \\
XTOL &= \quad 0.01 \\
FTOL &= \quad 0.0001 \\
NLIM &= \quad 20
\end{aligned}
$$

The output is shown with the program. We conclude that quite long ladders can negotiate this turn in the mine shaft; the calculated length is over 33 feet.

For our second program example, we exhibit a program to solve for the roots of a polynomial up to the tenth degree, using Bairstow's method to get the quadratic factors. The algorithm, which is an adaptation of the equations of Section 1.7, can be written as shown below. Subscripts are advanced by two, so the equations for all the b's and c's become identical when $b_1 = b_2 = c_1 = c_2 = 0$. (See Program 2, Fig. 1.15.)

Bairstow's method

To determine a quadratic factor, $x^2 - rx - s$, of the nth-degree polynomial $a_3 x^n + a_4 x^{n-1} + \ldots + a_{n+2} x + a_{n+3}$, choose initial coefficients R and S of the quadratic factor, then

Set B(1) = 0, B(2) = 0, C(1) = 0, C(2) = 0.
DO   WHILE DELR $\geq$ tolerance value, or
                  DELS $\geq$ tolerance value,
     DO   FOR J = 3 to n + 3 step 1,
         Set B(J) = A(J) + R * B(J − 1) + S * B(J − 2).
         Set C(J) = B(J) + R * C(J − 1) + S * C(J −2).
         Set DENOM = C(N + 1) * C(N + 1) − C(N + 2) * C(N).
         IF   DENOM = 0:
             Set R = R + 1.
             Set S = S + 1.
             Repeat from beginning.

---

* The relative advantages of flow charts as compared to the statement of the algorithm in a structured form are well illustrated by this example. The flow chart is much more specific as regards details of the program but is also much more dependent on the implementation language. The structured statement is much more general and presents the essentials of the procedure more clearly.

```
        ENDIF.
        Set DELR = [−B(N + 2) * C(N + 1) + B(N + 3)
                                        * C(N)]/DENOM.
        Set DELS = [−C(N + 1) * B(N + 3) + C(N + 2)
                                        * B(N + 2)]/DENOM.

        Set R = R + DELR.
        Set S = S + DELS.
     ENDDO.
  ENDDO.
```

Because the program handles only one specific type of equation, there is no advantage to subdividing it into subroutines. The number of iterations needed to determine the quadratic factor within a given tolerance is arbitrarily limited to 20. In case the denominator during the Cramer's-rule step is zero, the trial values of R and S are arbitrarily increased by unity and the search for a quadratic factor is begun again—this will ordinarily avoid the attempted division by zero. When a quadratic factor is found successfully, the deflated polynomial (the b's are its coefficients) is used to find additional factors. (The algorithm applies only to obtaining each quadratic factor.) The program could be used for polynomials of degree higher than 10 by changing the DIMENSION statement, but this is not advisable because accumulated errors will often be too large.

When the program was tested with the polynomial $x^5 − 17.8x^4 + 99.41x^3 − 261.218x^2 + 352.611x − 134.106$, the results were as shown below the program. The exact factors are

$$x^2 − 4.2x + 2.1, \qquad x^2 − 3.3x + 6.2, \qquad \text{and} \qquad x − 10.3,$$

and the program finds them with excellent precision.

The input data were:

$$N = 5 \text{ (degree)}$$
$$R = 0.0$$
$$S = 0.0$$
$$\text{TOL} = 0.0001$$

Coefficients = 1.0, −17.8, 99.41, −261.218, 352.611, −134.106

Programs 3 through 7 (Figs. 1.16 through 1.20) are FORTRAN implementations of other algorithms of this chapter:

Program	Method	Subroutine name
3	Interval Halving (Bisection)	INTHV
4	Linear Interpolation (*Regula falsi*)	LNINTP
5	Newton's Method	NEWTN
6	Iteration with $x = g(x)$	XGXIT
7	Muller's Method	MULLR

Each of the subroutines is tested with the function $f(x) = e^x − 3x = 0$.

Figure 1.13

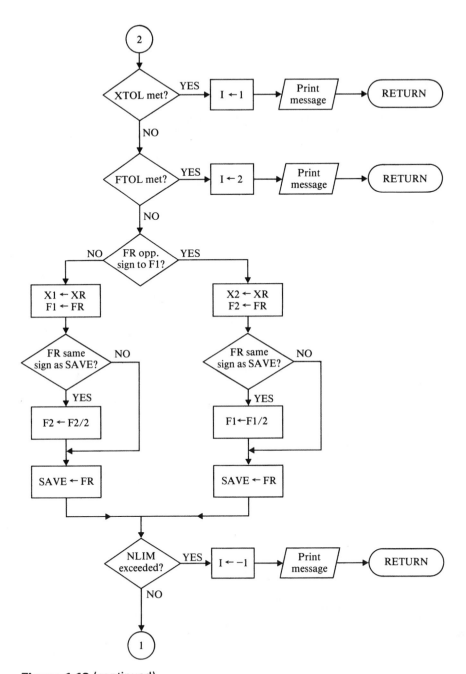

Figure 1.13 (continued)

```
      REAL LMAX
      EXTERNAL FCN
      COMMON W1, W2, A
      W1 = 7.0
      W2 = 9.0
      A = 123.0
      READ (5,100) X1, X2, XTOL, FTOL, NLIM
100   FORMAT (4F10.0, I3)
      I = 0
      CALL MOLNIN (FCN, X1, X2, XR, XTOL, FTOL, NLIM, I)
C     THE ROOT IS RETURNED IN XR.  USE THIS TO CALCULATE LMAX.
      PI = 3.141593
      LMAX = W2/SIN((180. - A - XR)/180.*PI) +
     +   W1/SIN(XR/180.*PI)
      WRITE (6,200) LMAX
200   FORMAT (1H0, 38H MAXIMUM LENGTH OF LADDER, IN FEET, IS ,F6.2)
      STOP
      END


      FUNCTION FCN (X)
      COMMON W1, W2, A
      ANG1 = (180. - A - X)/180.*3.141593
      ANG2 = X/180.*3.141593
      FCN = W2*COS(ANG1)/SIN(ANG1)**2 - W1*COS(ANG2)/SIN(ANG2)**2
      RETURN
      END


      SUBROUTINE MOLNIN (FCN, X1, X2, XR, XTOL, FTOL, NLIM, I)
C     SUBROUTINE FOR ROOT FINDING BY MODIFIED LINEAR INTERPOLATION
C     PARAMETERS ARE -
C     FCN        FUNCTION THAT COMPUTES VALUES FOR F(X).  MUST BE DECLARED
C                EXTERNAL IN CALLING PROGRAM.  IT HAS ONE ARGUMENT, X.
C     X1, X2     INITIAL VALUES OF X.  F(X) MUST CHANGE SIGN AT THESE POINTS.
C     XR         RETURNS THE ROOT TO MAIN PROGRAM.
C     XTOL, FTOL  TOLERANCE VALUES FOR X, F(X) TO TERMINATE INTERATIONS.
C     NLIM       LIMIT TO NUMBER OF ITERATIONS.
C     I          A SIGNAL FOR HOW ROUTINE TERMINATED.
C       I = 1    MEETS TOLERANCE FOR X VALUES.
C       I = 2    MEETS TOLERANCE FOR F(X).
C       I = -1   NLIM EXCEEDED.
C       I = -2   F(X1) NOT OPPOSITE IN SIGN TO F(X2).
C     WHEN THE SUBROUTINE IS CALLED, THE VALUE OF I INDICATES WHETHER TO
C        PRINT EACH VALUE OR NOT.  I = 0 MEANS PRINT THEM, .NE. 0 MEANS DON'T.
      LOGICAL PRINT
      PRINT = .TRUE.
      IF (I .NE. 0)  PRINT = .FALSE.
      F1 = FCN(X1)
      F2 = FCN(X2)
      IF (F1*F2 .GT. 0)  GO TO 50
      FSAVE = F1
      DO 20 J = 1,NLIM
      XR = X2 - F2*(X2 - X1)/(F2 - F1)
      FR = FCN(XR)
      XERR = ABS(X1 - X2)/2.
      IF (.NOT. PRINT)  GO TO 5
      WRITE (6,199) J, XR, FR
```

Figure 1.14 Program 1.   Solution to the ladder-in-the-mine problem, using modified linear interpolation.

```
199     FORMAT (1H , 13HAT ITERATION , I4, 5H X = ,E12.5,
    +   9H, F(X) = ,E12.5)
5       IF (XERR .LE. XTOL)  GO TO 60
        IF (ABS(FR) .LE. FTOL)  GO TO 70
        IF (FR*F1 .LT. 0)  GO TO 10
        X1 = XR
        F1 = FR
        IF (FR*FSAVE .GT. 0)  F2 = F2/2.
        FSAVE = FR
        GO TO 20
10      X2 = XR
        F2 = FR
        IF (FR*FSAVE .GT. 0)  F1 = F1/2.
        FSAVE = FR
20      CONTINUE
C   WHEN LOOP IS NORMALLY COMPLETED, NLIM IS EXCEEDED.
        I = -1
        WRITE (6,200) NLIM, XR, FR
200     FORMAT (1H0, 26HTOLERANCE NOT MET. AFTER , I4,
    +   15H ITERATIONS X = , E12.5,  12H AND F(X) = , E12.5)
        RETURN
C   THIS SECTION FOR RETURN WHEN F(X1) AND F(X2) NOT OPPOSITE IN SIGN.
50      I = -2
        WRITE (6,201)
201     FORMAT (1H0, 35HFUNCTION HAS SAME SIGN AT X1 AND X2)
        RETURN
C   THIS SECTION RETURNS AFTER MEETING XTOL.
60      I = 1
        WRITE (6,202) J, XR, FR
202     FORMAT (1H0, 19HX TOLERANCE MET IN ,I4, 18H ITERATIONS.  X = ,
    +   E12.5, 8H F(X) = , E12.5)
        RETURN
C   THIS SECTION RETURNS AFTER MEETING F(X) TOLERANCE.
70      I = 2
        WRITE (6,203) J, XR, FR
203     FORMAT (1H0, 19HF TOLERANCE MET IN , I4,18H ITERATIONS.  X = ,
    +   E12.5, 8H F(X) = , E12.5)
        RETURN
        END


        PROGRAM 1 OUTPUT

        AT ITERATION    1 X =    0.11678E 03, F(X) =   0.10024E 02
        AT ITERATION    2 X =    0.11556E 03, F(X) =   0.10160E 02
        AT ITERATION    3 X =    0.11314E 03, F(X) =   0.10524E 02
        AT ITERATION    4 X =    0.10836E 03, F(X) =   0.11660E 02
        AT ITERATION    5 X =    0.98739E 02, F(X) =   0.16241E 02
        AT ITERATION    6 X =    0.77902E 02, F(X) =   0.64523E 02
        AT ITERATION    7 X =    0.27286E 02, F(X) =   0.22131E 01
        AT ITERATION    8 X =    0.24282E 02, F(X) = -0.11813E 02
        AT ITERATION    9 X =    0.26812E 02, F(X) =   0.61096E-01
        AT ITERATION   10 X =    0.26799E 02, F(X) =   0.18463E-02
        AT ITERATION   11 X =    0.26798E 02, F(X) = -0.17853E-02
        AT ITERATION   12 X =    0.26799E 02, F(X) = -0.91553E-04

        X TOLERANCE MET IN    12 ITERATIONS.   X = 0.26799E 02 F(X) = -0.91553E-04

        MAXIMUM LENGTH OF LADDER, IN FEET, IS 33.42
```

Figure 1.14 (continued)

```
C    READ IN A CARD WITH THE DEGREE OF THE ORIGINAL POLYNOMIAL AND
C    ESTIMATES R AND S FOR THE QUADRATIC FACTOR. (IF ESTIMATES NOT AVAILABLE,
C    R AND S ARE TAKEN AS ZERO)  ALSO READ IN ERROR TOLERANCE.
     DIMENSION A(13), B(13), C(13)
     READ (5, 100) N, R, S, TOL
100  FORMAT (I10, 3F10.0)
C    READ IN COEFFICIENTS OF THE POLYNOMIAL.
     NP3 = N + 3
     READ (5, 101) (A(I), I = 3,NP3)
101  FORMAT (8F10.0)
C    PRINT HEADING AND ECHO PRINT THE COEFFICIENTS.
     WRITE (6,102)
102  FORMAT (1H1, 37HQUADRATIC FACTORS BY BAIRSTOW METHOD. /
    +        1H , 25HORIGINAL POLYNOMIAL IS -  //
    +        1H , 10HPOWER OF X , 5X, 11HCOEFFICIENT / )
     DO 5 I = 3,NP3
     J = NP3 - I
5    WRITE (6,103) J, A(I)
103  FORMAT (I6, 10X, F10.3)
     WRITE (6,104) TOL
104  FORMAT (1H0, 31HFACTORS ARE, WITH TOLERANCE OF , E14.6X)
C    COMPUTE B AND C ARRAYS.  LIMIT NUMBER OF ITERATIONS TO 20.
     B(1) = 0.
     B(2) = 0.
     C(1) = 0.
     C(2) = 0.
8    DO 20 I = 1,20
     DO 10 J = 3,NP3
     B(J) = A(J) + R*B(J-1) + S*B(J-2)
10   C(J) = B(J) + R*C(J-1) + S*C(J-2)
C    COMPUTE DENOMINATOR AND CHECK IF ZERO. IF ZERO,
C    PERTURB R AND S VALUES.  THEN START OVER.
     DENOM = C(N+1)*C(N+1) - C(N+2)*C(N)
     IF(DENOM  .NE. 0.)  GO TO 12
     R = R + 1.
     S = S + 1.
     GO TO 8
C    COMPUTE NEW R AND S VALUES.  CHECK IF ACCURACY IS OK.
12   DELR = (-B(N+2)*C(N+1) + B(N+3)*C(N))/DENOM
     DELS = (-C(N+1)*B(N+3) + C(N+2)*B(N+2))/DENOM
     R = R + DELR
     S = S + DELS
     IF (ABS(DELR) + ABS(DELS) .LE. TOL)  GO TO 21
20   CONTINUE
C    TOLERANCE NOT MET,  PRINT FACTOR AND QUIT.
     WRITE (6,105) R, S
105  FORMAT (1H0, 48HTOLERANCE NOT MET IN 20 ITERATIONS. LAST FACTOR
    +        9HFOUND WAS / 9H  X**2 - , F10.5, 5H X - , F10.5)
     STOP
C    PRINT FACTOR, TEST DEGREE OF REDUCED POLYNOMIAL.
C    AND CONTINUE IF GREATER THAN TWO.
21   WRITE (6,106) R, S
106  FORMAT (1H , 7HX**2 - , F10.5, 5H X - , F10.5)
     N = N - 2
     IF (N - 2)  22, 23, 24
```

Figure 1.15 Program 2:   Bairstow's method.

```
C   REDUCED EQUATION IS OF DEGREE ONE - PRINT IT.
22      WRITE (6,107) B(N+2), B(N+3)         CHECK THESE COEFF.
107     FORMAT (1H , F10.5, 4HX + , F10.5)
        STOP
C   REDUCED EQUATION IS OF DEGREE TWO - PRINT IT.
23      WRITE (6,108) B(N+1), B(N+2), B(N+3)
108     FORMAT (1H , F10.5, 7HX**2 + , F10.5, 4HX + , F10.5)
        STOP
C   DEGREE OF REDUCED EQUATION MORE THAN TWO.  SET COEFFICIENTS
C   INTO A ARRAY AND GET NEXT FACTOR.
24      NP3 = N + 3
        DO 30 I = 3,NP3
30      A(I) = B(I)
        GO TO 8
        END
```

PROGRAM 2 OUTPUT

QUADRATIC FACTORS BY BAIRSTOW METHOD.
ORIGINAL POLYNOMIAL IS —

POWER OF X	COEFFICIENT
5	1.000
4	−17.800
3	99.410
2	−261.218
1	352.611
0	−134.106

FACTORS ARE, WITH TOLERANCE OF   0.100000E−03
X**2 −   4.20000 X −   −2.10000
X**2 −   3.30001 X −   −6.20008
        1.00000X +   −10.30001

Figure 1.15 (continued)

```
      EXTERNAL FCN
      X1 = 0.
      X2 = 1.
      XTOL = .0001
      FTOL = .00001
      NLIM = 50
      I = 0
      CALL INTHV (FCN, X1, X2, XR, XTOL, FTOL, NLIM, I)
      STOP
      END


      SUBROUTINE INTHV (FCN, X1, X2, XR, XTOL, FTOL, NLIM, I)
C   SUBROUTINE FOR INTERVAL HALVING
C   PARAMETERS ARE -
C     FCN       FUNCTION THAT COMPUTES VALUES FOR F(X).  MUST BE DECLARED
C                 EXTERNAL IN CALLING PROGRAM.  IT HAS ONE ARGUMENT, X.
C     X1, X2    INITIAL VALUES OF X.  F(X) MUST CHANGE SIGN AT THESE POINTS.
C     XR        RETURNS THE ROOT TO MAIN PROGRAM.
C     XTOL, FTOL  TOLERANCE VALUES FOR X, F(X) TO TERMINATE INTERATIONS.
C     NLIM      LIMIT TO NUMBER OF ITERATIONS.
C     I         A SIGNAL FOR HOW ROUTINE TERMINATED.
C       I = 1   MEETS TOLERANCE FOR X VALUES.
C       I = 2   MEETS TOLERANCE FOR F(X).
C       I = -1  NLIM EXCEEDED.
C       I = -2  F(X1) NOT OPPOSITE IN SIGN TO F(X2).
C   WHEN THE SUBROUTINE IS CALLED, THE VALUE OF I INDICATES WHETHER TO
C     PRINT EACH VALUE OR NOT.  I = 0 MEANS PRINT THEM, .NE. 0 MEANS DON'T.
      LOGICAL PRINT
      PRINT = .TRUE.
      IF (I .NE. 0)  PRINT = .FALSE.
      F1 = FCN(X1)
      F2 = FCN(X2)
      IF (F1*F2 .GT. 0)  GO TO 50
      DO 20 J = 1,NLIM
      XR = (X1 + X2)/2.
      FR = FCN(XR)
      XERR = ABS(X1 - X2)/2.
      IF (.NOT. PRINT)  GO TO 5
      WRITE (6,199) J, XR, FR
199   FORMAT (1H , 13HAT ITERATION , I4, 5H X = ,E12.5,
     + 9H, F(X) = ,E12.5)
5     IF (XERR .LE. XTOL)  GO TO 60
      IF (ABS(FR) .LE. FTOL)  GO TO 70
      IF (FR*F1 .LT. 0)  GO TO 10
      X1 = XR
      F1 = FR
      GO TO 20
10    X2 = XR
      F2 = FR
20    CONTINUE
```

Figure 1.16 Program 3:   Interval halving.

```
C   WHEN LOOP IS NORMALLY COMPLETED, NLIM IS EXCEEDED.
      I = -1
      WRITE (6,200) NLIM, XR, FR
200   FORMAT (1H0, 26HTOLERANCE NOT MET.  AFTER , I4,
     +   15H ITERATIONS X = , E12.5,  12H AND F(X) = , E12.5)
      RETURN
C   THIS SECTION FOR RETURN WHEN F(X1) AND F(X2) NOT OPPOSITE IN SIGN.
50    I = -2
      WRITE (6,201)
201   FORMAT (1H0, 35HFUNCTION HAS SAME SIGN AT X1 AND X2)
      RETURN
C   THIS SECTION RETURNS AFTER MEETING XTOL.
60    I = 1
      WRITE (6,202) J, XR, FR
202   FORMAT (1H0, 19HX TOLERANCE MET IN ,I4, 18H ITERATIONS.  X = ,
     +   E12.5, 8H F(X) = , E12.5)
      RETURN
C   THIS SECTION RETURNS AFTER MEETING F(X) TOLERANCE.
70    I = 2
      WRITE (6,203) J, XR, FR
203   FORMAT (1H0, 19HF TOLERANCE MET IN , I4,18H ITERATIONS.  X = ,
     +   E12.5, 8H F(X) = , E12.5)
      RETURN
      END


      FUNCTION FCN(X)
      FCN = EXP(X) - 3.*X
      RETURN
      END


      PROGRAM 3 OUTPUT
      AT ITERATION    1 X =   0.50000E 00, F(X) =   0.14872E 00
      AT ITERATION    2 X =   0.75000E 00, F(X) =  -0.13300E 00
      AT ITERATION    3 X =   0.62500E 00, F(X) =  -0.67549E-02
      AT ITERATION    4 X =   0.56250E 00, F(X) =   0.67554E-01
      AT ITERATION    5 X =   0.59375E 00, F(X) =   0.29515E-01
      AT ITERATION    6 X =   0.60938E 00, F(X) =   0.11156E-01
      AT ITERATION    7 X =   0.61719E 00, F(X) =   0.21439E-02
      AT ITERATION    8 X =   0.62109E 00, F(X) =  -0.23193E-02
      AT ITERATION    9 X =   0.61914E 00, F(X) =  -0.91553E-04
      AT ITERATION   10 X =   0.61816E 00, F(X) =   0.10252E-02
      AT ITERATION   11 X =   0.61865E 00, F(X) =   0.46730E-03
      AT ITERATION   12 X =   0.61890E 00, F(X) =   0.18787E-03
      AT ITERATION   13 X =   0.61902E 00, F(X) =   0.48637E-04
      AT ITERATION   14 X =   0.61908E 00, F(X) =   0.20981E-04

      X TOLERANCE MET IN   14 ITERATIONS.    X =   0.61908E 00 F(X) = -0.20981E-04
```

Figure 1.16 (continued)

```
      EXTERNAL FCN
      X1 = 0.
      X2 = 1.
      XTOL = .0001
      FTOL = .00001
      NLIM = 50
      I = 0
      CALL LNINTP (FCN, X1, X2, XR, XTOL, FTOL, NLIM, I)
      STOP
      END


      SUBROUTINE LNINTP (FCN, X1, X2, XR, XTOL, FTOL, NLIM, I)
C  SUBROUTINE FOR LINEAR INTERPOLATION TO FIND A ROOT OF F(X) = 0.
C     PARAMETERS ARE -
C     FCN          FUNCTION THAT COMPUTES VALUES FOR F(X).  MUST BE DECLARED
C                     EXTERNAL IN CALLING PROGRAM.  IT HAO ONE ARGUMENT, X.
C     X1, X2    INITIAL VALUES OF X.  F(X) MUST CHANGE SIGN AT THESE POINTS.
C     XR           RETURNS THE ROOT TO MAIN PROGRAM.
C     XTOL, FTOL   TOLERANCE VALUES FOR X, F(X) TO TERMINATE INTERATIONS.
C     NLIM         LIMIT TO NUMBER OF ITERATIONS.
C     I            A SIGNAL FOR HOW ROUTINE TERMINATED.
C     I = 1     MEETS TOLERANCE FOR X VALUES.
C     I = 2     MEETS TOLERANCE FOR F(X).
C     I = -1    NLIM EXCEEDED.
C     I = -2    F(X1) NOT OPPOSITE IN SIGN TO F(X2).
C  WHEN THE SUBROUTINE IS CALLED, THE VALUE OF I INDICATES WHETHER TO
C     PRINT EACH VALUE OR NOT.  I = 0 MEANS PRINT THEM, .NE. 0 MEANS DON'T.
      LOGICAL PRINT
      PRINT = .TRUE.
      IF (I .NE. 0)  PRINT = .FALSE.
      F1 = FCN(X1)
      F2 = FCN(X2)
      IF (F1*F2 .GT. 0)  GO TO 50
      DO 20 J = 1,NLIM
      XR = X2 - F2*(X2 - X1)/(F2 - F1)
      FR = FCN(XR)
      XERR = ABS(X1 - X2)/2.
      IF (.NOT. PRINT)  GO TO 5
      WRITE (6,199) J, XR, FR
199   FORMAT (1H , 13HAT ITERATION , I4, 5H X = ,E12.5,
     +  9H, F(X) = ,E12.5)
5     IF (XERR .LE. XTOL)  GO TO 60
      IF (ABS(FR) .LE. FTOL)  GO TO 70
      IF (FR*F1 .LT. 0)  GO TO 10
      X1 = XR
      F1 = FR
      GO TO 20
10    X2 = XR
      F2 = FR
20    CONTINUE
C  WHEN LOOP IS NORMALLY COMPLETED, NLIM IS EXCEEDED.
      I = -1
      WRITE (6,200) NLIM, XR, FR
200   FORMAT (1H0, 26HTOLERANCE NOT MET.  AFTER , I4,
     +  15H ITERATIONS X = , E12.5,  12H AND F(X) = , E12.5)
      RETURN
```

Figure 1.17 Program 4:   Linear interpolation.

```
C   THIS SECTION FOR RETURN WHEN F(X1) AND F(X2) NOT OPPOSITE IN SIGN.
50      I = -2
        WRITE (6,201)
201     FORMAT (1H0, 35HFUNCTION HAS SAME SIGN AT X1 AND X2)
        RETURN
C   THIS SECTION RETURNS AFTER MEETING XTOL.
60      I = 1
        WRITE (6,202) J, XR, FR
202     FORMAT (1H0, 19HX TOLERANCE MET IN ,I4, 18H ITERATIONS.  X = ,
     +  E12.5, 8H F(X) = , E12.5)
        RETURN
C   THIS SECTION RETURNS AFTER MEETING F(X) TOLERANCE.
70      I = 2
        WRITE (6,203) J, XR, FR
203     FORMAT (1H0, 19HF TOLERANCE MET IN , I4,18H ITERATIONS.  X = ,
     +  E12.5, 8H F(X) = , E12.5)
        RETURN
        END


        FUNCTION FCN(X)
        FCN = EXP(X) - 3.*X
        RETURN
        END


        PROGRAM 4 OUTPUT

        AT ITERATION     1 X =   0.78020E 00, F(X) = -0.15869E 00
        AT ITERATION     2 X =   0.67335E 00, F(X) = -0.59252E-01
        AT ITERATION     3 X =   0.63568E 00, F(X) = -0.18736E-01
        AT ITERATION     4 X =   0.62399E 00, F(X) = -0.56114E-02
        AT ITERATION     5 X =   0.62051E 00, F(X) = -0.16518E-02
        AT ITERATION     6 X =   0.61949E 00, F(X) = -0.48447E-03
        AT ITERATION     7 X =   0.61919E 00, F(X) = -0.14210E-03
        AT ITERATION     8 X =   0.61910E 00, F(X) = -0.41962E-04
        AT ITERATION     9 X =   0.61907E 00, F(X) = -0.11444E-04
        AT ITERATION    10 X =   0.61906E 00, F(X) = -0.38147E-05

        F TOLERANCE MET IN    10 ITERATIONS.   X =   0.61906E 00 F(X) = -0.38147E-05
```

Figure 1.17 (continued)

```
      EXTERNAL FCN, FDER
      X = 0.5
      XTOL = .0001
      FTOL = .00001
      NLIM = 50
      I = 0
      CALL NEWTN (FCN, FDER, X, XTOL, FTOL, NLIM, I)
      STOP
      END


      SUBROUTINE NEWTN (FCN, FDER, X, XTOL, FTOL, NLIM, I)
C  SUBROUTINE FOR ROOT FINDING BY NEWTON'S METHOD
C   PARAMETERS ARE -
C    FCN        FUNCTION THAT COMPUTES F(X)
C    FDER       FUNCTION THAT COMPUTES THE DERIVATIVE OF F
C               FCN, FDER BOTH MUST BE DECLARED EXTERNAL IN CALLING PROGRAM.
C               EACH FUNCTION HAS ONE ARGUMENT, X.
C    X          INITIAL VALUE, SHOULD BE NEAR ROOT.  X ALSO RETURNS
C               THE VALUE OF THE ROOT TO THE CALLER.
C    XTOL, FTOL   TOLERANCE VALUES FOR X, F(X) TO TERMINATE ITERATIONS.
C    NLIM       LIMIT TO NUMBER OF ITERATIONS.
C    I          A SIGNAL FOR HOW ROUTINE TERMINATED.
C      I = 1    MEETS TOLERANCE FOR X VALUES.
C      I = 2    MEETS TOLERANCE FOR F(X).
C      I = -1   NLIM EXCEEDED.
C    WHEN THE SUBROUTINE IS CALLED, THE VALUE OF I INDICATES WHETHER TO
C      PRINT EACH VALUE OR NOT.  I = 0 MEANS PRINT THEM, .NE. 0 MEANS DON'T.
      LOGICAL PRINT
      PRINT = .TRUE.
      IF (I .NE. 0)  PRINT = .FALSE.
      FX = FCN(X)
      DO 20 J = 1,NLIM
      DELX = FX/FDER(X)
      X = X - DELX
      FX = FCN(X)
      IF (.NOT. PRINT)  GO TO 5
      WRITE (6,199) J, X, FX
199   FORMAT (1H , 13HAT ITERATION , I4, 5H X = , E12.5,
     +  9H, F(X) = , E12.5)
5     IF (ABS(DELX) .LE. XTOL)  GO TO 60
      IF (ABS(FX) .LE. FTOL)  GO TO 70
20    CONTINUE
C  WHEN LOOP IS NORMALLY COMPLETED, NLIM IS EXCEEDED.
      I = -1
      WRITE (6,200) NLIM, X, FX
200   FORMAT (1H0, 26HTOLERANCE NOT MET.  AFTER , I4,
     +  15H ITERATIONS X = , E12.5,  12H AND F(X) = , E12.5)
      RETURN
C  THIS SECTION RETURNS AFTER MEETING XTOL.
60    I = 1
      WRITE (6,202) J, X, FX
202   FORMAT (1H0, 19HX TOLERANCE MET IN ,I4, 18H ITERATIONS.  X = ,
     +  E12.5, 8H F(X) = , E12.5)
      RETURN
```

Figure 1.18 Program 5:   Newton's method.

```
C  THIS SECTION RETURNS AFTER MEETING F(X) TOLERANCE.
70    I = 2
      WRITE (6,203) J, X, FX
203   FORMAT (1H0, 19HF TOLERANCE MET IN , I4,18H ITERATIONS.   X = ,
     +  E12.5, 8H F(X) = , E12.5)
      RETURN
      END


      FUNCTION FCN(X)
      FCN = EXP(X) - 3.*X
      RETURN
      END


      FUNCTION FDER (X)
      FDER = EXP(X)   - 3.
      RETURN
      END


      PROGRAM 5 OUTPUT

      AT ITERATION     1 X =   0.61006E 00, F(X) =   0.10363E-01
      AT ITERATION     2 X =   0.61900E 00, F(X) =   0.73433E-04
      AT ITERATION     3 X =   0.61906E 00, F(X) =   0.95367E-06

      X TOLERANCE MET IN    3 ITERATIONS.    X =   0.61906E 00 F(X) =   0.95367E-06
```

Figure 1.18 (continued)

```
      EXTERNAL GFCN
      X = 0.6
      I = 0
      CALL XGXIT (GFCN, X, 0.0001, 50, I)
      STOP
      END


      SUBROUTINE XGXIT (GFCN, X, XTOL, NLIM, I)
C   SUBROUTINE FOR ROOT FINDING BY ITERATION WITH X = G(X).
C     PARAMETERS ARE -
C       GFCN        FUNCTION TO COMPUTE G(X), MUST BE DECLARED EXTERNAL IN
C                     CALLING PROGRAM.  IT HAS ONE ARGUMENT, X.
C       X           INITIAL VALUE TO BEGIN ITERATIONS.  X ALSO RETURNS
C                     THE VALUE OF THE ROOT TO THE CALLER.
C       XTOL        TOLERANCE VALUE FOR CHANGE IN X TO TERMINATE ITERATIONS.
C       NLIM        LIMIT TO NUMBER OF ITERATIONS.
C       I           A SIGNAL FOR HOW TERMINATED.
C       I = 1     MEETS TOLERANCE FOR X VALUES.
C       I = -1    NLIM EXCEEDED.
C       I = -2    ITERATIONS APPEAR TO DIVERGE IN INITIAL CALCULATIONS.
C     WHEN THE SUBROUTINE IS CALLED, THE VALUE OF I INDICATES WHETHER TO
C       PRINT EACH VALUE OR NOT.  I = 0 MEANS PRINT, .NE. 0 MEANS DON'T.
      LOGICAL PRINT
      PRINT = .TRUE.
      IF (I .NE. 0)  PRINT = .FALSE.
      J = 1
      SAVEX = X
      X = GFCN(X)
      DEL1 = ABS(X - SAVEX)
      IF (DEL1 .LE. XTOL)  GO TO 60
      IF (.NOT. PRINT)  GO TO 5
      WRITE (6,199) J, SAVEX, X
199   FORMAT (1H , 13HAT ITERATION , I4, 5H X = , F12.5,
     +  8H G(X) = , F12.5)
5     DO 20 J = 2,NLIM
      SAVEX = X
      X = GFCN(X)
      DEL2 = ABS(X - SAVEX)
      IF (J .GT. 2)  GO TO 10
      IF (DEL1 .LE. DEL2)  GO TO 50
10    IF (DEL2 .LE. XTOL)  GO TO 60
      IF (.NOT. PRINT)  GO TO 20
      WRITE (6,199) J, SAVEX, X
20    CONTINUE
C   WHEN LOOP IS NORMALLY TERMINATED, NLIM IS EXCEEDED.
      I = -1
      WRITE (6,200) NLIM, SAVEX, X
200   FORMAT(1H0, 27HMETHOD DID NOT CONVERGE IN , I4,
     +  35H ITERATIONS.  FINAL VALUES ARE X = , F12.5,
     +  9H, G(X) = , F12.5)
      RETURN
C   THIS SECTION RETURNS AFTER FINDING DIVERGENCE AT START.
50    I = -2
      WRITE (6,201)
```

Figure 1.19 Program 6:   Iteration with $X = G(X)$.

```
201    FORMAT (1H0, 41HFIRST THREE X VALUES INDICATE DIVERGENCE.)
       RETURN
C  THIS SECTION FOR RETURN AFTER MEETING XTOL.
60     I = 1
       WRITE (6,202) J, SAVEX, X
202    FORMAT (1H0, 12HXTOL MET IN , I4, 18H ITERATIONS.   X = ,
     +   F12.5, 9H, G(X) = , F12.5)
       RETURN
       END


       FUNCTION GFCN (X)
       GFCN = (EXP(X) - X)/2.
       RETURN
       END


       PROGRAM 6 OUTPUT

             $ ENTRY
       AT ITERATION      1 X =     0.60000 G(X) =       0.61106
       AT ITERATION      2 X =     0.61106 G(X) =       0.61566
       AT ITERATION      3 X =     0.61566 G(X) =       0.61761
       AT ITERATION      4 X =     0.61761 G(X) =       0.61844
       AT ITERATION      5 X =     0.61844 G(X) =       0.61879
       AT ITERATION      6 X =     0.61879 G(X) =       0.61895

       XTOL MET IN    7 ITERATIONS.    X =    0.61895,    G(X) =   0.61901
```

Figure 1.19 (continued)

```
      EXTERNAL FCN
      XR = 0.5
      H = 0.2
      XTOL = .0001
      FTOL = .00001
      NLIM = 50
      I = 0
      CALL MULLR (FCN, XR, H, XTOL, FTOL, NLIM, I)
      STOP
      END


      SUBROUTINE MULLR (FCN, XR, H, XTOL, FTOL, NLIM, I)
      DIMENSION Y(3)
C   SUBROUTINE FOR FINDING ROOTS BY MULLER'S METHOD
C     PARAMETERS ARE -
C     FCN          FUNCTION THAT COMPUTES F(X).  MUST BE DECLARED
C                  EXTERNAL IN CALLING PROGRAM.  IT HAS ONE ARGUMENT, X.
C     XR           INITIAL APPROXIMATION TO THE ROOT.  USED TO BEGIN THE
C                  ITERATIONS.  XR ALSO RETURNS THE ROOT TO CALLER.
C     H            DISPLACEMENT FROM X USED TO BEGIN CALCULATIONS.  THE
C                  FIRST QUADRATIC IS FITTED AT F(X), F(X+H), F(X-H).
C     XTOL, FTOL   TOLERANCE VALUES FOR X, F(X) TO TERMINATE ITERATIONS.
C     I            A SIGNAL FOR HOW THE ROUTINE TERMINATED.
C     I = 1    MEETS TOLERANCE FOR X VALUES.
C     I = 2    MEETS TOLERANCE FOR F(X).
C     I = -1   NLIM EXCEEDED.
C   WHEN THE SUBROUTINE IS CALLED, THE VALUE OF I INDICATES WHETHER TO
C     PRINT EACH VALUE OR NOT.  I = 0 MEANS PRINT THEM, .NE. 0 MEANS DON'T.
      LOGICAL PRINT
      PRINT = .TRUE.
      IF (I .NE. 0)  PRINT = .FALSE.
C   SET INITIAL VALUES INTO Y ARRAY AND EVALUATE F AT THESE PTS.
      Y(1) = XR - H
      Y(2) = XR
      Y(3) = XR + H
      F1 = FCN (Y(1))
      F2 = FCN (Y(2))
      F3 = FCN (Y(3))
C   BEGIN THE ITERATIONS.
      DO 20 J = 1,NLIM
      H1 = Y(2) - Y(1)
      H2 = Y(3) - Y(2)
      G = H1/H2
      A = (F3*G - F2*(1. + G) + F1)/(H1*(H1 + H2))
      B = (F3 - F2 - A*H2*H2)/H2
      C = F2
      DISC = SQRT(B*B - 4.*A*C)
      IF (B .LT. 0)  DISC = -DISC
      DELX = -2.*C/(B + DISC)
      XR = Y(2) + DELX
      FR = FCN (XR)
      IF (.NOT. PRINT)  GO TO 5
      WRITE (6,199) J, XR, FR
199   FORMAT (1H , 13HAT ITERATION , I4, 5H X = , E12.5,
     + 9H, F(X) = ,E12.5)
5     IF (ABS(DELX) .LE. XTOL)  GO TO 60
      IF (ABS(FR) .LE. FTOL)  GO TO 70
```

Figure 1.20 Program 7:   Muller's method.

```
C   SELECT THE THREE POINTS FOR THE NEXT ITERATION.  WHEN DELX .GT. 0,
C    CHOSE Y(2), Y(3), AND XR, BUT WHEN DELX .LT. 0, CHOOSE
C    Y(1), Y(2), AND XR.
C   ENTER THE PROPER SET INTO Y ARRAY SO THEY ARE IN ASCENDING ORDER.
        IF (DELX .LT. 0)  GO TO 7
        Y(1) = Y(2)
        F1 = F2
        IF (DELX .LT. H2)  GO TO 6
        Y(2) = Y(3)
        F2 = F3
        Y(3) = XR
        F3 = FR
        GO TO 20
6       Y(2) = XR
        F2 = FR
        GO TO 20
7       Y(3) = Y(2)
        F3 = F2
        IF (ABS(DELX).GT. H1)  GO TO 8
        Y(2) = XR
        F2 = FR
        GO TO 20
8       Y(2) = Y(1)
        F2 = F1
        Y(1) = XR
        F1 = FR
20      CONTINUE
C   WHEN LOOP IS NORMALLY TERMINATED, NLIM IS EXCEEDED.
        I = -1
        WRITE (6,200) NLIM, XR, FR
200     FORMAT (1H0, 26HTOLERANCE NOT MET.  AFTER , I4,
       +  15H ITERATIONS X = , E12.5, 12H AND F(X) = , E12.5)
        RETURN
C   THIS SECTION RETURNS AFTER MEETING XTOL.
60      I = 1
        WRITE (6,202) J, XR, FR
202     FORMAT (1H0, 19HX TOLERANCE MET IN , I4, 18H ITERATIONS.  X = ,
       +  E12.5, 8H F(X) = , E12.5)
        RETURN
C   THIS SECTION RETURNS AFTER MEETING F(X) TOLERANCE.
70      I = 2
        WRITE (6,203) J, XR, FR
203     FORMAT (1H0, 19HF TOLERANCE MET IN , I4, 18H ITERATIONS.  X = ,
       +  E12.5, 8H F(X) = , E12.5)
        RETURN
        END


        FUNCTION FCN (X)
        FCN = EXP(X) - 3.*X
        RETURN
        END


        PROGRAM 7 OUTPUT

        AT ITERATION      1 X =    0.61982E 00, F(X) = -0.87166E-03
        AT ITERATION      2 X =    0.61906E 00, F(X) =  0.19073E-05

        F TOLERANCE MET IN    2 ITERATIONS.  X =    0.61906E 00 F(X) =   0.19073E-05
```

Figure 1.20 (continued)

These subroutines might appropriately be placed in the FORTRAN library at your installation so that they may be called automatically by your programs.

EXERCISES*

Section 1.2

▶1. *The equation $x^2 - 2 = 0$ has the obvious roots* $\pm\sqrt{2} = \pm 1.414214$. Use six iterations of the method of halving the interval to evaluate the positive root beginning with the interval $[1, 2]$. How many iterations would be required to evaluate the root correct to five significant figures? After this many iterations, what is the bound to the error as measured by one-half the last interval? (You do not have to perform the iterations to answer the last two questions.)

2. The quadratic $(x - 0.4)(x - 0.6) = x^2 - x + 0.24$ has zeros at $x = 0.4$ and $x = 0.6$, of course. Observe that the endpoints of the interval $[0, 1]$ are not satisfactory to begin the interval-halving method. Graph the function, and from this deduce the boundaries of intervals which will converge to each of the zeros. If the endpoints of the interval $[0.5, 1.0]$ are used to begin the search, what is a bound to the error after five iterations? What is the actual error after five repetitions of interval-halving?

3. Interval-halving applies to any continuous function, not just to polynomials. Find where the graphs of $y = 3x$ and $y = e^x$ intersect by finding the root of $e^x - 3x = 0$ correct to four decimals.

4. Use interval-halving to find the smallest positive root of these equations. In each case first determine a suitable interval, then compute the root with relative accuracy of 0.5%.

a) $\tan x - x - 1 = 0$ ▶b) $x^3 - x^2 - 2x + 1 = 0$

▶c) $2e^{-x} - \sin x = 0$ d) $3x^3 + 4x^2 - 8x - 1 = 0$

Section 1.3

5. The polynomial $x^3 + x^2 - 3x - 3 = 0$, used as an example in Sections 1.2 and 1.3 where the root at $x = \sqrt{3}$ was approximated, has its other roots at $x = -1$ and $x = -\sqrt{3}$. Beginning with two suitable values that bracket the value $-\sqrt{3}$, show that the method of linear interpolation converges to that root.

6. In Exercise 5, if one tried as starting values $x = -1.5$ and $x = -1.7$, the function would not change sign, and, hence, they do not qualify for beginning the method of linear interpolation. However, the secant method can begin with these values. Use them to begin the secant method. How many iterations are needed to estimate the root correct to four decimals? Suppose the starting values are -1.5 and -1.1, which root is obtained by the secant method? What root if we begin with -1.5 and -1.25?

---

* Answers are given at the end of the text for exercises marked by ▶.

▶7. Find where the cubic $y = x^3 - x + 1$ intersects the parabola $y = 2x^2$. Make a sketch of the two curves to locate the intersections, and then use linear interpolation and/or the secant method to evaluate the x-values of the points of intersection.

8. a) Use the method of linear interpolation to solve the equations in Exercise 4. b) Use Modified Linear Interpolation to solve these equations and compare the rates of convergence with those obtained in part (a).

9. Write the algorithm for the secant method, following the model of the other algorithms of this chapter.

Section 1.4

▶10. Find a root near $x = -0.5$ of $e^x - 3x^2 = 0$ by Newton's method, to six-digit accuracy.

11. The equation $e^x - 3x^2 = 0$ not only has a root near $x = -0.5$, but also near $x = 4.0$. Find the positive root by Newton's method, accurate to 0.01% relative accuracy.

12. Use Newton's method to solve the equations in Exercise 4. Use synthetic division for evaluating the polynomials and their derivatives.

13. Use Newton's method on the equation $x^2 = N$ to derive the algorithm for the square root of N:

$$x_{i+1} = \frac{1}{2}\left(x_i + \frac{N}{x_i}\right),$$

where x_0 is an initial approximation to $\sqrt{N}$.

14. a) If the algorithm of Exercise 13 is applied twice, show that

$$\sqrt{N} \doteq \frac{A + B}{4} + \frac{N}{A + B}, \qquad \text{where} \quad N = AB.$$

b) Show also that the relative error (error/true value) in (a) is approximately

$$\frac{1}{8}\left(\frac{A - B}{A + B}\right)^4.$$

15. Use Newton's method to find the roots of the following polynomials:
   a) $x^3 - 2x + 1 = 0$ ▶b) $x^4 + x^3 - 4x^2 - 3x + 3 = 0$
   c) $x^4 - 4.4x^3 + 9.43x^2 - 14.86x + 7.15 = 0$

16. $(x - 1)^3(x - 2) = x^4 - 5x^3 + 9x^2 - 7x + 2 = 0$ obviously has a root at $x = 2$, and a triple root at $x = 1$. Beginning with $x = 2.1$, use Newton's method once, and observe the degree of improvement. Then start with $x = 0.9$, and note the much slower convergence to the triple root even though the initial error is only 0.1 in each case. Use the secant method beginning with $f(0.9)$ and $f(1.1)$, and observe that just one application brings one quite close to the root in contrast to Newton's method. Explain.

Section 1.5

17. $f(x) = e^x - 3x^2 = 0$ has three roots. An obvious rearrangement is

$$x = \pm\sqrt{e^x/3}.$$

Show, beginning with $x_0 = 0$, that this will converge to a root near -0.5 if the negative value is used, and that it converges to a root near 1.0 if the positive value is used. Show, however, that this form does not converge to the third root near 4.0 even when a nearly exact starting value is used. Find another form which will converge to the root near 4.0.

►18. One root of the quadratic $x^2 + x - 1 = 0 = x(x + 1) - 1$ is at $x = 0.6180$. The equivalent form $x = 1/(x + 1)$ converges to this root beginning at $x_0 = 1$. Carrying four or five decimals, how many steps are required to reach the root (correct to four decimals) by linear iteration? If Aitken acceleration is used after three approximations are available, how many iterations are required?

19. The form $x = 1/(x + 1)$ of Exercise 18 will converge to a root of the quadratic for many starting values in addition to $x_0 = 1$. For what starting values will it not converge to a root? (For this problem do not stop with a division by zero, i.e., for $x_0 = -1$, $x_1 = 1/0$. Use $x_3 = \lim_{x_2 \to \infty} 1/(x_2 + 1) = 0$.)

►20. The cubic $2x^3 + 4x^2 - 2x - 5 = 0$ has a root near $x = 1$. Find at least three rearrangements that will converge to this root beginning with $x_0 = 1.0$.

21. Use the method of iteration to solve the equations of Exercise 4.

Section 1.6

22. a) Show that if $P_n(x)$ has a double root at $x = r$, then $P'_n(r) = 0$.
 b) Show that if $P_n(x)$ has a root of multiplicity m at $x = r$, then

$$P_n^{(i)}(r) = 0, \quad i = 1, 2, \ldots, m - 1.$$

23. If $P_n(x) = 0$ has a double root at $x = r$, so that $P'_n(r) = 0$, then the condition

$$\left| \frac{f(x)f''(x)}{[f'(x)]^2} \right| < 1$$

may not hold for any interval including r. According to Section 1.6, we therefore have, for Newton's method, no assurance of convergence to the root at r. The simple equation $(x - 2)^2 = 0 = x^2 - 4x + 4$ has such a double root at $x = 2$, and $f'(x) = 2x - 4$ is zero at $x = 2$. Still, beginning at any finite value of x_0, Newton's method will converge! Reconcile this fact with the convergence criterion of Section 1.6.

►24. For the quadratic $x^2 - 4x + 4 = 0$, which has a double root at $x = 2$, begin with $x_0 = 1$ and compute successive approximations to the root by Newton's method. Tabulate the errors at each step and compare each with the next. Is Newton's method quadratically or only linearly convergent in this case? How could one accelerate the convergence?

25. If $f(x)$, $f'(x)$, $f''(x)$ are continuous and bounded on a certain interval containing $x = r$, and if both $f(r) = 0$ and $f'(r) = 0$, but $f''(r) \neq 0$, show that the form

$$x_{n+1} = x_n - 2 \frac{f(x_n)}{f'(x_n)}$$

will converge quadratically if x_n is in the interval. (*Hint.* The algorithm is of the form $x_{n+1} = g(x_n)$. Show that $g'(r) = 0$, using L'Hospital's rule.)

26. The method suggested by Exercise 25 extends to a root of multiplicity m:

$$x_{n+1} = x_n - m \frac{f(x_n)}{f'(x_n)}.$$

Show, with suitable restrictions on $f(x)$, that this is quadratically convergent.

▶27. Two of the four roots of the quartic $x^4 + 2x^3 - 7x^2 + 3 = 0$ are positive. Find these by Newton's method correct to seven decimal places using an electronic calculator, and then determine the other roots from the reduced equation. Use synthetic division.

28. Investigate the convergence rate of the square-root algorithm of Exercise 14(a).

Section 1.7

▶29. Beginning with the trial factor $x^2 - 4x + 5$, improve by successive applications of Bairstow's method to find the quadratic factors of

$$x^4 - 3.1x^3 + 2.1x^2 + 1.1x + 5.2.$$

What are the four zeros of this polynomial?

30. Solve Exercise 27 by Bairstow's method to get quadratic factors.

31. None of the roots of this fourth-degree polynomial are real. Find these complex roots by resolving into quadratic factors:

$$x^4 + 4x^3 + 21x^2 + 4x + 20 = 0.$$

32. When the modulus of one pair of complex roots is the same as for another pair of complex roots, the Bairstow method is slow to converge. Try your patience on

$$x^4 - x^3 + x^2 - x + 1 = 0,$$

which has as factors $(x^2 + 0.618034x + 1)(x^2 - 1.618034x + 1)$. Start with $x^2 + 0.6x + 1$. What is the modulus of the roots?

Section 1.8

33. Use the Q–D algorithm to approximate the roots of:

▶a) $x^3 - x^2 - 2x + 1 = 0$ b) $3x^3 + 3x^2 - 8x - 1 = 0$
c) $x^4 - 5x^3 + 9x^2 - 7x + 2 = 0$ ▶d) $x^4 - 3.1x^3 + 2.1x^2 + 1.1x + 5.2 = 0$

34. The Q–D algorithm is also not very efficient on Exercise 32. Solve that problem by Q–D.

Section 1.9

35. Use Muller's method to solve the problems below. Apply four iterations and determine how the successive errors are related to each other. Use starting values that differ by 0.2 from each other.

    a) $2x^3 + 4x^2 - 2x - 5 = 0$, root near 1.0 (exact value is 1.07816259);
    b) $e^x - 3x^2 = 0$, root near 4.0 (exact value is 3.73307903);
    c) $e^x - 3x^2 = 0$, root near 1.0;
    d) $\tan x - x - 1 = 0$, root near 1.1.

36. Muller's method is sometimes used in a "self-starting" form. Instead of specifying starting values which are near a root, the algorithm arbitrarily begins with $x_0 = 0$, $x_1 = 0.5$, $x_2 = -0.5$. Use these starting values on the equations of Exercise 35. One is supposed to obtain the root nearest the origin by this technique. Will this always be true?

37. An extension of the self-starting principle in Exercise 36 is to deflate the function after a first root is found. (Deflating means to form a new function that has the same roots as the original function except the root $x = r$, by dividing $f(x)$ by $(x - r)$.) Test this by comparing the graphs of $f(x) = x(x - 3)(x - 1)$, $g(x) = x(x - 3)$, $h(x) = x(x - 1)$, and $j(x) = (x - 1)(x - 3)$. Use this deflating technique to obtain all the roots of the equations in Exercise 35, beginning as in Exercise 36.

38. Since Muller's method subtracts function values that may be very nearly the same, there is a possibility of large relative errors. Investigate this by using starting values that are very close to each other (but not near a root) in some of the equations in Exercise 35. The difficulty should be exaggerated when $f(x)$ changes very slowly near a root. See if this exaggerated effect is exhibited by a polynomial function that has a triple root.

▶ 39. Muller's method works on complex roots if complex arithmetic is used. Find a complex root of $x^4 + 4x^3 + 21x^2 + 4x + 20 = 0$. Compare this method to Bairstow's method as used in Exercise 31. (Newton's method can also find complex roots; you may wish to make the comparison with Newton's method also. Complex arithmetic is laborious by hand; you probably will want to employ computer programs in this exercise.)

Section 1.10

40. In the simplified floating-point system of Section 1.10 (3-decimal-digit fractions, exponents on base of 10 from ±0 to ±9), how many different numbers are there?

In Exercises 41 through 44 below, use the floating-point system of Exercise 40.

41. Find the absolute error of each result, both with "chopping" and with rounding to three digits.

    a) $3.26 \times 10^{-3} + 2.07 \times 10^{-4}$
    ▶ b) $1.96 \times 10^5 - 1.94 \times 10^4$
    c) $(3.26 \times 10^{-3} + 2.07 \times 10^{-4}) - 2.01 \times 10^{-4}$
    d) $3.26 \times 10^{-3} + (2.07 \times 10^{-4} - 2.01 \times 10^{-4})$

42. Find the relative errors both with "chopping" and with rounding to three digits.
   a) $3.28 \times 10^{-2} * 6.98 \times 10^3$
   b) $3.28 \times 10^{-8} * 6.98 \times 10^{-7}$
   c) $(3.28 \times 10^{-2} * 6.98 * 10^3) \div 4.82 \times 10^{-8}$
   d) $3.28 \times 10^{-2} * (6.98 \times 10^3 \div 4.82 \times 10^{-8})$

43. Find the absolute and relative errors with "chopping" to three digits.
   a) $4.82 \times 10^2 \div 8.81 \times 10^8$
   b) $1.06 \times 10^{-9} \div 4.06 \times 10^2$
   ▶c) $4.82 \times 10^2 \div (8.81 \times 10^8 * 4.06 \times 10^{-2})$
   d) $(4.82 \times 10^2 \div 8.81 \times 10^8) \div 4.06 \times 10^{-2}$

44. Find examples where
   a) $(a + b) + c \neq a + (b + c)$
   ▶b) $(a * b) * c \neq a * (b * c)$
   c) $a * (b + c) \neq a * b + a * c$

45. Evaluate the polynomial $2.75x^3 - 2.95x^2 + 3.16x - 4.67$ for $x = 1.07$, using both chopping after three digits and rounding to three digits. What are the absolute and relative errors? Does nested multiplication differ from evaluation in "standard form"? What is the error of $2.75x^3 - (2.95x^2 - 3.16x) - 4.67$?

46. Determine how floating-point numbers are stored in the computer system available to you. How many decimal digits of accuracy for single-precision numbers? for double-precision? What are the execution times for each of the arithmetic operations for single-precision? for double-precision?

▶47. Write a computer program to determine the effect of the sums in (a), (b), and (c):
   a) 0.01 added 100 times;
   b) 0.001 added 1000 times;
   c) 0.0001 added 10,000 times.
   Print out values for sums that should equal 0.1, 0.2, ..., 1.0.
   d) Evaluate the divergent infinite series, $\Sigma 1 + \frac{1}{2} + \frac{1}{3} + \frac{1}{4} + \ldots$, with a computer program. Why is it not divergent in a computer?
   e) Compare the sum in part (d) when evaluated from right to left with the sum evaluated from left to right. Explain the differences.

APPLIED PROBLEMS AND PROJECTS

48. A sphere of density d and radius r weighs $\frac{4}{3}\pi r^3 d$. The volume of a spherical segment is $\frac{1}{3}\pi(3rh^2 - h^3)$. Find the depth to which a sphere of density 0.6 sinks in water as a fraction of its radius. (See Fig. 1.21.)

Figure 1.21

49. Given

$$x'' + x + 2y' + y = f(t),$$
$$x'' - x + y = g(t), \qquad x(0) = x'(0) = y(0) = 0.$$

In solving this pair of simultaneous second-order differential equations by the Laplace transform method, it becomes necessary to factor the expression

$$(S^2 + 1)(S) - (2S + 1)(S^2 - 1) = -S^3 - S^2 + 3S + 1,$$

so that partial fractions can be used in getting the inverse transform. What are the factors?

50. The solution of boundary-value problems by an analytical (Fourier series) method often involves finding the roots of transcendental equations to evaluate the coefficients. For example,

$$y'' + \lambda y = 0, \qquad y(0) = 0, \qquad y(1) = y'(1),$$

involves solving $\tan z = z$. Find three values of z other than $z = 0$.

51. In Chapter 4, a particularly efficient method for numerical integration of a function, called *Gaussian quadrature*, is discussed. In the development of formulas for this method it is necessary to evaluate the zeros of Legendre polynomials. Find the zeros of the Legendre polynomial of sixth order:

$$P_6(x) = \tfrac{1}{48}(693x^6 - 945x^4 + 315x^2 - 15).$$

(*Note:* All the zeros of the Legendre polynomials are less than one in magnitude and, for polynomials of even order, are symmetrical about the origin.)

52. The Legendre polynomials of Problem 51 are one set of a class of polynomials known as *orthogonal* polynomials. Another set are the *Laguerre* polynomials. Find the zeros of the following:

a) $L_3(x) = x^3 - 9x^2 + 18x - 6$ b) $L_4(x) = x^4 - 16x^3 + 72x^2 - 96x + 24$

53. Still another set of orthogonal polynomials are the *Chebyshev* polynomials. (We will use these in Chapter 10.) Find the roots of

$$T_6(x) = 32x^6 - 48x^4 + 18x^2 - 1 = 0.$$

(Note the symmetry of this function. All the roots of Chebyshev polynomials are also less than one in magnitude.)

54. R. DeSantis has derived a relationship for the compressibility factor of real gases (*I.E.C. Fundamentals*, **15**, 184 (1976)) of the form

$$z = \frac{1 + y + y^2 - y^3}{(1 - y)^3},$$

where $y = b/(4v)$, b being the van der Waals correction and v the molar volume. If $z = 0.892$, what is the value of y?

55. In studies of solar-energy collection by focusing a field of plane mirrors on a central collector, L. L. Vant-Hull (*Solar Energy*, **18**, 33 (1976)) derives an equation for the geometrical concentration factor, C:

$$C = \frac{\pi(h/\cos A)^2 F}{0.5\pi D^2(1 + \sin A - 0.5 \cos A)}$$

where A is the rim angle of the field, F is the fractional coverage of the field with mirrors, D is the diameter of the collector, and h is the height of the collector. Find A if $h = 300$, $C = 1200$, $F = 0.8$, and $D = 14$.

56. Lee and Duffy (*A.I.Ch.E. Journal*, July, 1976) relate the friction factor for flow of a suspension of fibrous particles to the Reynoids number by this empirical equation:

$$\frac{1}{\sqrt{f}} = \left(\frac{1}{k}\right) \ell n\,(RE\sqrt{f}) + \left(14 - \frac{5.6}{k}\right).$$

In their relation, f is the friction factor, RE is the Reynolds number, and k is a constant determined by the concentration of the suspension. For a suspension with 0.08% concentration, $k = 0.28$. What is the value of f if RE $= 3750$?

57. The Redlich–Kwong equation is

$$P = \frac{RT}{v - b} - \frac{A(T)}{v(v + b)}.$$

Measurements of $P = 87.3$, $T = 486.9$, and $v = 2.005$ have been made. It is known that $A(T) = 0.0837$ under these conditions. $R = 1.98$, a constant. Find the value of b that satisfies the equation.

58. Based on the work of Frank–Kamenetski in 1955, temperatures in the interior of a material with imbedded heat sources can be determined if we solve this equation:

$$e^{-(1/2)t} \cosh^{-1}(e^{(1/2)t}) = \sqrt{\tfrac{1}{2}L_{cr}}.$$

Given that $L_{cr} = 0.88$, find t.

59. In cooperation with another student (or your instructor), find the roots of a function whose form is unknown to you. (The function is installed in the computer disk file where it can be called as a subprogram, but it is against the rules to study the function itself, although values can be obtained.) Find all the roots in a given interval with the least number of function evaluations, getting the roots to a precision of 0.01% relative error.

60. Modify one or more of the computer subroutines given in this chapter so they find n roots of a function (n is a parameter in the call statement) in a self-starting manner, as described in Exercises 36 and 37. For some programs, it may be preferred to begin with a search to find where $f(x)$ changes sign.

Chapter Two

Solving Sets of Equations

2.1 POTENTIALS AND CURRENTS IN AN ELECTRICAL NETWORK

Eight resistances of varying size are connected together in the complex network shown in Fig. 2.1. The potentials at points a and b are specified. Can we find the voltage at each numbered node? Can we find the current through each resistor?

The two questions are interrelated. If we know the voltage at each node, Ohm's law lets us calculate the currents. If we know the currents, we can similarly calculate the voltages. We elect to analyze the network in terms of voltages. Kirkhoff's Law is fundamental for us—the sum of the currents leaving (or entering) each node is zero.

We apply this law to each node in turn, employing the relation $\Delta V * (1/R) =$ current. We simplify the problem somewhat by observing that $v_5 = v_6$ because there is no resistance between points ⑤ and ⑥. Similarly, $v_6 = v_b$.

At node ①: $(v_1 - v_a) \cdot \frac{1}{3} + (v_1 - v_2) \cdot \frac{1}{5} + (v_1 - v_b) \cdot \frac{1}{6}$
$$+ (v_1 - v_b) \cdot \frac{1}{2} = 0;$$

At node ②: $(v_2 - v_1) \cdot \frac{1}{5} + (v_2 - v_3) \cdot \frac{1}{10} + (v_2 - v_4) \cdot 1 = 0;$

At node ③: $(v_3 - v_2) \cdot \frac{1}{10} + (v_3 - v_4) \cdot \frac{1}{7} = 0;$

At node ④: $(v_4 - v_2) \cdot 1 + (v_4 - v_3) \cdot \frac{1}{7} + (v_4 - v_b) \cdot \frac{1}{4} = 0.$

Collecting terms and simplifying, we get:

$$36v_1 - 6v_2 \qquad\qquad = 10v_a + 20v_b,$$
$$-2v_1 + 13v_2 - v_3 - 10v_4 = 0,$$
$$- 7v_2 + 17v_3 - 10v_4 = 0,$$
$$- 28v_2 - 4v_3 + 39v_4 = 7v_b.$$

Figure 2.1

If we substitute the specified values for v_a and v_b, the problem reduces to solving four equations simultaneously for the four unknown voltages. We observe that the equations are linear in each of the unknown variables. The methods of this chapter constitute ways to solve such systems in a systematic way.

(The alternative method of solving the above problem is called the *loop method*. In that method the algebraic sum of the potential drops around each loop is equated to zero. This would result in four simultaneous linear equations in the four loop currents as unknowns. In both methods of analysis, the same mathematical problem occurs—solve a set of simultaneous linear equations. For this problem, one can also replace combinations of resistors in series or in parallel with equivalent resistances and so simplify the problem that its solution is easy. Our method of setting up a set of simultaneous equations using Kirkhoff's Law is more general, however, and can solve any circuit, no matter how complicated.)

A variety of engineering problems may be solved through network analysis, in addition to determining the potentials in certain electrical networks, such as finding the stresses in a building frame or a bridge truss, computing the flow rates in a hydraulic system with interconnected branches, or estimating the concentrations of reactants subject to simultaneous chemical reactions. The mathematical problem in all these cases reduces to solving a set of equations simultaneously. We shall also find, when we study the numerical methods of solving boundary-value problems and partial-differential equations, that these also require the solution of sets of equations, often very large in number.

The solution of sets of equations is much more difficult when the equations are nonlinear. Fortunately the majority of applications involve only linear equations, although when such systems are large, careful choice of methods is required to avoid overly long procedures and to preserve maximum accuracy. In this chapter we shall first discuss some of the basic ways of solving sets of linear equations, and later take up the harder problem of nonlinear equations.

2.2 MATRIX NOTATION

Our discussion will be facilitated by some of the concepts and notation of matrix algebra. Only the more elementary ideas will be needed.

A *matrix* is a rectangular array of numbers in which not only the value of the number is important but also its position in the array. The size of the matrix is described by the number of its rows and columns. A matrix of n rows and m columns is said to be $n \times m$. The elements of the matrix are generally enclosed in brackets, and double-subscripting is the common way of indexing the elements. The first subscript always denotes the row and the second denotes the column in which the element occurs. Capital letters are used to refer to matrices. For example,

$$A = \begin{bmatrix} a_{11} & a_{12} \ldots a_{1m} \\ a_{21} & a_{22} \ldots a_{2m} \\ \cdot \\ \cdot \\ \cdot \\ a_{n1} & a_{n2} \ldots a_{nm} \end{bmatrix} = [a_{ij}], \qquad i = 1, 2, \ldots, n, \qquad j = 1, 2, \ldots, m.$$

Enclosing the general element, a_{ij}, in brackets is another way of representing matrix A, as shown above.

Two matrices of the same size may be added or subtracted. The sum of

$$A = [a_{ij}] \qquad \text{and} \qquad B = [b_{ij}]$$

is the matrix whose elements are the sum of the corresponding elements of A and B,

$$C = A + B = [a_{ij} + b_{ij}] = [c_{ij}].$$

Similarly we get the *difference* of two equal-sized matrices by subtracting corresponding elements. If two matrices are not equal in size, they cannot be added or subtracted. Two matrices are equal if and only if each element of one is the same as the corresponding element of the other. Obviously, equal matrices must be of the same size. Some examples will help make this clear.

If

$$A = \begin{bmatrix} 3 & 2 & -1 \\ -4 & 0 & 2 \end{bmatrix} \qquad \text{and} \qquad B = \begin{bmatrix} 0 & 3 & 2 \\ 4 & -2 & 1 \end{bmatrix},$$

we say that A is 2×3 because it has two rows and three columns. B is also 2×3. Their sum, C, is also 2×3:

$$C = A + B = \begin{bmatrix} 3 & 5 & 1 \\ 0 & -2 & 3 \end{bmatrix}.$$

The difference, D, of A and B is:

$$D = A - B = \begin{bmatrix} 3 & -1 & -3 \\ -8 & 2 & 1 \end{bmatrix}.$$

Multiplication of two matrices is defined as follows, when A is $n \times m$ and B is $m \times r$:

$$[a_{ij}][b_{ij}] = [c_{ij}] = \begin{bmatrix} (a_{11}b_{11} + a_{12}b_{21} + \ldots + a_{1m}b_{m1}) \ldots (a_{11}b_{1r} + \ldots + a_{1m}b_{mr}) \\ (a_{21}b_{11} + a_{22}b_{21} + \ldots + a_{2m}b_{m1}) \ldots (a_{21}b_{1r} + \ldots + a_{2m}b_{mr}) \\ \cdot \\ \cdot \\ \cdot \\ (a_{n1}b_{11} + a_{n2}b_{21} + \ldots + a_{nm}b_{m1}) \ldots (a_{n1}b_{1r} + \ldots + a_{nm}b_{mr}) \end{bmatrix},$$

$$c_{ij} = \sum_{k=1}^{m} a_{ik}b_{kj}, \quad i = 1, 2, \ldots, n, \quad j = 1, 2, \ldots, r.$$

It is simplest to select the proper elements if one counts across the rows of A with the left hand while counting down the columns of B with the right. Unless the number of columns of A equals the number of rows of B (so the counting comes out even), the matrices cannot be multiplied. Hence if A is $n \times m$, B must have m rows or else they are said to be "nonconformable for multiplication" and their product is undefined. In general $AB \neq BA$, so the order of factors must be preserved in matrix multiplication.

If a matrix is multiplied by a scalar (a pure number), the product is a matrix, each element of which is the scalar times the original element. We can write:

$$\text{If } kA = C, \quad c_{ij} = ka_{ij}.$$

A matrix with only one column, $n \times 1$ in size, is termed a *column vector*, and one of only one row, $1 \times m$ in size, is called a *row vector*. When the *unqualified* term "vector" is used, it nearly always means a *column* vector. Frequently the elements of vectors are only singly subscripted.

Some examples of matrix multiplication are:

$$\text{Suppose } A = \begin{bmatrix} 2 & 4 & 0 \\ -1 & 2 & 3 \end{bmatrix}, \quad B = \begin{bmatrix} -1 & 2 \\ 0 & 1 \\ 4 & -1 \end{bmatrix}, \quad x = \begin{bmatrix} 4 \\ -2 \\ 1 \end{bmatrix}, \quad y = \begin{bmatrix} y_1 \\ y_2 \\ y_3 \end{bmatrix}.$$

$$A \cdot B = \begin{bmatrix} -2 & 8 \\ 13 & -3 \end{bmatrix}; \quad B \cdot A = \begin{bmatrix} -4 & 0 & 6 \\ -1 & 2 & 3 \\ 9 & 14 & -3 \end{bmatrix};$$

$$Ax = \begin{bmatrix} 0 \\ -5 \end{bmatrix}; \quad Ay = \begin{bmatrix} 2y_1 + 4y_2 \\ -y_1 + 2y_2 + 3y_3 \end{bmatrix}.$$

Since A is 2×3 and B is 3×2, they are conformable for multiplication and their product is 2×2. When we form the product of $B \cdot A$, it is 3×3. Observe that not only is $AB \neq BA$; AB and BA are not even the same size. The product of A and the vector x (a 3×1 matrix) is another vector, one with two components.

Similarly, Ay has two components. We cannot multiply B times x or B times y; they are nonconformable.

The product of the scalar number 2 and A is

$$2A = \begin{bmatrix} 4 & 8 & 0 \\ -2 & 4 & 6 \end{bmatrix}.$$

Since a vector is just a special case of a matrix, a column vector can be multiplied by a matrix, so long as they are conformable in that the number of columns of the matrix equals the number of elements (rows) in the vector. The product, in this case will be another column vector. The size of a product of two matrices, the first $m \times n$ and the second $n \times r$, is $m \times r$. An $m \times n$ matrix times an $n \times 1$ vector gives an $m \times 1$ product.

The general relation for $Ax = b$ is

$$b_i = \sum_{k=1}^{\text{No of cols.}} a_{ik}x_k, \qquad i = 1, 2, \ldots, \text{No of rows.}$$

Two vectors, each with the same number of components, may be added or subtracted. Two vectors are equal if each component of one equals the corresponding component of the other.

This definition of matrix multiplication permits us to write the set of linear equations,

$$a_{11}x_1 + a_{12}x_2 + \ldots + a_{1n}x_n = b_1,$$
$$a_{21}x_1 + a_{22}x_2 + \ldots + a_{2n}x_n = b_2,$$
$$\vdots$$
$$a_{n1}x_1 + a_{n2}x_2 + \ldots + a_{nn}x_n = b_n,$$

much more simply in matrix notation, as $Ax = b$, where

$$A = \begin{bmatrix} a_{11} & a_{12} & \ldots & a_{1n} \\ a_{21} & a_{22} & \ldots & a_{2n} \\ & & & \\ & & & \\ a_{n1} & a_{n2} & \ldots & a_{nn} \end{bmatrix}, \quad x = \begin{bmatrix} x_1 \\ x_2 \\ \cdot \\ \cdot \\ \cdot \\ x_n \end{bmatrix}, \quad b = \begin{bmatrix} b_1 \\ b_2 \\ \cdot \\ \cdot \\ \cdot \\ b_n \end{bmatrix}.$$

For example,

$$\begin{bmatrix} 3 & 2 & 4 \\ 1 & -2 & 0 \\ -1 & 3 & 2 \end{bmatrix} \cdot x = \begin{bmatrix} 14 \\ -7 \\ 2 \end{bmatrix}$$

is the same as the set of equations

$$3x_1 + 2x_2 + 4x_3 = 14,$$
$$x_1 - 2x_2 \qquad = -7,$$
$$-x_1 + 3x_2 + 2x_3 = 2.$$

A very important special case is the multiplication of two vectors. The first must be a row vector if the second is a column vector, and each must have the same number of components. For example,

$$[1 \quad 3 \quad -2] \cdot \begin{bmatrix} 4 \\ -1 \\ 3 \end{bmatrix} = [-5]$$

gives a "matrix" of one row and one column. The result is a pure number, a scalar. This product is called the *scalar product* of the vectors, also called the *inner product*.

Certain square matrices have special properties. The diagonal elements are the line of elements a_{ii} from upper left to lower right of the matrix. If only the diagonal terms are nonzero, the matrix is called a *diagonal matrix*. When the diagonal elements are each equal to unity while all off-diagonal elements are zero, the matrix is said to be the *identity matrix of order n*. The usual symbol for such a matrix is I_n, and it has properties similar to unity. For example, the order-four identity matrix is

$$\begin{bmatrix} 1 & 0 & 0 & 0 \\ 0 & 1 & 0 & 0 \\ 0 & 0 & 1 & 0 \\ 0 & 0 & 0 & 1 \end{bmatrix} = I_4.$$

The subscript is omitted when the order is clear from the context.

A vector that has all its elements equal to zero except one element, which has a value of unity, is called a *unit vector*. There are three distinct unit vectors for order-three vectors; they are

$$\begin{bmatrix} 1 \\ 0 \\ 0 \end{bmatrix}, \quad \begin{bmatrix} 0 \\ 1 \\ 0 \end{bmatrix}, \quad \text{and} \quad \begin{bmatrix} 0 \\ 0 \\ 1 \end{bmatrix}.$$

If all the elements above the diagonal are zero, the matrix is called *lower triangular*; it is called *upper triangular* when all the elements below the diagonal are zero. For example, these order-three matrices are lower and upper triangular:

$$L = \begin{bmatrix} 1 & 0 & 0 \\ 4 & 6 & 0 \\ -2 & 1 & -4 \end{bmatrix}, \quad U = \begin{bmatrix} 1 & -3 & 3 \\ 0 & -1 & 0 \\ 0 & 0 & 1 \end{bmatrix}.$$

Tridiagonal matrices are those which have nonzero elements only on the diagonal and in the positions adjacent to the diagonal; they will be of special importance in certain partial-differential equations. An example of a tridiagonal matrix is

$$\begin{bmatrix} -4 & 2 & 0 & 0 & 0 \\ 1 & -4 & 1 & 0 & 0 \\ 0 & 1 & -4 & 1 & 0 \\ 0 & 0 & 1 & -4 & 1 \\ 0 & 0 & 0 & 2 & -4 \end{bmatrix}$$

The *transpose* of a matrix is the matrix that results when the rows are written as columns (or, alternatively, when the columns are written as rows). The symbol A^T is used for the *transpose of A*.

Example

$$A = \begin{bmatrix} 3 & -1 & 4 \\ 0 & 2 & -3 \\ 1 & 1 & 2 \end{bmatrix}; \qquad A^T = \begin{bmatrix} 3 & 0 & 1 \\ -1 & 2 & 1 \\ 4 & -3 & 2 \end{bmatrix}.$$

It should be obvious that the transpose of A^T is just A itself.

When a matrix is square, a quantity called its *trace* is defined. The trace of a square matrix is the sum of the elements on its main diagonal. For example, the traces of the above matrices are:

$$\text{tr}(A) = 3 + 2 + 2 = 7; \qquad \text{tr}(A^T) = 3 + 2 + 2 = 7.$$

It should be obvious that the trace remains the same if a square matrix is transposed.

We present here some additional examples of arithmetic operations with matrices.

$$(3) \cdot \begin{bmatrix} 1 & 2 \\ 3 & 4 \end{bmatrix} = \begin{bmatrix} 3 & 6 \\ 9 & 12 \end{bmatrix}.$$

$$\begin{bmatrix} 1 & 3 & 2 \\ -1 & 0 & 4 \end{bmatrix} + \begin{bmatrix} -1 & 0 & 2 \\ 4 & 1 & -3 \end{bmatrix} = \begin{bmatrix} 0 & 3 & 4 \\ 3 & 1 & 1 \end{bmatrix};$$

$$\begin{bmatrix} 2 & 1 \\ 0 & -4 \\ 7 & 2 \end{bmatrix} - \begin{bmatrix} 3 & -2 \\ 4 & 1 \\ 0 & -2 \end{bmatrix} = \begin{bmatrix} -1 & 3 \\ -4 & -5 \\ 7 & 4 \end{bmatrix}.$$

$$\begin{bmatrix} 2 & 0 & -1 \\ 3 & 2 & 6 \end{bmatrix} \cdot \begin{bmatrix} -1 \\ 2 \\ 1 \end{bmatrix} = \begin{bmatrix} -3 \\ 7 \end{bmatrix} \quad \text{but} \quad \begin{bmatrix} 6 & 1 \\ 3 & -2 \end{bmatrix} \cdot \begin{bmatrix} -1 \\ 2 \\ 1 \end{bmatrix} = \text{not defined.}$$

$$\begin{bmatrix} 1 & 3 \\ 2 & -1 \end{bmatrix} \cdot \begin{bmatrix} 0 & 3 \\ -1 & 1 \end{bmatrix} = \begin{bmatrix} -3 & 6 \\ 1 & 5 \end{bmatrix} \quad \text{but} \quad \begin{bmatrix} 0 & 3 \\ -1 & 1 \end{bmatrix} \cdot \begin{bmatrix} 1 & 3 \\ 2 & -1 \end{bmatrix} = \begin{bmatrix} 6 & -3 \\ 1 & -4 \end{bmatrix}$$

Division of a matrix by another matrix is not defined, but we shall discuss the *inverse* of a matrix later in this chapter.

The *determinant* of a square matrix is a number. For a 2×2 matrix, the determinant is computed by subtracting the product of the elements on the minor diagonal (from upper right to lower left) from the product of terms on the major diagonal. For example

$$A = \begin{bmatrix} 3 & -1 \\ 2 & 4 \end{bmatrix}, \quad \det(A) = (3)(4) - (-1)(2) = 14;$$

$\det(A)$ is the usual notation for the determinant of A. Sometimes the determinant is symbolized by writing the elements of the matrix between vertical lines (similar to representing the absolute value of a number.)

For a three-by-three matrix, you may have learned a crisscross way of forming products of terms (I call it the "spaghetti rule") that probably had best be forgotten for it applies only to the special case of a 3×3 matrix; it won't work for larger systems. The general rule that applies in all cases is to expand in terms of the minors of some row or column. The minor of any term is the matrix of lower order formed by striking out the row and column in which the term is found. The determinant is found by adding the product of each term in any row or column by the determinant of its minor, with signs alternating $+$ and $-$. We expand each of the determinants of the minor until we reach 2×2 matrices. For example:

$$\text{Given } A = \begin{bmatrix} 3 & 0 & -1 & 2 \\ 4 & 1 & 3 & -2 \\ 0 & 2 & -1 & 3 \\ 1 & 0 & 1 & 4 \end{bmatrix},$$

$$\det(A) = 3\begin{vmatrix} 1 & 3 & -2 \\ 2 & -1 & 3 \\ 0 & 1 & 4 \end{vmatrix} - 0\begin{vmatrix} 4 & 3 & -2 \\ 0 & -1 & 3 \\ 1 & 1 & 4 \end{vmatrix} + (-1)\begin{vmatrix} 4 & 1 & -2 \\ 0 & 2 & 3 \\ 1 & 0 & 4 \end{vmatrix}$$
$$- 2\begin{vmatrix} 4 & 1 & 3 \\ 0 & 2 & -1 \\ 1 & 0 & 1 \end{vmatrix}$$

$$= 3\left\{(1)\begin{vmatrix} -1 & 3 \\ 1 & 4 \end{vmatrix} - (3)\begin{vmatrix} 2 & 3 \\ 0 & 4 \end{vmatrix} + (-2)\begin{vmatrix} 2 & -1 \\ 0 & 1 \end{vmatrix}\right\}$$

$$+ (-1)\left\{(4)\begin{vmatrix} 2 & 3 \\ 0 & 4 \end{vmatrix} - (1)\begin{vmatrix} 0 & 3 \\ 1 & 4 \end{vmatrix} + (-2)\begin{vmatrix} 0 & 2 \\ 1 & 0 \end{vmatrix}\right\}$$

$$- 2\left\{(4)\begin{vmatrix} 2 & -1 \\ 0 & 1 \end{vmatrix} - (1)\begin{vmatrix} 0 & -1 \\ 1 & 1 \end{vmatrix} + (3)\begin{vmatrix} 0 & 2 \\ 1 & 0 \end{vmatrix}\right\}$$

$$= 3\{(1)(-7) - (3)(8) + (-2)(2)\} + (-1)\{(4)(8) - (1)(-3) + (-2)(-2)\}$$
$$- 2\{(4)(2) - (1)(1) + (3)(-2)\}$$

$$= 3(-7 - 24 - 4) + (-1)(32 + 3 + 4) - 2(8 - 1 - 6)$$

$$= 3(-35) + (-1)(39) - 2(1) = -146.$$

In computing the determinant, the expansion can be about the elements of any row or column. To get the signs, give the first term a plus sign if the sum of its column number and row number is even; give it a minus if the sum is odd, with alternating signs thereafter. (For example, in expanding about the elements of the third row we begin with a plus; the first element a_{31} has $3 + 1 = 4$, an even number.) Judicious selection of rows or columns with many zeros can hasten the process, but this method of calculating determinants is a lot of work if the matrix is of large size.

2.3 ELIMINATION METHOD

The first method we shall study for the solution of a set of equations is just an enlargement of the familiar method of eliminating one unknown between a pair of simultaneous equations. It is generally called Gaussian elimination and is the basic pattern of a large number of methods that can be classed as *direct methods*. (This is to distinguish them from indirect, or iterative, methods, which we shall discuss later.)

Consider the simple example of three equations

$$3x_1 - x_2 + 2x_3 = 12,$$
$$x_1 + 2x_2 + 3x_3 = 11,$$
$$2x_1 - 2x_2 - x_3 = 2.$$

Multiplying the first equation by -1 and the second by 3 and adding will eliminate x_1. Similarly, multiplying the first by -2 and the third by 3 and adding also eliminates x_1. (We prefer, in hand calculations, to multiply by the negative values and add, to avoid making mistakes when subtracting quantities of unlike sign.) The result is

$$3x_1 - x_2 + 2x_3 = 12,$$
$$7x_2 + 7x_3 = 21,$$
$$-4x_2 - 7x_3 = -18.$$

We eliminate x_2 between the second and third equations by multiplying the second by 4 and the third by 7 and adding. (Of course, just adding them as they stand would eliminate x_3, which is equally satisfactory, but we wish to keep our method systematic to lead up to an algorithm that can be readily programmed.) After this operation we have the upper triangular system

$$3x_1 - x_2 + 2x_3 = 12,$$
$$7x_2 + 7x_3 = 21,$$
$$-21x_3 = -42.$$

Obviously $x_3 = 2$ from the third equation, and back-substitution gives $x_2 = 1$, (from the second equation), and $x_1 = 3$ (from the first equation).

We now present the same problem, solved in exactly the same way, in matrix notation:

$$\begin{bmatrix} 3 & -1 & 2 \\ 1 & 2 & 3 \\ 2 & -2 & -1 \end{bmatrix} \begin{bmatrix} x_1 \\ x_2 \\ x_3 \end{bmatrix} = \begin{bmatrix} 12 \\ 11 \\ 2 \end{bmatrix}.$$

The arithmetic operations we have performed affect only the coefficients and the constant terms, of course, so we work with the matrix of coefficients "augmented" with the b-vector:

$$A \mathbin{\vdots} b = \begin{bmatrix} 3 & -1 & 2 & \vdots & 12 \\ 1 & 2 & 3 & \vdots & 11 \\ 2 & -2 & -1 & \vdots & 2 \end{bmatrix}.$$

(The dotted line is usually omitted.)

We perform elementary row transformations* to convert A to upper triangular:

$$\begin{bmatrix} 3 & -1 & 2 & 12 \\ 1 & 2 & 3 & 11 \\ 2 & -2 & -1 & 2 \end{bmatrix} \begin{smallmatrix} \\ 3R_2 + (-1)R_1 \\ 3R_3 + (-2)R_1 \end{smallmatrix} \to \begin{bmatrix} 3 & -1 & 2 & 12 \\ 0 & 7 & 7 & 21 \\ 0 & -4 & -7 & -18 \end{bmatrix} 7R_3 + 4R_2 \to \begin{bmatrix} 3 & -1 & 2 & 12 \\ 0 & 7 & 7 & 21 \\ 0 & 0 & -21 & -42 \end{bmatrix}. \quad (2.1)$$

The steps are to add 3 times the second row to -1 times the first row and 3 times the third row to -2 times the first row. The next phase adds 7 times the third row to 4 times the second row.

We are now ready for back-substitution. Note that, except for notation and terminology, there is nothing new here. We depend on our memory to know which numbers in the converted augmented matrix are coefficients and which are the righthand sides (the constant terms).

The back-substitution step can be performed quite mechanically by eliminating the coefficients above the diagonal in (2.1). Adding the third row of (2.1) to 3 times the second row, and adding twice the third row to 21 times the first row gives

$$\begin{bmatrix} 63 & -21 & 0 & 168 \\ 0 & 21 & 0 & 21 \\ 0 & 0 & -21 & -42 \end{bmatrix}.$$

We finish the elimination of off-diagonal elements by adding the second row to the first:

$$\begin{bmatrix} 63 & 0 & 0 & 189 \\ 0 & 21 & 0 & 21 \\ 0 & 0 & -21 & -42 \end{bmatrix}.$$

---

* Elementary row operations are arithmetic operations which are obviously valid rearrangements of a set of equations: 1. Any equation can be multiplied by a constant. 2. The order of the equations can be changed. 3. Any equation can be replaced by its sum with another of the equations.

If we divide each row by the diagonal element, we get a form in which the elements of x, the vector whose components are the unknowns x_i, $i = 1, 2, \ldots, n$, are equal to the components of the transformed b-vector:

$$\begin{bmatrix} 1 & 0 & 0 & 3 \\ 0 & 1 & 0 & 1 \\ 0 & 0 & 1 & 2 \end{bmatrix}, \quad x = \begin{bmatrix} 3 \\ 1 \\ 2 \end{bmatrix}, \quad x_1 = 3, \quad x_2 = 1, \quad x_3 = 2.$$

Thinking of this procedure in terms of matrix operations, we transform the augmented coefficient matrix by elementary row operations until the identity matrix is created on the left. The x-vector then stands as the rightmost column.

Note that there exists the possibility that the set of equations has no solution, or that the above procedure will fail to find it. During the triangularization step, if a zero is encountered on the diagonal, we cannot use that row to eliminate coefficients below that zero element. However, in that case, we can continue by interchanging rows and eventually achieve an upper triangular matrix of coefficients. The real stumbling block is finding a zero on the diagonal after we have triangularized. If that occurs, the back-substitution fails for we cannot divide by zero.

It is worthwhile to explain in more detail what we mean by the *elementary row operations* that we have used above, and to see why they can be used in solving a linear system. There are three of these operations:

1.   We may multiply any row of the augmented coefficient matrix by a constant.

2.   We can add a multiple of one row to a multiple of any other row.

3.   We can interchange the order of any two rows (this was not used above.)

The validity of these row operations is intuitively obvious if we think of them applied to a set of linear equations. Certainly, multiplying one equation through by a constant does not change the truth of the equality. Adding equal quantities to both sides of an equality results in an equality, and this is the equivalent of the second transformation. Obviously the order of the set is arbitrary, so rule three is valid.

These operations, which do not change the relationships represented by a set of equations, can be applied to an augmented matrix, for this is only a different notation for the equations. (We need to add one proviso to the above. Since round-off error is related to the magnitude of the values when we express them in fixed-word-length computer representations, some of the above operations may have an effect on the accuracy of the computed solution.)

We should also observe that the "back-substitution" phase, when it is done by making the coefficients above the diagonal zero and then reducing the coefficient matrix to the identity matrix, is exactly the same as back-substitution in the more explicit sense. The order of operations is changed but each of the steps is identical.

2.4 GAUSS AND GAUSS–JORDAN METHODS

While the procedure of the previous section is satisfactory for hand calculations on small systems, there are several objections that we should eliminate before we write a computer program to perform Gaussian elimination. In a large set of equations, and that is the situation we must prepare for, the multiplications will give very large and unwieldy numbers that may overflow the computer's registers. We therefore will eliminate the first coefficient in the ith row by subtracting a_{i1}/a_{11} times the first equation from the ith equation. (this is equivalent to making the leading coefficient 1 in the equation that retains that leading term.) We use similar ratios of coefficients in eliminating coefficients in the other columns.

We must also guard against dividing by zero. Observe that zeros may be created in the diagonal positions even if they are not present in the original matrix of coefficients. A useful strategy to avoid (if possible) such zero divisors is to rearrange the equations so as to put the coefficient of largest magnitude on the diagonal at each step. This is called *pivoting*. Complete pivoting may require both row and column interchanges. This is not frequently done. Partial pivoting, which places a coefficient of larger magnitude on the diagonal by row interchanges only, will guarantee a nonzero divisor if there is a solution to the set of equations, and will have the added advantage of giving improved arithmetic precision. The diagonal elements that result are called *pivot elements*. (When there are large differences in magnitude of coefficients in one equation compared to the other equations, we may need to *scale* the values; we consider this later.)

We repeat the example of the previous section incorporating these ideas, and carrying four significant digits in our work. We begin with the augmented matrix.

$$\begin{bmatrix} 3 & -1 & 2 & 12 \\ 1 & 2 & 3 & 11 \\ 2 & -2 & -1 & 2 \end{bmatrix}$$

$$\begin{matrix} \text{Row } 2 - (\tfrac{1}{3})\text{Row } 1 \rightarrow \\ \text{Row } 3 - (\tfrac{2}{3})\text{Row } 1 \rightarrow \end{matrix} \begin{bmatrix} 3 & -1 & 2 & 12 \\ 0 & 2.333 & 2.334 & 7.004 \\ 0 & -1.334 & -2.332 & -5.992 \end{bmatrix}$$

$$\text{Row } 3 - \left(\frac{-1.334}{2.333}\right)\text{Row } 2 \rightarrow \begin{bmatrix} 3 & -1 & 2 & 12 \\ 0 & 2.333 & 2.334 & 7.004 \\ 0 & 0 & -1.000 & -1.993 \end{bmatrix}$$

The method we have just illustrated is called *Gaussion Elimination*. (In this example, no pivoting was required to make the largest coefficients be on the diagonal.) Back-substitution, beginning with the third equation and then moving "backward" to the second and first equations, gives $x_3 = 1.993$, $x_2 = 1.008$, $x_1 = 3.007$. The differences of these values from 2, 1, and 3 are due to the effects of round-off error. In this example, we have truncated after the fourth digit rather than round to the nearest fourth digit, which is the same as the arithmetic performed in most computers. In either case, the errors similarly affect the accuracy of the results. When there are many equations, the effects of round-off (the term is applied to the error due to chopping as well as when rounding is used) may cause large effects. In certain cases, the coefficients are such that the results are particularly sensitive to round-off; such systems are called *ill-conditioned.*

Let us summarize the operations of Gaussian elimination in a form that will facilitate writing a computer program. We use a less formal method of stating this algorithm.

Gaussian elimination

To solve a system of linear equations,

1. Augment the $n \times n$ coefficient matrix with the vector of righthand sides to form a $n \times (n + 1)$ matrix.

2. Interchange rows if necessary to make the value of a_{11} the largest magnitude of any coefficient in the first column.

3. Create zeros in the second through nth rows in the first column by subtracting a_{i1}/a_{11} times the first row from the ith row.

4. Repeat steps (2) and (3) for the second through the $(n - 1)$st rows, putting the largest-magnitude coefficient on the diagonal by interchanging rows (considering only rows j to n), and then subtracting a_{ij}/a_{jj} times the jth row from the ith row so as to create zeros in all positions of the jth column below the diagonal. At the conclusion of this step, the system is upper-triangular.

5. Solve for x_n from the nth equation by

$$x_n = a_{n,n+1}/a_{nn}.$$

6. Solve for $x_{n-1}, x_{n-2}, \ldots, x_1$ from the $(n - 1)$st through the first equation in turn, by

$$x_i = \frac{a_{i,n+1} - \sum_{j=i+1}^{n} a_{ij}x_j}{a_{ii}}.$$

Some computer programs do not actually interchange all the elements of the rows when pivoting. In these programs, one keeps track of the order in which the rows are to be used in a vector whose elements represent row order. When an interchange is indicated, only the elements of this ordering vector are changed. These numbers are then used to locate the positions of the elements in the matrix of coefficients that are to be operated on, both during the reduction step and during the back-substitution. This can reduce the computer time for large systems, but adds to the complexity of the program.

The algorithm for Gaussian elimination will be clarified by an additional numerical example.

Solve:

$$
\begin{aligned}
2x_2 \quad\quad + \ x_4 &= \ 0, \\
2x_1 + 2x_2 + 3x_3 + 2x_4 &= -2, \\
4x_1 - 3x_2 \quad\quad + \ x_4 &= -7, \\
6x_1 + \ x_2 - 6x_3 - 5x_4 &= \ 6.
\end{aligned}
$$

The augmented coefficient matrix is

$$
\begin{bmatrix}
0 & 2 & 0 & 1 & 0 \\
2 & 2 & 3 & 2 & -2 \\
4 & -3 & 0 & 1 & -7 \\
6 & 1 & -6 & -5 & 6
\end{bmatrix}.
$$

We cannot permit a zero in the a_{11} position because that element is the pivot in reducing the first column. We could interchange the first row with any of the other rows to merely avoid a zero divisor, but interchanging the first and fourth rows is our best choice. This gives

$$
\begin{bmatrix}
6 & 1 & -6 & -5 & 6 \\
2 & 2 & 3 & 2 & -2 \\
4 & -3 & 0 & 1 & -7 \\
0 & 2 & 0 & 1 & 0
\end{bmatrix}.
$$

We make all the elements in the first column zero by subtracting the appropriate multiple of row one:

$$
\begin{bmatrix}
6 & 1 & -6 & -5 & 6 \\
0 & 1.6667 & 5 & 3.6667 & -4 \\
0 & -3.6667 & 4 & 4.3333 & -11 \\
0 & 2 & 0 & 1 & 0
\end{bmatrix}.
$$

We again interchange before reducing the second column, not because we have a zero divisor, but because we want to preserve accuracy.* Interchanging the second and third rows puts the element of largest magnitude on the diagonal. (We could

---

* A numerical example that demonstrates the improved accuracy when partial pivoting is used will be found in Section 2.9.

also interchange the fourth column with the second, giving an even larger diagonal element, but we do not do this.) After the interchange, we have

$$\begin{bmatrix} 6 & 1 & -6 & -5 & 6 \\ 0 & -3.6667 & 4 & 4.3333 & -11 \\ 0 & 1.6667 & 5 & 3.6667 & -4 \\ 0 & 2 & 0 & 1 & 0 \end{bmatrix}.$$

Now we reduce in the second column:

$$\begin{bmatrix} 6 & 1 & -6 & -5 & 6 \\ 0 & -3.6667 & 4 & 4.333 & -11 \\ 0 & 0 & 6.8182 & 5.6364 & -9.0001 \\ 0 & 0 & 2.1818 & 3.3636 & -5.9999 \end{bmatrix}$$

No interchange is indicated in the third column. Reducing, we get

$$\begin{bmatrix} 6 & 1 & -6 & -5 & 6 \\ 0 & -3.6667 & 4 & 4.3333 & -11 \\ 0 & 0 & 6.8182 & 5.6364 & -9.0001 \\ 0 & 0 & 0 & 1.5600 & -3.1199 \end{bmatrix}$$

Back-substitution gives

$$x_4 = \frac{-3.1199}{1.5600} = -1.9999,$$

$$x_3 = \frac{-9.0001 - 5.6364(-1.9999)}{6.8182} = 0.33325,$$

$$x_2 = \frac{-11 - 4.3333(-1.9999) - 4(0.33325)}{-3.6667} = 1.0000,$$

$$x_1 = \frac{6 - (-5)(-1.9999) - (-6)(0.33325) - (1)(1.0000)}{6} = -0.50000.$$

The correct answers are -2, $\frac{1}{3}$, 1, and $-\frac{1}{2}$ for x_4, x_3, x_2, and x_1. In this calculation we have carried five significant figures and rounded each calculation. Even so, we do not have five-digit accuracy in the answers. The discrepancy is due to round-off. The question of the accuracy of the computed solution to a set of equations is a most important one, and at several points in the following discussion we will discuss how to minimize the effects of round-off and avoid conditions that can cause round-off errors to be magnified.

There are many variants to the Gaussian elimination scheme. The back-substitution step can be performed by eliminating the above-diagonal elements, using elementary row operations and proceeding upward from the last row, after the triangularization has been finished. This is similar to an example presented in the previous section. The diagonal elements all may be made unity as a first step before creating zeros in their column; this does the divisions of the back-substitution phase at an earlier time.

One variant that is sometimes used is the Gauss–Jordan scheme. In it, the elements above the diagonal are made zero at the *same time* that zeros are created below the diagonal. Usually the diagonal elements are made unity at the same time that the reduction is performed; this transforms the coefficient matrix into the identity matrix. When this has been accomplished, the column of righthand sides has been transformed into the solution vector. Pivoting is normally employed to preserve arithmetic accuracy.

The previous example, solved by the Gauss–Jordan method, gives this succession of calculations:

The original augmented matrix is:

$$\begin{bmatrix} 0 & 2 & 0 & 1 & 0 \\ 2 & 2 & 3 & 2 & -2 \\ 4 & -3 & 0 & 1 & -7 \\ 6 & 1 & -6 & -5 & 6 \end{bmatrix}.$$

Interchanging rows one and four, dividing the first row by 6, and reducing the first column gives:

$$\begin{bmatrix} 1 & 0.16667 & -1 & -0.83335 & 1 \\ 0 & 1.6667 & 5 & 3.6667 & -4 \\ 0 & -3.6667 & 4 & 4.3334 & -11 \\ 0 & 2 & 0 & 1 & 0 \end{bmatrix}.$$

Interchanging rows two and three, dividing the second row by -3.6667, and reducing the second column (operating above the diagonal as well as below) gives:

$$\begin{bmatrix} 1 & 0 & -1.5000 & -1.2000 & 1.4000 \\ 0 & 1 & 2.9999 & 2.2000 & -2.4000 \\ 0 & 0 & 15.000 & 12.400 & -19.800 \\ 0 & 0 & -5.9998 & -3.4000 & 4.8000 \end{bmatrix}.$$

No interchanges are required for the next step. We divide the third row by 15.000 and make the other elements in the third column into zeros:

$$\begin{bmatrix} 1 & 0 & 0 & 0.040005 & -0.58000 \\ 0 & 1 & 0 & -0.27993 & 1.5599 \\ 0 & 0 & 1 & 0.82667 & -1.3200 \\ 0 & 0 & 0 & 1.5599 & -3.1197 \end{bmatrix}.$$

We now divide the fourth row by 1.5599 and create zeros above the diagonal in the fourth column:

$$\begin{bmatrix} 1 & 0 & 0 & 0 & -0.49999 \\ 0 & 1 & 0 & 0 & 1.0001 \\ 0 & 0 & 1 & 0 & 0.33326 \\ 0 & 0 & 0 & 1 & -1.9999 \end{bmatrix}.$$

The solution is essentially the same as with the usual Gaussian method; round-off errors have created inaccuracies in a slightly different way than did the previous computation.

While the above Gauss–Jordan scheme appears to be a duplicate of the work done in the standard procedure, a count of the arithmetic operations shows that the Gauss–Jordan method requires almost 50% more operations. We therefore do not recommend its use.

A frequently occurring situation is to have to solve a set of equations with the same coefficient matrix but with a number of different righthand sides. For example, in the case of interconnected resistances at the beginning of this chapter, we may desire to study the effects of different voltages applied at points a and b. In the design of a truss, one usually wishes to determine the stresses under a variety of external loads—this causes only the righthand-side terms to vary.

If all of the different righthand-side vectors are known in advance, the multiple solutions of the system can be obtained simultaneously using our Gaussian elimination method. One augments the coefficient matrix with all of the righthand-side vectors, and treats each augmentation column in the same way as a single added column. In the back-substitution phase, each of the columns is employed to give a solution vector. At the end of this chapter, a subroutine is exhibited that does this.

Example Solve the system $Ax = b$, with multiple values of b, by Gaussian elimination:

$$A = \begin{bmatrix} 3 & 2 & -1 & 2 \\ 1 & 4 & 0 & 2 \\ 2 & 1 & 2 & -1 \\ 1 & 1 & -1 & 3 \end{bmatrix}, \quad b^{(1)} = \begin{bmatrix} 0 \\ 0 \\ 1 \\ 0 \end{bmatrix}, \quad b^{(2)} = \begin{bmatrix} -2 \\ 1 \\ 3 \\ 4 \end{bmatrix}, \quad b^{(3)} = \begin{bmatrix} 2 \\ 2 \\ 0 \\ 0 \end{bmatrix}.$$

We augment A with all of the b's, then triangularize:

$$\begin{bmatrix} 3 & 2 & -1 & 2 & 0 & -2 & 2 \\ 1 & 4 & 0 & 2 & 0 & 1 & 2 \\ 2 & 1 & 2 & -1 & 1 & 3 & 0 \\ 1 & 1 & -1 & 3 & 0 & 4 & 0 \end{bmatrix} \rightarrow$$

$$
\begin{bmatrix}
3 & 2 & -1 & 2 & 0 & -2 & 2 \\
0 & 3.333 & 0.333 & 1.333 & 0 & 1.667 & 1.333 \\
0 & -0.333 & 2.667 & -2.333 & 1 & 4.333 & -1.333 \\
0 & 0.333 & -0.667 & 2.333 & 0 & 4.667 & -0.667
\end{bmatrix} \rightarrow
$$

$$
\begin{bmatrix}
3 & 2 & -1 & 2 & 0 & -2 & 2 \\
0 & 3.333 & 0.333 & 1.333 & 0 & 1.667 & 1.333 \\
0 & 0 & 2.700 & -2.200 & 1 & 4.500 & -1.200 \\
0 & 0 & -0.700 & 2.200 & 0 & 4.500 & -0.800
\end{bmatrix} \rightarrow
$$

$$
\begin{bmatrix}
3 & 2 & -1 & 2 & 0 & -2 & 2 \\
0 & 3.333 & 0.333 & 1.333 & 0 & 1.667 & 1.333 \\
0 & 0 & 2.700 & -2.200 & 1 & 4.500 & -1.200 \\
0 & 0 & 0 & 1.630 & 0.259 & 5.667 & -1.111
\end{bmatrix}
$$

$$
\uparrow \qquad\qquad\qquad\qquad
$$
$$
c^{(1)} \qquad c^{(2)} \qquad c^{(3)}
$$

We obtain the three solution vectors by back-substitution, employing the proper b' vector. (These are indicated above as $c^{(i)}$.)

$$
x^{(1)} = \begin{bmatrix} 0.137 \\ -0.114 \\ 0.500 \\ 0.159 \end{bmatrix}, \qquad
x^{(2)} = \begin{bmatrix} -0.591 \\ -1.340 \\ 4.500 \\ 3.477 \end{bmatrix}, \qquad
x^{(3)} = \begin{bmatrix} 0.273 \\ 0.773 \\ -1.000 \\ -0.682 \end{bmatrix}.
$$

Scaling

We have mentioned that the rows of the augmented coefficient matrix may need to scaled before a proper choice of pivot element can be made. *Scaling* is the operation of adjusting the coefficients of a set of equations so that they are all of the same order of magnitude. In some instances, a set of equations may involve relationships between quantities measured in widely different units (microvolts versus kilovolts, for example, or nanoseconds versus years). This may result in some of the equations having very large numbers and others very small. If we select the pivot elements without scaling, pivoting may put numbers on the diagonal that are not large in comparison to others in their row; this can actually create the round-off errors that pivoting was supposed to avoid. An example will clarify the concept.

$$
\text{Given:} \quad \begin{bmatrix} 3 & 2 & 100 \\ -1 & 3 & 100 \\ 1 & 2 & -1 \end{bmatrix} x = \begin{bmatrix} 105 \\ 102 \\ 2 \end{bmatrix}.
$$

Carrying only three digits to emphasize round-off, and using partial pivoting, we find that the triangularized system is:

$$\begin{bmatrix} 3 & 2 & 100 & 105 \\ 0 & 3.66 & 133 & 137 \\ 0 & 0 & -82.6 & -82.7 \end{bmatrix}$$

from which $x_3 = 1.00$, $x_2 = 1.09$, $x_1 = 0.94$; the exact solution vector should be [1.00, 1.00, 1.00].

If we scale the values before reduction by dividing each row by the magnitude of the largest coefficient, so that the system is

$$\begin{bmatrix} 0.03 & -0.02 & 1.00 \\ -0.01 & 0.03 & 1.00 \\ 0.50 & 1.00 & -0.50 \end{bmatrix} x = \begin{bmatrix} 1.05 \\ 1.02 \\ 1.00 \end{bmatrix},$$

we get, with the same arithmetic precision, the triangularized system

$$\begin{bmatrix} 0.50 & 1.00 & -0.50 & 1.00 \\ 0 & 0.05 & 0.99 & 1.04 \\ 0 & 0 & 1.82 & 1.82 \end{bmatrix},$$

from which $x_3 = 1.00$, $x_2 = 1.00$, $x_1 = 1.00$. The reason for the improvement is that rows are interchanged after scaling has been done. No interchanges are indicated in the unscaled equations.

Whenever the coefficients in one row are widely different from those in another row, scaling is beneficial. When all values are about the same order of magnitude, scaling should be avoided, for the additional round-off error incurred during the scaling operation itself may adversely affect the accuracy. The usual way to scale is as we have done here, by dividing each row by the magnitude of the largest term. Some authorities recommend scaling so that the sum of the magnitudes of the coefficients in each row is the same. This is probably very slightly more economical to do in a computer program.

2.5 *LU* DECOMPOSITION

A modification of the elimination method, called the *LU* decomposition method (also named Crout reduction, or, after another discoverer, Cholesky), is frequently used in computer programs. In it the matrix of coefficients A is transformed into the product of two matrices L and U, where L is a lower triangular and U is an upper triangular matrix with 1's on its diagonal.*

---

* An equivalent method transforms A into an LU pair in which L has 1's on its diagonal. This is called *Doolittle's method*.

Any matrix can be written as a product of a lower triangular and an upper triangular matrix in an infinity of ways. For example,

$$\underset{A}{\begin{bmatrix} 2 & -1 & -1 \\ 0 & -4 & 2 \\ 6 & -3 & 0 \end{bmatrix}} = \underset{L_1}{\begin{bmatrix} 2 & 0 & 0 \\ 0 & -4 & 0 \\ 6 & 0 & 3 \end{bmatrix}} \underset{U_1}{\begin{bmatrix} 1 & -\frac{1}{2} & -\frac{1}{2} \\ 0 & 1 & -\frac{1}{2} \\ 0 & 0 & 1 \end{bmatrix}}$$

$$= \underset{L_2}{\begin{bmatrix} 1 & 0 & 0 \\ 0 & 1 & 0 \\ 3 & 0 & 1 \end{bmatrix}} \underset{U_2}{\begin{bmatrix} 2 & -1 & -1 \\ 0 & -4 & 2 \\ 0 & 0 & 3 \end{bmatrix}}$$

$$= \underset{L_3}{\begin{bmatrix} 1 & 0 & 0 \\ 0 & 2 & 0 \\ 3 & 0 & 1 \end{bmatrix}} \underset{U_3}{\begin{bmatrix} 2 & -1 & -1 \\ 0 & -2 & 1 \\ 0 & 0 & 3 \end{bmatrix}}$$

$$= \text{etc., etc.}$$

Of the entire set of *LU*'s whose product equals matrix *A*, we choose the pair in which *U* has only ones on its diagonal, as in the first pair above. We get the rules for such an *LU* decomposition from the relationship that $LU = A$. In the case of a 4 × 4 matrix:

$$\begin{bmatrix} \ell_{11} & 0 & 0 & 0 \\ \ell_{21} & \ell_{22} & 0 & 0 \\ \ell_{31} & \ell_{32} & \ell_{33} & 0 \\ \ell_{41} & \ell_{42} & \ell_{43} & \ell_{44} \end{bmatrix} \begin{bmatrix} 1 & u_{12} & u_{13} & u_{14} \\ 0 & 1 & u_{23} & u_{24} \\ 0 & 0 & 1 & u_{34} \\ 0 & 0 & 0 & 1 \end{bmatrix} = \begin{bmatrix} a_{11} & a_{12} & a_{13} & a_{14} \\ a_{21} & a_{22} & a_{23} & a_{24} \\ a_{31} & a_{32} & a_{33} & a_{34} \\ a_{41} & a_{42} & a_{43} & a_{44} \end{bmatrix}$$

Multiplying the rows of *L* by the first column of *U*, we get $\ell_{11} = a_{11}$, $\ell_{21} = a_{21}$, $\ell_{31} = a_{31}$, $\ell_{41} = a_{41}$; the first column of *L* is the same as the first column of *A*. We now multiply the first row of *L* by the columns of *U*:

$$\ell_{11}u_{12} = a_{12}, \qquad \ell_{11}u_{13} = a_{13}, \qquad \ell_{11}u_{14} = a_{14}, \tag{2.3}$$

from which

$$u_{12} = \frac{a_{12}}{\ell_{11}}, \qquad u_{13} = \frac{a_{13}}{\ell_{11}}, \qquad u_{14} = \frac{a_{14}}{\ell_{11}}. \tag{2.4}$$

Thus the first row of *U* is determined.

In this method we alternate between getting a column of *L* and a row of *U*, so we next get the equations for the second column of *L* by multiplying the rows of L by the second column of U:

$$\ell_{21}u_{12} + \ell_{22} = a_{22},$$
$$\ell_{31}u_{12} + \ell_{32} = a_{32} \tag{2.5}$$
$$\ell_{41}u_{12} + \ell_{42} = a_{42},$$

which gives

$$\ell_{22} = a_{22} - \ell_{21}u_{12},$$
$$\ell_{32} = a_{32} - \ell_{31}u_{12}, \tag{2.6}$$
$$\ell_{42} = a_{42} - \ell_{41}u_{12}.$$

Proceeding in the same fashion, the equations we need are

$$u_{23} = \frac{a_{23} - \ell_{21}u_{13}}{\ell_{22}}, \qquad u_{24} = \frac{a_{24} - \ell_{21}u_{14}}{\ell_{22}},$$

$$\ell_{33} = a_{33} - \ell_{31}u_{13} - \ell_{32}u_{23}, \qquad \ell_{43} = a_{43} - \ell_{41}u_{13} - \ell_{42}u_{23}, \tag{2.7}$$

$$u_{34} = \frac{a_{34} - \ell_{31}u_{14} - \ell_{32}u_{24}}{\ell_{33}},$$

$$\ell_{44} = a_{44} - \ell_{41}u_{14} - \ell_{42}u_{24} - \ell_{43}u_{34}.$$

The general formula for getting elements of L and U corresponding to the coefficient matrix for n simultaneous equations can be written

$$\ell_{ij} = a_{ij} - \sum_{k=1}^{j-1} \ell_{ik}u_{kj}, \qquad j \le i, \qquad i = 1, 2, \ldots, n,$$

$$u_{ij} = \frac{a_{ij} - \sum_{k=1}^{i-1} \ell_{ik}u_{kj}}{\ell_{ii}}, \qquad i > j, \qquad j = 2, 3, \ldots, n.$$

$\left(\vphantom{\sum}\right.$ For $j = 1$, the rule for ℓ reduces to

$$\ell_{i1} = a_{i1}.$$

For $i = 1$, the rule for u reduces to

$$u_{1j} = \frac{a_{1j}}{\ell_{11}} = \frac{a_{1j}}{a_{11}}.\left.\vphantom{\sum}\right)$$

The reason this method is popular in programs is that storage space may be economized. There is no need to store the 0's in either L or U, and the 1's on the diagonal of U can also be omitted. (Since these values are always the same and are always known, it is redundant to record them.) One can then store the essential elements of U where the 0's appear in the L array. Examination of Eqs. (2.2) through (2.7) shows that after any element of A, a_{ij}, is once used, it never again appears in the equations. Hence its place in the original $n \times n$ array A can be used to store an element of either L or U. In other words, the A array can be transformed by the above equations and becomes

$$
\begin{bmatrix}
a_{11} & a_{12} & a_{13} & a_{14} \\
a_{21} & a_{22} & a_{23} & a_{24} \\
a_{31} & a_{32} & a_{33} & a_{34} \\
a_{41} & a_{42} & a_{43} & a_{44}
\end{bmatrix}
\rightarrow
\begin{bmatrix}
\ell_{11} & u_{12} & u_{13} & u_{14} \\
\ell_{21} & \ell_{22} & u_{23} & u_{24} \\
\ell_{31} & \ell_{32} & \ell_{33} & u_{34} \\
\ell_{41} & \ell_{42} & \ell_{43} & \ell_{44}
\end{bmatrix}
$$

Because we can condense the L and U matrices into one array and store their elements in the space of A, this method is often called a *compact scheme*.

Example Consider the matrix A,

$$
A = \begin{bmatrix}
3 & -1 & 2 \\
1 & 2 & 3 \\
2 & -2 & -1
\end{bmatrix}.
$$

Applying the equations for the ℓ's and u's, we obtain:

$$\ell_{11} = 3, \qquad \ell_{21} = 1, \qquad \ell_{31} = 2; \qquad u_{12} = -\tfrac{1}{3}, \qquad u_{13} = \tfrac{2}{3}.$$

$$\ell_{22} = 2 - (1)(-\tfrac{1}{3}) = \tfrac{7}{3}, \qquad \ell_{32} = -2 - (2)(-\tfrac{1}{3}) = -\tfrac{4}{3}.$$

$$u_{23} = \frac{3 - (1)(\tfrac{2}{3})}{\tfrac{7}{3}} = 1, \qquad \ell_{33} = -1 - (2)(\tfrac{2}{3}) - (-\tfrac{4}{3})(1) = -1.$$

$$
L = \begin{bmatrix}
3 & 0 & 0 \\
1 & \tfrac{7}{3} & 0 \\
2 & -\tfrac{4}{3} & -1
\end{bmatrix}, \qquad
U = \begin{bmatrix}
1 & -\tfrac{1}{3} & \tfrac{2}{3} \\
0 & 1 & 1 \\
0 & 0 & 1
\end{bmatrix}.
$$

If the quantities are written in the compact form as they are computed, we have:

$$
LU = \begin{bmatrix}
3 & -\tfrac{1}{3} & \tfrac{2}{3} \\
1 & \tfrac{7}{3} & 1 \\
2 & -\tfrac{4}{3} & -1
\end{bmatrix}
\begin{matrix} \leftarrow ② \\ \leftarrow ④ \\ \ \end{matrix}
$$
$$
\begin{matrix} \uparrow & \uparrow & \uparrow \\ ① & ③ & ⑤ \end{matrix}
$$

The circled numbers show the order in which columns and rows of the new matrix are obtained.

Here is an algorithm for LU decomposition. It does not compute the L and U matrices in place, but sets them up as separate matrices.

LU decomposition

To transform an $n \times n$ matrix A into the equivalent product of L and U matrices,

```
DO FOR I = 1 to n step 1,
    L(I, 1) = A(I, 1).
ENDDO
DO FOR J = 1 to n step 1,
    U(1, J) = A(1, J)/L(1, 1).
ENDDO.
DO FOR J = 2 to n step 1,
    DO FOR I = J to n step 1,
        DO FOR K = 1 to J-1 step 1,
            Accumulate sum of L(I, K) * U(K, J).
        ENDDO.
        L(I, J) = A(I, J) − SUM.
    ENDDO.
    U(J, J) = 1.
    DO FOR I = J + 1 to n step 1,
        DO FOR K = 1 to J − 1 step 1,
            Accumulate sum of L(J, K) * U(K, J).
        ENDDO
        U(J, I) = [A(J, I) − SUM]/L(J, J).
    ENDDO.
ENDDO.
```

The solution of the set of equations $Ax = b$ is readily obtained with the L and U matrices. Once the coefficient matrix has been converted to its LU equivalent, we are prepared to find the solution to the set of equations that corresponds to any given righthand-side vector b. The L matrix is really a record of the operations required to make the coefficient matrix, A, into the upper triangular matrix U. We apply these same transformations to the RHS vector, b, converting it to a new vector, b'. If we augment b' to U and back-substitute, the solution appears.

The general equation for the reduction of b (it is exactly the same as the rule for forming the elements of U) is:

$$b'_1 = b_1/\ell_{11}$$

$$b'_i = \frac{b_i - \sum_{k=1}^{i-1} \ell_{ik} b'_k}{\ell_{ii}}, \qquad i = 2, 3, \ldots, n.$$

The equations for the back-substitution are:

$$x_n = b'_n,$$

$$x_j = b'_j - \sum_{k=j+1}^{n} u_{jk}b'_k, \qquad j = n-1, n-2, \ldots, 1.$$

For example, if

$$A = \begin{bmatrix} 3 & -1 & 2 \\ 1 & 2 & 3 \\ 2 & -2 & -1 \end{bmatrix},$$

we get

$$L = \begin{bmatrix} 3 & 0 & 0 \\ 1 & \frac{7}{3} & 0 \\ 2 & -\frac{4}{3} & -1 \end{bmatrix} \quad \text{and} \quad U = \begin{bmatrix} 1 & -\frac{1}{3} & \frac{2}{3} \\ 0 & 1 & 1 \\ 0 & 0 & +1 \end{bmatrix}.$$

For $b = \begin{bmatrix} 12 \\ 11 \\ 2 \end{bmatrix}$,

See below * .

$$b' = \begin{bmatrix} 4 \\ 3 \\ 2 \end{bmatrix}, \qquad \boxed{L \cdot b' = b}$$

because

$$b'_1 = \frac{12}{3} = 4,$$

$$b'_2 = \frac{11 - (1)(4)}{\frac{7}{3}} = 3,$$

$$b'_3 = \frac{2 - (2)(4) - (-\frac{4}{3})(3)}{-1} = 2.$$

Augmenting b' to U and back-substituting,

$$\begin{bmatrix} 1 & -\frac{1}{3} & \frac{2}{3} & \vdots & 4 \\ 0 & 1 & 1 & \vdots & 3 \\ 0 & 0 & 1 & \vdots & 2 \end{bmatrix} \qquad \begin{bmatrix} U \vdots b' \end{bmatrix}$$

$(U \qquad \vdots b')$

we get

$$x_3 = 2,$$

$$x_2 = 3 - 1(2) = 1,$$

$$x_1 = 4 - \frac{2}{3}(2) - (-\frac{1}{3})(1) = 3.$$

Examination of the above operations reveals that we can get b' by augmenting L with b and solving that triangular system (a kind of "forward" substitution):

$$* \qquad L \vdots b = \begin{bmatrix} 3 & 0 & 0 & \vdots & 12 \\ 1 & \frac{7}{3} & 0 & \vdots & 11 \\ 2 & -\frac{4}{3} & -1 & \vdots & 2 \end{bmatrix}$$

and

$$b_1' = \tfrac{12}{3} = 4$$
$$b_2' = [11 - (1)(4)]/(\tfrac{7}{3}) = 3$$
$$b_3' = [2 - (2)(4) - (-\tfrac{4}{3})(3)]/(-1) = 2.$$

A special advantage of the LU method is its ability to solve the system with new righthand-side vectors with great economy of effort.* We note in particular that these new righthand sides do not have to be known in advance as is the case with Gaussion elimination as described in Section 2.4. The number of arithmetic operations to get the solution corresponding to each b turns out to be exactly the same as to multiply an $n \times n$ matrix by an n-component vector.

Pivoting with the LU method is somewhat more complicated than with Gaussion elimination because we do not usually handle the righthand-side vector simultaneously with our reduction of the A matrix. This means we must keep a record of any row interchanges made during the formation of L and U so that the elements of the righthand-side vector can be similarly interchanged. We do the interchanges immediately after computing each column of L, choosing the value to appear on the diagonal so as to have the one of largest magnitude. We illustrate with an example:

$$\text{Given } A = \begin{bmatrix} 0 & 2 & 1 \\ 1 & 0 & 0 \\ 3 & 0 & 1 \end{bmatrix}.$$

We will keep a record of row order in a vector: $O = [1, 2, 3]$, representing the original ordering.

The first column of L is

$$\begin{bmatrix} 0 \\ 1 \\ 3 \end{bmatrix};$$

we need to interchange rows 3 and 1. To keep track of this, we interchange the first and third elements of O, so O becomes $[3, 2, 1]$.

Interchange the rows of A and compute the first row of the U matrix. (We use the compact scheme.)

$$\begin{bmatrix} 3 & 0 & \tfrac{1}{3} \\ 1 & 0 & 0 \\ 0 & 2 & 1 \end{bmatrix}, \quad \text{with } O = [3, 2, 1].$$

---

* A numerical example that illustrates using the LU method with multiple righthand sides will be found in Section 2.7.

Now compute the second column of L; it is $\begin{bmatrix} 0 \\ 0 \\ 2 \end{bmatrix}$.

We must interchange again, the second row with the third, and O becomes $[3, 1, 2]$. Making the interchange of rows and computing the second row of U gives

$$\begin{bmatrix} 3 & 0 & \frac{1}{3} \\ 0 & 2 & \frac{1}{2} \\ 1 & 0 & 0 \end{bmatrix}, \quad \text{with } O = [3, 1, 2].$$

Completing the reduction, we find $\ell_{33} = 0 - (1)(\frac{1}{3}) = -\frac{1}{3}$, giving

$$LU = \begin{bmatrix} 3 & 0 & \frac{1}{3} \\ 0 & 2 & \frac{1}{2} \\ 1 & 0 & -\frac{1}{3} \end{bmatrix}, \quad \text{with } O = [3, 1, 2].$$

To solve the problem $Ax = b$, with $b^T = [5, -1, -2]$, we rearrange the elements of b in the order given by O and compute b':

$$L \vdots b = \begin{bmatrix} 3 & 0 & 0 & \vdots & -2 \\ 0 & 2 & 0 & \vdots & 5 \\ 1 & 0 & -\frac{1}{3} & \vdots & -1 \end{bmatrix}, \quad b' = \begin{bmatrix} -\frac{2}{3} \\ \frac{5}{2} \\ 1 \end{bmatrix},$$

so

$$U \vdots b' = \begin{bmatrix} 1 & 0 & \frac{1}{3} & \vdots & -\frac{2}{3} \\ 0 & 1 & \frac{1}{2} & \vdots & \frac{5}{2} \\ 0 & 0 & 1 & \vdots & 1 \end{bmatrix}, \quad \text{giving } x = \begin{bmatrix} -1 \\ 2 \\ 1 \end{bmatrix}.$$

2.6 PATHOLOGICAL SITUATIONS WITH LINEAR SYSTEMS— SINGULAR MATRICES

When a real physical situation is modelled by a set of linear equations, one can anticipate that the set of equations will have a solution that matches the values of the quantities in the physical problem, at least so far as the equations truly do represent it.* Because of round-off errors, the solution vector which is calculated may imperfectly predict the physical quantity, but there is assurance that a solution exists, at least in principle. Consequently, it must always be theoretically possible to avoid divisions by zero when the set of equations has a solution.

An arbitrary set of equations may not have such a guaranteed solution, however. There are several such possible situations, which we term "pathological." In each case, there is *no unique solution* to the set of equations.

---

* There are certain problems for which values of interest are determined from a set of equations that do not have a unique solution; these are called *eigenvalue* problems and are discussed in another chapter.

First, if the number of equations relating the variables is less than the number of *unknowns*, we certainly cannot solve for unique values of the unknown variables. It turns out, in this case, that there is an infinite set of solutions, for we may arrange the n equations with all but n of the variables on the righthand sides, grouped with the constant terms. We may assign almost any desired values to these segregated variables (combining them with the constant terms) and then solve for the n remaining variables.* Assigning new values to the variables on the righthand sides gives another set of values for the unknowns, etc., etc. For example:

$$\begin{cases} x_1 - 2x_2 + x_3 = 4, \\ x_1 - x_2 + x_3 = 5. \end{cases}$$

Rewrite as

$$x_1 - 2x_2 = 4 - x_3,$$
$$x_1 - x_2 = 5 - x_3.$$

If $x_3 = 0$,

$$x_1 = 6, \qquad x_2 = 1.$$

If $x_3 = 1$,

$$x_1 = 5, \qquad x_2 = 1.$$

If $x_3 = -1$,

$$x_1 = 7, \qquad x_2 = 1, \qquad \text{etc., etc.}$$

A second situation where we might not expect a set of equations to have a solution is that in which the number of equations is greater than the number of unknowns. If there are n unknowns, we can normally find a subset of the equations which can be solved for the unknowns. There are two subcases to consider:

If the remaining equations are satisfied by the values of the unknowns we have just determined (we would say these equations are *consistent* with the others), there exists a unique solution to the set of equations. Really, of course, there are not truly more equations than unknowns in this case; the extra equations are *redundant*.

The other subcase is a pathological one. If the solution to the first n equations does not satisfy the remaining ones, the set is clearly *inconsistent*, and *no* solution exists that satisfies the system.

Realizing that an equation may be redundant in the above situation makes us re-examine our more standard case of n equations in n unknowns. What if there

---

* An important instance of this situation is the solution of a linear programming problem by the simplex method. In this method, the segregated variables are all assigned the value of zero.

is redundancy there? How can we recognize redundancy when it is present? In this example it is obvious:

$$x + y = 3, \qquad 2x + 2y = 6.$$

The second equation is clearly redundant and contains no information not already given by the first. This system will then have an *infinity* of values for x and y; it is an example of fewer equations than unknowns.

Inconsistency may also be present:

$$x + y = 3, \qquad 2x + 2y = 7.$$

In this case there is *no* solution.

In $n \times n$ systems that do not have a unique solution, they have a (square) coefficient matrix that is called *singular*. If the coefficient matrix can be triangularized without having zeros on the diagonal (hence the set of equations has a solution), the matrix is said to be *nonsingular*.

Larger systems may have redundancy or inconsistency even though it is not obvious at a glance. Even in a 3×3 system, it is not easy to tell:

$$
\begin{aligned}
x_1 - 2x_2 + 3x_3 &= 5, \\
2x_1 + 4x_2 - x_3 &= 7, \\
-x_1 - 14x_2 + 11x_3 &= 2.
\end{aligned}
$$

Are these inconsistent or redundant? (In other words, is the coefficient matrix singular?) Or do they have a unique solution? (That is, is the matrix nonsingular?) Is there a rule that we can apply, especially one that works for large systems? The answer is *yes*, there is a rule, or rather, there are several tests we can apply. The standard response from mathematics is to determine the *rank* of the coefficient matrix. If this value is less than n, the number of equations, then no unique solution exists; the equations are either inconsistent or one or more is redundant, depending on the righthand-side values.

But how does one determine the rank? The only practical method is to triangularize by Gaussian elimination: If no zeros show up on the diagonal of the final triangularized coefficient matrix, the rank is equal to n (the matrix is said to be of *full rank*) and a unique solution exists. If, in spite of pivoting, one or more zeros occur on the final diagonal, there is no unique solution. The set will be consistent (and have redundancy) if back-substitution gives $(0/0)$ indeterminate forms. When one would need to divide a nonzero term by zero in back-substituting, inconsistency occurs. Let us apply this to our earlier 3×3 example.

$$
\begin{bmatrix}
1 & -2 & 3 & \vdots & 5 \\
2 & 4 & -1 & \vdots & 7 \\
-1 & -14 & 11 & \vdots & 2
\end{bmatrix}
$$

On reduction we get:

$$\begin{bmatrix} 1 & -2 & 3 & \vdots & 5 \\ 0 & 8 & -7 & \vdots & -3 \\ 0 & 0 & 0 & \vdots & 1 \end{bmatrix}$$

We see that there is no solution and the equations are inconsistent. If the constant term in the third equation were 1 rather than 2, reduction gives:

$$\begin{bmatrix} 1 & -2 & 3 & \vdots & 5 \\ 0 & 8 & -7 & \vdots & -3 \\ 0 & 0 & 0 & \vdots & 0 \end{bmatrix}$$

This system is redundant.

Another way to find whether a set of equations has a unique solution is to test the rows or columns of the coefficient matrix for *linear dependency*. Vectors are called linearly dependent if a linear combination of them can be found that equals the zero vector (one with all components equal to zero). (Of course the linear combination $a\bar{x} + b\bar{y} + c\bar{z}$ always equals zero if all the coefficients are zero; we rule out this possibility in our test for linear dependency. When $a = b = c = 0$, we say we have the *trivial case*.)

If the vectors are linearly independent, the only way a weighted sum of them can equal the zero vector is to weight each of them with a zero coefficient.

Our singular 3×3 system has columns that form vectors that are linearly dependent:

$$(-10)\begin{bmatrix} 1 \\ 2 \\ -1 \end{bmatrix} + (7)\begin{bmatrix} -2 \\ 4 \\ -14 \end{bmatrix} + (8)\begin{bmatrix} 3 \\ -1 \\ 11 \end{bmatrix} = \begin{bmatrix} 0 \\ 0 \\ 0 \end{bmatrix}.$$

Similarly, the rows form linearly dependent vectors:

$$(-3)[1 \quad -2 \quad 3] + (2)[2 \quad 4 \quad -1] + (1)[-1 \quad -14 \quad 11] = [0 \quad 0 \quad 0].$$

In the general case, we say that vectors $\bar{x}_1, \bar{x}_2, \bar{x}_3, \ldots, \bar{x}_n$ are *linearly dependent* if we can find scalar coefficients, $a_1, a_2, \ldots, a_n$ (with not all the a_i simultaneously zero,) for which

$$\sum_{i=1}^{n} a_i \bar{x}_i = \bar{0}. \tag{2.8}$$

If the only linear combination of the $\bar{x}_i$ that equals the zero vector requires that all the a_i be zero, the set of vectors is called linearly independent. It follows that, if a set of vectors is linearly dependent, at least one of the vectors can be written as a linear combination of the others. If the set is linearly independent, none of the vectors can be written as a linear combination of the others. As a practical matter, we do not usually test the columns (or rows) for linear dependency to determine whether a matrix is singular.

If we are interested in determining the coefficients $a_1, a_2, \ldots, a_n$ that appear in the linear combination of Eq. (2.8), it turns out that we have to solve a set of linear equations to obtain them.

It is worth while to summarize the concepts and terminology of this section. The following lists of terms are all equivalent expressions. If a square matrix can be shown to have one property, it has all the others.

Equivalent properties of singular or nonsingular matrices

The matrix is singular.	The matrix is nonsingular.
A set of equations with these coefficients has no unique solution.	A set of equations with these coefficients has a unique solution.
Gaussian elimination cannot avoid a zero on the diagonal.	Gaussian elimination proceeds without a zero on the diagonal.
The rank of the matrix is less than n.	The rank of the matrix equals n.
The rows form linearly dependent vectors.	The rows form linearly independent vectors.
The columns form linearly dependent vectors.	The columns form linearly independent vectors.

In the next section we will consider two other properties of the matrix: its determinant and its inverse. This adds two more attributes to our lists: A singular matrix has a zero determinant and a nonsingular matrix has a nonzero determinant. A singular matrix has no inverse and a nonsingular matrix does have an inverse.

2.7 DETERMINANT OF A MATRIX AND MATRIX INVERSION

The student has perhaps wondered why there has been no reference heretofore in this chapter to the solution of linear equations by deteminants (Cramer's rule). The reason for this is that, except for systems of only two or three equations, the determinant method is too inefficient. For example, for a set of 10 simultaneous equations, about 70,000,000 multiplications and divisions are required if the usual method of expansion in terms of minors is used. A more efficient method of evaluating the determinants can reduce this to about 3000 multiplications, but even this is inefficient compared to Gaussian elimination, which would require about 380. (The Gauss–Jordan scheme requires about 50% more arithmetic operations than Gauss with back-substitution.)

In fact, the evaluation of a determinant can perhaps best be done by adapting the Gauss elimination procedure. Its utility derives from the fact that the determinant of a triangular matrix (either upper- or lower-triangular) is just the product of its diagonal elements. This is easily seen, in the case of an upper-triangular matrix, by expansion by minors of the first column at each step. For example,

$$
\begin{vmatrix} a_{11} & a_{12} & a_{13} & a_{14} \\ 0 & a_{22} & a_{23} & a_{24} \\ 0 & 0 & a_{33} & a_{34} \\ 0 & 0 & 0 & a_{44} \end{vmatrix} = a_{11} \begin{vmatrix} a_{22} & a_{23} & a_{24} \\ 0 & a_{33} & a_{34} \\ 0 & 0 & a_{44} \end{vmatrix} - 0 + 0 - 0
$$

$$
= a_{11}(a_{22} \begin{vmatrix} a_{33} & a_{34} \\ 0 & a_{44} \end{vmatrix} - 0 + 0)
$$

$$
= a_{11}a_{22}(a_{33}a_{44} - 0) = a_{11}a_{22}a_{33}a_{44}.
$$

Since elementary row transformations do not change the value of a determinant (except that interchanging two rows changes its sign), using the procedure of Gaussian elimination to convert to upper-triangular is a simple way to evaluate determinants. If row interchanges are made to take advantage of pivoting, the value of the determinant is multiplied by -1 for each such interchange. If a row is multiplied by a constant, the value of the determinant is also multiplied, so scaling must be considered as well.

Example Find the value of the determinant by using elementary row transformations to make it upper-triangular.

$$
\begin{vmatrix} 1 & 4 & -2 & 3 \\ 2 & 2 & 0 & 4 \\ 3 & 0 & -1 & 2 \\ 1 & 2 & 2 & -3 \end{vmatrix} = \begin{vmatrix} 1 & 4 & -2 & 3 \\ 0 & -6 & 4 & -2 \\ 0 & -12 & 5 & -7 \\ 0 & -2 & 4 & -6 \end{vmatrix} = \begin{vmatrix} 1 & 4 & -2 & 3 \\ 0 & -6 & 4 & -2 \\ 0 & 0 & -3 & -3 \\ 0 & 0 & \frac{8}{3} & -\frac{16}{3} \end{vmatrix}
$$

$$
= \begin{vmatrix} 1 & 4 & -2 & 3 \\ 0 & -6 & 4 & -2 \\ 0 & 0 & -3 & -3 \\ 0 & 0 & 0 & -8 \end{vmatrix} = (1)(-6)(-3)(-8) = -144.
$$

While division of matrices is not defined, the *matrix inverse* gives the equivalent result. If the product of two square matrices is the identity matrix, the matrices are said to be *inverses*. If $AB = I$, we write $B = A^{-1}$; also $A = B^{-1}$. Inverses *commute* on multiplication, which is not true for matrices in general: $AB = BA = I$. Not all square matrices have an inverse. Singular matrices do not have an inverse, and these are of extreme importance in connection with the coefficient matrix of a set of equations, as discussed above.

The inverse of a matrix can be defined in terms of the matrix of the minors of its determinant, but this is not a useful way to find an inverse. Gaussian elimination can be adapted to provide a practical way to invert a matrix. The procedure is to augment the given matrix with the identity matrix of the same order. One then reduces the original matrix to the identity matrix by elementary row transformations, performing the same operations on the augmentation columns. When the identity matrix stands as the left half of the augmented matrix, the inverse of the original stands as the right half. It should be apparent that this is equivalent to solving a set of equations with n different righthand sides; each of the righthand sides is a *unit vector*, in which the position of the element whose value is unity changes from row 1 to row 2 to row 3 ... to row n.

Example   Find the inverse of

$$A = \begin{bmatrix} 1 & -1 & 2 \\ 3 & 0 & 1 \\ 1 & 0 & 2 \end{bmatrix}.$$

Augment A with the identity matrix and then reduce:

$$\begin{bmatrix} 1 & -1 & 2 & 1 & 0 & 0 \\ 3 & 0 & 1 & 0 & 1 & 0 \\ 1 & 0 & 2 & 0 & 0 & 1 \end{bmatrix} \rightarrow \begin{bmatrix} 1 & -1 & 2 & 1 & 0 & 0 \\ 0 & 3 & -5 & -3 & 1 & 0 \\ 0 & 1 & 0 & -1 & 0 & 1 \end{bmatrix}$$

$$\overset{(1)}{\rightarrow} \begin{bmatrix} 1 & -1 & 2 & 1 & 0 & 0 \\ 0 & 1 & 0 & -1 & 0 & 1 \\ 0 & 0 & -5 & 0 & 1 & -3 \end{bmatrix} \overset{(2)}{\rightarrow} \begin{bmatrix} 1 & -1 & 0 & 1 & \frac{2}{5} & -\frac{6}{5} \\ 0 & 1 & 0 & -1 & 0 & 1 \\ 0 & 0 & 1 & 0 & -\frac{1}{5} & \frac{3}{5} \end{bmatrix}$$

$$\rightarrow \begin{bmatrix} 1 & 0 & 0 & 0 & \frac{2}{5} & -\frac{1}{5} \\ 0 & 1 & 0 & -1 & 0 & 1 \\ 0 & 0 & 1 & 0 & -\frac{1}{5} & \frac{3}{5} \end{bmatrix}.$$

(1) Interchange the third and second rows before eliminating from the third row.
(2) Divide the third row by -5 before eliminating from the first row.

We confirm the fact that we have found the inverse by multiplication:

$$\begin{bmatrix} 1 & -1 & 2 \\ 3 & 0 & 1 \\ 1 & 0 & 2 \end{bmatrix} \begin{bmatrix} 0 & \frac{2}{5} & -\frac{1}{5} \\ -1 & 0 & 1 \\ 0 & -\frac{1}{5} & \frac{3}{5} \end{bmatrix} = \begin{bmatrix} 1 & 0 & 0 \\ 0 & 1 & 0 \\ 0 & 0 & 1 \end{bmatrix}.$$

If we have the LU equivalent to a matrix, we can compute its inverse by solving the system n times, with successive unit vectors as the righthand sides. For the matrix above,

$$LU \text{ (compact form)} = \begin{bmatrix} 1 & -1 & 2 \\ 3 & 3 & -\frac{5}{3} \\ 1 & 1 & \frac{5}{3} \end{bmatrix}.$$

We append the righthand sides, back-substitute to get the b' vectors, then back-substitute with the elements of U:

$$
\begin{bmatrix}
1 & -1 & 2 \\
3 & 3 & -\frac{5}{3} \\
1 & 1 & \frac{5}{3}
\end{bmatrix}
\underbrace{\begin{bmatrix}
1 & 0 & 0 \\
0 & 1 & 0 \\
0 & 0 & 1
\end{bmatrix}}_{b \text{ vectors}}
\rightarrow
\begin{bmatrix}
1 & 0 & 0 \\
-1 & \frac{1}{3} & 0 \\
0 & -\frac{1}{5} & \frac{3}{5}
\end{bmatrix}
\underbrace{}_{b' \text{ vectors}}
\rightarrow
\begin{bmatrix}
0 & \frac{2}{5} & -\frac{1}{5} \\
-1 & 0 & 1 \\
0 & -\frac{1}{5} & \frac{3}{5}
\end{bmatrix}
$$

Solution vectors; gives A^{-1}

The inverse of the coefficient matrix provides a way of solving the set of equations $Ax = b$ because, when we multiply both sides of the relation by A^{-1}, we get

$$A^{-1}Ax = A^{-1}b,$$
$$x = A^{-1}b.$$

The second equation follows because the product $A^{-1}A = I$, the identity matrix, and $Ix = x$. If we know the inverse of A, we can solve the system for any righthand side b simply by multiplying the b vector by A^{-1}. This would seem like a good way to solve systems of equations, and one finds frequent references to it.

If we care about the efficiency of our method of solving the equations, however, this is not the preferred method, because solving the system with the LU decomposition of A, and doing the equivalent of two back-substitutions, requires exactly the same effort as multiplying b by the matrix. We compare the efficiency of the two schemes, then, by comparing the work needed to get the inverse and that to get the LU equivalent. Getting the inverse is more work, because it is the equivalent of solving the system with n righthand sides, while getting the LU is the equivalent of doing only the reduction to triangular form.

Even though the inverse is not the most efficient way to solve a set of simultaneous equations, the inverse is very important for theoretical reasons, and is essential to the understanding of many situations in applied mathematics. The use of the inverse concept and notation often simplifies the development of some fundamental relationships. We illustrate this by considering again the LU decomposition.

Find a pair of matrices such that $LU = A$.
Then $Ax = b$ can be written as

$$LUx = b.$$

Multiply both sides by L^{-1},

$$(L^{-1}L)Ux = L^{-1}b,$$

so $Ux = L^{-1}b$ because $L^{-1}L = I$.

We see that Ux (which is a vector because the product of a matrix times a vector always yields a vector) is equal to the vector formed by $L^{-1}b$. Call this Ux vector

b'. Then

$$Ux = b', \qquad b' = L^{-1}b, \qquad \text{or} \qquad Lb' = b.$$

We can get the vector b' by solving the system $Lb' = b$. This is particularly easy to do because L is triangular; all we need to do is the back-substitution phase (actually a forward substitution because L is lower-triangular.)

Once we have b', we can solve for x from the system $Ux = b'$. This is also easy because U is triangular.

Observe how using the concept and notation of inverses helps to clarify and prove the validity of the LU method.

2.8 MATRIX AND VECTOR NORMS

When we discuss multicomponent entities like matrices and vectors, we frequently need a way to express their magnitude—some measure of "bigness" or "smallness." For ordinary numbers, the absolute value tells us how large the number is; but for a matrix there are many components, each of which may be large or small in magnitude. (We are not talking about the *size* of a matrix, meaning the number of elements it contains.)

Any good measure of the magnitude of a matrix (the technical term is *norm*) must have four properties that are intuitively essential:

1. The norm must always have a value greater than or equal to zero, and must be zero only when the matrix is the zero matrix (one with all elements equal to zero.)

2. The norm must be multiplied by k if the matrix is multiplied by the scalar k.

3. The norm of the sum of two matrices must not exceed the sum of the norms.

4. The norm of the product of two matrices must not exceed the product of the norms.

More formally, we can state these conditions, using $\|A\|$ to represent the *norm of matrix A*:

1. $\|A\| \geqslant 0$ and $\|A\| = 0$ if and only if $A = 0$.
2. $\|kA\| = k\|A\|$.
3. $\|A + B\| \leqslant \|A\| + \|B\|$.
4. $\|AB\| \leqslant \|A\| \|B\|$.

$$(2.9)$$

The third relationship is called the *triangle inequality*. The fourth is important when we deal with the product of matrices.

For the special kind of matrices that we call vectors, our past experience can help us. For vectors in two- or three-space, the length satisfies all four requirements and is a good value to use for the norm of a vector. This norm is called the *Euclidean norm*, and is computed by $\sqrt{x_1^2 + x_2^2 + x_3^2}$.

We compute the Euclidean norm of vectors with more than three components by generalizing:

$$\|x\|_e = \sqrt{x_1^2 + x_2^2 + \ldots + x_n^2} = \left(\sum_{i=1}^{n} x_i^2\right)^{1/2}.$$

This is not the only way to compute a vector norm, however. The sum of the absolute values of the x_i can be used as a norm; the maximum value of the magnitudes of the x_i will also serve. These three norms can be interrelated by defining the p-norm as

$$\|x\|_p = \left(\sum_{i=1}^{n} |x_i|^p\right)^{1/p}.$$

From this it is readily seen that:

$$\|x\|_1 = \sum_{i=1}^{n} |x_i| = \text{Sum of magnitudes};$$

$$\|x\|_2 = \left(\sum_{i=1}^{n} x_i^2\right)^{1/2} = \text{Euclidean norm};$$

$$\|x\|_\infty = \max_{1 \leq i \leq n} |x_i| = \text{maximum-magnitude norm}.$$

Which of these vector norms is best to use may depend on the problem. In most cases, satisfactory results are obtained with any of these measures of the "size" of a vector.

Example Compute the 1-, 2-, and ∞-norms of the vector x, if $x = (1.25, 0.02, -5.15, 0)$.

$$\|x\|_1 = |1.25| + |0.02| + |-5.15| + |0| = 6.42;$$
$$\|x\|_2 = [(1.25)^2 + (0.02)^2 + (-5.15)^2 + (0)^2]^{1/2} = 5.2996;$$
$$\|x\|_\infty = |-5.15| = 5.15.$$

The norms of a matrix are developed by a correspondence to vector norms. Matrix norms that correspond to the above, for matrix A, can be shown to be:

$$\|A\|_1 = \max_{1 \leq j \leq n} \sum_{i=1}^{n} |a_{ij}| = \text{Maximum column-sum};$$

$$\|A\|_\infty = \max_{1 \leq i \leq n} \sum_{j=1}^{n} |a_{ij}| = \text{Maximum row-sum}.$$

The matrix norm $\|A\|_2$ that corresponds to the 2-norm of a vector is not readily computed. It is related to the eigenvalues of the matrix (which we discuss in a later chapter). It sometimes has special utility because no other norm is smaller than this norm. It therefore provides the "tightest" measure of the "size" of a matrix, but is also the most difficult to compute. This norm is also called the *spectral norm*.

For an $m \times n$ matrix, we can paraphase the Euclidean norm,* and write

$$\|A\|_e = \left(\sum_{i=1}^m \sum_{j=1}^n a_{ij}^2 \right)^{1/2}.$$

Example Compute the Euclidean norms of A, B, and C, and the ∞-norms, given that:

$$A = \begin{bmatrix} 5 & 9 \\ -2 & 1 \end{bmatrix}; \quad B = \begin{bmatrix} 0.1 & 0 \\ 0.2 & -0.1 \end{bmatrix}; \quad \text{and} \quad C = \begin{bmatrix} 0.2 & 0.1 \\ 0.1 & 0 \end{bmatrix}.$$

$\|A\|_e = \sqrt{25 + 81 + 4 + 1} = \sqrt{111} = 10.53;$ $\|A\|_\infty = 14.$

$\|B\|_e = \sqrt{0.01 + 0 + 0.04 + 0.01} = \sqrt{0.06} = 0.2449;$ $\|B\|_\infty = 0.3$

$\|C\|_e = \sqrt{0.04 + 0.01 + 0.01 + 0} = \sqrt{0.06} = 0.2449;$ $\|C\|_\infty = 0.3.$

The results of our examples look quite reasonable; certainly A is "larger" than B or C. While $B \neq C$, both are equally "small." The Euclidean norm is a good measure of the magnitude of a matrix. (Note that $\|A\|_e \neq \|A\|_2$ for a matrix although the 2-norm for a vector is the same as its Euclidean norm.)

We see then that there are a number of ways that the norm of a matrix can be expressed. Which way is preferred? There are certainly differences in their cost; e.g., some will require more extensive arithmetic than others. The spectral norm is usually the most "expensive." Which norm is best? The answer to this question depends in part on the use for the norm. In most instances, we want the norm that puts the smallest upper bound on the magnitude of the matrix. In this sense, the spectral norm is "best." We observe, in the next example, that not all the norms give the same value for the magnitude of a matrix.

Example

$$A = \begin{bmatrix} 5 & -4 & -7 \\ -4 & 2 & -4 \\ -7 & -4 & 5 \end{bmatrix}.$$

$\|A\|_e = $ Euclidean norm $= 14.69;$
$\|A\|_\infty = 16;$
$\|A\|_1 = 16;$
$\|A\|_2 = $ Spectral norm $= 12.$

If a matrix is a diagonal matrix, all p-norms have the same value, however.

Why are norms important? For one thing, they let us express the accuracy of the solution to a set of equations in quantitative terms by stating the norm of the

---

* Be alert to a situation that may be confusing: For a vector, the 2-norm is the same as the Euclidean norm, but for a matrix, the 2-norm is not the same as the Euclidean norm.

error vector (the true solution minus the approximate solution vector). Norms are also used to study quantitatively the convergence of iterative methods of solving linear systems (which we will cover in a later section).

2.9 ERRORS IN THE SOLUTION AND CONDITION NUMBERS

When we solve a set of linear equations, $Ax = b$, we hope that the calculated vector $\bar{x}$ is a close representation of the true solution vector x. In the previous examples we have seen how round-off can make the computed solution differ from the exact solution; and several times the use of pivoting has been recommended as a way to minimize the effect. The next example will demonstrate how much the accuracy can be improved by pivoting. We will exaggerate the effect by using only four-digit arithmetic (rounding to four digits after each operation). The same effects are observed with more precise arithmetic in large systems.

Given $Ax = b$ with

$$A = \begin{bmatrix} -0.002 & 4.000 & 4.000 \\ -2.000 & 2.906 & -5.387 \\ 3.000 & -4.031 & -3.112 \end{bmatrix}$$

and

$$b = (7.998, -4.481, -4.143)^T.$$

After Gaussian elimination without pivoting, the triangularized matrix (augmented with the b vector) is

$$\begin{bmatrix} -0.002 & 4.000 & 4.000 & 7.998 \\ 0 & -3.997 & -4.005 & -8.002 \\ 0 & 0 & -10.00 & 0.000 \end{bmatrix},$$

which gives the computed solution

$$\bar{x} = (-1496, 2.000, 0.000)^T.$$

The exact solution is $x = (1.000, 1.000, 1.000)^T$, and we see that round-off errors (particularly obvious in the last row of the triangularized system) have given a completely incorrect result. In this example the very small value of a_{11} has been the source of difficulty.

If the equations are reordered to give:

$$A = \begin{bmatrix} 3.000 & -4.031 & -3.112 \\ -0.002 & 4.000 & 4.000 \\ -2.000 & 2.906 & -5.387 \end{bmatrix}, \quad \text{and} \quad b = \begin{bmatrix} -4.413 \\ 7.998 \\ -4.481 \end{bmatrix},$$

and we repeat the four-digit computations, the triangularized system becomes:

$$\begin{bmatrix} 3.000 & -4.031 & -3.112 & -4.143 \\ 0 & 3.997 & 3.998 & 7.995 \\ 0 & 0 & -7.681 & -7.681 \end{bmatrix},$$

which gives $\bar{x} = (1.000, 1.000, 1.000)^T$, whose error is nil.

We conclude from the above that the algorithm that we employ can have a very significant effect on the accuracy.

Unfortunately, the error due to round-off is sometimes large even with the best available algorithm, because the problem itself may be very sensitive to the effects of small errors. Consider this example:

Given $Ax = b$, with

$$A = \begin{bmatrix} 3.02 & -1.05 & 2.53 \\ 4.33 & 0.56 & -1.78 \\ -0.83 & -0.54 & 1.47 \end{bmatrix}, \quad \text{and} \quad b = \begin{bmatrix} -1.61 \\ 7.23 \\ -3.38 \end{bmatrix}.$$

If we solve the system by Gaussian elimination, using pivoting and carrying three digits rounded, the triangularized system is:

$$\begin{bmatrix} 4.33 & 0.56 & -1.78 & 7.23 \\ 0 & -1.44 & 3.77 & -6.65 \\ 0 & 0 & -0.00362 & 0.00962 \end{bmatrix}.$$

The computed solution is $\bar{x} = (0.880, -2.35, -2.66)^T$ while the true solution vector is $x = (1, 2, -1)^T$. The very small number on the diagonal in the third row is a sign of such inherent sensitivity to round-off. Such a system is called "ill-conditioned." One strategy to use with ill-conditioned systems is greater precision in the arithmetic operations. For the above example, when six-digit computations are used, we get marked improvement:

$$x = (0.9998, 1.9995, -1.002)^T;$$

but note that there is still an appreciable error. A large ill-conditioned system would require even more digits of accuracy than six if we wished to compute a solution anywhere near to the exact answer.

The occurrence of an unavoidable small value on the diagonal in an ill-conditioned system indicates that the determinant of the original coefficient matrix tends to be a small number. (We can make this statement because we can evaluate the determinant of a matrix by triangularizing it and then multiplying the elements on the diagonal.) Since a zero value for the determinant occurs only for a singular matrix, we can say that an ill-conditioned matrix is "nearly singular." This is hardly a quantitative measure of the property, however, so we will look for a different way to express the degree of ill-conditioning.

There are some important problems involving linear systems in which the coefficient matrix is always nearly singular, and hence ill-conditioned. One example is fitting a polynomial of relatively high degree to a set of points by the least-squares method. This is covered in another chapter.

Another way to look at ill-conditioned systems is to examine the effect of small changes in the coefficients. In the above system, changing the a_{11} coefficient from 3.02 to 3.00 gives a solution (exact) of

$$x_3 = 0.05856, \qquad x_2 = 4.75637, \qquad x_1 = 1.07968.$$

This property of ill-conditioned systems—that their solution is extremely sensitive to small changes in the coefficients—explains why they are also so sensitive to round-off error. The small inaccuracies in values as the computation proceeds, caused by round-off errors, are equivalent to numbers we would encounter in using exact arithmetic on a problem with slightly altered coefficients.

A more vivid conception of ill-conditioning is obtained by examining a system of two equations with two unknowns.

Consider the system:

$$\begin{bmatrix} 1.01 & 0.99 \\ 0.99 & 1.01 \end{bmatrix} \begin{bmatrix} x \\ y \end{bmatrix} = \begin{bmatrix} 2 \\ 2 \end{bmatrix},$$

which has the obvious solution $x = 1$, $y = 1$. The set of equations can be represented by two straight lines that intersect at the point $(1, 1)$. The lines are very nearly parallel, as Fig. 2.2(a) shows. Suppose there is some uncertainty in the coefficients: The position of the lines might be anywhere in the fuzzy region as indicated by Fig. 2.2(b), and a corresponding amplified uncertainty in the point of intersection results. On the other hand, as indicated by Fig. 2.2(c), lines that are not almost parallel do not show an amplified uncertainty in their point of intersection when the position of the lines is not known precisely. If two lines are perpendicular, the uncertainty in the point of intersection is no greater than the uncertainty in the lines themselves.

One can consider linearly dependent vectors of order 2 as representing parallel lines. The analogy of a consistent set of righthand sides, with a corresponding redundancy in the set of equations, is two coincident straight lines. There is an infinite set of solutions because the lines "intersect" everywhere.

Two noncoincident parallel lines are analogous to an inconsistent set of equations with a singular coefficient matrix; there is no solution and no intersection of the lines. The visual picture of linear dependence in three space would involve parallel planes. We need to think of parallel hyperplanes in higher-dimensional spaces.

A simple way to detect ill-conditioning would be to make a deliberate change in some of the coefficients and determine the degree of change in the solution. An even simpler way is to examine the values on the diagonal of the triangularized system, but a computer subroutine may not always exhibit these values. When

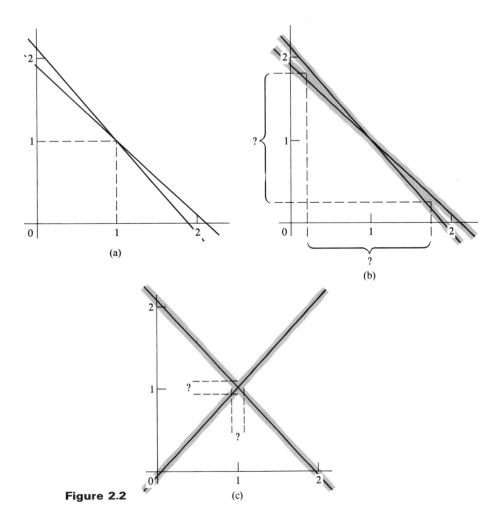

Figure 2.2

ill-conditioning is suspected, one might compare the solution obtained using single precision with that when double precision is used. Then both the presence of ill-conditioning is observed and a better approximation of the solution is obtained.

In some situations, one can combat ill-conditioning by transforming the problem into an equivalent set of equations that are not ill-conditioned. The efficiency of this scheme is related to the relative amount of computation required for the transformation, compared to the cost of doing the calculations in higher precision.*

---

* Double precision is not required throughout the computations. When the system is solved through LU decomposition, the accumulation of inner products in double precision is sufficient.

An interesting phenomenon of an ill-conditioned system is that we cannot test for the accuracy of the computed solution merely by substituting it into the equations to see whether the righthand sides are reproduced. Consider again the ill-conditioned example we have previously examined:

$$A = \begin{bmatrix} 3.02 & -1.05 & 2.53 \\ 4.33 & 0.56 & -1.78 \\ -0.83 & -0.54 & 1.47 \end{bmatrix}, \quad b = \begin{bmatrix} -1.61 \\ 7.23 \\ -3.38 \end{bmatrix}.$$

If we compute the vector Ax, using the exact solution $x = (1, 2, -1)^T$, we of course get

$$Ax = (-1.61, 7.23, -3.38)^T = b.$$

But if we substitute a clearly erroneous vector

$$\bar{x} = (0.880, -2.35, -2.66)^T,$$

we get $A\bar{x} = (-1.6047, 7.2292, -3.3716)^T$, which is very close to b.

We define the residual of a solution vector as the difference between b and $A\bar{x}$, where $\bar{x}$ is the computed solution:

$$r = b - A\bar{x}.$$

Our example shows that the norm of r is not a good measure of the norm of the error vector ($e = x - \bar{x}$) for an ill-conditioned system.

Because the degree of ill-condition of the coefficient matrix is so important in determining the magnitude of round-off effects, it is valuable to have a quantitative measure. The condition number is normally defined as the product of two matrix norms:

$$\text{cond}\,(A) = \|A\|\,\|A^{-1}\|.$$

Unfortunately this is not an inexpensive quantity to compute, for it requires us to invert A. Because inverting a matrix amounts to solving a linear system (solving it with n different righthand sides, actually), and the computed solution for an ill-conditioned system may be inexact, we will not compute A^{-1} very accurately. This suggests that the condition number will not be computed very exactly either. Ordinarily this causes no great difficulty; if the condition number is large, we know we are in deep trouble. Observe that condition numbers will always be at least unity, which corresponds to the condition number of the identity matrix.

For our previous example, we have:

$$A = \begin{bmatrix} 3.02 & -1.05 & 2.53 \\ 4.33 & 0.56 & -1.78 \\ -0.83 & -0.54 & 1.47 \end{bmatrix}, \quad A^{-1} = \begin{bmatrix} 5.661 & -7.273 & -18.55 \\ 200.5 & -268.3 & -669.9 \\ 76.85 & -102.6 & -255.9 \end{bmatrix}.$$

Using matrix ∞-norms, we find that the condition number is

$$\|A\|\,\|A^{-1}\| = (6.67)(1138.7) = 7595.$$

The elements of A^{-1} will be large relative to the elements of A when A is ill-conditioned. However, this can also be true when the elements of A are small, even in the absence of ill-conditioning. Multiplying the two norms has a normalizing effect, so the condition number is large only for an ill-conditioned system.

The condition number lets us relate the magnitude of the error in the computed solution to the magnitude of the residual. We use norms to express the size of the vectors.

Let $e = x - \bar{x}$, where x is the exact solution to $Ax = b$ and $\bar{x}$ is an approximate solution. Let $r = b - A\bar{x}$, the residual. Since $Ax = b$, we have

$$r = b - A\bar{x} = Ax - A\bar{x} = A(x - \bar{x}) = Ae. \qquad (2.10)$$

Hence,

$$e = A^{-1}r.$$

Taking norms and recalling Eq. (2.9), line 4, for a product, we write:

$$\|e\| \leq \|A^{-1}\|\,\|r\|. \qquad (2.11)$$

From $r = Ae$, we also have $\|r\| \leq \|A\|\,\|e\|$, which combines with Eq. (2.11) to give

$$\frac{\|r\|}{\|A\|} \leq \|e\| \leq \|A^{-1}\|\,\|r\|. \qquad (2.12)$$

Applying the same reasoning to $Ax = b$ and $x = A^{-1}b$, we get

$$\frac{\|b\|}{\|A\|} \leq \|x\| \leq \|A^{-1}\|\,\|b\|. \qquad (2.13)$$

Taking Eqs. (2.12) and (2.13) together, we reach a most important relationship:

$$\frac{1}{\|A\|\,\|A^{-1}\|}\frac{\|r\|}{\|b\|} \leq \frac{\|e\|}{\|x\|} \leq \|A\|\,\|A^{-1}\|\frac{\|r\|}{\|b\|}, \qquad (2.14)$$

or

$$\frac{1}{(\text{Cond. no.})}\frac{\|r\|}{\|b\|} \leq \frac{\|e\|}{\|x\|} \leq (\text{Cond. no.})\frac{\|r\|}{\|b\|}.$$

Equation (2.14) shows that the relative error in the computed solution vector $\bar{x}$ can be as great as the relative residual multiplied by the condition number. Of course it can be also be as small as the relative residual divided by the condition number. Therefore, when the condition number is large, the residual gives little information about the accuracy of $\bar{x}$. Conversely, when the condition number is near unity, the relative residual is a good measure of the relative error of $\bar{x}$.

When we solve a linear system, we are normally doing so to determine values for a physical system for which the set of equations is a model. We use the measured values of the parameters of the physical system to evaluate the coefficients of the equations, so we expect these coefficients to be known only as precisely as the measurements. When these are in error, the solution of the equations will reflect these errors. We have already seen that an ill-conditioned system is extremely sensitive to small changes in the coefficients. The condition number lets us relate the change in the solution vector to such errors in the coefficients of the set of equations $Ax = b$.

Assume that the errors in measuring the parameters cause errors in the coefficients of A so that the actual set of equations being solved is $(A + E)\bar{x} = b$, where $\bar{x}$ represents the solution of the perturbed system and A represents the true (but unknown) coefficients. We let $\bar{A} = A + E$ represent the perturbed coefficient matrix. We desire to know how large $x - \bar{x}$ is.

Using $Ax = b$ and $\bar{A}\bar{x} = b$, we can write:

$$x = A^{-1}b = A^{-1}(\bar{A}\bar{x}) = A^{-1}(A + \bar{A} - A)\bar{x}$$
$$= [I + A^{-1}(\bar{A} - A)]\bar{x}$$
$$= \bar{x} + A^{-1}(\bar{A} - A)\bar{x}.$$

Since $\bar{A} - A = E$, we have:

$$x - \bar{x} = A^{-1}E\bar{x}.$$

Taking norms, we get:

$$\|x - \bar{x}\| \leq \|A^{-1}\|\|E\|\|\bar{x}\| = \|A^{-1}\|\|A\|\frac{\|E\|}{\|A\|}\|\bar{x}\|,$$

so that

$$\frac{\|x - \bar{x}\|}{\|\bar{x}\|} \leq (\text{Cond. no.})\frac{\|E\|}{\|A\|}. \qquad (2.15)$$

This says that the error of the solution relative to the norm of the computed solution can be as large as the relative error in the coefficients of A multiplied by the condition number. The net effect is that, if the coefficients of A are known to only four-digit precision and the condition number is 1000, the computed vector x may have only one digit of accuracy.

When the solution to the system $Ax = b$ has been computed, and, because of round-off error, we obtain the approximate solution vector $\bar{x}$, it is possible to apply iterative improvement to correct $\bar{x}$ so that it more closely agrees with x. Define $e = x - \bar{x}$. Define $r = b - A\bar{x}$. As shown above [Eq. (2.10)],

$$Ae = r. \qquad (2.16)$$

If we could solve this equation for e, we could apply this as a correction to $\bar{x}$. Furthermore, if $\|e\|/\|\bar{x}\|$ is small, it means that $\bar{x}$ should be close to x. In fact, if the value of $\|e\|/\|\bar{x}\|$ is 10^{-p}, we know that $\bar{x}$ is probably correct to p digits.

The process of iterative improvement is based on solving Eq. (2.16). Of course this is also subject to the same round-off error as the original solution of the system for $\bar{x}$; so we actually get $\bar{e}$, an approximation to the true error vector. Even so, unless the system is so ill-conditioned that $\bar{e}$ is not a reasonable approximation to e, we will get an improved estimate of x from $\bar{x} + e$. One special caution is important to observe: The computation of the residual vector, r, must be as precise as possible. One always uses double-precision arithmetic; otherwise iterative improvement will not be successful. An example will make this clear.

Given

$$A = \begin{bmatrix} 4.23 & -1.06 & 2.11 \\ -2.53 & 6.77 & 0.98 \\ 1.85 & -2.11 & -2.32 \end{bmatrix}, \quad b = \begin{bmatrix} 5.28 \\ 5.22 \\ -2.58 \end{bmatrix},$$

whose true solution is

$$x = \begin{bmatrix} 1.000 \\ 1.000 \\ 1.000 \end{bmatrix}.$$

If three-digit chopped arithmetic is used, the approximate solution vector is $\bar{x} = (0.991, 0.997, 1.000)^t$. Using double precision, we compute $A\bar{x}$ and the residual as

$$\bar{A}x = \begin{bmatrix} 5.24511 \\ 5.22246 \\ -2.59032 \end{bmatrix}, \quad r = \begin{bmatrix} 0.0348 \\ -0.00246 \\ 0.0103 \end{bmatrix}.$$

We now solve $A\bar{e} = r$, again using three-digit precision, and get

$$\bar{e} = \begin{bmatrix} 0.00822 \\ 0.00300 \\ -0.00000757 \end{bmatrix}.$$

Finally, correcting $\bar{x}$ with $\bar{x} + \bar{e}$ gives almost exactly the correct solution:

$$\bar{x} + \bar{e} = \begin{bmatrix} 0.999 \\ 1.000 \\ 1.000 \end{bmatrix}.$$

In the general case, the iterations are repeated until the corrections are negligible. Since we want to make the solution of Eq. (2.16) as economical as possible, we should use an LU method to solve the original system and apply the LU to Eq. (2.16).

2.10 SOLVING LINEAR SYSTEMS BY ITERATION

As opposed to the direct method of solving a set of linear equations by elimination, we now discuss iterative methods. In certain cases, these methods are preferred over the direct methods—when the coefficient matrix is sparse (has many zeros) they may be more rapid.* They may be more economical in core-storage requirements of a computer. For hand computation they have the distinct advantage that they are self-correcting if an error is made; they may sometimes be used to reduce round-off error in the solutions computed by direct methods, as discussed above. They can also be applied to sets of nonlinear equations.

We illustrate the method by a simple example:

$$8x_1 + x_2 - x_3 = 8,$$
$$2x_1 + x_2 + 9x_3 = 12, \tag{2.17}$$
$$x_1 - 7x_2 + 2x_3 = -4.$$

The solution is $x_1 = 1$, $x_2 = 1$, $x_3 = 1$. We begin our iterative scheme by solving each equation for one of the variables, choosing, when possible, to solve for the variable with largest coefficient:

$x_1 = 1 \qquad\qquad - 0.125x_2 + 0.125x_3$	(from first equation),	
$x_2 = 0.571 + 0.143x_1 \qquad\qquad + 0.286x_3$	(from third equation),	(2.18)
$x_3 = 1.333 - 0.222x_1 - 0.111x_2$	(from second equation).	

We begin with some initial approximation to the value of the variables. (Each component might be taken equal to zero if no better initial estimates are at hand.) Substituting these into the righthand sides of the set of equations generates new approximations which are closer to the true value. The new values are substituted into the righthand sides to generate a second approximation, and the process is repeated until successive values of each of the variables are sufficiently alike. For the set of equations given above we get:

Successive Estimates of Solution (Jacobi Method)

	First	Second	Third	Fourth	Fifth	Sixth	Seventh	Eighth
x_1	0	1.000	1.095	0.995	0.993	1.002	1.001	1.000
x_2	0	0.571	1.095	1.026	0.990	0.998	1.001	1.000
x_3	0	1.333	1.048	0.969	1.000	1.004	1.001	1.000

Note that this method is exactly the same as the method of iteration for a single equation that was discussed in Chapter 1, but applied to a *set* of equations, for we

---

* When the occurrence of zeros follows some easy pattern, elimination methods can take advantage of this. See the program for a tridiagonal system at the end of the chapter.

may write Eq. (2.18) in the form

$$x^{(n+1)} = Gx^{(n)} = b' - Bx^{(n)},$$

which is identical in form to $x_{n+1} = g(x_n)$ as used in Chapter 1.

In the present context, of course, $x^{(n)}$ and $x^{(n+1)}$ refer to the nth and $(n + 1)$st iterates of a vector rather than a simple variable, and G is a linear transformation rather than a nonlinear function. For the example above,

$$b' = \begin{bmatrix} 1.000 \\ 0.571 \\ 1.333 \end{bmatrix} \quad \text{and} \quad B = \begin{bmatrix} 0 & 0.125 & -0.125 \\ -0.143 & 0 & -0.286 \\ 0.222 & 0.111 & 0 \end{bmatrix}.$$

The procedure we have just described is known as the *Jacobi method*, also called "the method of simultaneous displacements," because each of the equations is simultaneously changed by using the most recent set of x-values.

We can write the algorithm for the Jacobi iterative method as follows:

Algorithm for Jacobi Iteration

To solve a system of N linear equations, rearrange the rows so that the diagonal elements have magnitudes as large as possible relative to the magnitudes of other coefficients in the same row. Define the rearranged system as $Ax = b$. Beginning with an initial approximation to the solution vector, $x^{(1)}$, compute each component of $x^{(n+1)}$ for $i = 1, 2, \ldots, N$, by:

$$x_i^{(n+1)} = \frac{b_i}{a_{ii}} - \sum_{\substack{j=1, \\ j \neq i}}^{N} \frac{a_{ij}}{a_{ii}} x_j^{(n)}, \qquad n = 1, 2, \ldots$$

A sufficient condition for convergence is that

$$|a_{ii}| > \sum_{\substack{j=1, \\ j \neq i}}^{N} |a_{ij}|, \qquad i = 1, 2, \ldots, N.$$

When this is true, $x^{(n)}$ will converge to the solution no matter what initial vector is used.

Actually, the x-values of the next trial are not all calculated "simultaneously" when we perform the Jacobi method. In the above, we calculated the second estimate of x_1 before we did the x_2, and new values of both x_1 and x_2 were available before we improved the value of x_3. In nearly all cases the new values are better than the old, and should be used in preference to the poorer values.*

---

* There may be times when this obvious move is not beneficial, due to cancellation of errors, but in general this is good strategy in numerical analysis.

When this is done, the method is known by the name *Gauss–Seidel*. In this method our first step is to rearrange the set of equations by solving each equation for one of the variables in terms of the others, exactly as we have done above in the Jacobi method. One then proceeds to improve each x-value in turn, using always the most recent approximations to the values of the other variables. The rate of convergence is more rapid, as shown by reworking the same example as above (Eqs. (2.18), rearranged form of (2.17)):

Successive Estimates of Solution (Gauss–Seidel Method)

	First	Second	Third	Fourth	Fifth	Sixth
x_1	0	1.000	1.041	0.997	1.001	1.000
x_2	0	0.714	1.014	0.996	1.000	1.000
x_3	0	1.032	0.990	1.002	1.000	1.000

These values were computed by using Eq. (2.18) in the following way:

$$x_1^{(n+1)} = 1 \qquad\qquad - 0.125x_2^{(n)} + 0.125x_3^{(n)},$$
$$x_2^{(n+1)} = 0.571 + 0.143x_1^{(n+1)} \qquad\qquad + 0.286x_3^{(n)},$$
$$x_3^{(n+1)} = 1.333 - 0.222x_1^{(n+1)} - 0.111x_2^{(n+1)},$$

beginning with $x^{(1)} = (0, 0, 0)^T$.

An algorithm for Gauss–Seidel iteration is as follows:

Algorithm for Gauss–Seidel Iteration

To solve a system of N linear equations, rearrange the rows so that the diagonal elements have magnitudes as large as possible relative to the magnitudes of other coefficients in the same row. Define the rearranged system as $Ax = b$. Beginning with an initial approximation to the solution vector, $x^{(1)}$, compute each component of $x^{(n+1)}$, for $i = 1, 2, \ldots, N$, by

$$x_i^{(n+1)} = \frac{b_i}{a_{ii}} - \sum_{j=1}^{i-1} \frac{a_{ij}}{a_{ii}} x_j^{(n+1)} - \sum_{j=i+1}^{N} \frac{a_{ij}}{a_{ii}} x_j^{(n)}, \qquad n = 1, 2, \ldots$$

A sufficient condition for convergence is that

$$|a_{ii}| > \sum_{\substack{j=1, \\ j \neq i}}^{N} |a_{ij}|, \qquad i = 1, 2, \ldots, N.$$

When this is true, $x^{(n)}$ will converge to the solution no matter what initial vector is used.

It can be shown that the Gauss–Seidel method will always converge if the Jacobi method converges, and will do so more rapidly. (If Jacobi diverges, so does Gauss–Seidel.) The Gauss–Seidel method should always be used because of this faster convergence.

These iteration methods will not converge for all sets of equations, nor for all possible rearrangements of the equations. When the equations can be ordered so that each diagonal entry is larger in magnitude than the sum of the magnitudes of the other coefficients in that row (such a system is called *diagonally dominant*), the iteration will converge for any starting values. This is easy to visualize, because all the equations can be put in the form

$$ x_i = \frac{b_i}{a_{ii}} - \frac{a_1}{a_{ii}} x_1 - \frac{a_2}{a_{ii}} x_2 - \dots \tag{2.19} $$

The error in the next value of x_i will be the sum of the errors in all other x's multiplied by the coefficients of Eq. (2.19), and if the sum of the magnitudes of the coefficients is less than unity, the error will decrease as the iteration proceeds. The above convergence condition is a sufficient condition only; i.e., if the condition holds, the system always converges, but sometimes the system converges even if the condition is violated.*

The speed with which the iterations converge is obviously related to the degree of dominance of the diagonal terms, for the coefficients in Eq. (2.19) are then smaller and x_i is less affected by the errors in the other components. When the initial approximation is close† to the solution vector, relatively few iterations are needed to get an acceptable solution.

The next section shows how the Gauss–Seidel method can be accelerated.

2.11 RELAXATION METHOD

There is an iteration method which is more rapidly convergent than Gauss–Seidel and which can be used to advantage for hand calculations. It is unfortunately not well adapted to computer application. The method is due to a British engineer, Richard Southwell, and has been applied to a wide variety of problems. (Allen (1954) is an excellent reference.) We discuss the method because of its historical importance and because it leads to an important acceleration technique called *over-relaxation*.

If we consider the Gauss–Seidel scheme, we realize that the order in which the equations are used is important. We should improve that x which is most in

---

* A more quantitative analysis of the convergence of iterative schemes considers the iterative method in the form $x^{(n+1)} = Bx^{(n)} + c$, where $x^{(i)}$ is the ith iterate. If the norm of B is less than unity, the scheme will converge. A good treatment is found in Stewart (1973).

† "Closeness" could be expressed by the norm of the difference in the vectors.

error, since, in the rearranged form, that variable does not appear on the right, and hence its own error will not affect the next iterate. By using that equation, then, we introduce lesser errors into the computation of the next iterate. The *method of relaxation* is a scheme which permits one to select the best equation to be used for maximum rate of convergence.

We illustrate the method by the same example we solved in Section 2.10. The original equations are

$$\begin{aligned}
8x_1 + x_2 - x_3 &= 8, \\
2x_1 + x_2 + 9x_3 &= 12, \\
x_1 - 7x_2 + 2x_3 &= -4.
\end{aligned} \tag{2.20}$$

We again begin by a rearrangement of the equations, but different from that for the Gauss–Seidel or Jacobi methods. We transpose all the terms to one side, and then divide by the negative of the largest coefficient. Equations (2.20) become

$$\begin{aligned}
-x_1 - 0.125x_2 + 0.125x_3 + 1 &= 0, \\
-0.222x_1 - 0.111x_2 - x_3 + 1.333 &= 0, \\
0.143x_1 - x_2 + 0.286x_3 + 0.571 &= 0.
\end{aligned} \tag{2.21}$$

If we begin with some initial set of values, and substitute in Eqs. (2.21), the equations will not be satisfied (unless, by chance, we have stumbled onto the solution); the left sides will not be zero, but some other value which we call the *residual* and denote by R_i. It is also convenient to reorder the equation so the -1 coefficients are on the diagonal. Equations (2.21) become, with these rearrangements,

$$\begin{aligned}
-x_1 - 0.125x_2 + 0.125x_3 + 1 &= R_1, \\
0.143x_1 - x_2 + 0.286x_3 + 0.571 &= R_2, \\
-0.222x_1 - 0.111x_2 - x_3 + 1.333 &= R_3.
\end{aligned}$$

For example, with $x_1 = 0$, $x_2 = 0$, $x_3 = 0$, we have

$$R_1 = 1, \qquad R_2 = 0.571, \qquad R_3 = 1.333.$$

The largest residual in magnitude, R_3, tells us that the third equation is most in error and should be improved first. The method gets its name "relaxation" from the fact that we make a change in x_3 to relax R_3 (the greatest residual) so as to make it zero. Observing the coefficients of the various equations, we see that increasing the value of x_3 by one, say, will decrease R_3 by one, will increase R_1 by 0.125, and increase R_2 by 0.286. To change R_3 from its initial value of 1.333 to zero, we should increase x_3 by that same amount.

We then select the new residual of greatest magnitude, and relax it to zero. We continue until all residuals are zero and when this is true, the values of the x's will be at the exact solution. In implementing this method, there are some modifications that make the work easier. We illustrate in Fig. 2.3.

Eq. No.	x_1	x_2	x_3
I	−1	−0.125	0.125
II	0.143	−1	0.286
III	−0.222	−0.111	−1

x_1	R_I	x_2	R_{II}	x_3	R_{III}
0	1000	0	571	0	1333
	+ 167		+ 381		
	1167		952	1333	0
			+ 167		− 259
1167	0		1119	−259	
	− 140				− 124
	−140	1119		−383	
	− 48		− 109		
	−189		−109	−38.	0
			− 27		+ 42
−189	0		−136	42	
	+ 17				+ 15
	17	−136		57	
	+ 7		+ 16		
	24		16	57	0
			+ 3		− 5
24	0		19		−5
	− 2				− 2
	−2	19			−7
	− 1		− 2		
	−3		−2	−7	0
			0		+ 1
−3	0		−2	1	
	0				0
	0	−2	0		1
	0		0	1	0
999.		1000		1001	

Check residuals: 1   −1   0

Figure 2.3 Solving a set of linear equations by relaxation.

We make three double columns, one for each variable and for the residual of the equation in which that variable appears with −1 coefficient. The initial x values and the initial residuals are entered as the first row of the table. It is convenient to work entirely with integers by multiplying the initial x-values and residuals by 1000, and then to scale down the solution by dividing by 1000 at the end of the computations. We avoid fractions; if a fractional change in a variable is needed to relax to zero we only relax to near zero.

In Fig. 2.3, we set down the increments to the x's but record the cumulative effect on the residuals. (The old values of the residuals are crossed out when replaced by a new value.) When the residuals are zero, we add the various increments to the initial value to get the final value. In this example, round-off errors cause an error of one in the third decimal.

It is important to make a final check by recomputing residuals at the end of the calculation to check for mistakes in arithmetic. The method is not usually programmed because searching on the computer for the largest residual is slow, and adds enough execution time that the acceleration gives no net benefit. The search can be done rapidly by scanning the residuals in a hand calculation, however.

Southwell and his co-workers observed, for many situations, that relaxing the residuals to zero was less efficient than relaxing beyond zero (*over-relaxing*) or relaxing short of zero (*under-relaxing*). The reason this is an improved strategy is that a zero residual doesn't stay zero; relaxing the residual of another equation affects the first residual, so it is appropriate to anticipate and allow for this by an appropriate under- or over-relaxation.

Table 2.1 shows that a significant improvement in the speed of convergence is obtained if R_1 is under-relaxed by 10% and R_3 is under-relaxed by 25%.

Table 2.1 Accelerated solution of linear equations by relaxation.

Eq. No.	x_1	x_2	x_3
I	−1	−0.125	0.125
II	0.143	−1	0.286
III	−0.222	−0.111	−1

x_1	R_I	x_2	R_{II}	x_3	R_{III}
0	1000	0	571	0	1333
	+ 125		+ 286		
	1125		857	1000	333
			+ 144		− 225
1013	112		1001		108
	− 125				− 111
	−13	1001	0		−3
			− 2		+ 3
−12	−1		−2		0
	+ 0				+ 0
	−1	−2	0		0
			+ 0		+ 0
−1	0		0		0
1000		999		1000	

Unfortunately the optimum degree of under- or over-relaxation is not easily determined. In many problems, acceleration is obtained by over-relaxing rather than under-relaxing.

Even though Southwell's relaxation method is not often used today, there is one aspect of it that has influence on the iterative solution of linear equations by computer. In using the Gauss–Seidel method, we can speed up the convergence by "over-relaxation," that is, by making the residuals go to the other side of zero instead of just relaxing to zero as in the first example. We can apply this to Gauss–Seidel iteration by modifying the algorithm.

The standard relationship for Gauss–Seidel iteration for the set of equations $Ax = b$, for variable x_i, can be written:

$$x_i^{(k+1)} = \frac{1}{a_{ii}}\left(b_i - \sum_{j=1}^{i-1} a_{ij}x_j^{(k+1)} - \sum_{j=i+1}^{n} a_{ij}x_j^{(k)} \right), \qquad (2.22)$$

where the superscript $(k + 1)$ indicates that this is the $(k + 1)$st iterate. On the right side we use the most recent estimates of the x_j, which will be either $x_j^{(k)}$ or $x_j^{(k+1)}$.

An algebraically equivalent form for Eq. (2.22) is:

$$x_i^{(k+1)} = x_i^{(k)} + \frac{1}{a_{ii}}\left(b_i - \sum_{j=1}^{i-1} a_{ij}x_j^{(k+1)} - \sum_{j=i}^{n} a_{ij}x_j^{(k)} \right),$$

because $x_i^{(k)}$ is both added to and subtracted from the right side. In this form, we see that Gauss–Seidel and Southwell's relaxation can have identical arithmetic: The term we add to $x_i^{(k)}$ to get $x_i^{(k+1)}$ is exactly the increment that relaxes the residual to zero. (Of course, we apply the relaxation to the x_i's in a different sequence in the two methods.) Over-relaxation can be applied to Gauss–Seidel if we will add to $x_i^{(k)}$ some multiple of the second term. It can be shown that this multiple should never be more than 2 in magnitude (to avoid divergence) and the optimum over-relaxation factor lies between 1.0 and 2.0. Our iteration equations take this form, where w is the *over-relaxation factor*:

$$x_i^{(k+1)} = x_i^{(k)} + \frac{w}{a_{ii}}\left(b_i - \sum_{j=1}^{i-1} a_{ij}x_j^{(k+1)} - \sum_{j=1}^{n} a_{ij}x_j^{(k)} \right). \qquad (2.23)$$

Table 2.2 shows how the convergence rate is influenced by the value of w for the system

$$\begin{bmatrix} -4 & 1 & 1 & 1 \\ 1 & -4 & 1 & 1 \\ 1 & 1 & -4 & 1 \\ 1 & 1 & 1 & -4 \end{bmatrix} x = \begin{bmatrix} 1 \\ 1 \\ 1 \\ 1 \end{bmatrix},$$

starting with an initial estimate of $x = 0$. The exact solution is

$$x_1 = -1, \qquad x_2 = -1, \qquad x_3 = -1, \qquad x_4 = -1.$$

Table 2.2 Acceleration of convergence of Gauss–Seidel iteration.

w, the over-relaxation factor	Number of iterations to reach error $< 1 \times 10^{-5}$
1.0	24
1.1	18
1.2	13
1.3	11← Minimum
1.4	14 of iterations
1.5	18
1.6	24
1.7	35
1.8	55
1.9	100+

We see that the optimum value for the over-relaxation factor is about $w = 1.3$ for this example. The optimum value will vary between 1.0 and 2.0 depending on the size of the coefficient matrix and the values of the coefficients. Over-relaxation is considered further in Chapter 7 in connection with methods to solve partial-differential equations.

2.12 SETS OF NONLINEAR EQUATIONS

As mentioned previously, the problem of finding the solution of a set of non-linear equations is much more difficult than for linear equations. (In fact, some sets have no real solutions.) Consider the example of a pair of nonlinear equations:

$$\begin{align} x^2 + y^2 &= 4, \\ e^x + y &= 1. \end{align} \tag{2.24}$$

Graphically, the solution to this system is represented by the intersections of the circle $x^2 + y^2 = 4$ with the curve $y = 1 - e^x$. Figure 2.4 shows that these are near $(-1.8, 0.8)$ and $(1, -1.7)$. We can use the method of iteration to improve these approximations. Just as in Section 5, Chapter 1, we rearrange both equations to a form of the pattern $x = f(x, y)$, $y = g(x, y)$, and use the method of iteration on each equation in turn. Under proper conditions, these will converge. For example, if we rearrange Eqs. (2.24) in the form

$$\begin{align} x &= \pm\sqrt{4 - y^2}, \quad (- \text{ sign for leftmost root}), \\ y &= 1 - e^x, \end{align}$$

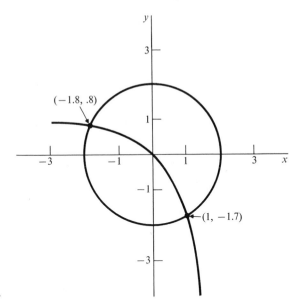

Figure 2.4

we get the following successive values, beginning with $y_1 = 0.8$ in the first equation:

x-values: −1.83   −1.815   −1.8163   −1.8162

y-values: 0.8   0.84   0.8372   0.8374   0.8374

When we begin at $y = -1.7$ to find the root to the right of the origin, we get

x: 1.05   0.743   1.669   Imaginary value

y: −1.7   −1.857   −1.102   −4.307

The equations diverge! (Beginning with $x = 1.0$ in the second equation is no help. This also diverges.) However, with a different rearrangement of the original equations, such as

$$x = \ell n\,(1 - y),$$
$$y = \pm\sqrt{4 - x^2}, \quad (-\text{ sign for rightmost root}),$$

(2.25)

we get

x: 0.993   1.006   1.0038   1.0042   1.0042

y: −1.7   −1.736   −1.7286   −1.7299   −1.7296

The pair of rearranged equations in (2.25) converges.

Some of the difficulties with sets of nonlinear equations are apparent from this simple example. If there are more than two equations in the system, finding a convergent form of the equations is increasingly difficult. A criterion for convergence (sufficiency condition only) is as follows:

The set of equations

$$x = f(x, y, z, \ldots), \qquad y = g(x, y, z, \ldots), \qquad z = h(x, y, z, \ldots), \qquad \ldots$$

will converge if, in an interval about the root,

$$|f_x| + |f_y| + |f_z| + \ldots < 1,$$
$$|g_x| + |g_y| + |g_z| + \ldots < 1,$$
$$|h_x| + |h_y| + |h_z| + \ldots < 1,$$
$$\cdot \quad \cdot \quad \cdot \quad \cdot \quad \cdot \quad \cdot \quad \cdot \quad < 1.$$

In the above inequalities, the subscript notation designates partial derivatives. Computing all the partial derivatives and knowing where the root is are major problems. Getting starting values for the multidimensioned system is also correspondingly difficult.

Newton's method can also be applied to systems. We begin with the forms

$$f(x, y) = 0,$$
$$g(x, y) = 0.$$

Let $x = r$, $y = s$ be a root, and expand both functions as a Taylor series about the point (x_1, y_1) in terms of $(r - x_1)$, $(s - y_1)$, where (x_1, y_1) is a point near the root:

$$f(r, s) = 0 = f(x_1, y_1) + f_x(x_1, y_1)(r - x_1) + f_y(x_1, y_1)(s - y_1) + \ldots ,$$
$$g(r, s) = 0 = g(x_1, y_1) + g_x(x_1, y_1)(r - x_1) + g_y(x_1, y_1)(s - y_1) + \ldots \tag{2.26}$$

If (x_1, y_1) is near enough to (r, s), we can truncate after the first-derivative terms and solve for the unknowns $(r - x_1)$, $(s - y_1)$. We exhibit the solution in determinant form in Eq. (2.27). Because we have truncated the series, we get only an approximation:

$$r - x_1 \doteq \frac{\begin{vmatrix} -f & f_y \\ -g & g_y \end{vmatrix}}{\begin{vmatrix} f_x & f_y \\ g_x & g_y \end{vmatrix}} \tag{2.27}$$

$$s - y_1 \doteq \frac{\begin{vmatrix} f_x & -f \\ g_x & -g \end{vmatrix}}{\text{Denominator}}$$

(*Note.* All functions are evaluated at (x_1, y_1).)

The righthand sides of Eq. (2.27) are increments to add to x_1, y_1 to get an improved estimate of the root (r, s). The extension to more than two simultaneous equations is straightforward in principle, but the effort soon becomes tremendous.

We illustrate by repeating the previous example:

$$f(x, y) = 4 - x^2 - y^2 = 0,$$
$$g(x, y) = 1 - e^x - y = 0.$$

The partials are

$$f_x = -2x, \qquad f_y = -2y,$$
$$g_x = -e^x, \qquad g_y = -1.$$

Beginning at $x_1 = 1$, $y_1 = -1.7$, we get

$$r = \quad 1 + \frac{\begin{vmatrix} -0.110 & 3.4 \\ 0.0183 & -1 \end{vmatrix}}{\begin{vmatrix} -2 & 3.4 \\ -2.7183 & -1 \end{vmatrix}} = 1 + \frac{0.04778}{11.2422} = 1.0042,$$

$$s = -1.7 + \frac{\begin{vmatrix} -2 & -0.110 \\ -2.7183 & 0.0183 \end{vmatrix}}{11.2422} = -1.7 + \frac{-0.3356}{11.2422} = -1.7298.$$

The results already agree with the true value of the root within two in the fourth decimal place.

We can write Newton's method for a system of n equations by expanding on Eq. (2.26):

$$0 = (f_1) + (f_1)_x(r - x_0) + (f_1)_y(s - y_0) + (f_1)_z(t - z_0) + \ldots,$$
$$0 = (f_2) + (f_2)_x(r - x_0) + (f_2)_y(s - y_0) + (f_2)_z(t - z_0) + \ldots,$$
$$0 = (f_3) + (f_3)_x(r - x_0) + (f_3)_y(s - y_0) + (f_3)_z(t - z_0) + \ldots,$$
$$\vdots \tag{2.28}$$
$$0 = (f_n) + (f_n)_x(r - x_0) + (f_n)_y(s - y_0) + (f_n)_z(t - z_0) + \ldots$$

In these equations, each function is evaluated at the approximate root $(x_0, y_0, z_0, \ldots)$ Note that we have reduced the problem from solving a set of n nonlinear equations to solving a set of n linear equations. The unknowns are the improvements in each estimated variable $(r - x_0)$, $(s - y_0)$, $(t - z_0)$, ...

In a computer program, it is awkward to introduce each of the partial-derivative functions (which must be developed by hand) in order to implement

Eqs. (2.28). An alternative technique is to approximate these partials by recalculating the function with a small perturbation to each of the variables in turn:

$$(f_1)_x \doteq \frac{f_1(x + \delta, y, z, \ldots) - f_1(x, y, z, \ldots)}{\delta},$$

$$(f_1)_y \doteq \frac{f_1(x, y + \delta, z, \ldots) - f_1(x, y, z, \ldots)}{\delta}, \qquad (2.29)$$

$$\vdots$$

$$(f_i)_{x_i} \doteq \frac{f_i(x, y, z, \ldots, x_j + \delta, \ldots) - f_i(x, y, z, \ldots, x_j, \ldots)}{\delta}.$$

Similar relations are used for each variable in each function.* A computer program in the next section exploits this idea.

It is interesting to observe that Newton's method, as applied to a set of nonlinear equations, reduces the problem to solving a set of *linear equations* in order to determine the values that improve the accuracy of the estimates. This points out quite dramatically how linear and nonlinear problems vary in difficulty.

Newton's method has the advantage of converging quadratically, at least when we are near a root, but it is expensive in terms of function evaluations. For the 2×2 system above, there are six function evaluations at each step, while for a 3×3 system, there are twelve. For n simultaneous equations, the number of function evaluations is $n^2 + n$. One can see why this is rarely applied to large systems. It is good strategy, in all cases of simultaneous nonlinear equations, to reduce the number of equations as much as possible by solving for one variable in terms of the others and eliminating that one by substituting for it in the other equations. For example, we should attack the previous examples as follows:

$$\begin{cases} 4 - x^2 - y^2 = 0 \\ 1 - e^x - y = 0 \end{cases} \qquad \text{Solve for } y: y = 1 - e^x.$$

Substituting in the first equation, we get:

$$4 - x^2 - (1 - e^x)^2 = 0,$$
$$3 - x^2 + 2e^x - e^{2x} = 0.$$

We then use the methods of Chapter 1.

When we must solve a larger system of nonlinear equations, a modification of Newton's method is often used. It converges less than quadratically but usually faster than linearly. Unfortunately, it may *diverge* unless we start fairly near to a

---

* Approximation of derivatives by such difference quotients is discussed in Chapter 4. If the limiting value (as $\delta \to 0$) of this ratio were used, we would have exactly the definition of a derivative. Since limits are not used, we have approximate values.

root. In this method we use the simpler equations:

$$r - x_0 = -\frac{f_1(x_0, y_0, z_0, \ldots)}{\dfrac{\partial f_1}{\partial x}(x_0, y_0, z_0, \ldots)},$$

$$s - y_0 = -\frac{f_2(x_0, y_0, z_0, \ldots)}{\dfrac{\partial f_2}{\partial y}(x_0, y_0, z_0, \ldots)},$$ (2.30)

$$t - z_0 = -\frac{f_3(x_0, y_0, z_0, \ldots)}{\dfrac{\partial f_3}{\partial z}(x_0, y_0, z_0, \ldots)},$$

.
.
.

Comparing Eqs. (2.30) to Eqs. (2.28), we see that the number of function evaluations has been reduced to $2n$ from $n^2 + n$. We illustrate with the same 2×2 system above (although this does not fully reveal the saving in work at each iteration):

$$f(x, y) = 4 - x^2 - y^2 = 0, \qquad f_x(x, y) = -2x,$$
$$g(x, y) = 1 - e^x - y = 0, \qquad g_y(x, y) = -1.$$

Beginning at $x_0 = 1$, $y_0 = -1.7$, we get:

$$x_1 = 1 - \frac{0.110}{-2} = 1.005, \qquad y_1 = -1.7 - \frac{-0.0183}{-1} = -1.7183;$$

$$x_2 = 1.055 - \frac{-0.0655}{-2.110} = 1.023, \qquad y_2 = -1.7183 - \frac{-0.1536}{-1} = -1.8719;$$

$$x_3 = 1.023 - \frac{-0.5505}{-2.046} = 0.754, \qquad y_3 = -1.8719 - \frac{0.0903}{-1} = -1.7815;$$

$$x_4 = 0.754 - \frac{0.2577}{-1.508} = 0.925, \qquad y_4 = -1.7815 - \frac{0.6560}{-1} = -1.1255.$$

This is certainly not converging very fast. Reversing the order of the equations, however, gives no improvement; the values diverge.

On the other hand, the system

$$e^x - y = 0,$$
$$xy - e^x = 0$$

for which

$$x_{n+1} = x_n - \frac{e^{x_n} y_n}{e^{x_n}}, \qquad y_{n+1} = y_n - \frac{x_n y_n - e^{x_n}}{x_n},$$

gives these results, with $x_0 = 0.95$, $y_0 = 2.7$:

$$x_1 = 0.9942, \qquad y_1 = 2.7218;$$
$$x_2 = 1.0013, \qquad y_2 = 2.7183.$$

These are rapidly converging to the root at $(1, e)$.

2.13 COMPUTER PROGRAMS FOR SETS OF EQUATIONS

Several examples of computer programs utilizing the methods of this chapter are exhibited. These are grouped near the end of the chapter, as Figs. 2.6 through 2.13.) The first two employ the Gaussian elimination method—Program 1 for a general $n \times n$ array of coefficients and Program 2 for a tridiagonal system. The latter embodies an ingenious scheme to compress the matrix so that less storage is used by it for 500 equations than is required by the former for only 40. Programs 3, 4, and 5 are a combination of routines that form the LU equivalent of the coefficient matrix and then use it to solve the set of equations with one or several righthand sides. Program 6 is a subroutine that can be used to scale the coefficient matrix when values that differ widely in magnitude are involved. Program 7 uses iteration to solve a set of equations which must have the diagonal coefficients dominant. The last program solves n simultaneous nonlinear equations by Newton's method.

The first Gaussian elimination program proceeds quite conventionally, utilizing the straightforward method that we discussed earlier in the chapter (see Sec. 2.4). To use subroutine ELIM (Program 1, Fig. 2.6) the coefficient matrix is augmented with the constant vector and passed to the subprogram. The set of equations can be solved with multiple righthand sides: They are all augmented to the coefficient matrix and all the solutions are obtained at once and returned in the space of the augmentation columns. The coefficient matrix is made upper-triangular (although the elements below the diagonal are not converted to zeros because we know that the row reduction would create a zero; hence it need not actually be done.) In order to minimize round-off error as well as to avoid division by zero, we scan each column to find the element of largest magnitude on or below the diagonal, and interchange rows to put this largest element on the diagonal. In other words, partial pivoting is performed. If the largest element is very small (less than 10^{-5}), the program terminates with a message, concluding that the matrix is either singular or nearly so. In the case of a small divisor, the accuracy will be poor. (It is, of course, true that if all the coefficients are very small in magnitude, this is not the correct conclusion. It is assumed that the values have been scaled so that they are not all small.)

(The pair of subprograms below, LUDCMP and SOLVLU, in Programs 3 and 5, also can be used to solve equations with multiple righthand sides. The difference between these and Program 1 is that with ELIM all the righthand sides must be known in advance. The other subroutines can solve with a new righthand-side vector at any time after LUDCMP has reduced the coefficient matrix without repeating the reduction of the coefficient matrix.)

Subroutine TRIDG (Program 2) in Fig. 2.7 solves a system with a tridiagonal coefficient matrix. To conserve storage in the computer, the tridiagonal matrix of coefficients is compressed, together with the constant terms, into an $N \times 4$ matrix. Column one holds the coefficients to the left of the diagonal, column two holds the diagonal terms, column three holds the coefficient to the right of the diagonal, and column four holds the constant terms. The ith and $(i - 1)$st rows of the compressed matrix correspond to elements in the uncompressed matrix, as shown:

$$
\begin{bmatrix}
\cdot & & \cdot & & \cdot & \\
& \cdot & & \cdot & & \cdot \\
a_{i-1,1} & a_{i-1,2} & a_{i-1,3} & & & \\
& a_{i,1} & a_{i,2} & a_{i,3} & & \\
& & \cdot & \cdot & \cdot & \\
& & \cdot & \cdot & & \cdot
\end{bmatrix}
\begin{bmatrix}
\vdots \\
a_{i-1,4} \\
a_{i,4} \\
\vdots
\end{bmatrix}
$$

We can eliminate $a_{i,1}$ by subtracting $a_{i,1} / a_{i-1,2}$ times the $(i - 1)$st row from the ith row. Since we know a zero will replace $a_{i,1}$ we don't need actually to perform the arithmetic. The values of $a_{i,2}$ and $a_{i,4}$ change as follows:

$$
a_{i,2} = a_{i,2} - \frac{a_{i,1}}{a_{i-1,2}} * a_{i-1,3},
$$

$$
a_{i,4} = a_{i,4} - \frac{a_{i,1}}{a_{i-1,2}} * a_{i-1,4}, \qquad \text{for } i = 2, 3, 4, \ldots, n.
$$

When the ith row is reduced, $a_{i,3}$ is unaffected because there is a zero above it. Elements a_{11} and $a_{n,3}$ are never referred to and their values can be anything, or they can be left undefined.

After reduction, a back-substitution is performed. The elements of the solution vector replace the constant vector in the fourth column of the matrix. The equations for back-substitution are:

$$
a_{n,4} = \frac{a_{n,4}}{a_{n,2}},
$$

$$
a_{i,4} = \frac{a_{i,4} - a_{i,3} * a_{i+1,4}}{a_{i,2}}, \qquad i = n - 1, n - 2, \ldots, 1.
$$

The third, fourth and fifth subroutines (Figs. 2.8 through 2.10) will ordinarily be used together to solve a linear system. The first, LUDCMP, accepts the square matrix of coefficients and forms the LU decomposition which is equivalent to the coefficient matrix. This is returned in place of the coefficient matrix by compacting the L and U matrices as described in Section 2.5. The algorithm is a straight-forward implementation of the formulas of Section 2.5. In this routine, partial pivoting is employed by a call to Program 4. This APVT subroutine does partial pivoting after a column of L's has been obtained. The order of the rows, as affected by pivoting, is recorded in a vector named ORDER, so that the elements of the b vector can be rearranged into the same order as the coefficient matrix. A very small diagonal element is signalled by printing a message.

After the LU decomposition has been obtained from LUDCMP, the solution to the system with a given righthand-side vector is obtained by calling the SOLVLU subroutine, Program 5. In this program, the L-values are used to reduce the elements of the constant terms in the same way that the coefficient matrix was triangularized. The subroutine then does a back-substitution, using the reduced coefficients in the U matrix, and returns the solution vector to the caller.

Additional calls to SOLVLU will give the solutions corresponding to new righthand sides with economy of computational effort, since the LU decomposition is done only once. This scheme is recommended when a given set of equations must be solved with a series of constant terms.

Program 6 (Fig. 2.11) is a utility subroutine that scales the values of the coefficient matrix so that the largest magnitude in each row is unity. This subroutine should be called before any of the previous routines are used if the coefficient matrix has values that are widely different in some rows than in other rows, as discussed in Section 2.3.

Program 7 (Fig. 2.12) is a subroutine named GSITRN, which finds the solution of a linear system by iteration. To make the convergence as rapid as possible, the coefficients must be arranged so that the terms of largest magnitude are on the diagonal. In some systems this cannot be totally achieved; in that case one should approach the condition as closely as possible. An initial approximation to the solution is passed to the subroutine as a parameter.

The first step in the subroutine is to divide each row by the diagonal term, including the corresponding element in the righthand-side vector; this avoids later divisions. Iteration continues until each component of the approximate solution vector changes less than the input parameter TOL. If this does not occur within the number of iterations specified by NITER, the program prints a message and returns the last approximation.

Iteration is best adapted to sparse matrices.* The subroutine GSITRN does not illustrate this case. In fact, writing a general-purpose subroutine for iterative solutions to a sparse matrix is impossible, for this would require a repetitive pattern in the coefficients; such patterns cannot be anticipated. Many texts illustrate iteration with a tridiagonal system, but for these, iteration is not the best

---

* A major area of application is in solving certain partial-differential equations. This is discussed in later chapters.

choice. A program like TRIDG, above, has a minimum storage requirement and runs much faster than an iterative scheme.

Program 8 (Fig. 2.13), named NLSYST, is a subroutine that solves a nonlinear system. Newton's method (see Eqs. (2.28)) is employed. The partial derivatives are computed by making a small change in the value of a variable and dividing the change in value of the function by the change in value of the variable; this estimates the derivative by a difference quotient (these are discussed in detail in Chapter 4). After all the partials have been approximated, the set of equations equivalent to Eq. (2.28) is solved by a call to subroutine ELIM, giving values for increments to the initial guesses for the variables that should make them closer to a solution of the nonlinear system. The subroutine repeats the procedure until convergence is obtained.

The calling program passes initial guesses for the variables, a value for DELTA (the change in variable value used to estimate the partial derivative), and tolerance values (FTOL and XTOL) to terminate iterations when the function values are all sufficiently small or the changes in the x-values are all sufficiently small. The maximum number of iterations is limited by MAXIT. Function values are supplied through an external subroutine.

Some judgment is needed to use NLSYST. First, the initial guesses for values of the variables must be near enough to a solution to give convergence. Second, the value for DELTA should be small enough to give a reasonable approximation to the partial derivative but not so small as to lead to excessive round-off. (DELTA $= 0.01$ seems to give good results in an IBM 360 system.)

As an illustration of how the subroutine NLSYST is used, the program in Fig. 2.5, together with the function subroutine, will determine a root of the pair of equations:

$$f(x, y) = x^2 + y^2 - 5 = 0,$$
$$g(x, y) = y - e^x - 1 = 0.$$

The starting value $(2, 4)$ is supplied, along with the values for DELTA, FTOL, XTOL, and MAXIT. After six iterations, the root at $(0.20437, 2.22672)$ was found.

```
DIMENSION X(2) , F(2)
DATA N, MAXIT, DELTA / 2, 10, 0.01/
DATA X, XTOL, FTOL /   2.,4.,   0.0001, 0.0005/
EXTERNAL FCN
I = 0
CALL NLSYST (FCN, N, MAXIT, X, F, DELTA, XTOL, FTOL, I)
STOP
END

SUBROUTINE FCN (X, F)
DIMENSION X(2), F(2)
F(1) = X(1)*X(1) + X(2)*X(2) - 5.
F(2) = X(2) - EXP(X(1)) - 1.
RETURN
END
```

Figure 2.5

```
      SUBROUTINE ELIM (AB, N, NP, NDIM)
      DIMENSION AB(NDIM,NP)
C  THIS SUBROUTINE SOLVES A SET OF LINEAR EQUATIONS.
C  THE GAUSS ELIMINATION METHOD IS USED, WITH PARTIAL PIVOTING.
C  MULTIPLE RIGHT HAND SIDES ARE PERMITTED, THEY SHOULD BE SUPPLIED
C  AS COLUMNS THAT AUGMENT THE COEFFICIENT MATRIX.
C  PARAMETERS ARE -
C     AB     COEFFICIENT MATRIX AUGMENTED WITH R.H.S. VECTORS
C     N      NUMBER OF EQUATIONS
C     NP     TOTAL NUMBER OF COLUMNS IN THE AUGMENTED MATRIX.
C     NDIM   FIRST DIMENSION OF MATRIX AB IN THE CALLING PROGRAM.
C  THE SOLUTION VECTOR(S) ARE RETURNED IN THE AUGMENTATION
C  COLUMNS OF AB.
C
C  BEGIN THE REDUCTION
      NM1 = N - 1
      DO 35 I = 1,NM1
C  FIND THE ROW NUMBER OF THE PIVOT ROW.  WE WILL THEN
C  INTERCHANGE ROWS TO PUT THE PIVOT ELEMENT ON THE DIAGONAL.
         IPVT = I
         IP1 = I + 1
         DO 10 J = IP1,N
            IF (ABS(AB(IPVT,I)) .LT. ABS(AB(J,I)))  IPVT = J
10       CONTINUE
C  CHECK TO BE SURE THE PIVOT ELEMENT IS NOT TOO SMALL, IF SO
C  PRINT A MESSAGE AND RETURN.
         IF ((ABS(AB(IPVT,I)) .LT. 1.E-5))  GO TO 99
C  NOW INTERCHANGE, EXCEPT IF THE PIVOT ELEMENT IS ALREADY ON
C  THE DIAGONAL, DON'T NEED TO.
         IF (IPVT .EQ. I)  GO TO 25
         DO 20 JCOL = I,NP
            SAVE = AB(I,JCOL)
            AB(I,JCOL) = AB(IPVT,JCOL)
            AB(IPVT,JCOL) = SAVE
20       CONTINUE
C  NOW REDUCE ALL ELEMENTS BELOW THE DIAGONAL IN THE I-TH ROW.  CHECK
C  FIRST TO SEE IF A ZERO ALREADY PRESENT.  IF SO,
C  CAN SKIP REDUCTION FOR THAT ROW.
25       DO 32 JROW = IP1,N
            IF (AB(JROW,I) .EQ. 0)  GO TO 32
            RATIO = AB(JROW,I)/AB(I,I)
            DO 30 KCOL = IP1,NP
               AB(JROW,KCOL) = AB(JROW,KCOL) - RATIO*AB(I,KCOL)
30          CONTINUE
32       CONTINUE
35    CONTINUE
C  WE STILL NEED TO CHECK A(N,N) FOR SIZE.
      IF (ABS(AB(N,N)) .LT. 1.E-5)  GO TO 99
C  NOW WE BACK SUBSTITUTE
      NP1 = N + 1
      DO 50 KCOL = NP1,NP
         AB(N,KCOL) = AB(N,KCOL)/AB(N,N)
         DO 45 J = 2,N
            NVBL = NP1 - J
            L = NVBL + 1
            VALUE = AB(NVBL,KCOL)
            DO 40 K = L,N
               VALUE = VALUE - AB(NVBL,K)*AB(K,KCOL)
40          CONTINUE
            AB(NVBL,KCOL) = VALUE/AB(NVBL,NVBL)
45       CONTINUE
50    CONTINUE
      RETURN
C  MESSAGE FOR A NEAR SINGULAR MATRIX
99    WRITE (6,100)
100   FORMAT(1H0, 57HSOLUTION NOT FEASIBLE. A NEAR ZERO PIVOT WAS ENCOUN
     +TERED.)
      RETURN
      END
```

Figure 2.6 Program 1.

```
      SUBROUTINE TRIDG (A, N, NDIM)
C  THIS SUBROUTINE PERFORMS GAUSSIAN ELIMINATION ON A TRIDIAGONAL
C  SYSTEM.  THE COEFFICIENTS OF N EQUATIONS ARE STORED IN THE
C  N X 4 ARRAY, A.  THE FIRST COLUMN OF A HOLDS THE ELEMENTS TO THE
C  LEFT OF THE DIAGONAL, THE SECOND HOLDS THE DIAGONAL ELEMENTS, AND
C  THE THIRD HOLDS THE ELEMENTS TO THE RIGHT.  THE FOURTH COLUMN HOLDS THE
C  RIGHT HAND SIDE TERMS..
C  THE SOLUTION VECTOR IS RETURNED IN THE FOURTH COLUMN OF A.
C  PARAMETERS ARE -
C    A      MATRIX OF COEFFICIENTS AND R.H.S. AS DESCRIBED
C    N      NUMBER OF EQUATIONS
C    NDIM   FIRST DIMENSION OF  A IN THE CALLING PROGRAM.
      DIMENSION A(NDIM,4)
C  FIRST WE ELIMINATE ALL BELOW DIAGONAL TERMS
      DO 10 I = 2,N
         A(I,2) = A(I,2) - A(I,1)/A(I-1,2)*A(I-1,3)
         A(I,4) = A(I,4) - A(I,1)/A(I-1,2)*A(I-1,4)
10    CONTINUE
C  NOW WE DO THE BACK SUBSTITUTING
      NM1 = N - 1
      A(N,4) = A(N,4)/A(N,2)
      DO 20 I = 1,NM1
C  THE INDEX M WILL COUNT UP THE ROWS
      M = N - I
         A(M,4) = (A(M,4) - A(M,3)*A(M+1,4))/A(M,2)
20    CONTINUE
      RETURN
      END
```

Figure 2.7 Program 2.

```
          SUBROUTINE LUDCMP (A, N, NDIM, ORDER)
C   THIS SUBROUTINE COMPUTES THE L AND U TRIANGULAR MATRICES
C   EQUIVALENT TO THE A MATRIX, SUCH THAT LU = A.  THESE MATRICES
C   ARE RETURNED IN THE SPACE OF A, IN COMPACT FORM.  THE U MATRIX
C   HAS ONES ON ITS DIAGONAL.
C   PARTIAL PIVOTING IS USED TO GIVE MAXIMUM VALUED ELEMENTS ON THE
C   DIAGONAL OF L.  THE ORDER OF THE ROWS AFTER PIVOTING IS RETURNED IN
C   THE INTEGER VECTOR ORDER.  THIS SHOULD BE USED TO REORDER R.H.S.
C   VECTORS BEFORE SOLVING THE SYSTEM AX = B.
C   PARAMETERS ARE -
C    A       THE N X N MATRIX OF COEFFICIENTS
C    N       THE NUMBER OF EQUATIONS
C    NDIM    THE FIRST DIMENSION OF A IN THE CALLING PROGRAM
C    ORDER INTEGER VECTOR HOLDING ROW ORDER AFTER PIVOTING
C   THIS ROUTINE CALLS A SUBROUTINE APVT TO LOCATE THE PIVOT ROW AND
C   MAKE INTERCHANGES
          DIMENSION A(NDIM,N)
          INTEGER ORDER (N)
C   ESTABLISH INITIAL ORDERING IN ORDER VECTOR.
          DO 10 I = 1,N
            ORDER(I) = I
10        CONTINUE
C   DO PIVOTING FOR FIRST COLUMN BY CALL TO SUBROUTINE APVT
          CALL APVT (A, N, NDIM, ORDER, 1)
C   IF PIVOT ELEMENT VERY SMALL, PRINT ERROR MESSAGE AND RETURN
          IF (ABS(A(1,1)).LT. 1.E-5)  GO TO 99
C   NOW COMPUTE ELEMENTS FOR FIRST ROW OF U
          DO 20 KCOL = 2,N
            A(1,KCOL) = A(1,KCOL)/A(1,1)
20        CONTINUE
C   COMPLETE THE COMPUTING OF L AND U ELEMENTS.  THE GENERAL PLAN IS TO
C   COMPUTE A COLUMN OF L'S, THEN CALL APVT TO INTERCHANGE ROWS,
C   AND THEN GET A ROW OF U'S
          NM1 = N - 1
          DO 80 JCOL = 2,NM1
C   FIRST COMPUTE A COLUMN OF L'S
            JM1 = JCOL - 1
            DO 50 IROW = JCOL,N
              SUM = 0.
              DO 40 KCOL = 1,JM1
                SUM = SUM + A(IROW,KCOL)*A(KCOL,JCOL)
40            CONTINUE
              A(IROW,JCOL) = A(IROW,JCOL) - SUM
50          CONTINUE
C   NOW INTERCHANGE ROWS IF NEED BE, THEN TEST FOR TOO SMALL PIVOT.
            CALL APVT (A, N, NDIM, ORDER, JCOL)
            IF (ABS(A(JCOL,JCOL)).LT. 1.E-5)  GO TO 99
C   NOW WE GET A ROW OF U'S
            JP1 = JCOL + 1
            DO 70 KCOL = JP1,N
              SUM = 0.
              DO 60 IROW = 1,JM1
                SUM = SUM + A(JCOL,IROW)*A(IROW,KCOL)
60            CONTINUE
              A(JCOL,KCOL) = (A(JCOL,KCOL) - SUM)/A(JCOL,JCOL)
70          CONTINUE
80        CONTINUE
C   STILL NEED TO GET LAST ELEMENT IN L MATRIX
          SUM = 0.
          DO 90 KCOL = 1,NM1
            SUM = SUM + A(N,KCOL)*A(KCOL,N)
90        CONTINUE
          A(N,N) = A(N,N) - SUM
          RETURN
C   WE COME HERE IF A PIVOT ELEMENT IS .LT. 1.E-5, INDICATING THAT
C   THE A MATRIX IS NEARLY SINGULAR
99        WRITE (6,100)
100       FORMAT (1H0, 60H VERY SMALL PIVOT ELEMENT INDICATES A NEARLY SINGU
         +LAR MATRIX )
          RETURN
          END
```

Figure 2.8 Program 3.

```
        SUBROUTINE APVT (A, N, NDIM, ORDER, JCOL)
C  THIS SUBROUTINE FINDS THE LARGEST ELEMENT FOR PIVOT IN JCOL OF
C  MATRIX A, PERFORMS INTERCHANGES OF ELEMENTS IN A AND ALSO
C  INTERCHANGES THE ELEMENTS IN THE ORDER VECTOR.
C  PARAMETERS ARE -
C     A     MATRX OF COEFFICIENTS WHOSE ROWS ARE TO BE INTERCHANGED
C     N     NUMBER OF EQUATIONS
C     NDIM  FIRST DIMENSION OF A IN THE MAIN PROGRAM
C     ORDER INTEGER VECTOR TO HOLD ROW ORDERING
C     JCOL  COLUMN OF A BEING SEARCHED FOR PIVOT ELEMENT
        DIMENSION A(NDIM, N)
        INTEGER ORDER(N)
C  FIND PIVOT ROW, CONSIDERING ONLY THE ELEMENTS ON AND BELOW DIAGONAL
        IPVT = JCOL
        BIG = ABS(A(JCOL,JCOL))
        JP1 = JCOL + 1
        DO 10 IROW = JP1,N
          ANEXT = ABS(A(IROW,JCOL))
          IF (ANEXT .GT. BIG)  IPVT = IROW
10      CONTINUE
C  NOW INTERCHANGE ROW ELEMENTS IN THE ROW WHOSE NUMBER EQUALS JCOL
C  WITH THE PIVOT ROW UNLESS PIVOT ROW IS JCOL
        IF (IPVT .EQ. JCOL)  GO TO 25
        DO 20 KCOL = 1,N
          SAVE = A(JCOL,KCOL)
          A(JCOL,KCOL) = A(IPVT,KCOL)
          A(IPVT,KCOL) = SAVE
20      CONTINUE
C  NOW SWITCH ELEMENTS IN THE ORDER VECTOR
        ISAVE = ORDER(JCOL)
        ORDER(JCOL) = ORDER(IPVT)
        ORDER(IPVT) = ISAVE
25      RETURN
        END
```

Figure 2.9 Program 4.

```
      SUBROUTINE SOLVLU (LU, B, X, N, NDIM, ORDER)
C  THIS SUBROUTINE IS USED TO FIND THE SOLUTION TO A SYSTEM
C  OF EQUATIONS, AX = B, AFTER THE LU EQUIVALENT OF A HAS BEEN FOUND.
C  BEFORE USING THIS ROUTINE, THE VECTOR B SHOULD BE SCALED IF MATRIX A WAS
C  SCALED, USING THE SAME SCALE FACTORS.  WITHIN THIS ROUTINE, THE ELEMENTS
C  OF B ARE REARRANGED IN THE SAME WAY THAT THE ROWS OF A WERE
C  INTERCHANGED, USING THE ORDER VECTOR WHICH HOLDS THE ROW ORDERINGS.
C  THE SOLUTION IS RETURNED IN X.
C  PARAMETERS ARE -
C     LU     THE LU EQUIVALENT OF THE COEFFICIENT MATRIX
C     B      THE VECTOR OF RIGHT HAND SIDES
C     X      SOLUTION VECTOR
C     N      NUMBER OF EQUATIONS
C     NDIM   FIRST DIMENSION OF A IN THE MAIN PROGRAM
C     ORDER  INTEGER ARRAY OF ROW ORDER AS REARRANGED DURING PIVOTING
      REAL LU(NDIM,N)
      DIMENSION B(N), X(N)
      INTEGER ORDER(N)
C  REARRANGE THE ELEMENTS OF THE B VECTOR.  X IS USED TO HOLD THEM.
      DO 10 I = 1,N
         J = ORDER(I)
         X(I) = B(J)
10    CONTINUE
      WRITE (6,200) X
C  COMPUTE THE B' VECTOR, STORING BACK IN X
      X(1) = X(1)/LU(1,1)
      DO 50 IROW = 2,N
         IM1 = IROW - 1
         SUM = 0.
         DO 40 JCOL = 1,IM1
            SUM = SUM + LU(IROW,JCOL)*X(JCOL)
40       CONTINUE
         X(IROW) = (X(IROW) - SUM)/LU(IROW,IROW)
50    CONTINUE
C  NOW GET THE SOLUTION VECTOR.  X(N) = X(N) ALREADY.
      DO 70 IROW = 2,N
         NVBL = N - IROW + 1
         SUM = 0.
         NP1 = NVBL + 1
         DO 60 JCOL = NP1,N
            SUM = SUM + LU(NVBL,JCOL)*X(JCOL)
60       CONTINUE
         X(NVBL) = X(NVBL) - SUM
70    CONTINUE
200   FORMAT (1H0, 3F10.3)
      RETURN
      END
```

Figure 2.10 Program 5.

```
      SUBROUTINE SCALES (A, N, NDIM, SCFAC)
C THIS ROUTINE SCALES THE VALUES IN AN N X N COEFFICIENT MATRIX
C SO THAT THE LARGEST ELEMENT IN EACH ROW IS UNITY.  THE SCALED
C VALUES OF A ARE RETURNED IN THE A MATRIX AND THE SCALE FACTORS
C FOR EACH ROW ARE RETURNED IN SCFAC VECTOR.  USE SCFAC TO SCALE THE
C ELEMENTS IN THE B VECTOR (R.H. SIDES( BEFORE SOLVING THE SET OF
C EQUATIONS AX = B.
C PARAMETERS ARE -
C    A        MATRIX OF COEFFICIENTS
C    N        NUMBER OF EQUATIONS
C    NDIM   FIRST DIMENSION OF A IN THE CALLING PROGRAM
C    SCFAC ARRAY TO HOLD THE SCALE FACTORS
      DIMENSION A(NDIM,NDIM), SCFAC(N)
C FIND LARGEST VALUE IN EACH ROW.  IF ANY ROW HAS ONLY ZEROS,
C PRINT MESSAGE AND RETURN.
      DO 20 I = 1,N
        BIG = ABS(A(I,1))
        DO 10 J = 2,N
          ANEXT = ABS(A(I,J))
          IF (ANEXT .GT. BIG)  BIG = ANEXT
10      CONTINUE
        IF (BIG .EQ. 0)  GO TO 99
        SCFAC (I) = 1./BIG
20    CONTINUE
C NOW SCALE THE A VALUES
      DO 40 I = 1,N
        DO 50 J = 1,N
          A(I,J) = A(I,J)*SCFAC(I)
50      CONTINUE
40    CONTINUE
      RETURN
C WE COME HERE WHEN ALL ELEMENTS IN ANY ROW ARE ZERO
99    WRITE (6,200) I
200   FORMAT (1H0, 21H ALL ELEMENTS IN ROW , I3, 9H ARE ZERO)
      RETURN
      END
```

Figure 2.11 Program 6.

```
      SUBROUTINE GSITRN (A, B, X, N, NDIM, NITER, TOL)
C     THIS SUBROUTINE OBTAINS THE SOLUTION TO N LINEAR EQUATIONS
C     BY GAUSS-SEIDEL ITERATION.  AN INITIAL APPROXIMATION IS
C     SENT TO THE SUBROUTINE IN THE VECTOR X.  THE SOLTUION, AS
C     APPROXIMATED BY THE SUBROUTINE , IS RETURNED IN X.
C     THE ITERATIONS ARE CONTINUED UNTIL THE MAXIMUM CHANGE IN
C     ANY X COMPONENT IS LESS THAN TOL.  IF THIS CANNOT
C     BE ACCOMPLISHED IN NITER ITERATIONS, A MESSAGE IS PRINTED
C     AND THE CURRENT APPROXIMATION IS RETURNED.
C     THE COEFFICIENTS ARE TO BE ARRANGED IN A SO AS TO HAVE THE
C     LARGEST VALUES ON THE DIAGONAL.
C     PARAMETERS ARE -
C     A      COEFFICIENT MATRIX WITH LARGEST VALUES ON THE DIAGONAL
C     B      RIGHT HAND SIDE VECTOR
C     X      INITIAL APPROXIMATION TO THE SOLUTION, ALSO RETURNS THE RESULT
C     N      THE NUMBER OF EQUATIONS
C     NDIM   FIRST DIMENSION OF A IN THE CALLING PROGRAM
C     NITER  LIMIT TO THE NUMBER OF ITERATIONS
C     TOL    TEST VALUE TO STOP ITERATING THE SOLUTION
      DIMENSION A(NDIM,N), B(N), X(N)
C     WE CAN SAVE SOME DIVISIONS BY MAKING THE DIAGONAL  ELEMENTS
C     EQUAL TO UNITY
      DO 10 I = 1,N
         SAVE = A(I,I)
         B(I) = B(I)/SAVE
         DO 5 J = 1,N
            A(I,J) = A(I,J)/SAVE
5        CONTINUE
10    CONTINUE
C     NOW WE PERFORM THE ITERATIONS.  STORE MAX CHANGE IN X VALUES
C     FOR TESTING AGAINST TOL.  OUTER LOOP LIMITS ITERATIONS TO NITER.
      DO 40 ITER = 1,NITER
         XMAX = 0.
         DO 30 I = 1,N
            SAVE = X(I)
            X(I) = B(I)
            DO 20 J = 1,N
               IF (J .EQ. I)  GO TO 20
               X(I) = X(I) + A(I,J)*X(J)
20          CONTINUE
            IF (ABS(X(I) - SAVE) .GT. XMAX)  XMAX = ABS(X(I) - SAVE)
30       CONTINUE
         IF (XMAX .LE. TOL)  RETURN
40    CONTINUE
C     NORMAL EXIT FROM THE LOOP MEANS NON-CONVERGENT IN NITER
C     ITERATIONS.  PRINT MESSAGE AND RETURN.
      WRITE (6,200) TOL, NITER
200   FORMAT (1H0, 26HDID NOT MEET TOLERANCE OF , E14.6,
     +  4H IN , I4, 11H ITERATIONS, /, 1H , 38HLAST VALUE OF X WAS RETUR
     +NED TO CALLER)
      RETURN
      END
```

Figure 2.12 Program 7.

```
      SUBROUTINE NLSYST (FCN, N, MAXIT, X, F, DELTA, XTOL, FTOL, I)
C  THIS SUBROUTINE SOLVES A SYSTEM OF N NON-LINEAR EQUATIONS BY
C  NEWTON'S METHOD.  THE PARTIAL DERIVATIVES OF THE FUNCTIONS
C  ARE ESTIMATED BY DIFFERENCE QUOTIENTS WHEN A VARIABLE IS PERTURBED
C  BY AN AMOUNT EQUAL TO DELTA (DELTA IS ADDED.)  THIS IS DONE FOR EACH
C  VARIABLE IN EACH FUNCTION.  INCREMENTS TO IMPROVE THE ESTIMATES
C  FOR THE X-VALUES ARE COMPUTED FROM A SYSTEM OF EQUATIONS USING
C  SUBROUTINE ELIM.
C  PARAMETERS ARE
C     FCN    SUBROUTINE THAT COMPUTES VALUES OF THE FUNCTIONS.  MUST
C              BE DECLARED EXTERNAL IN THE CALLING PROGRAM
C     N      THE NUMBER OF EQUATIONS
C     MAXIT  LIMIT TO THE NUMBER OF ITERATIONS THAT WILL BE USED
C     X      ARRAY TO HOLD THE X VALUES.  INITIALLY THIS ARRAY HOLDS THE
C              INITIAL GUESSES.  IT RETURNS THE FINAL VALUES.
C     F      AN ARRAY THAT HOLDS VALUES OF THE FUNCTIONS
C     DELTA  A SMALL VALUE USED TO PERTURB THE X VALUES SO PARTIAL DERIVATIVES
C              CAN BE COMPUTED BY DIFFERENCE QUOTIENT
C     XTOL   TOLERANCE VALUE FOR CHANGE IN X VALUES TO STOP ITERATIONS.
C              WHEN THE LARGEST CHANGE IN ANY X MEETS XTOL, THE
C              SUBROUTINE TERMINATES.
C     FTOL   TOLERANCE VALUE ON F TO TERMINATE.  WHEN THE LARGEST F
C              VALUE IS LESS THAN FTOL, SUBROUTINE TERMINATES.
C     I      RETURNS VALUES TO INDICATE HOW THE ROUTINE TERMINATED.
C              I = 1   XTOL WAS MET
C              I = 2   FTOL WAS MET
C              I = -1  MAXIT EXCEEDED BUT TOLERANCES NOT MET
C              I = -2  VERY SMALL PIVOT ENCOUNTERED IN GAUSSIAN
C                      ELIMINATION STEP.  NO RESULTS OBTAINED
C              I = -3  INCORRECT VALUE OF N WAS SUPPLIED.  N MUST BE
C                      BE BETWEEN 2 AND 10
      DIMENSION X(N), F(N), A(10,11), XSAVE(10), FSAVE(10)
      LOGICAL PRINT
C  CHECK VALIDITY OF VALUE OF N
      IF (N .LT. 2 .OR. N .GT. 10)  GO TO 999
      PRINT = .TRUE.
      IF (I .NE. 0)  PRINT = .FALSE.
C  BEGIN ITERATIONS
C  SAVE X VALUES, THEN GET F VALUES
      NP = N + 1
      DO 100 IT = 1,MAXIT
        DO 10 IVBL = 1,N
          XSAVE(IVBL) = X(IVBL)
10      CONTINUE
        CALL FCN (X, F)
C  TEST F VALUES AND SAVE THEM
        ITEST = 0
        DO 20 IFCN = 1,N
          IF (ABS(F(IFCN)) .GT. FTOL)  ITEST = ITEST + 1
          FSAVE(IFCN) = F(IFCN)
20      CONTINUE
C  PRINT CURRENT VALUES IF PRINT IS .TRUE.
        IF (.NOT. PRINT)  GO TO 30
          WRITE (6,1000) IT, X
1000      FORMAT (1H0, 23HAFTER ITERATION NUMBER , I3,
     +                 19H X AND F VALUES ARE  / 1H0, 10F13.5)
```

Figure 2.13 Program 8.

```
            WRITE (6,1001) F
1001        FORMAT (1H0, 10F13.5)
C  SEE IF FTOL IS MET.  IF NOT, CONTINUE.  IF SO, SET I = 2 AND RETURN.
30      IF (ITEST .NE. 0)  GO TO 35
        I = 2
        RETURN
C  THIS DOUBLE LOOP COMPUTES THE PARTIAL DERIVATIVES OF EACH FUNCTION
C  FOR EACH VARIABLE AND STORES THEM IN A COEFFICIENT ARRAY
35      DO 50 JCOL = 1,N
        X(JCOL) = XSAVE(JCOL) + DELTA
        CALL FCN (X, F)
        DO 40 IROW = 1,N
           A(IROW,JCOL) = (F(IROW) - FSAVE(IROW))/DELTA
40      CONTINUE
C  RESET X VALUES FOR NEXT COLUMN OF PARTIALS
        X(JCOL) = XSAVE(JCOL)
50      CONTINUE
C  NOW WE PUT NEGATIVE OF F VALUES AS RIGHT HAND SIDES AND CALL ELIM
        DO 60 IROW = 1,N
        A(IROW,NP) = -FSAVE(IROW)
60      CONTINUE
        CALL ELIM (A, N, NP, 10)
C  BE SURE THAT COEFFICIENT MATRIX NOT TOO ILL-CONDITIONED
        DO 70 IROW = 1,N
        IF (ABS(A(IROW,IROW)) .LE. 1.E-5)  GO TO 998
70      CONTINUE
C  APPLY THE CORRECTIONS TO THE X VALUES, ALSO SEE IF XTOL IS MET.
        ITEST = 0
        DO 80 IVBL = 1,N
        X(IVBL) = XSAVE(IVBL) + A(IVBL,NP)
        IF (ABS(A(IVBL,NP)) .GT. XTOL)  ITEST = ITEST + 1
80      CONTINUE
C  IF XTOL IS MET, PRINT LAST VALUES AND RETURN, ELSE DO
C  ANOTHER ITERATION
        IF (ITEST .EQ. 0)  GO TO 997
100     CONTINUE
C  WHEN WE HAVE DONE MAXIT ITERATIONS, SET I = -1 AND RETURN
        I = -1
        RETURN
C  WE COME HERE WHEN XTOL IS MET.  PRINT LAST VALUES, SET I = 1, RETURN
997     I = 1
        IF (.NOT. PRINT)  GO TO 110
        WRITE (6,1002) IT, X
1002    FORMAT (1H0, 23HAFTER ITERATION NUMBER , I3,
     +          29H X VALUES (MEETING XTOL) ARE / 1H0, 10F13.5)
110     RETURN
C  WE COME HERE WHEN PARTIALS FORM A NEARLY SINGULAR MATRIX.  SET I = -2
C  PRINT MESSAGE, RETURN
998     I = -2
        WRITE (6,1003)
1003    FORMAT (1H0, 44HCANNOT SOLVE SYSTEM.  MATRIX NEARLY SINGULAR )
        RETURN
C  WE COME HERE WHEN NUMBER OF EQUATIONS PASSED TO NLSYST IS INVALID.
C  PRINT MESSAGE, RETURN
999     I = -3
        WRITE (6,1004) N
1004    FORMAT (1H0, 81HNUMBER OF EQUATIONS PASSED TO NLSYST IS INVALID.
     + MUST BE 1 < N < 11.  VALUE WAS , I3)
        RETURN
        END
```

Figure 2.13 (continued)

EXERCISES

Section 2.2

1. Given matrices A, B, and C and vector x:

$$A = \begin{bmatrix} 1 & 2 & -1 & 3 \\ 0 & 4 & -2 & 1 \\ 3 & -1 & 1 & 1 \end{bmatrix}, \quad B = \begin{bmatrix} 0 & 1 & 0 \\ 1 & 2 & -1 \\ 1 & -1 & 3 \\ 2 & -2 & 1 \end{bmatrix},$$

$$C = \begin{bmatrix} 0 & 1 & 4 & 2 \\ -1 & -2 & 3 & 1 \\ 0 & 1 & 2 & -1 \end{bmatrix}, \quad x = \begin{bmatrix} -1.2 \\ 3.7 \\ -2.0 \end{bmatrix}.$$

   a) Find $A + B$, $A + C$, $A - 2C$.
   ▶b) Find AB, BA, AC, Bx, Cx.
   c) What is the size of ABC? of Bx?
   d) Show that $(AB)C = A(BC)$.
   e) Find A^T, B^T.

2. Given the square matrix M:

$$M = \begin{bmatrix} 1 & 1 & -1 \\ -1 & 0 & 1 \\ 1 & -1 & -1 \end{bmatrix}.$$

   ▶a) Find M^2, M^3.
   b) Show that $MI = IM = M$, where I is the order-3 identity matrix.
   c) A square matrix can always be expressed as a sum of an upper-triangular and a lower-triangular matrix. Find a U and an L such that $U + L = M$.
   d) A square matrix can also be expressed as a sum of a lower- and upper-triangular and a diagonal matrix. Find $L + D + U = M$.
   ▶e) Find $\det(M)$.
   f) Find $\operatorname{tr}(M)$.

3. Write the following set of equations in matrix notation:

$$a_1 x + b_1 y + c_1 z = d_1,$$
$$a_2 x + b_2 y + c_2 z = d_2,$$
$$a_3 x + b_3 y + c_3 z = d_\gamma.$$

4. Write as a set of equations:

$$\begin{bmatrix} 1 & 3 & 1 & -1 \\ 2 & 0 & 1 & 1 \\ 0 & -1 & 4 & 1 \\ 0 & 1 & 1 & -5 \end{bmatrix} \begin{bmatrix} x_1 \\ x_2 \\ x_3 \\ x_4 \end{bmatrix} = \begin{bmatrix} 3 \\ 1 \\ 6 \\ 16 \end{bmatrix}.$$

Section 2.3

5. Solve by elimination:

$$\begin{bmatrix} 5 & -1 & 0 \\ -1 & 5 & -1 \\ 0 & -1 & 5 \end{bmatrix} \begin{bmatrix} x_1 \\ x_2 \\ x_3 \end{bmatrix} = \begin{bmatrix} 9 \\ 4 \\ -6 \end{bmatrix}.$$

▶ 6. Solve by elimination:

$$\begin{bmatrix} 5 & -1 & 1 \\ 2 & 4 & 0 \\ 1 & 1 & 5 \end{bmatrix} x = \begin{bmatrix} 10 \\ 12 \\ -1 \end{bmatrix}.$$

7. Solve the set of equations in Exercise 4.

8. Show that the following set of equations does not have a solution:

$$\begin{aligned} 3x_1 + 2x_2 - \ x_3 - 4x_4 &= 10, \\ x_1 - \ x_2 + 3x_3 - \ x_4 &= -4, \\ 2x_1 + \ x_2 - 3x_3 \quad\quad &= 16, \\ - \ x_2 + 8x_3 - 5x_4 &= 3. \end{aligned}$$

9. If the constant vector in Exercise 8 has 2, 3, 1, 3 as components, show that the equations have an infinite set of solutions.

Section 2.4

10. Solve Exercise 6 by Gaussian elimination. Carry four decimals in your work.

11. Reorder the equations in Exercise 6 by putting the first equation last, and then solve using the Gauss–Jordan method. Are your answers the same as in Exercise 10?

▶12. a) Solve the system

$$\begin{aligned} 2.51x_1 + 1.48x_2 + 4.53x_3 &= 0.05, \\ 1.48x_1 + 0.93x_2 - 1.30x_3 &= 1.03, \\ 2.68x_1 + 3.04x_2 - 1.48x_3 &= -0.53, \end{aligned}$$

by Gaussian elimination, carrying just three significant digits and chopping. Do not interchange rows. Note that one divides by a small coefficient in reducing the third equation.

b) Solve the system again, using elimination with pivoting. Note now that the small coefficient is no longer a divisor.

c) Substitute each set of answers in the original equations and observe that the left- and righthand sides match better with the (b) answers. The solution, when seven places are carried to eliminate round-off, is

$$x_1 = 1.45310, \quad x_2 = -1.58919, \quad x_3 = -0.27489.$$

13. Solve the systems in Exercises 5, 6, and 12 by the Gauss–Jordan method.

▶14. By augmenting the coefficient matrix with each of the b vectors simultaneously, solve $Ax = b$ with multiple righthand sides, given that:

$$A = \begin{bmatrix} 4 & 0 & -1 & 3 \\ 2 & 1 & -2 & 0 \\ 0 & 3 & 2 & -2 \\ 1 & 1 & 0 & 5 \end{bmatrix}, \quad b_1 = \begin{bmatrix} 0 \\ 1 \\ 4 \\ -2 \end{bmatrix}, \quad b_2 = \begin{bmatrix} 0 \\ 0 \\ -2 \\ 4 \end{bmatrix}, \quad b_3 = \begin{bmatrix} 7 \\ -1 \\ 4 \\ -2 \end{bmatrix}.$$

15. Restate the algorithm given in Section 2.4 for Gaussian elimination in the structured format of Chapter 1.

16. Write an algorithm for the Gauss–Jordan method incorporating partial pivoting.

Section 2.5

17. Use LU decomposition to solve the equations in Exercise 6.

18. Use LU decomposition to solve the set of equations in Exercise 4.

19. Solve the system in Exercise 12(a) by the LU method. Note that the small divisor again appears and causes round-off error to be severe. Pivoting can be used with this scheme; one must interchange corresponding rows in both the working matrix (LU) and the original matrix (A).

20. a) Solve the tridiagonal system

$$\begin{bmatrix} 4 & -1 & 0 & 0 & 0 \\ -1 & 4 & -1 & 0 & 0 \\ 0 & -1 & 4 & -1 & 0 \\ 0 & 0 & -1 & 4 & -1 \\ 0 & 0 & 0 & -1 & 4 \end{bmatrix} x = \begin{bmatrix} 100 \\ 200 \\ 200 \\ 200 \\ 100 \end{bmatrix}$$

by finding the LU equivalent to the coefficient matrix. Note how readily the operations proceed for this particular case. The coefficient matrix occurs in the solution of partial-differential equations as discussed in Chapter 8.

▶b) Solve the equations in part (a) with a new righthand-side vector, $b = (100, 0, 0, 0, 200)^T$.

Section 2.6

21 Which of these matrices are singular?

a)
$$A = \begin{bmatrix} 3 & 2 & -1 \\ 0 & -1 & 4 \\ 6 & 3 & 2 \end{bmatrix}$$

b)
$$B = \begin{bmatrix} 1 & 0 & -2 & 3 \\ 3 & 1 & 1 & 4 \\ -1 & 0 & 2 & -1 \\ 4 & 2 & 6 & 0 \end{bmatrix}$$

c)
$$C = \begin{bmatrix} 1 & 0 & -2 & 3 \\ 3 & 1 & 1 & 4 \\ -1 & 0 & 2 & -1 \\ 4 & 3 & 6 & 0 \end{bmatrix}$$

22. a) Find values of x and y that make A singular:

$$A = \begin{bmatrix} 4 & 1 & 0 \\ 2 & -1 & 2 \\ x & y & -1 \end{bmatrix}$$

b) Find values for x and y that make A nonsingular.

23. a) Matrix A in Exercise 21 is singular. Do its rows form linearly independent vectors? Find the values for the a_i in Eq. (2.8).

▶b) Repeat part (a) for the elements of A considered as column vectors.

24. Do these sets of equations have a solution? Find a solution if it exists.

a)

$$\begin{cases} 3x - 2y + z = 2, \\ x - 3y + z = 5, \\ x + y - z = -5, \\ 3x \quad\quad + z = 0. \end{cases}$$

b)

$$\begin{bmatrix} 1 & 1 & 0 \\ 0 & 1 & 1 \\ 1 & 0 & 1 \\ 1 & 1 & 1 \end{bmatrix} x = \begin{bmatrix} 1 \\ -2 \\ 0 \\ 4 \end{bmatrix}$$

c)

$$\begin{bmatrix} 2 & 1 & 3 \\ -1 & 0 & 2 \\ 6 & 2 & 2 \end{bmatrix} x = \begin{bmatrix} 1 \\ 0 \\ 0 \end{bmatrix}$$

d)

$$\begin{bmatrix} 2 & 1 & 3 \\ -1 & 0 & 2 \\ 6 & 2 & 2 \end{bmatrix} x = \begin{bmatrix} 1 \\ 0 \\ 2 \end{bmatrix}$$

▶25. The Hilbert matrix is a classical case of the pathological situation called "ill-conditioning." The 4×4 Hilbert matrix is

$$H = \begin{bmatrix} 1 & \frac{1}{2} & \frac{1}{3} & \frac{1}{4} \\ \frac{1}{2} & \frac{1}{3} & \frac{1}{4} & \frac{1}{5} \\ \frac{1}{3} & \frac{1}{4} & \frac{1}{5} & \frac{1}{6} \\ \frac{1}{4} & \frac{1}{5} & \frac{1}{6} & \frac{1}{7} \end{bmatrix}$$

For the system $Hx = b$, with $b^T = [25/12, 77/60, 57/60, 319/420]$, the exact solution is $x^T = [1, 1, 1, 1]$.

a) Show that the matrix is ill-conditioned by showing that it is nearly singular.

b) Using only three significant digits (chopped) in your arithmetic, find the solution to $Hx = b$.

c) Using only three significant digits, but rounding, again find the solution and compare it to that obtained in part (a).

d) Using five significant digits in your arithmetic, again find the solution and compare it to those found in parts (b) and (c).

Section 2.7

26. Find the determinant of the matrix

$$\begin{bmatrix} 0 & 1 & -1 \\ 3 & 1 & -4 \\ 2 & 1 & 1 \end{bmatrix}$$

by row operations to make it (a) upper-triangular, (b) lower-triangular.

▶27. Find the determinant of the matrix

$$\begin{bmatrix} 1 & 4 & -2 & 1 \\ -1 & 2 & -1 & 1 \\ 3 & 3 & 0 & 4 \\ 4 & -4 & 2 & 3 \end{bmatrix}.$$

28. Invert the coefficient matrix in Exercise 5, and then use the inverse to generate the solution. Note that a symmetric matrix has a symmetric inverse.

29. If the constant vector in Exercise 5 is changed to one with components $(13, 11, -22)$, what is now the solution? Observe that the inverse obtained in Exercise 28 gives the answer readily.

30. Attempt to find the inverse of the coefficient matrix in Exercise 8. Note that a singular matrix has no inverse.

31. Find the determinant of the Hilbert matrix in Exercise 25. A small value of the determinant (when the matrix has elements of the order of unity) indicates ill-conditioning.

▶32. Find the inverse of the Hilbert matrix in Exercise 25. The inverse of an ill-conditioned matrix has some very large elements in comparison to the elements of the original matrix.

Section 2.8

33. Find the Euclidean norm of these vectors and matrices.

a) $x = [1.06, -2.15, 14.05, 0.0]$.

b) $y = [1, 0, 0, 0]$.

c) $z^T = [1, 1, 1]$.

d)
$$A = \begin{bmatrix} 14 & 0 & 0 \\ 0 & 3 & 0 \\ 0 & 0 & -2 \end{bmatrix}.$$

e)
$$B = \begin{bmatrix} 1 & 0 & 1 \\ 1 & 1 & 1 \\ 0 & 1 & 1 \end{bmatrix}.$$

f) Find the Euclidean norm of B^2, of B^3, of B^4.

g) Find the norms of AB; of Az.

h) Does the triangle inequality of Eq. (2.9) hold for $A + B$? for $B + B$? for $x + y$?

34. Find the Euclidean norm of the Hilbert matrix of Exercise 25.

▶35. Find the Euclidean norm of the inverse of the Hilbert matrix of order 4 (Exercise 32).

Section 2.9

36. Consider the system $Ax = b$, where

$$A = \begin{bmatrix} 3.01 & 6.03 & 1.99 \\ 1.27 & 4.16 & -1.23 \\ 0.987 & -4.81 & 9.34 \end{bmatrix}, \qquad b = \begin{bmatrix} 1 \\ 1 \\ 1 \end{bmatrix}.$$

▶a) Using double precision (or a calculator with 10 or more digits of accuracy), solve for x.

b) Solve the system using 3-digit (chopped) arithmetic for each arithmetic operation; call this solution $\bar{x}$.

c) Compare x and $\bar{x}$ and compute $e = x - \bar{x}$. What is $\|e\|_2$?

d) Is the system ill-conditioned? What evidence is there to support your conclusion?

37. Repeat Exercise 36, except change the element a_{33} to -9.34.

▶38. Suppose, in Exercise 36, that uncertainties of measurement give slight changes in some of the elements of A. Specifically, suppose a_{11} is 3.00 instead of 3.01 and a_{31} is 0.99 instead of 0.987. What change does this cause in the solution vector (using precise arithmetic)?

39. Compute the residuals for the imperfect solutions in 36(b) and 37(b). Use double precision in this computation.

40. What are the condition numbers of the coefficient matrices in Exercises 36 and 37? Use the 1-norms.

▶41. Verify Eq. (2.14), using the results in Exercises 36, 39, and 40.

42. Verify Eq. (2.14), using the results in Exercises 37, 39, and 40.

43. Verify Eq. (2.15) for the results of Exercise 38.

44. Apply iterative improvement to the imperfect solution of Exercise 36.

▶45. Apply iterative improvement to the imperfect solution of Exercise 37.

Section 2.10

46. Solve Exercise 5 by the Jacobi method, beginning with the initial vector $(0, 0, 0)$. Compare the rate of convergence when Gauss–Seidel is used with the same starting vector.

47. Solve Exercise 6 by Gauss–Seidel iteration, beginning with approximate solution $(2, 2, -1)$.

▶48. The pair of equations

$$\begin{aligned} x_1 + 2x_2 &= 3, \\ 3x_1 + x_2 &= 4, \end{aligned}$$

can be rearranged to give $x_1 = 3 - 2x_2$, $x_2 = 4 - 3x_1$. Apply the Jacobi method to this rearrangement, beginning with a vector very close to the solution: $x^{(1)} = (1.01, 1.01)^T$ and observe divergence. Now apply Gauss–Seidel. Which method diverges most rapidly?

49. Solve Exercise 20(a) by Gauss–Seidel iteration.

Section 2.11

▶50. Beginning with $(0, 0, 0)$, use relaxation to solve the system
$$6x_1 - 3x_2 + x_3 = 11,$$
$$2x_1 + x_2 - 8x_3 = -15,$$
$$x_1 - 7x_2 + x_3 = 10.$$

51. Solve Exercise 6 by relaxation. Begin with the approximate solution $(2, 2, -1)$.

52. Relaxation is especially well adapted to problems like Exercise 20. Solve by the relaxation method, starting with the vector $(45, 80, 90, 80, 45)$ which one obtains by inspection.

53. Solve the system of equations in Exercise 20 by Gauss–Seidel iteration with over-relaxation (Eq. (2.13)). Vary the over-relaxation factor, w, to find its optimum value. (You will probably want to write a computer program for this.)

Section 2.12

▶54. Find the two intersections nearest the origin of the curves $x^2 + x - y^2 = 1$ and $y - \sin x^2 = 0$. Use the method of iteration.

55. Solve the system
$$x^2 + y^2 + z^2 = 9,$$
$$xyz = 1,$$
$$x + y - z^2 = 0,$$
by iteration to obtain the solution near $(2.5, 0.2, 1.6)$.

56. Solve Exercise 54 by Newton's method.

▶57. Solve by using Newton's method.
$$x^3 + 3y^2 = 21,$$
$$x^2 + 2y + 2 = 0.$$
Make sketches of the graphs to locate approximate values of the intersections.

58. Apply Eq. (2.29) to compute partials and solve this system by Newton's method:
$$xyz - x^2 + y^2 = 1.34,$$
$$xy - z^2 = 0.09,$$
$$e^x - e^y + z = 0.41.$$
There should be a solution near $(1, 1, 1)$.

59. Use the method illustrated by Eqs. (2.30) to solve the systems in Exercises 54 and 57.

APPLIED PROBLEMS AND PROJECTS

60. In considering the movement of space vehicles, it is frequently necessary to transform coordinate systems. The standard inertial coordinate system has the N-axis pointed to the north, the E-axis pointed east, and the D-axis pointed toward the center of the earth. A second system is the vehicle's local coordinate system (with the i-axis straight ahead of the vehicle, the j-axis to the right, and the k-axis downward.) We can transform the vector whose local coordinates are (i, j, k) to the inertial system by multiplying by transformation matrices:

$$
\begin{bmatrix} n \\ e \\ d \end{bmatrix} = \begin{bmatrix} \cos a & -\sin a & 0 \\ \sin a & \cos a & 0 \\ 0 & 0 & 1 \end{bmatrix} \begin{bmatrix} \cos b & 0 & \sin b \\ 0 & 1 & 0 \\ -\sin b & 0 & \cos b \end{bmatrix} \begin{bmatrix} 1 & 0 & 0 \\ 0 & \cos c & -\sin c \\ 0 & \sin c & \cos c \end{bmatrix} \begin{bmatrix} i \\ j \\ k \end{bmatrix}
$$

Transform the vector $(2.06, -2.44, -0,47)^T$ to the inertial system if $a = 27°$, $b = 5°$, $c = 72°$.

61. In this statically determinate trust with pin joints, the tension (F_i) in each member can be obtained from the matrix equation below (the equations result from setting the sum of all forces acting horizontally or vertically at each pin equal to zero):

$$
\begin{bmatrix}
0.7071 & 0 & 0 & -1 & -0.8660 & 0 & 0 & 0 & 0 \\
0.7071 & 0 & 1 & 0 & 0.5 & 0 & 0 & 0 & 0 \\
0 & 1 & 0 & 0 & 0 & -1 & 0 & 0 & 0 \\
0 & 0 & -1 & 0 & 0 & 0 & 0 & 0 & 0 \\
0 & 0 & 0 & 0 & 0 & 0 & 1 & 0 & 0.7071 \\
0 & 0 & 0 & 1 & 0 & 0 & 0 & 0 & -0.7071 \\
0 & 0 & 0 & 0 & 0.8660 & 1 & 0 & -1 & 0 \\
0 & 0 & 0 & 0 & -0.5 & 0 & -1 & 0 & 0 \\
0 & 0 & 0 & 0 & 0 & 0 & 0 & 1 & 0.7071
\end{bmatrix}
F =
\begin{bmatrix}
0 \\ -1000 \\ 0 \\ 0 \\ 500 \\ 0 \\ 0 \\ -500 \\ 0
\end{bmatrix}
$$

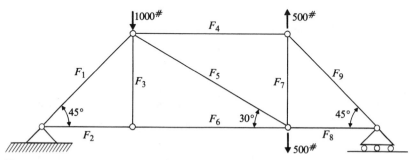

Figure 2.14

a) Solve the system by Gauss elimination.

b) The matrix is quite sparse, so it is a candidate for Gauss–Seidel iteration. Can it be arranged into a diagonally dominant form? Is the system convergent for a starting vector with all elements 0?

62. A bridge network has resistances, in ohms, as shown in Fig. 2.15. If $E = V_1 - V_6 = 5.0$ volts, what is the current through R_4?

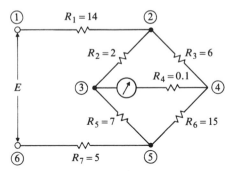

Figure 2.15

Using Kirkhoff's law, we can set up equations as given below. Rearrange them so that you can solve for $V_4 - V_3$.

$$\frac{1}{R_1}(V_1 - V_2) + \frac{1}{R_2}(V_3 - V_2) + \frac{1}{R_3}(V_4 - V_2) = 0,$$

$$\frac{1}{R_2}(V_2 - V_3) + \frac{1}{R_4}(V_4 - V_3) + \frac{1}{R_5}(V_5 - V_3) = 0,$$

$$\frac{1}{R_3}(V_2 - V_4) + \frac{1}{R_4}(V_3 - V_4) + \frac{1}{R_6}(V_5 - V_4) = 0,$$

$$\frac{1}{R_7}(V_6 - V_5) + \frac{1}{R_5}(V_3 - V_5) + \frac{1}{R_6}(V_4 - V_5) = 0.$$

63. Mass spectrometry analysis gives a series of peak height readings for various ion masses. For each peak, the height, h_j, is contributed to by the various constituents. These make different contributions c_{ij} per unit concentration p_i so that the relation

$$h_j = \sum_{i=1}^{n} c_{ij} p_i$$

holds, with n being the number of components present. Carnahan (Ph.D. Thesis, Univ. of Mich.) gives the values shown in Table 2.3 for c_{ij}:

Table 2.3

Peak number	Component				
	CH_4	C_2H_4	C_2H_6	C_3H_6	C_3H_8
1	0.165	0.202	0.317	0.234	0.182
2	27.7	0.862	0.062	0.073	0.131
3		22.35	13.05	4.420	6.001
4			11.28	0	1.110
5				9.850	1.684
6					15.94

If a sample had measured peak heights of $h_1 = 5.20$, $h_2 = 61.7$, $h_3 = 149.2$, $h_4 = 79.4$, $h_5 = 89.3$, and $h_6 = 69.3$, calculate the values of p_i for each component. The total of all the p_i values was 21.53.

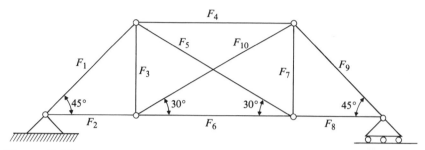

Figure 2.16

64. The truss in Exercise 61 is called *statically determinate* because nine linearly independent equations can be established to relate the nine unknown values of the tensions in the members. If an additional cross brace is added, as sketched in Fig. 2.16, we have ten unknowns but still only nine equations can be written; we now have a statically *indeterminate* system. Consideration of the stretching or compression of the members permits a solution, however. We need to solve a set of equations that gives the displacements x of each pin, which is of the form $ASA^Tx = P$. We then get the tensions f by matrix multiplication: $SA^Tx = f$. The necessary matrices and vectors are:

$$A = \begin{bmatrix} 0.7071 & 0 & 0 & -1 & -0.8660 & 0 & 0 & 0 & 0 & 0 \\ 0.7071 & 0 & 1 & 0 & 0.5 & 0 & 0 & 0 & 0 & 0 \\ 0 & 1 & 0 & 0 & 0 & -1 & 0 & 0 & 0 & -0.8660 \\ 0 & 0 & -1 & 0 & 0 & 0 & 0 & 0 & 0 & -0.5 \\ 0 & 0 & 0 & 0 & 0 & 0 & 1 & 0 & 0.7071 & 0.5 \\ 0 & 0 & 0 & 1 & 0 & 0 & 0 & 0 & -0.7071 & 0.8660 \\ 0 & 0 & 0 & 0 & 0.8660 & 1 & 0 & -1 & 0 & 0 \\ 0 & 0 & 0 & 0 & -0.5 & 0 & -1 & 0 & 0 & 0 \\ 0 & 0 & 0 & 0 & 0 & 0 & 0 & 1 & 0.7071 & 0 \end{bmatrix}.$$

S is a diagonal matrix with values (from upper left to lower right) of

$$\begin{array}{ccccc} 4255, & 6000, & 6000, & 3670, & 3000, \\ 3670, & 6000, & 6000, & 4255, & 3000. \end{array}$$

(These quantities are the values of aE/L, where a is the cross-sectional area of a member, E is the Young's modulus for the material, and L is the length.)

Solve the system of equations to determine the values of f for each of three loading vectors:

$$P_1 = [0, -1000, 0, 0, 500, 0, 0, -500, 0]^T;$$
$$P_2 = [1000, 0, 0, -500, 0, 1000, 0, -500, 0]^T;$$
$$P_3 = [0, 0, 0, -500, 0, 0, 0, -500, 0]^T.$$

65. For turbulent flow of fluids in an interconnected network (see Fig. 2.17), the flow rate V from one node to another is about proportional to the square root of the difference in pressures at the nodes. (Thus, fluid flow differs from flow of electrical current in a network in that nonlinear equations result.) For the conduits sketched below, we are to find the pressure at each node. The values of b represent conduc-

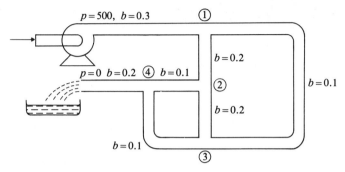

Figure 2.17

tance factors in the relation $v_{ij} = b_{ij}(p_i - p_j)^{1/2}$. These equations can be set up for the pressures at each node:*

$$\text{At node 1:} \quad 0.3\sqrt{500 - p_1} = 0.2\sqrt{p_1 - p_2} + 0.2\sqrt{p_1 - p_3};$$
$$\text{node 2:} \quad 0.2\sqrt{p_1 - p_2} = 0.1\sqrt{p_2 - p_4} + 0.2\sqrt{p_2 - p_3};$$
$$\text{node 3:} \quad 0.1\sqrt{p_1 - p_3} = 0.2\sqrt{p_3 - p_2} + 0.1\sqrt{p_3 - p_4};$$
$$\text{node 4:} \quad 0.1\sqrt{p_2 - p_4} + 0.1\sqrt{p_3 - p_4} = 0.2\sqrt{p_4 - 0}.$$

66. For a tridiagonal matrix of coefficients, the subroutine TRIDG is much more economical of memory space than the subroutine ELIM. It should also be much faster in execution. Analyze each program to determine how the numbers of arithmetic operations vary with size of the matrix. Perform tests to find how the execution times with your computer system compare for systems of various sizes.

67. Make similar comparison between subroutine ELIM and LUDCMP (plus APVT and SOLVLU) as described in Exercise 66. Be sure to consider both sparse and dense matrices.

68. Compare the execution times to obtain the solution to some large sparse systems when using subroutine GSITRN in comparison to ELIM or LUDCMP. Vary the TOL criterion that is input to find the effect of this parameter.

---

* These assume $p_3 > p_2$. If this is not true, modification is necessary.

Chapter Three

Interpolating Polynomials

In chapters one and two, we examined the question: "Given an explicit function of the independent variable x, what is the value of x corresponding to a certain value of the function?" In chapter one, x was a simple variable. In chapter two, x was a vector. We now want to consider a question somewhat the reverse of this: "Given values of an unknown function corresponding to certain values of x, what is the behavior of the function?" We would like to answer the question "What is the function?", but this is always impossible to determine with a limited amount of data.

Our purpose in determining the behavior of the function, as evidenced by the sample of data pairs $[x, f(x)]$, is several-fold. We will wish to approximate other values of the function at values of x not tabulated (interpolation or extrapolation) and to estimate the integral of $f(x)$, and its derivative. The latter objectives will lead us into ways of solving ordinary and partial differential equations.

The strategy we will use in approximating unknown values of the function is straightforward. We will find a polynomial that fits a selected set of points $(x_i, f(x_i))$ and assume that the polynomial and the function behave nearly the same over the interval in question. Values of the polynomial then should be reasonable estimates of the values of the unknown function. When the polynomial is of the first degree, this leads to the familiar linear interpolation. We will be interested in polynomials of degree higher than the first, so we can approximate functions that are far from linear, or so we can get good values from a table with wider spacing.

Interpolating in tables of data was once a very important topic. Today, with computers and calculators that recalculate values for most functions at electronic speeds, we look up values from published tables less frequently. A major reason for discussing this subject, apart from its historical value, is to lay the ground work for numerical integration and differentiation.

3.1 DIFFERENCE TABLES

If we desire to find a polynomial that passes through the same points as our unknown function, we could set up a system of equations involving the coefficients of the polynomial. For example, suppose we want to fit a cubic to this data:

x	$f(x)$
1.0	14.2
2.7	17.8
3.2	22.0
4.8	38.3
5.6	51.7

First, we need to select four points to determine our polynomial. (The maximum degree of the polynomial is always one less than the number of points.) Suppose we choose the first four points. If the cubic is $ax^3 + bx^2 + cx + d$, we can write four equations involving the unknown coefficients, a, b, c, and d:

when $x = 1$: $a(1.0)^3 + b(1.0)^2 + c(1.0) + d = 14.2$,

if $x = 2.7$: $a(2.7)^3 + b(2.7)^2 + c(2.7) + d = 17.8$,

if $x = 3.2$: $a(3.2)^3 + b(3.2)^2 + c(3.2) + d = 22.0$,

if $x = 4.8$: $a(4.8)^3 + b(4.8)^2 + c(4.8) + d = 38.3$.

Solving these equations by the methods of the previous chapter gives us the polynomial. We can then estimate the values of the function at some value of x, say $x = 3.0$, by substituting 3.0 for x in the polynomial.

We seek a better and simpler way of finding such interpolating polynomials. The above procedure is awkward, especially if we want a new polynomial that is made to fit also at the point $(5.6, 51.7)$, or to see what difference it would make to use a quadratic instead of a cubic. Furthermore, this technique leads to an ill-conditioned system of equations.*

The problem at hand is considerably simplified if the values of the function are given at evenly spaced intervals of the independent variable, so we first consider this case. It is convenient to arrange the data in a table with x-values in ascending order. In addition to columns for x and $f(x)$, we shall tabulate differences of the functional values. Table 3.1 is a typical difference table.

Each of the columns to the right of the $f(x)$ column is computed by calculating the difference between two values in the column to its left.

---

* For example, the coefficient matrix for a quartic through five equispaced points has a condition number of the order of 3000.

Table 3.1 A difference table

x	$f(x)$	$\Delta f(x)$	$\Delta^2 f(x)$	$\Delta^3 f(x)$	$\Delta^4 f(x)$
0.0	0.000				
		0.203			
0.2	0.203		0.017		
		0.220		0.024	
0.4	0.423		0.041		0.020
		0.261		0.044	
0.6	0.684		0.085		0.052
		0.346		0.096	
0.8	1.030		0.181		0.211
		0.527		0.307	
1.0	1.557		0.488		
		1.015			
1.2	2.572				

Symbols that represent the entries in a difference table will be useful in using the table of differences to determine coefficients in interpolating polynomials. It is conventional to let the letter h stand for the uniform difference in the x-values, $h = \Delta x$. Using subscripts to represent the order of the x and $f(x)$ values, we define the first differences of the function as

$$\Delta f_1 = f_2 - f_1, \qquad \Delta f_2 = f_3 - f_2, \qquad \ldots, \qquad \Delta f_i = f_{i+1} - f_i.$$

The second- and higher-order differences are similarly defined:*

$$\Delta^2 f_1 = \Delta(\Delta f_1) = \Delta(f_2 - f_1) = \Delta f_2 - \Delta f_1 = (f_3 - f_2) - (f_2 - f_1)$$
$$= f_3 - 2f_2 + f_1,$$
$$\Delta^2 f_i = f_{i+2} - 2f_{i+1} + f_i,$$
$$\Delta^3 f_1 = \Delta(\Delta^2 f_1) = f_4 - 3f_3 + 3f_2 - f_1, \tag{3.1}$$
$$\Delta^3 f_i = f_{i+3} - 3f_{i+2} + 3f_{i+1} - f_i,$$

$$\vdots \qquad\qquad \vdots$$

$$\Delta^n f_i = f_{i+n} - nf_{i+n-1} + \frac{n(n-1)}{2!}f_{i+n-2} - \frac{n(n-1)(n-2)}{3!}f_{i+n-3} + \ldots.$$

In Eqs. (3.1) and throughout this chapter, a subscript on f indicates the x-value at which it is evaluated: viz., $f_3 = f(x_3)$. The pattern of the coefficients in

---

* The differences that we define here are called forward differences. Some texts define differences that are called backward differences (written as ∇f_i, defined as $\nabla f_i = f_i - f_{i-1}$) and central differences (written δf_i). Our treatment uses only the forward difference.

Eqs. (3.1) is the familiar array of coefficients in the binomial expansion. This fact we can prove most readily by symbolic methods, which we postpone until a later section. The second- and higher-order differences are generally obtained by differencing the previous differences, but Eq. (3.1) shows how any difference can be calculated directly from the functional values.

Table 3.2 Difference table, using symbols.

s	x	$f(x)$	Δf	$\Delta^2 f$	$\Delta^3 f$	$\Delta^4 f$
-2	x_{-2}	f_{-2}				
			Δf_{-2}			
-1	x_{-1}	f_{-1}		$\Delta^2 f_{-2}$		
			Δf_{-1}		$\Delta^3 f_{-2}$	
0	x_0	f_0		$\Delta^2 f_{-1}$		$\Delta^4 f_{-2}$
			Δf_0		$\Delta^3 f_{-1}$	
1	x_1	f_1		$\Delta^2 f_0$		$\Delta^4 f_{-1}$
			Δf_1		$\Delta^3 f_0$	
2	x_2	f_2		$\Delta^2 f_1$		$\Delta^4 f_0$
			Δf_2		$\Delta^3 f_1$	
3	x_3	f_3		$\Delta^2 f_2$		
			Δf_3			
4	x_4	f_4				

Table 3.2 illustrates the formation of a difference table using symbolic representation of the entries. While it might be natural to refer to the first x in the table as x_1, we shall frequently wish to refer to only a portion of the data pairs of the table, so the "first x" loses its significance. We then arbitrarily choose the origin for the subscripted variable, because by the use of negative and zero values we can refer to x-entries that precede x_1. The variable s is equal to the subscript on x and serves to index the x-values.

In hand computation of a difference table, great care should be exercised to avoid arithmetic errors in the subtractions—the fact that we subtract the upper entry from the lower adds a real source of confusion. One of the best ways to check for mistakes is to add the sum of the numbers in each column to the top entry in the column to its left. This sum should equal the bottom entry in the column to the left.

Since each entry in the difference table is the difference of a pair of numbers in the column to its left, one could recompute one of this pair if it should be erased. As a consequence of this, the entire table could be reproduced, given only one value in each column, if the table is extended to the highest possible order of differences.

When $f(x)$ behaves like a polynomial for the set of data given, the difference table has special properties. In Table 3.3 a function is tabulated over the domain $x = 1$ to $x = 6$, and $f(x)$ obviously behaves the same as x^3. [Note carefully that this does *not* imply that $f(x) \equiv x^3$; the value of $f(x)$ at $x = 7$ might well be 17 instead of $7^3 = 343$. We *only* know the values of $f(x)$ as given in the table.]

Table 3.3 Difference table for a function behaving like x^3.

x	$f(x)$	Δf	$\Delta^2 f$	$\Delta^3 f$	$\Delta^4 f$
0	0				
		1			
1	1		6		
		7		6	
2	8		12		0
		19		6	
3	27		18		0
		37		6	
4	64		24		0
		61		6	
5	125		30		
		91			
6	216				

We observe that the third differences are constant. Consequently, the fourth and all higher differences will be zero. The fact that the nth-order differences of any nth-degree polynomial are constant is readily shown. To prove that these nth order differences are constant, we first examine the differences of ax^n:

$$\Delta(ax^n) = a(x + h)^n - ax^n$$
$$= ax^n + anx^{n-1}h + \ldots + ah^n - ax^n$$
$$= (anh)x^{n-1} + \text{terms of lower degree in } x,$$
$$\Delta(anh\, x^{n-1}) = an(n - 1)h^2 x^{n-2} + \text{terms of lower degree.}$$

Noting that every time a difference is taken the leading term has a power of x one less than originally and that the difference of a constant term will be zero, we have for a polynomial

$$\Delta P_n(x) = \Delta(a_1 x^n + a_2 x^{n-1} + \ldots + a_n x + a_{n+1})$$
$$= a_1 nh x^{n-1} + \text{terms of lower degree,}$$
$$\Delta^2 P_n(x) = a_1 n(n - 1)h^2 x^{n-2} + \text{terms of lower degree,}$$

$$\vdots \tag{3.2}$$

$$\Delta^n P_n(x) = a_1 n(n - 1)(n - 2) \cdots (1)h^n x^{n-n} = a_1 n! h^n.$$

This shows not only that the nth difference is a constant, but that its value is $a_1 n! h^n$. For $P_3(x) = x^3$,

$$\Delta^3 P_3(x) = (1)(3!)(1)^3 = 6 \qquad \text{when } h = 1.$$

This is exactly what was found in Table 3.3. Note a similarity between the nth difference and the nth derivative of $P_n(x)$:

$$P_n^{(n)}(x) = a_1 n!.$$

3.2 EFFECT OF ERRORS IN TABLE

Suppose there is an error in one of the $f(x)$ values in the table. Table 3.4 shows how the differences are affected. The data are the same as in Table 3.3, except that the value of $f(3)$ has been changed by 1. We show Table 3.4 as a perturbation of Table 3.3 to more clearly show the effect. Observe, in the second difference column, that changes of $-1, 2, -1$ occur. In the third difference column, they are $-1, 3, -3, 1$. The fourth differences are not extensive enough to show all of the expected pattern of $-1, 4, -6, 4, -1$.

Table 3.4 Effect of errors on a difference table.

x	$f(x)$	Δf	$\Delta^2 f$	$\Delta^3 f$	$\Delta^4 f$
0	0				
		1			
1	1				
		7		~~6~~ 5	
2	8		~~12~~ 11		~~0~~ 4
		~~19~~ 18		~~6~~ 9	
3	~~27~~ 26		~~18~~ 20		~~0~~ −6
		~~37~~ 38		~~6~~ 3	
4	64		~~24~~ 23		~~0~~ 4
		61		~~6~~ 7	
5	125		30		
		91			
6	216				

If the table is long enough and the effect of errors does not overlap, one can use this principle of error propagation to detect and correct errors. Today, with so much computation being done on a computer, correcting of tables is not as important as it once was. However, it is of major concern to us that a single error has an effect on many entries in the difference columns as it fans out to the right, and these increase in magnitude. We have shown a gross error in Table 3.4. What one must always expect in all tables that are expressed decimally are the little errors due to round-off of our entries—they too fan out and affect the entries with increasing magnitude. This growth and spreading of round-off will be an important factor affecting the accuracy of interpolation using polynomials.

3.3 INTERPOLATING POLYNOMIALS

When the function that is tabulated behaves like a polynomial (and this we can tell by observing that its nth-order differences are constant or nearly so), we can approximate it by the polynomial it resembles. Our problem is to find the simplest means of writing the nth-degree polynomial that passes through $n + 1$ pairs of points, $(x_i, f_i) = 1, 2, \ldots, n + 1$. Note that such a polynomial is unique—there is only one polynomial of degree n passing through $n + 1$ points.*

Perhaps the easiest way to write a polynomial that passes through a group of equispaced points is the Newton–Gregory forward polynomial:

$$P_n(x_s) = f_0 + s\Delta f_0 + \frac{s(s - 1)}{2!} \Delta^2 f_0 + \frac{s(s - 1)(s - 2)}{3!} \Delta^3 f_0 + \ldots$$

$$= f_0 + \binom{s}{1}\Delta f_0 + \binom{s}{2}\Delta^2 f_0 + \binom{s}{3}\Delta^3 f_0 + \binom{s}{4}\Delta^4 f_0 + \ldots . \qquad (3.3)$$

In this equation we have used the notation $\binom{s}{n}$, the number of combinations of s things taken n at a time, which is the same as the factorial ratios also shown. Referring to Table 3.2, we now observe that $P_n(x)$ does match the table at all the data pairs (x_i, f_i), $i = 0, 1, 2, \ldots, n$. When $s = 0$, $P_n(x_0) = f_0$. If $s = 1$,

$$P_n(x_1) = f_0 + \Delta f_0 = f_0 + f_1 - f_0 = f_1.$$

If $s = 2$,

$$P_n(x_2) = f_0 + 2\Delta f_0 + \Delta^2 f_0 = f_2.$$

Similarly we can demonstrate that $P_n(x)$ formed according to Eq. (3.3) matches at all $n + 1$ points.†

---

* This seems intuitively true, since there are $n + 1$ constants in the polynomial and the $n + 1$ data pairs are exactly enough to determine them. More formally we can reason thus, the proof being by contradiction:

Suppose there are two different polynomials of degree n which are alike at the $n + 1$ points. Call these $P_n(x)$ and $Q_n(x)$ and write their difference:

$$D(x) = P_n(x) - Q_n(x),$$

where $D(x)$ is a polynomial of at most degree n. But since P and Q match at the $n + 1$ pairs of points, their difference, $D(x)$, is equal to zero for all $n + 1$ of these x-values; i.e., it is a polynomial of degree n at most but has $n + 1$ distinct zeros. This is impossible unless $D(x)$ is identically zero. Hence $P_n(x)$ and $Q_n(x)$ are not different—they must be the same polynomial.

† This demonstration is not a proof, of course. The section on symbolic methods gives perhaps the neatest proof.

If, over the domain from x_0 to x_n, $P_n(x)$ and $f(x)$ have the same values at tabulated values of x, it is perhaps reasonable to assume they will be nearly the same at intermediate x-values. This assumption is the basis for the use of $P_n(x)$ as an interpolating polynomial. We again emphasize that $f(x)$ and $P_n(x)$ will, in general, not be the same function. Hence there is some error to be expected in the estimate from such interpolation. We use the polynomial in Eq. (3.3) as an interpolation polynomial by letting s take on nonintegral values. Observe that, for any value of x,

$$s = \frac{x - x_0}{h}.$$

Example Write a Newton–Gregory forward polynomial of degree three that fits Table 3.1 for the four points at $x = 0.4$ to $x = 1.0$. Use it to interpolate for $f(0.73)$.

To make the polynomial fit as specified, we must index the x's so that $x_0 = 0.4$. It follows then that $f_0 = 0.423$, $\Delta f_0 = 0.261$, $\Delta^2 f_0 = 0.085$, and $\Delta^3 f_0 = 0.096$. We compute s:

$$s = \frac{x - x_0}{h} = \frac{0.73 - 0.4}{0.2} = 1.65.$$

Applying these to Eq. (3.3) with terms through $\Delta^3 f_0$ to give a cubic, we have

$$f(0.73) = 0.423 + (1.65)(0.261) + \frac{(1.65)(0.65)}{2}(0.085)$$

$$+ \frac{(1.65)(0.65)(-0.35)}{6}(0.096)$$

$$= 0.423 + 0.4306 + 0.0456 - 0.0060$$

$$= 0.893. \tag{3.4}$$

The function tabulated in Table 3.1 is $\tan x$, for which the true value is 0.895 at $x = 0.73$. We hence see that there is an error in the third decimal place. We should anticipate some error because, since the third differences are far from constant, our cubic polynomial is not a perfect representation of the function. Even so, our interpolating polynomial gave a fair estimate, certainly better than linear interpolation.

Even though the fourth differences are also not constant, we would hope for some improvement if we approximated $f(x)$ by a fourth-degree polynomial. We can just add one more term to Eq. (3.4) to do this:

$$\binom{s}{4}\Delta^4 f_0 = \frac{(1.65)(0.65)(-0.35)(-1.35)}{4!}(0.211) = 0.0044,$$

$$f(0.73) = 0.893 + 0.0044 = 0.898.$$

Normally the higher-degree polynomial is better, but in this instance, adding another term does not improve our estimate; in fact it worsens it slightly. This is due to round-off errors in the original values in this instance. We shall study the errors of interpolating polynomials in a later section.

The domain over which an interpolating polynomial agrees with the function is most readily found by working backward from the last difference that is included in Eq. (3.3), drawing imaginary diagonals to the left between the entries. The x-values included between this fan of diagonals is the domain of the interpolating polynomial. It will always be found to include one more x-entry than the degree of the polynomial.

3.4 OTHER INTERPOLATING POLYNOMIALS

It is sometimes convenient to write the interpolating polynomial in other forms. The Newton–Gregory backward polynomial is

$$P_n(x) = f_0 + \binom{s}{1}\Delta f_{-1} + \binom{s+1}{2}\Delta^2 f_{-2} + \binom{s+2}{3}\Delta^3 f_{-3} + \binom{s+3}{4}\Delta^4 f_{-4} + \ldots .$$

$$(3.5)$$

It is seen that the differences used here form a diagonal row going upward and to the right,* in contrast to the downward sloping diagonal row of differences used in the Newton–Gregory forward formula. Trial with various negative integer values of s demonstrates that Eq. (3.5) also matches with data pairs in the table from $x = x_0$ to $x = x_{-n}$.

If the subscripts are suitably chosen, the points where $P_n(x)$ matches the table will be the same as for the Newton–Gregory forward formula, however. When this is done, the two polynomials are really identical though of a different form. We illustrate by reworking the same problem as in Section 3.3.

Choosing $x_0 = 1.0$, so that

$$s = \frac{0.73 - 1.0}{0.2} = -1.35,$$

gives the cubic that fits Table 3.1 between $x = 0.4$ to $x = 1.0$; hence

$$f(0.73) = 1.557 + (-1.35)(0.527) + \frac{(-0.35)(-1.35)}{2}(0.181)$$

$$+ \frac{(0.65)(-0.35)(-1.35)}{6}(0.096)$$

$$= 1.557 - 0.7114 + 0.0428 + 0.0049$$

$$= 0.893$$

One observes that the identical result is obtained as before.

---

* This ascending diagonal row of differences starting at f_i is equal to the backward differences of f_i; those in the downward sloping row are forward differences.

If we again add one more term to make our interpolation correspond to a fourth-degree polynomial, we have

$$f(0.73) = 0.893 + \frac{(1.65)(0.65)(-0.35)(-1.35)}{24}(0.052) = 0.894.$$

In this instance we do improve the estimate, coming closer to the true value of 0.895. Why did this fourth-degree polynomial not match the fourth-degree one of Section 3.3? In the present case, the domain is from $x = 0.2$ to $x = 1.0$, as is found by going back diagonally from the last difference, 0.052, in contrast to the domain of $x = 0.4$ to $x = 1.2$. The two fourth-degree polynomials are not identical.

There is much nonsense in many books about the application of the Newton–Gregory forward polynomial only at the beginning of the table, and the backward formula only at the end. As our examples clearly show, we may use either formula anywhere in the table by suitably subscripting the x's. Furthermore, identically the same results are given by *any* interpolating polynomial that ends on the *same* difference entry. The reason why books tell us to use the forward polynomial at the beginning of a table is so we can increase the degree merely by adding a term. The backward polynomial can similarly be increased in degree very easily when applied to points near the end of the table.

There is a rich variety of interpolation formulas beyond the two we have so far discussed. They differ in the paths taken through the difference table. For example, the Gauss forward goes through the table in a zigzag path, the first step being a forward one. Stirling's and Bessel's formulas proceed horizontally, using averages of differences, one starting with f_0 and the other halfway between f_0 and f_1. In the next section we present a handy device that permits us to write any of these, knowing only the path that is taken through the difference table. The utility and convenience of that device, called a *lozenge diagram*, is apparent when we look at the formulas for the various forms of interpolating polynomials. We repeat the Newton–Gregory forward and backward polynomials for convenience.

Formulas for interpolation of equispaced data

Newton–Gregory Forward (fits at x_0 to x_n):

$$P_n(x) = f_0 + \binom{s}{1}\Delta f_0 + \binom{s}{2}\Delta^2 f_0 + \binom{s}{3}\Delta^3 f_0 + \binom{s}{4}\Delta^4 f_0 + \ldots + \binom{s}{n}\Delta^n f_0.$$

Newton–Gregory Backward (fits at x_{-n} to x_0):

$$P_n(x) = f_0 + \binom{s}{1}\Delta f_{-1} + \binom{s+1}{2}\Delta^2 f_{-2} + \binom{s+2}{3}\Delta^3 f_{-3} + \binom{s+3}{4}\Delta^4 f_{-4}$$

$$+ \ldots + \binom{s+n-1}{n}\Delta^n f_{-n}.$$

Gauss Forward (fits at $x_{-[n/2]}$ to $x_{[(n+1)/2]}$):

$$P_n(x) = f_0 + \binom{s}{1}\Delta f_0 + \binom{s}{2}\Delta^2 f_{-1} + \binom{s+1}{3}\Delta^3 f_{-1} + \binom{s+1}{4}\Delta^4 f_{-2}$$

$$+ \ldots + \binom{s + [\![(n-1)/2]\!]}{n}\Delta^n f_{-[n/2]}.$$

Gauss Backward (fits at $x_{-[(n+1)/2]}$ to $x_{[n/2]}$):

$$P_n(x) = f_0 + \binom{s}{1}\Delta f_{-1} + \binom{s+1}{2}\Delta^2 f_{-1} + \binom{s+1}{3}\Delta^3 f_{-2} + \binom{s+2}{4}\Delta^4 f_{-2}$$

$$+ \ldots + \binom{s + [\![n/2]\!]}{n}\Delta^n f_{-[(n+1)/2]}.$$

Stirling (fits at $x_{-[n/2]}$ to $x_{[n/2]}$ if n is even):

$$P_n(x) = f_0 + \binom{s}{1}\frac{\Delta f_{-1} + \Delta f_0}{2} + \frac{\binom{s+1}{2} + \binom{s}{2}}{2}\Delta^2 f_{-1} + \binom{s+1}{3}\frac{\Delta^3 f_{-2} + \Delta^3 f_{-1}}{2}$$

$$+ \frac{\binom{s+2}{4} + \binom{s+1}{4}}{2}\Delta^4 f_{-2} + \ldots$$

$$+ \begin{cases} \dfrac{\binom{s + [\![n/2]\!]}{n} + \binom{s + [\![(n-1)/2]\!]}{n}}{2}\Delta^n f_{-[n/2]} & \text{if } n \text{ is even;} \\[3ex] \binom{s + [\![(n-1)/2]\!]}{n}\dfrac{\Delta^n f_{-[(n+1)/2]} + \Delta^n f_{-[n/2]}}{2} & \text{if } n \text{ is odd.} \end{cases}$$

Bessel (fits at $x_{-[n/2]}$ to $x_{[(n+1)/2]}$ if n is odd):

$$P_n(x) = \frac{f_0 + f_1}{2} + \frac{\binom{s-1}{1} + \binom{s}{1}}{2}\Delta f_0 + \binom{s}{2}\frac{\Delta^2 f_0 + \Delta^2 f_{-1}}{2} + \frac{\binom{s}{3} + \binom{s+1}{3}}{2}\Delta^3 f_{-1}$$

$$+ \binom{s+1}{4}\frac{\Delta^4 f_{-1} + \Delta^4 f_{-2}}{2} + \ldots$$

$$+ \begin{cases} \binom{s + [\![(n-2)/2]\!]}{n}\dfrac{\Delta^n f_{-[\![(n-2)/2]\!]} + \Delta^n f_{-[n/2]}}{2} & \text{if } n \text{ is even;} \\[3ex] \dfrac{\binom{s + [\![(n-2)]\!]}{n} + \binom{s + [\![n/2]\!]}{n}}{2}\Delta^n f_{-[n/2]} & \text{if } n \text{ is odd.} \end{cases}$$

In the above,

$$\binom{s + j}{k} = \frac{(s + j)(s + j - 1)(s + j - 2) \cdots (s + j - k + 1)}{k!},$$

$\left[\!\left[\dfrac{m}{2}\right]\!\right]$ is greatest integer in $\dfrac{m}{2}$.

To use the formulas, choose x_0, compute $s = (x - x_0)/h$, and substitute values.

3.5 LOZENGE DIAGRAM FOR INTERPOLATION

Since all polynomials that fit a table at the same points are identical, it follows that there is a relationship between the coefficients to adjust for the different f-differences that are employed. Figure 3.1 is lozenge diagram, so called because of the diamond pattern. Any of the various forms of interpolating polynomial can be written from this. One may even invent his own polynomial, different in form from any of the named ones.

Examination of the lozenge diagram, Fig. 3.1, shows it to be a difference table in symbolic form (similar to Table 3.2), with factorial ratios interspersed. These factorial ratios are used as coefficients of the differences of f in forming the interpolating polynomial, and are selected by tracing a pattern through the difference table.

Certain rules must be followed:

1. One begins at the column of f's usually at f_0, but not necessarily so. The point of beginning dictates the subscripting, however, and hence the value of x_0 must be consistent with the subscripts of the f's.

2. One proceeds from left to right, either diagonally upward or downward to the next difference column, or alternatively horizontally.* A term is added for every column crossed.

3. The term that we add is the entry in the lozenge which is traversed, multiplied by the entry above if the last step was diagonally downward, by the entry below if the last step was diagonally upward, or by the average of entries above and below if it was horizontal. The coefficient of the f-term is always unity, as the interspersed figures indicate.

Our familiar Newton–Gregory forward polynomial results from a diagonally downward path. Beginning at f_0, and using the rules given, we have, identically with Eq. (3.3):

$$P_n(x_s) = f_0 + \binom{s}{1}\Delta f_0 + \binom{s}{2}\Delta^2 f_0 + \binom{s}{3}\Delta^3 f_0 + \cdots$$

---

* Kunz (1957) discusses the lozenge in detail, giving the rule if the path progresses from right to left (a term is subtracted), but such variations seem only of academic importance.

f_{-4} $\Delta^2 f_{-5}$ $\Delta^4 f_{-6}$ $\Delta^6 f_{-7}$

1   Δf_{-4} $\binom{s+4}{2}$ $\Delta^3 f_{-5}$ $\binom{s+5}{4}$ $\Delta^5 f_{-6}$ $\binom{s+6}{6}$

f_{-3} $\binom{s+3}{1}$ $\Delta^2 f_{-4}$ $\binom{s+4}{3}$ $\Delta^4 f_{-5}$ $\binom{s+5}{5}$ $\Delta^6 f_{-6}$

1   Δf_{-3} $\binom{s+3}{2}$ $\Delta^3 f_{-4}$ $\binom{s+4}{4}$ $\Delta^5 f_{-5}$ $\binom{s+5}{6}$

f_{-2} $\binom{s+2}{1}$ $\Delta^2 f_{-3}$ $\binom{s+3}{3}$ $\Delta^4 f_{-4}$ $\binom{s+4}{5}$ $\Delta^6 f_{-5}$

1   Δf_{-2} $\binom{s+2}{2}$ $\Delta^3 f_{-3}$ $\binom{s+3}{4}$ $\Delta^5 f_{-4}$ $\binom{s+4}{6}$

f_{-1} $\binom{s+1}{1}$ $\Delta^2 f_{-2}$ $\binom{s+2}{3}$ $\Delta^4 f_{-3}$ $\binom{s+3}{5}$ $\Delta^6 f_{-4}$

1   Δf_{-1} $\binom{s+1}{2}$ $\Delta^3 f_{-2}$ $\binom{s+2}{4}$ $\Delta^5 f_{-3}$ $\binom{s+3}{6}$

f_0 $\binom{s}{1}$ $\Delta^2 f_{-1}$ $\binom{s+1}{3}$ $\Delta^4 f_{-2}$ $\binom{s+2}{5}$ $\Delta^6 f_{-3}$

1   Δf_0 $\binom{s}{2}$ $\Delta^3 f_{-1}$ $\binom{s+1}{4}$ $\Delta^5 f_{-2}$ $\binom{s+2}{6}$

f_1 $\binom{s-1}{1}$ $\Delta^2 f_0$ $\binom{s}{3}$ $\Delta^4 f_{-1}$ $\binom{s+1}{5}$ $\Delta^6 f_{-2}$

1   Δf_1 $\binom{s-1}{2}$ $\Delta^3 f_0$ $\binom{s}{4}$ $\Delta^5 f_{-1}$ $\binom{s+1}{6}$

f_2 $\binom{s-2}{1}$ $\Delta^2 f_1$ $\binom{s-1}{3}$ $\Delta^4 f_0$ $\binom{s}{5}$ $\Delta^6 f_{-1}$

1   Δf_2 $\binom{s-2}{2}$ $\Delta^3 f_1$ $\binom{s-1}{4}$ $\Delta^5 f_0$ $\binom{s}{6}$

f_3 $\binom{s-3}{1}$ $\Delta^2 f_2$ $\binom{s-2}{3}$ $\Delta^4 f_1$ $\binom{s-1}{5}$ $\Delta^6 f_0$

1   Δf_3 $\binom{s-3}{2}$ $\Delta^3 f_2$ $\binom{s-2}{4}$ $\Delta^5 f_1$ $\binom{s-1}{6}$

f_4 $\binom{s-4}{1}$ $\Delta^2 f_3$ $\binom{s-3}{3}$ $\Delta^4 f_2$ $\binom{s-2}{5}$ $\Delta^6 f_1$

1   Δf_4 $\binom{s-4}{2}$ $\Delta^3 f_3$ $\binom{s-3}{4}$ $\Delta^5 f_2$ $\binom{s-2}{6}$

f_5 $\Delta^2 f_4$ $\Delta^4 f_3$ $\Delta^6 f_2$

Coefficients for each term depend on the path:

⟋ Use term
( ) below

( ) Use term
⟍ above

½( ) Use average
⟶ of above and
½( ) below

Paths for certain polynomials:
  Newton–Gregory forward—diagonally downward
  Newton–Gregory backward—diagonally upward
  Gauss forward—zig-zag, first step downward
  Gauss backward—zig-zag, first step upward
  Stirling—horizontally, beginning at y_0
  Bessel—horizontally, beginning between y_0 and y_1

Figure 3.1 Lozenge diagram for interpolating polynomials.

The Newton–Gregory backward polynomial results from a diagonally upward path. If we begin at f_0, we have:

$$P_n(x_s) = f_0 + \binom{s}{1}\Delta f_{-1} + \binom{s+1}{2}\Delta^2 f_{-2} + \binom{s+2}{3}\Delta^3 f_{-3} + \cdots$$

The coefficients of all the other polynomials of the last section are also given by the lozenge diagram. For these, we only need to know the proper path.

Gauss Forward   The path is a zigzag one, taking a first step downward, then a step upward, then downward, then upward.

Gauss Backward   This is also a zigzag path, except that the first step is upward, followed by downward and upward steps alternating.

Stirling   The path is a horizontal one; beginning at f_0 we move directly to the right. For this polynomial we use averages of entries above and below to multiply the entry that is crossed by our path.

Bessel   This is also a horizontal path, involving multiplication by averages of entries above and below. It differs from the Stirling in that we begin at a point between two f-values.

We illustrate the use of the lozenge diagram by writing several interpolating polynomials for the values in Table 3.5.

Table 3.5

x	$f(x)$	Δf	$\Delta^2 f$	$\Delta^3 f$	$\Delta^4 f$
0.2	1.06894				
		0.11242			
0.5	1.18136		0.01183		
		0.12425		0.00123	
0.8	1.30561		0.01306		0.00015
		0.13731		0.00138	
1.1	1.44292		0.01444		0.00014
		0.15175		0.00152	
1.4	1.59467		0.01596		
		0.16771			
1.7	1.76238				

Newton–Gregory forward fitting   at $x = 0.5$ through $x = 1.4$ ($x_0 = 0.5$):

$$P_3(x) = 1.18136 + s(0.12425) + \binom{s}{2}(0.01306) + \binom{s}{3}(0.00138).$$

Newton–Gregory backward fitting   at $x = 1.1$ through $x = 0.2$ ($x_0 = 1.1$):

$$P_3(x) = 1.44292 + s(0.13731) + \binom{s+1}{2}(0.01306) + \binom{s+2}{3}(0.00123).$$

We could, however also define $x_0 = 0.2$ and write:

$$P_3(x) = 1.44292 + \binom{s-3}{1}(0.13731) + \binom{s-2}{2}(0.01306) + \binom{s-1}{3}(0.00123).$$

These last two are really identical because the values of s are computed with different values of x_0 so that the coefficients turn out to be the same. For example, at $x = 0.35$, $s = -2.5$ for the first, but $s = 0.5$ for the second.

Gauss forward fitting   at $x = 0.5$ through $x = 1.1$ ($x_0 = 0.8$):

$$P_2(x) = 1.30561 + s(0.13731) + \binom{s}{2}0.01306.$$

Gauss backward fitting   at $x = 0.8$ through $x = 1.7$ ($x_0 = 1.4$):

$$P_3(x) = 1.59467 + s(0.15175) + \binom{s+1}{2}0.01596 + \binom{s+1}{3}0.00152.$$

Stirling fitting   at $x = 0.2$ through $x = 1.4$ ($x_0 = 0.8$):

$$P_4(x) = 1.30561 + \binom{s}{1}\frac{0.12425 + 0.13731}{2} + \frac{\binom{s}{2} + \binom{s+1}{2}}{2}(0.01306)$$

$$+ \binom{s+1}{3}\frac{0.00123 + 0.00138}{2}.$$

Bessel fitting   at $x = 0.8$ through $x = 1.1$ ($x_0 = 0.8$):

$$P_1(x) = \frac{1.30561 + 1.44292}{2} + \frac{\binom{s-1}{1} + \binom{s}{1}}{2}(0.13731).$$

(This is a pretty elaborate way to write a linear polynomial! It reduces to $1.30561 + s(0.13731)$, as given directly by the Newton–Gregory forward formula.)

3.6 ERROR TERMS AND ERROR OF INTERPOLATION

Since the interpolating polynomial is not, in general, identical with the unknown function $f(x)$ even though they match at certain points, predicting the values of the function at nontabulated points is subject to error. We now develop an expression for the error of $P_n(x)$, an nth-degree interpolating polynomial. We write the error function in a form which has the known property that it is zero at the $n + 1$ points, from x_0 through x_n, where $P_n(x)$ and $f(x)$ are the same. We call this function $E(x)$:

$$E(x) = f(x) - P_n(x) = (x - x_0)(x - x_1) \cdots (x - x_n)g(x).$$

The $n + 1$ linear factors give $E(x)$ the zeros we know it must have and $g(x)$ accounts for its behavior at values other than at $x_0, x_1, \cdots, x_n$. Obviously, $f(x) - P_n(x) - E(x) = 0$, so

$$f(x) - P_n(x) - (x - x_0)(x - x_1) \cdots (x - x_n)g(x) = 0. \tag{3.6}$$

In order to determine $g(x)$, we now use the interesting mathematical device of constructing an auxiliary function (the reason for its special form becomes apparent as the development proceeds). We call this auxiliary function $W(t)$, and define it as:

$$W(t) = f(t) - P_n(t) - (t - x_0)(t - x_1) \cdots (t - x_n)g(x).$$

Note in particular that x has *not* been replaced by t in the $g(x)$ portion. (W is really a function of both t and x, but we are only interested in variations of t.) We now examine the zeros of $W(t)$.

Certainly at $t = x_0, x_1, \ldots, x_n$, the W function is zero ($n + 1$ times), but it is also zero if $t = x$ by virtue of Eq. (3.6). There are then a total of $n + 2$ values of t that make $W(t) = 0$. We now impose the necessary requirements on $W(t)$ for the *law of mean value* to hold. $W(t)$ must be continuous and differentiable. If this is so, there is a zero to its derivative, $W'(t)$, between each of the $n + 2$ zeros of $W(t)$, a total of $n + 1$ zeros. If $W''(t)$ exists, and we suppose it does, there will be n zeros of $W''(t)$, and likewise $n - 1$ zeros of $W'''(t)$, etc., until we reach $W^{(n+1)}(t)$, which must have at least one zero in the interval that has x_0, x_n, or x as endpoints. Call this value of $t = \xi$. We then have

$$W^{(n+1)}(\xi) = 0 = \frac{d^{n+1}}{dt^{n+1}}[f(t) - P_n(t) - (t - x_0) \cdots (t - x_n)g(x)]_{t=\xi}$$

$$= f^{(n+1)}(\xi) - 0 - (n + 1)!g(x). \tag{3.7}$$

The righthand side of Eq. (3.7) occurs because of the following arguments. The $(n + 1)$st derivative of $f(t)$, evaluated at $t = \xi$, is obvious. The $(n + 1)$st derivative of $P_n(t)$ is zero because every time any polynomial is differentiated its degree is reduced by one, so that the nth derivative is of degree zero (a constant) and its $(n + 1)$st derivative is zero. We apply the same argument to the $(n + 1)$st degree polynomial in t that occurs in the last term—its $(n + 1)$st derivative is a constant and this constant results from the t^{n+1} term and is $(n + 1)!$. Of course

$g(x)$ is independent of t and goes through the differentiations unchanged. The form of $g(x)$ is now apparent:

$$g(x) = \frac{f^{(n+1)}(\xi)}{(n+1)!}, \qquad \xi \text{ between } (x_0, x_n, x).$$

The conditions on $W(t)$ that are required for this development (continuous and differentiable $n+1$ times) will be met *if* $f(x)$ has these same properties, because $P_n(x)$ is continuous and differentiable. We now have our error term:

$$E(x) = (x - x_0)(x - x_1) \cdots (x - x_n) \frac{f^{(n+1)}(\xi)}{(n+1)!}. \tag{3.8}$$

We wish to modify this, expressing it in terms of $s = (x - x_0)/h$, to make it more compatible with our interpolating polynomials. Remembering that

$$x_1 = x_0 + h, \qquad x_2 = x_0 + 2h, \qquad \ldots,$$

so that

$$(x - x_0) = sh, \qquad (x - x_1) = sh - h = (s - 1)h,$$
$$(x - x_2) = sh - 2h = (s - 2)h, \qquad \ldots,$$

we find that Eq. (3.8) becomes

$$E(x_s) = \frac{(s)(s - 1)(s - 2) \cdots (s - n)}{(n+1)!} h^{n+1} f^{(n+1)}(\xi) = \binom{s}{n+1} h^{n+1} f^{(n+1)}(\xi),$$
$$\tag{3.9}$$
$$\xi \text{ between } (x_0, x_n, x_s).$$

Referring again to Eq. (3.3) or to the lozenge diagram (Fig. 3.1), we observe that the next term after the last one included in an nth-degree Newton–Gregory forward interpolating polynomial is $\binom{s}{n+1} \Delta^{n+1} f_0$. One can get the error term of Eq. (3.9) by substituting $h^{n+1} f^{(n+1)}(\xi)$ for the $(n+1)$st difference. This is true for all interpolating polynomials, not just the Newton–Gregory forward one. This property extends the usefulness of the lozenge diagram to give the error term by this simple change in the next term after the last one included, no matter which interpolating polynomial is used.

Since all polynomials that pass through the same points are identical, this easy rule by which the error term can be written applies to any polynomial developed from the lozenge diagram—just find the next term and change the difference therein to the product of a power of h times the derivative of f of the same order as the difference replaced.

Example The data below is for sin x. Interpolate to estimate sin (0.8), using a quadratic through the first three points; also estimate the error using Eq. (3.9).

x	$f(x)$	Δf	$\Delta^2 f$	$\Delta^3 f$
0.1	0.09983			
		0.37960		
0.5	0.47943		−0.07570	
		0.30390		−0.04797
0.9	0.78333		−0.12367	
		0.18023		−0.02846
1.3	0.96356		−0.15213	
		0.02810		
1.7	0.99166			

We take $x_0 = 0.1$, and $s = (0.8 - 0.1)/0.4 = 1.75$. Then

$$f(0.8) = 0.09983 + 1.75(0.37960) + \frac{1.75(0.75)}{2}(-0.07570)$$

$$= 0.71445.$$

$n = 2$ 'cause why?

$$\text{Error} = \frac{1.75(0.75)(-0.25)}{3 \cdot 2 \cdot 6}(0.4)^3(-\cos \xi); \ \xi = ?$$

why?

We don't know what value to use for ξ. However, ξ lies in the interval bounded by $x_0 = 0.1$ and $x_2 = 0.9$. Because the cosine is monotonic in this interval, we can easily find maximum and minimum values for $\cos \xi$. (They occur at the endpoints.) This should bound the error:

$$\text{Error} \leqslant \frac{1.75(0.75)(-0.25)}{6}(0.4)^3[-\cos(0.1)] = 3.48 \times 10^{-3};$$

$$\text{Error} \geqslant \frac{1.75(0.75)(-0.25)}{66}(0.4)^3[-\cos(0.9)] = 2.18 \times 10^{-3}.$$

The actual error is 2.90×10^{-3}, falling between the two calculated values, as expected. Note carefully that the error estimate we have made is for the truncation error only (the error that arises because we do not use a polynomial of infinite degree.) The round-off error, which is negligible in this example, acts independently.

We now can determine which of several interpolating polynomials, all of the same degree, give the best estimates of $f(x)$. It makes no difference which form we use—any convenient one is satisfactory, but we will choose the domain where it fits the table in such a way as to minimize the error term. Changing the points where $P_n(x)$ and $f(x)$ match will cause two changes in the error term; the coefficient involving s will vary, and the value of $f^{(n+1)}(\xi)$ may vary because the interval in which ξ lies changes. Since the function $f(x)$ is, in general, unknown, we certainly do not know its derivatives, and further, the value of ξ is not known except for the intervals that contain it. We then choose the polynomial for which the coefficient involving s has the smallest value. This occurs if the value of x_s at which the polynomial is to be evaluated (the point we are interpolating) lies nearest the midpoint of the interval from x_0 to x_n. Note that this implies that extrapolation will normally be less accurate than interpolation, in accordance with our intuition.

In the above example where $f(x)$ is known, we reduce the error if we use the central three points to fit the polynomial.

The computations, with $x_0 = 0.5$, $s = 0.75$, give an estimate of $f(0.8) = 0.71895$. The error estimates are -0.7×10^{-3} and -2.19×10^{-3}, which bracket the actual error of -1.59×10^{-3}. We see that properly choosing the domain where the polynomial fits can reduce the error by nearly one half.

Choosing $x_0 = 0.9$, so that the last three points are the domain will be a bad choice, with the greatest error of all.

What if, as is usual, we do not know $f(x)$? In that situation, we cannot find bounds on $f^{(n+1)}(\xi)$. We shall later see (Chapter 4) that the nth derivative and the nth differences are related. Anticipating some results from that chapter, we have

$$f^{(n+1)}(x) \doteq \frac{\Delta^{n+1} f(x)}{h^{n+1}}.$$

In the absence of knowledge of the function, one may use this relation in the error-term computation, provided that the $(n + 1)$st differences do not vary too greatly. This accounts for the rough rule of thumb that the error of interpolation is about the magnitude of the next term beyond that included in the formula, for with this approximation to $h^{n+1} f^{(n+1)}(\xi)$, the error term is this next term of the polynomial.

In the above examples, this rule gives the following results:

For the quadratic that fits at $x_0 = 0.1$ to $x_2 = 0.9$, the next term gives:

$$\text{Error} \doteq \frac{1.75(0.75)(-0.25)}{6} (-0.04797)$$

$$= 2.62 \times 10^{-3}.$$

For the quadratic that fits at $x_0 = 0.5$ to $x_2 = 1.3$, we get:

$$\text{Error} \doteq \frac{0.75(-0.25)(-1.25)}{6} (-0.02846)$$

$$= -1.11 \times 10^{-3}.$$

While these are not the same as the exact errors, they are certainly close enough to give a good working value. We recommend this technique when the derivative needed in Eq. (3.8) or Eq. (3.9) is unknown.

To keep the error of our polynomial interpolation as small as possible, we have seen that we must choose the points to fit our polynomial so that the x-value at which we do the interpolation is as well centered as possible. The Gauss forward and backward polynomials make it simple to keep the x-value centered in the range of fit, even though the degree is increased. Stirling's and Bessel's formulas act similarly.

For even-degree polynomials, Newton and Gauss polynomials are poorer in this respect than are Stirling's. For odd degrees, Bessel's formulas are preferred. This is another reason why the lozenge diagram of Section 3.5 is presented, to make it easy to write polynomials that keep a well-centered x-value as the degree increases.

In addition to keeping the range symmetrical about the value for interpolation, a second most important decision must be made—the degree of the polynomial. The truncation error will decrease with the degree but the round-off error that perturbs the differences in the table will increase, as we have seen. This argues for an intermediate degree as being most accurate.

Another problem occurs with high-degree interpolation in certain cases. A local irregularity in the tabulated function, due either to a local "bump" in the function or to a relatively large experimental error at one point, can cause amplified distortions in the interpolating polynomial at points remote from the point of disturbance. Such amplification increases with the degree; our only recourse is to keep the degree small; otherwise we will have to approximate the function by a different technique. This phenomenon of amplified distortions and alternative techniques for fitting functions to data is discussed in another chapter.

To summarize, then, the error of polynomial interpolation is reduced by making its range as symmetrical as possible about the point of interpolation, and by choosing a higher degree of polynomial, up to the point where round-off or the effect of local irregularities cause off-setting errors. But another factor has the greatest importance of all—the step size, h. With a given set of tabulated data, we may not be able to do much about the step size. However, if we are designing a new set of tables, the step size may be open to our selection; it is then advantageous to make it small. (This makes our tables more bulky and adds proportionally to the computational effort, of course.)

This effect of h is so important that a special notation is often used to focus attention on it. We write

$$\text{Error} = O(h^n) \qquad \text{if } \lim_{h \to 0} (\text{error}) = Kh^n,$$

where K is some constant. The expression $O(h^n)$ is read "order of h to the nth power." For example, the error of a quadratic interpolating polynomial whose range is (x_0, x_2) would be of order h *cubed* because

$$\text{Error} = \frac{(s)(s-1)(s-2)}{6} h^3 f'''(\xi), \quad \text{or} \quad \text{Error} = O(h^3).$$

As h gets small, $f'''(\xi) \to f'''(x_0)$ because ξ is squeezed between x_0 and x_2, and hence approaches a constant value.

3.7 DERIVATION OF FORMULAS BY SYMBOLIC METHODS

In several instances we have presented formulas without proof, although we have demonstrated their validity in specific cases. Symbolic operator methods are a convenient way to establish these relations. We supplement the forward-differencing operator Δ with a backward-differencing operator ∇ and a *stepping* operator E. These are defined by their actions on a function:

$$\Delta f(x_0) = f(x_0 + h) - f(x_0),$$
$$\Delta^2 f(x_0) = \Delta[\Delta f(x_0)] = \Delta f(x_0 + h) - \Delta f(x_0),$$

$$\nabla f(x_0) = f(x_0) - f(x_0 - h)$$
$$\nabla^2 f(x_0) = \nabla[\nabla f(x_0)] = \Delta f(x_0) - \Delta f(x_0 - h), \quad (3.10)$$

$$Ef(x_0) = f(x_0 + h),$$
$$E^2 f(x_0) = E[Ef(x_0)] = Ef(x_0 + h) = f(x_0 + 2h),$$
$$E^n f(x_0) = f(x_0 + nh).$$

These obvious relationships exist:

$$\Delta f(x_0) = Ef(x_0) - f(x_0) = (E - 1)f(x_0),$$
$$\nabla f(x_0) = f(x_0) - E^{-1}f(x_0) = (1 - E^{-1})f(x_0). \quad (3.11)$$

We abstract from Eq. (3.11) the symbolic operator relations:

$$\Delta = E - 1,$$
$$\nabla = 1 - E^{-1}. \quad (3.12)$$

Equations (3.12) are really meaningless, for neither side is defined as they stand. What they signify is that the effect of Δ when operating on a function is the same as the effect of operating with $(E - 1)$, and that ∇ and $(1 - E^{-1})$ have the same effect on a function, though all these quantities are without significance standing alone. We must apply them to a function to interpret them. This is no different, really, from other operational symbols that we use, such as $\sqrt{}$ for square root, and d/dx for differentiation.

Since all the operators represented by Δ, ∇, and E are linear operators (the effect on a linear combination of functions is the same as the linear combination of the operator acting on the functions), the laws of algebra are obeyed in relationships between them. This means we can manipulate the relations of Eq.

(3.12) by algebraic transformations, and then interpret the results by letting them operate on a function. For example, by raising $\Delta = E - 1$ to the nth power, we have:

$$\Delta^n = (E - 1)^n = E^n - nE^{n-1} + \binom{n}{2}E^{n-2} - \binom{n}{3}E^{n-3} + \ldots,$$

so

$$\Delta^n f(x_0) = \left[E^n - nE^{n-1} + \binom{n}{2}E^{n-2} - \ldots\right]f(x_0)$$

$$= f(x_0 + nh) - nf[x_0 + (n-1)h] + \binom{n}{2}f[x_0 + (n-2)h] - \ldots,$$

or

$$\Delta^n f_0 = f_n - nf_{n-1} + \binom{n}{2}f_{n-2} - \ldots. \tag{3.13}$$

Equation (3.13) is a proof of the alternating-sign, binomial-coefficient formula given in Section 3.1.

We can develop interesting relations between Δ and ∇ such as:

$$E\nabla = E(1 - E^{-1}) = E - 1 = \Delta,$$
$$E^n\nabla^n = \nabla^n E^n = \Delta^n,$$
$$\Delta^n f_0 = \nabla^n E^n f_0 = \nabla^n f_n.$$

This illustrates the fact that a given difference entry in a table can be interpreted as either a forward or backward difference of the appropriate f-values. (This was already obvious from inspection of a difference table.)

We can very simply derive the Newton–Gregory forward formula:

$$E = 1 + \Delta, \qquad E^s = (1 + \Delta)^s,$$

$$f_s = E^s f_0 = (1 + \Delta)^s f_0 = \left[1 + s\Delta + \binom{s}{2}\Delta^2 + \binom{s}{3}\Delta^3 + \ldots\right]f_0$$

$$= f_0 + s\Delta f_0 + \binom{s}{2}\Delta^2 f_0 + \binom{s}{3}\Delta^3 f_0 + \ldots.$$

The Newton–Gregory backward formula is similarly easy:

$$E^{-1} = 1 - \nabla, \qquad E^s = (1 - \nabla)^{-s},$$

$$f_s = E^s f_0 = (1 - \nabla)^{-s} f_0 = \left[1 + s\nabla + \binom{s+1}{2}\nabla^2 + \binom{s+2}{3}\nabla^3 + \ldots\right]f_0$$

$$= f_0 + s\nabla f_0 + \binom{s+1}{2}\nabla^2 f_0 + \binom{s+2}{3}\nabla^3 f_0 + \ldots$$

$$= f_0 + s\Delta f_{-1} + \binom{s+1}{2}\Delta^2 f_{-2} + \binom{s+2}{3}\Delta^3 f_{-3} + \ldots. \tag{3.14}$$

In the last relation of Eq (3.14) we have expressed the differences as forward differences to agree with our lozenge-diagram notation, using $\nabla^n = E^{-n}\Delta^n$. We shall use symbolic methods to derive derivative formulas in the next chapter.

3.8   INTERPOLATION WITH NONUNIFORMLY SPACED x-VALUES

When the x-values are not evenly spaced, our methods of writing an interpolating polynomial fail. A parallel development using the concept of divided differences* is possible, but perhaps the simplest approach is the Lagrangian polynomial. Data where the x-values are not equispaced often occur as the result of experimental observations, or when historical data is examined.

Suppose we have a table of data, of x- and $f(x)$-values: r

x	$f(x)$
x_1	f_1
x_2	f_2
x_3	f_3
x_4	f_4

Here we do not assume uniform Δx, nor do we need the x-values arranged in a particular order. The x-values must all be distinct, however. Through these four data pairs we can pass a cubic. The Lagrangian form for this is:

$$P_3(x) = \frac{(x - x_2)(x - x_3)(x - x_4)}{(x_1 - x_2)(x_1 - x_3)(x_1 - x_4)}f_1 + \frac{(x - x_1)(x - x_3)(x - x_4)}{(x_2 - x_1)(x_2 - x_3)(x_2 - x_4)}f_2$$

$$+ \frac{(x - x_1)(x - x_2)(x - x_4)}{(x_3 - x_1)(x_3 - x_2)(x_3 - x_4)}f_3 + \frac{(x - x_1)(x - x_2)(x - x_3)}{(x_4 - x_1)(x_4 - x_2)(x_4 - x_3)}f_4.$$

Note that it is made up of four terms, each of which is a cubic in x; hence the sum is a cubic. The pattern of each term is to form the numerator as a product of linear factors of the form $(x - x_i)$, omitting one x_i in each term, the omitted value

---

* The first-order divided difference relative to x-values x_0 and x_1 is defined as $f[x_1, x_0] = [f(x_1) - f(x_0)]/(x_1 - x_0)$. The higher-order divided differences are defined in terms of the lower-order differences:

$$f[x_2, x_1, x_0] = \frac{f[x_2, x_1] - f[x_1, x_0]}{x_2 - x_0},$$

$$f[x_3, x_2, x_1, x_0] = \frac{f[x_3, x_2, x_1] - f[x_2, x_1, x_0]}{x_3 - x_0}.$$

The Newton interpolating polynomials can be written in terms of these divided differences in the fashion of Section 3.3. When the x-values are evenly spaced, the formulas are identical.

being used to form the denominator by replacing x in each of the numerator factors. In each term, we multiply by the f_i corresponding to the x_i omitted in the numerator factors. The Lagrangian polynomial for other degrees of interpolating polynomials employs this same pattern of forming a sum of polynomials all of the desired degree.

Example Interpolate for $f(2.3)$ from the table:

x	$f(x)$
1.1	10.6
1.7	15.2
3.0	20.3

With only three pairs of data, a quadratic is the highest-degree polynomial possible. It is:

$$P_2(x) = \frac{(x - 1.7)(x - 3.0)}{(1.1 - 1.7)(1.1 - 3.0)}(10.6) + \frac{(x - 1.1)(x - 3.0)}{(1.7 - 1.1)(1.7 - 3.0)}(15.2)$$
$$+ \frac{(x - 1.1)(x - 1.7)}{(3.0 - 1.1)(3.0 - 1.7)}(20.3).$$

At $x = 2.3$, we get $P_2(2.3) = 18.38$.

The arithmetic in this method is tedious, although electronic calculators are convenient for this type of computation. A computer program at the end of this chapter implements the method.

We develop the error term in the same way as in Section 3.6. Write $E(x)$ to exhibit the fact that the error of $P_n(x)$ is zero at the $n + 1$ values of x that are fitted exactly:

$$E(x) = f(x) - P_n(x) = (x - x_1)(x - x_2)\cdots(x - x_{n+1})g(x).$$

The auxiliary function $W(t)$ is

$$W(t) = f(t) - P_n(t) - (t - x_1)(t - x_2)\cdots(t - x_{n+1})g(x).$$

$W(t) = 0$ for $t = x_1, x_2, \ldots, x_{n+1}$, and at $t = x$, for a total of $n + 2$ zeros. Hence $W'(t)$ has n zeros, $W''(t)$ has $n - 1$ zeros, $\ldots$, $W^{(n+1)}(t)$ has one zero. Let ξ be the value of t at which $W^{(n+1)}(t) = 0$. Hence,

$$W^{(n+1)}(\xi) = 0 = f^{(n+1)}(\xi) - 0 - (n + 1)!g(x),$$

$$g(x) = \frac{f^{(n+1)}(\xi)}{(n + 1)!}.$$

The error is then

$$E(x) = (x - x_1)(x - x_2)\cdots(x - x_{n+1})\frac{f^{(n+1)}(\xi)}{(n + 1)!}. \tag{3.15}$$

The interval containing ξ lies between the largest and smallest of the x_i for interpolation. The error can be bracketed between a maximum and a minimum value only if we have information on the $(n + 1)$st derivative of the actual function $f(x)$.

3.9 INVERSE INTERPOLATION

Suppose we have a table of data such as in Table 3.6, and we are required to find the x-value corresponding to a certain value of the function, say at $y = 5.0$. We have two approaches. We can consider the y's to be the independent variable (unevenly spaced) and interpolate for x with a Lagrangian polynomial. Doing so gives $x = 2.312$. This technique is straightforward, but in some instances gives poor results, the reason being that x considered as a function of y may not be well approximated by a polynomial. This may be true even though y itself behaves quite like a polynomial. (Try inverse interpolation among three or four points on the function $y = x^2$, especially for y-values outside the given range.)

Table 3.6

x	y	Δy	$\Delta^2 y$	$\Delta^3 y$	$\Delta^4 y$
1.6	2.3756				
		0.8926			
1.9	3.2682		0.2963		
		1.1889		0.1079	
2.2	4.4571		0.4042		0.0365
		1.5931		0.1444	
2.5	6.0502		0.5486		
		2.1417			
2.8	8.1919				

The second method is to write y as a polynomial in x and then use the methods of Chapter 1. This will generally make it necessary to multiply out the interpolating polynomial so as to express the function in the usual polynomial form. If we use the Gauss forward polynomial, with $x_0 = 2.2$, we have

$$y_s = P_3(x_s) = 4.4571 + s(1.5931) + \frac{(s)(s - 1)}{2}(0.4042)$$

$$+ \frac{(s + 1)(s)(s - 1)}{6}(0.1444)$$

$$+ \frac{(s + 1)(s)(s - 1)(s - 2)}{24}(0.0365)$$

$$= 0.00152s^4 + 0.02103s^3 + 0.2006s^2 + 1.3700s + 4.4571.$$

[We have left the polynomial in terms of s to save work; after determining the value of s corresponding to $y = 5.0$, we get x from $s = (x - 2.2)/0.3$.] The polynomial of Eq. (3.16) could, of course, be determined through the Lagrangian polynomial, but the arithmetic effort is much greater. To complete the problem, we must find the root near zero of a fourth-degree polynomial in s, which is rather tedious.

The equivalent of this second technique for inverse interpolation can be accomplished more readily by the method of successive approximations. Our Gauss forward polynomial is

$$y_s = y_0 + s\Delta y_0 + \frac{s(s - 1)}{2}\Delta^2 y_{-1} + \frac{(s + 1)(s)(s - 1)}{6}\Delta^3 y_{-1}$$

$$+ \frac{(s + 1)(s)(s - 1)(s - 2)}{24}\Delta^4 y_{-2}.$$

We rearrange to solve for s in the second term:

$$s = \frac{1}{\Delta y_0}\left[y_s - y_0 - s(s - 1)\frac{\Delta^2 y_{-1}}{2} - (s + 1)(s)(s - 1)\frac{\Delta^3 y_{-1}}{6} \right.$$

$$\left. - (s + 1)(s)(s - 1)(s - 2)\frac{\Delta^4 y_{-2}}{24} \right]. \tag{3.17}$$

The method of successive approximations finds s by first neglecting all the terms in s on the right, to give

$$s_1 = \frac{1}{6.0502 - 4.4571}(5.0 - 4.4571) = 0.34.$$

The second approximation is obtained using s_1 on the right side of Eq. (3.17), including now one more term:

$$s_2 = \frac{1}{1.5931}[5.0 - 4.4571 - (0.34)(0.34 - 1)(0.2021)] = 0.369.$$

The next approximation uses s_2 on the right and picks up another term:

$$s_3 = \frac{1}{1.5931}[5.0 - 4.4571 - (0.369)(-0.631)(0.2021)$$

$$- (1.369)(0.369)(-0.631)(0.02407)] = 0.375.$$

In the same fashion, $s_4 = 0.3748$, giving

$$x = 2.2 + (0.3)(-0.3748) = 2.31244.$$

(The data in Table 3.6 are for $y = \sinh x$. Substitution in the hyperbolic sine function gives $\sinh(2.31244) = 5.00001$.)

3.10 POLYNOMIAL INTERPOLATION IN TWO DIMENSIONS

When a function u is a polynomial function* of two variables x and y, say of degree three in x and of degree two in y, we would have

$$u = f(x, y) = a_0 + a_1x + a_2y + a_3x^2 + a_4xy + a_5y^2 + a_6x^3$$
$$+ a_7x^2y + a_8xy^2 + a_9x^3y + a_{10}x^2y^2 + a_{11}x^3y^2. \quad (3.18)$$

The functional relation is seen to involve many terms. If we are concerned with four independent variables (three space dimensions plus time, say), even low-degree polynomials would be quite intractable. Except for special purposes, such as when we need an explicit representation, perhaps to permit ready differentiation at an arbitrary point, we can avoid such complications by handling each variable separately. We shall treat only this case.

Note the immediate simplification of Eq. (3.18) if we let y take on a constant value, say $y = y_1$. Combining the y factors with the coefficients, we get

$$u|_{y=c} = b_0 + b_1x + b_2x^2 + b_3x^3.$$

This will be our attack in interpolating at the point (a, b) in a table of two variables—hold one variable constant, say $y = y_1$, and the table becomes a single-variable problem. The above methods then apply to give $f(a, y_1)$. If we repeat this at various values of y, $y = y_2, y_3, \ldots, y_n$, we will get a table with x constant at the value $x = a$ and with y varying. We then interpolate at $y = b$.

Example Estimate $f(1.6, 0.33)$ from the values in Table 3.7. Use quadratic interpolation in the x-direction and cubic interpolation for y. We select one of the variables to hold constant, say x. [This choice is arbitrary since we would get the same result (except for differences due to round-off) if we had chosen to hold y constant.] We decide to interpolate for y within the three rows of the table at $x = 1.0$, 1.5, and 2.0 since the desired value at $x = 1.6$ is most nearly centered within this set. We choose y-values of 0.2, 0.3, 0.4, and 0.5 so that $y = 0.33$ is centralized. The shading in Table 3.7 shows the region of fit for our polynomials.

Table 3.7 Tabulation of a function of two variables, $u = f(x, y)$.

y \ x	0.1	0.2	0.3	0.4	0.5	0.6
0.5	0.165	0.428	0.687	0.942	1.190	1.431
1.0	0.271	0.640	1.003	1.359	1.703	2.035
1.5	0.447	0.990	1.524	2.045	2.549	3.031
2.0	0.738	1.568	2.384	3.177	3.943	4.672
2.5	1.216	2.520	3.800	5.044	6.241	7.379
3.0	2.005	4.090	6.136	8.122	10.030	11.841
3.5	3.306	6.679	9.986	13.196	16.277	19.198

---

* We approximate a nonpolynomial function by a polynomial that agrees with the function, just as we have done with a function of one variable.

We may either use Lagrangian interpolation or derive the interpolated values using the lozenge diagram, Fig. 3.1 after making difference tables. Let us use the later method:

	y	u	Δu	$\Delta^2 u$	$\Delta^3 u$
	0.2	0.640			
			0.363		
	0.3	1.003		−0.007	
$x = 1.0$			0.356		−0.005
	0.4	1.359		−0.012	
			0.344		
	0.5	1.703			
	0.2	0.990			
			0.534		
	0.3	1.524		−0.013	
$x = 1.5$			0.521		−0.004
	0.4	2.045		−0.017	
			0.504		
	0.5	2.549			
	0.2	1.568			
			0.816		
	0.3	2.384		−0.023	
$x = 2.0$			0.793		−0.004
	0.4	3.177		−0.027	
			0.766		
	0.5	3.943			

We need the subtables from $y = 0.2$ to $y = 0.5$ since, for a cubic interpolation, four points are required. Using any convenient path through the tables with coefficients given by the lozenge, we arrive at the results:

	x	u	Δu	$\Delta^2 u$
	1.0	1.1108		
			0.5710	
$y = 0.33$	1.5	1.6818		0.3717
			0.9427	
	2.0	2.6245		

In the last tabulation we carry one extra decimal to guard against round-off errors. Interpolating again, we get $u = 1.8406$, which we report as $u = 1.841$.

The function tabulated in Table 3.7 is $f(x, y) = e^x \sin y + y - 0.1$, so the true value is $f(1.6, 0.33) = 1.8350$. Our error of -0.006 occurs because quadratic interpolation for x is inadequate in view of the large second difference. In retrospect, it would have been better to use quadratic interpolation for y, since the third differences of the y-subtables are small, and let x take on a third-degree relationship.

It is instructive to observe which of the values in Table 3.7 entered into our computation. The shaded rectangle covers these values. This is the "region of fit" for the interpolating polynomial that we have used. The principle of choosing values so that the point at which the interpolating polynomial is used is centered in the region of fit obviously applies here in exact analogy to the one-way table situation. It also applies to tables of three and four variables in the same way. Of course, the labor of interpolating in such multidimensional cases soon becomes burdensome.

A rectangular region of fit is not the only possibility. We may change the degree of interpolation as we subtabulate the different rows or columns. Intuitively, it would seem best to use higher-degree polynomials for the rows near the interpolating point, decreasing the degree as we get farther away. The coefficient of the error term, when this is done, will be found to be minimized thereby, though for multidimensional interpolating polynomials the error term is quite complex. The region of fit will be diamond-shaped when such tapered degree functions are used.

We may adapt the Lagrangian form of interpolating polynomial to the multidimensional case also. It is perhaps easiest to employ a process similar to the above example. Holding one variable constant, we write a series of Lagrangian polynomials for interpolation at the given value of the other variable, and then combine these values in a final Lagrange form. The net result is a Lagrangian polynomial in which the function factors are replaced by Lagrangian polynomials. The resulting expression for the above example would be:

$$\frac{(y - 0.3)(y - 0.4)(y - 0.5)}{(0.2 - 0.3)(0.2 - 0.4)(0.2 - 0.5)}$$

$$\times \left[\frac{(x - 1.5)(x - 2.0)}{(1.0 - 1.5)(1.0 - 2.0)}(0.640) + \frac{(x - 1.0)(s - 2.0)}{(1.5 - 1.0)(1.5 - 2.0)}(0.990) + \frac{(x. - 1.0)(x - 1.5)}{(2.0 - 1.0)(2.0 - 1.5)}(1.568) \right]$$

$$+ \frac{(y - 0.2)(y - 0.4)(y - 0.5)}{(0.3 - 0.2)(0.3 - 0.4)(0.3 - 0.5)}$$

$$\times \left[\frac{(x - 1.5)(x - 2.0)}{(1.0 - 1.5)(1.0 - 2.0)}(1.003) + \frac{(x - 1.0)(x - 2.0)}{(1.5 - 1.0)(1.5 - 2.0)}(1.534) + \frac{(x - 1.0)(x - 1.5)}{(2.0 - 1.0)(2.0 - 1.5)}(2.384) \right]$$

$$+ \frac{(y - 0.2)(y - 0.3)(y - 0.5)}{(0.4 - 0.2)(0.4 - 0.3)(0.4 - 0.5)}$$

$$\times \left[\frac{(x - 1.5)(x - 2.0)}{(1.0 - 1.5)(1.0 - 2.0)}(1.359) + \frac{(x - 1.0)(x - 2.0)}{(1.5 - 1.0)(1.5 - 2.0)}(2.045) + \frac{(x - 1.0)(x - 1.5)}{(2.0 - 1.0)(2.0 - 1.5)}(3.177) \right]$$

$$+ \frac{(y - 0.2)(y - 0.3)(y - 0.4)}{(0.5 - 0.2)(0.5 - 0.3)(0.5 - 0.4)}$$

$$\times \left[\frac{(x - 1.5)(x - 2.0)}{(1.0 - 1.5)(1.0 - 2.0)}(1.703) + \frac{(x - 1.0)(x - 2.0)}{(1.5 - 1.0)(1.5 - 2.0)}(2.549) + \frac{(x - 1.0)(x - 1.5)}{(2.0 - 1.0)(2.0 - 1.5)}(3.943) \right].$$

The equation is easy to write, but its evaluation by hand is laborious. If one is writing a computer program for interpolation in such multivariate situations, the Lagrangian form is recommended. There is a special advantage in that equal spacing in the table is not required. The Lagrangian form is also perhaps the most straightforward way to write out the polynomial as an explicit function.

The cubic spline (described in Chapter 10) could also be used to interpolate in multivariate tables. Here again holding all but one variable fixed so that a series of one-way table problems is solved is perhaps the best approach. The computational effort to employ spline interpolation would be very severe.

3.11 INTERPOLATION IN A COMPUTER PROGRAM

The first of the following subroutines (Program 1, Fig. 3.2) interpolates using the Lagrangian form of interpolating polynomial. The coefficients of each term are built up as the ratio of $(x - x_j)/(x_i - x_j)$ with j varying from 1 to N but omitting the factor when $i = j$. Up to a ninth-degree polynomial can be employed. The x, y-pairs can be in any order, but the number must be just one more than the degree of the interpolating polynomial. In other words, a polynomial that fits exactly through all N points is used.

Program 2 (Fig. 3.3) employs the Newton–Gregory interpolating polynomial, again of degree up to nine, to interpolate in a function tabulated with uniform x-spacing. The differences are computed as they are needed, and a single array is employed both to compute and to store them, for maximum economy of memory use. The value of x_0 is automatically chosen to put the point of interpolation as near the center of the domain as possible.

```
        SUBROUTINE LAGINT (X, Y, N, XINT, YOUT)
C   THIS SUBROUTINE PERFORMS LAGRANGIAN INTERPOLATION WITHIN A SET
C   OF (X, Y) PAIRS TO GIVE THE Y VALUE CORRESPONDING TO XINT.  THE
C   DEGREE OF THE INTERPOLATING POLYNOMIAL IS ONE LESS THAN THE NUMBER OF
C   POINTS SUPPLIED
C   PARAMETERS ARE -
C      X      ARRAY OF VALUES OF THE INDEPENDENT VARIABLE
C      Y      ARRAY OF FUNCTION VALUES CORRESPONDING TO X
C      N      NUMBER OF POINTS
C      XINT   THE X-VALUE FOR WHICH ESTIMATE OF Y IS DESIRED
C      YOUT   THE Y-VALUE RETURNED TO CALLER
        DIMENSION X(N), Y(N)
        YOUT = 0.0
        DO 20 I = 1,N
          TERM = Y(I)
          DO 10 J = 1,N
            IF (I .EQ. J)   GO TO 10
            TERM = TERM*(XINT - X(J))/(X(I) - X(J))
10        CONTINUE
          YOUT = YOUT + TERM
20      CONTINUE
        RETURN
        END
```

Figure 3.2 Program 1.

```
      SUBROUTINE INTERP (X1, XN, H, N, Y, X, M, YOUT)
C  SUBROUTINE TO INTERPOLATE IN A TABLE OF UNIFORMLY SPACED VALUES.
C  PARAMETERS ARE -
C     X1, XN      BEGINNING AND ENDING X VALUES
C     H           DELTA X - THE UNIFORM SPACING
C     N           NUMBER OF ENTRIES IN THE TABLE
C     Y           ARRAY OF FUNCTION VALUES
C     X           VALUE AT WHICH Y IS TO BE INTERPOLATED
C     M           DEGREE OF INTERPOLATING POLYNOMIAL.  THE SUBROTUINE
C                    WILL HANDLE UP TO 10TH DEGREE, BUT USUSALLY THE
C                    DEGREE WILL BE LESS THAN 10 TO AVOID ROUND-OFF ERRORS.
C     YOUT        THE INTERPOLATED Y VALUE RETURNED TO THE CALLER.
C  AN ARRAY D IS USED IN THE SUBROUTINE TO HOLDS DELTA Y'S.
      DIMENSION Y(N), D(10)
C  FIRST FIND PROPER SUBSCRIPT FOR YO SO THAT X IS CENTERED IN
C  THE DOMAIN AS WELL AS POSSIBLE.  THIS SUBSCRIPT VALUE IS CALLED J.
      FM = M + 1
      J = (X - X1)/H - FM/2. + 2.
      IF (X .LE. X1 + FM/2.*H)   J = 1
      IF (X .GE. XN - FM/2.*H)   J = N - M
      FJ = J
      XO = X1 + (FJ - 1.)*H
      YO = Y(J)
C  COMPUTE THE DIFFERENCES THAT ARE NEEDED.
      DO 10 I = 1,M
         D(I) = Y(J+1) - Y(J)
         J = J + 1
10    CONTINUE
      IF (M .EQ. 1)  GO TO 25
      DO 20 J = 2,M
      DO 15 I = J,M
         K = M - I + J
         D(K) = D(K) - D(K-1)
15       CONTINUE
20       CONTINUE
C  COMPUTE S VALUE
25    S = (X - XO)/H
C  COMPUTE INTERPOLATED Y VALUE
      YOUT = YO
      FNUM = S
      DEN = 1.
      DO 30 I = 1,M
       FI = I
       YOUT = YOUT + FNUM/DEN*D(I)
       FNUM = FNUM*(S - FI)
       DEN = DEN*(FI + 1.)
30     CONTINUE
      RETURN
      END
```

Figure 3.3 Program 2.

EXERCISES

Section 3.1

1. Complete a difference table for the following data:

x	1.20	1.25	1.30	1.35	1.40	1.45	1.50
$f(x)$	0.1823	0.2231	0.2624	0.3001	0.3365	0.3716	0.4055

▶ 2. In Exercise 1, what degree of polynomial is required to fit exactly to all seven data pairs? What lesser-degree polynomial will nearly fit the data? Justify your answer.

3. In Exercise 1, assume that the third differences are constant and are equal to 0.0001 (which is about their average value). Working backward from this, extrapolate the table to $x = 1.15$ and to $x = 1.55$. The data in Problem 1 are for $\ell n\ x$. Compare your extrapolated values to a table of logarithms.

▶4. Look up values for sinh x in a handbook and tabulate for $x = 1.0(0.1)2.0$. (This notation means for $x = 1.0$ to $x = 2.0$ at intervals of 0.1.) Round the values to four decimals and make a difference table, computing differences up to and including the fourth. Assuming that the fourth differences are constant at the average value of those in your table, extrapolate to get values of sinh x at $x = 0.9$ and $x = 2.1$. How do these compare to the handbook values? The value at $x = 1.1$ is 1.33565, to five decimals. A common rule for rounding when the last figure is a five is to round to the nearest even digit (which here would give 1.3356) while another rule always rounds a five upwards. What difference does this make in this problem?

5. Form a difference table for $f(x) = x^3 - 3x^2 + 2x + 1$ for $x = -1(0.2)1$. (Recall that nested multiplication is more efficient, especially on a desk calculator.) Verify from your table the validity of Eq. (3.1).

6. The statement is made in Section 3.1 that one can recompute an entire difference table given only one entry from each column. Verify by completing this table:

x	y	Δy	$\Delta^2 y$	$\Delta^3 y$	$\Delta^4 y$	$\Delta^5 y$
0	___					
		___				
5	___		0.0013			
		0.0888				
10	___		___	___	0.0002	
				___		−0.0002
15	___		___		___	
		___		0.0017		
20	___		___			
		___				
25	0.4663					

Section 3.2

7. Form a difference table through fourth differences:

x	$f(x)$	x	$f(x)$
1	0.7	5	7.1
2	0.8	6	13.2
3	1.5	7	22.3
4	3.4	8	35.0

Suppose, in transcribing the above, the entry for $x = 4$ was incorrectly written as 4.3. What would the difference table then be?

Section 3.3

▶ 8. Use Newton–Gregory forward-interpolating polynomials of degree three to estimate $f(0.158)$ and $f(0.636)$, given the following table. In the first polynomial, choose $x_0 = 0.125$. In the second, let $x_0 = 0.375$.

x	$f(x)$	Δf	$\Delta^2 f$	$\Delta^3 f$	$\Delta^4 f$
0.125	0.79168				
		−0.01834			
0.250	0.77334		−0.01129		
		−0.02963		0.00134	
0.375	0.74371		−0.00995		0.00038
		−0.03958		0.00172	
0.500	0.70413		−0.00823		0.00028
		−0.04781		0.00200	
0.625	0.65632		−0.00623		
		−0.05404			
0.750	0.60228				

9. Add one term to the work of Exercise 8 to estimate $f(0.158)$ by a fourth-degree polynomial.

10. Use the data below to find the value of y at $x = 0.58$, using a cubic polynomial that fits the table at x-values of 0.3, 0.5, 0.7, and 0.9.

x	y	Δy	$\Delta^2 y$	$\Delta^3 y$
0.1	0.003			
		0.064		
0.3	0.067		0.017	
		0.081		0.002
0.5	0.148		0.019	
		0.100		0.003
0.7	0.248		0.022	
		0.122		0.004
0.9	0.370		0.026	
		0.148		0.005
1.1	0.518		0.031	
		0.179		
1.3	0.697			

▶ 11. Convert the cubic polynomial of Exercise 10 to standard form:

$$a_1 x^3 + a_2 x^2 + a_3 x + a_4.$$

12. What is the minimum-degree polynomial that will exactly fit all seven data pairs of Exercise 10? (Answer is *not* sixth-degree.)

Section 3.4

13. Using the data of Exercise 8, write a Newton–Gregory forward polynomial that fits the table at x-values of 0.500, 0.625, and 0.750. Then write the Newton–Gregory backward polynomial that fits the same three points. Demonstrate that these are two different forms of the same polynomial by rewriting both in the form $a_1x^2 + a_2x + a_3$.

14. Repeat Exercise 8, using Newton–Gregory backward polynomials, but choose $x_0 = 0.500$ in the first and $x_0 = 0.750$ in the second polynomial. Are the same results obtained?

▶15. Using the data of Exercise 8, estimate $f(0.385)$ by the following: Use quadratic polynomials in each case. (a) Newton–Gregory forward, $x_0 = 0.250$; (b) Newton–Gregory backward, $x_0 = 0.500$; (c) Gauss backward, $x_0 = 0.375$; (d) Stirling, $x_0 = 0.375$. Why are all these results identical?

Section 3.5

16. For the data of Exercise 10, write cubic interpolation polynomials that terminate on the third difference whose value is 0.004. Use the lozenge diagram to write polynomials of the following forms: (a) Newton–Gregory forward; (b) Newton–Gregory backward; (c) Gauss forward; (d) Gauss backward; (e) Bessel.

17. Can a Stirling polynomial be written which is the same as the other polynomials of Exercise 16? What about the set of quadratics that terminate on the second difference whose value is 0.022? Which polynomial cannot be included in this set?

▶18. Use each of the polynomials of Exercise 16 to interpolate at $x = 0.92$. For each polynomial, compare in a table the values of y_0, s, $y(0.92)$ that are involved.

19. Interpolate the data of Exercise 16 for $y(0.92)$ using a third-degree Stirling polynomial for which y_0 is 0.370. Should this value be identical to the results of Exercise 18?

Section 3.6

20. Write the error terms of each polynomial of Exercise 18. Evaluate the coefficients (the s polynomials).

21. Write the error term of the polynomial of Exercise 19. Evaluate the coefficient and compare to the result of Exercise 20.

22. Exercise 21 demonstrates that the coefficient of the error term is smallest when the point at which the interpolation polynomial is used is centered in the range of fit of the polynomial. On what third difference should the cubics of Exercise 18 terminate to give the best polynomial for interpolation at $x = 0.92$? How much is the coefficient of the error term reduced over that in Exercise 20?

▶23. What is the error term for linear interpolation? Show that the coefficient is of maximum magnitude at the midpoint between x_0 and x_1.

24. A table of $\tan x$ with the argument given in radians has $\Delta x = 0.01$. Near $x = \pi/2$, the tangent function increases very rapidly and hence linear interpolation is not very accurate. What degree of interpolating polynomial is needed to interpolate at $x = 1.506$ to three-decimal accuracy?

Section 3.7

25. An operator α is called a *linear* operator if $\alpha(f + g) = \alpha f + \alpha g$ and $\alpha(cf) = c\alpha f$, where $c = $ constant. Show that E, Δ, and ∇ are linear operators based on their definitions by Eq. (3.10).

26. Two operators, α and β, are said to commute if the result of operating on a function is not changed by changing the order of the operations, that is, $\alpha(\beta f) = \beta(\alpha f)$. Show that E, Δ, and ∇ commute.

27. Define D as the differentiation operator. Show that D commutes with E, Δ, and ∇.

▶28. Show that:

   a) $\Delta[f(x)g(x)] = f(x)\Delta g(x) + g(x + h)\Delta f(x)$
   b) $\Delta^n x^n = \nabla^n x^n = \nabla^n E^m x^n = n!$ when $h = 1$.

29. Express $\Delta^3 E^{-2} \nabla^2 y_4$ in terms of y_i entries of the table.

▶30. If $\Delta^n y_s = \nabla^n y_r$, express r in terms of s.

31. $\delta = E^{1/2} - E^{-1/2}$ defines the central difference operator. Show that

$$\delta^n y_i = \Delta^n E^{-n/2} y_i = \nabla^n E^{n/2} y_i.$$

32. Based on the relations of Exercise 31, where are the even-order central differences located in a difference table? Why is it appropriate to call Stirling's formula a "central difference formula"?

Section 3.8

33. Write an interpolating polynomial that passes through each point:

x	0.5	1.2	3.1
y	-3.2	1.6	-1.8

   Plot the points and sketch the parabola that passes through them.

▶34. Given the four points $(1, 0)$, $(-2, 15)$, $(-1, 0)$, $(2, 9)$, write the Lagrangian form of the cubic that passes through them. Multiply out each term to express in standard form, $ax^3 + bx^2 + cx + d$.

35. Given that $\ln 2 = 0.69315$ and $\ln 5 = 1.60944$, compute the natural logarithms of each integer from 1 to 10. Compare to a five-place table.

▶36. If $e^{0.2}$ is estimated by interpolation among the values of $e^0 = 1$, $e^{0.1} = 1.1052$, and $e^{0.3} = 1.3499$, find the maximum and minimum estimates of the error. Compare with the actual error.

37. Repeat Exercise 36, except for extrapolation to get $e^{0.4}$.

38. If the x-values are evenly spaced, the Lagrangian form of interpolating polynomial must be identical to the various interpolating polynomials written with the aid of the lozenge diagram (Fig. 3.1). Show that the error terms are also identical by comparing Eqs. (3.9) and (3.15) for this case.

Section 3.9

▶ 39. For the function $y = x^2$, obviously the points $(1, 1)$, $(2, 4)$, $(3, 9)$ are on its graph. Considering x as a function of y, we have $x = \sqrt{y}$, and for $y = 25$, $x = 5$. But one may also compute x corresponding to $y = 25$ by inverse interpolation from the first three points. Do this by Lagrangian interpolation (extrapolating, of course). What error estimates are available?

40. Repeat Exercise 39, but determine x at $y = 0$ by inverse interpolation. Are you surprised that the interpolated x-value is again 0.6? Sketch the curves for $x = \sqrt{y}$ and for the interpolating parabola.

41. Use the method of successive approximations to find the value of x corresponding to $y = 3.0$ from the data of Table 3.6. Compare to $\sinh^{-1} 3.0$.

Section 3.10

42. In Section 3.10, the assertion is made that the order of interpolation makes no difference, Demonstrate that this is true by interpolating within the data of Table 3.7 to find the values at $y = 0.33$ (within rows with x constant at 1.0, 1.5, and 2.0), using cubic interpolation formulas that fit the table at $y = 0.2$, 0.3, 0.4, and 0.5. Then interpolate within these three values to determine $f(1.6, 0.33)$, and compare to the value 1.841 obtained in the text.

▶ 43. After the example of Section 3.10 was finished, it was observed that a cubic in x and a quadratic in y would have been preferable. Do this to obtain $f(1.6, 0.33)$ and compare to the true value, 1.8350. Use the best "region of fit."

44. Interpolate in the function $f(x, y)$ tabulated in Section 3.10 to find $f(2.9, 0.21)$ employing a quartic formula in x and a quadratic in y.

45. Interpolation from a table of values for a function of three or more variables is done similarly to interpolation in a two-way table. Construct a table for $f(x, y, z) = xe^{y+z}$ for

$$x = 1.0(0.2)1.4, \qquad y = 0.2(0.2)0.6, \qquad z = 0.2(0.2)0.6.$$

Use these values to interpolate at $x = 1.3$, $y = 0.35$, $z = 0.46$, and compare to the exact value.

▶ 46. The example of Section 3.10 used a rectangular region of fit when a more nearly circular region would appear to have advantages. Interpolate from the data in Table 3.7 to evaluate $f(1.62, 0.31)$ by a set of polynomials that fits the table at $x = 1.5$ and 2.0 when $y = 0.2$ and when $y = 0.4$, and at $x = 0.5(0.5)2.5$ when $y = 0.3$. Do this by forming a series of difference tables. In this instance it is very awkward to interpolate first with x held constant, but there is no problem if we begin with y constant.

47. Repeat Exercise 46, but now use the Lagrangian form of interpolating polynomials.

APPLIED PROBLEMS

48. S. H. P. Chen and S. C. Saxena report experimental data for the emittance of tungsten as a function of temperature [*Ind. Eng. Chem. Fund.* **12,** 220 (1973)]. Their data are given below. They found that the equation

$$e(T) = 0.02424\left(\frac{T}{303.16}\right)^{1.27591}$$

correlated the data for all temperatures accurately to three digits. What degree of interpolating polynomial is required to match to their correlation at points midway between the tabulated temperatures? Discuss the pros and cons of polynomial interpolation in comparison to using their correlation.

$T, °K$	300	400	500	600	700	800	900	1000	1100
e	0.024	0.035	0.046	0.058	0.067	0.083	0.097	0.111	0.125
$T, °K$	1200	1300	1400	1500	1600	1700	1800	1900	2000
e	0.140	0.155	0.170	0.186	0.202	0.219	0.235	0.252	0.269

49. In studies of radiation-induced polymerization, a source of gamma rays was employed to give measured doses of radiation. However, the dosage varied with position in the apparatus, with these figures being recorded;

Position, in. from base point	0	0.5	1.0	1.5	2.0	3.0	3.5	4.0
Dosage, 10^5 rads/hr	1.90	2.39	2.71	2.98	3.20	3.20	2.98	2.74

For some reason, the reading at 2.5 inches was not reported, but the value of radiation there is needed. Fit interpolating polynomials of various degrees to the data to supply the missing information. What do you think is the best estimate for the dosage level at 2.5 inches?

50. M. S. Selim and R. C. Seagraves studied the kinetics of elution of copper compounds from ion-exchange resins. The normality of the leaching liquid was the most important factor in determining the diffusivity. Their data was obtained at convenient values of normality; we desire a table of D for integer values of normality ($N = 0.0, 1.0, 2.0, 3.0, 4.0, 5.0$). Use their data to construct such a table.

N	$D \times 10^6, cm^2/sec$	N	$D \times 10^6, cm^2/sec$
0.0521	1.65	0.9863	3.12
0.1028	2.10	1.9739	3.06
0.2036	2.27	2.443	2.92
0.4946	2.76	5.06	2.07

51. When the steady-state heat-flow equation is solved numerically, temperatures, $u(x, y)$, are calculated at the nodes of a gridwork constructed in the domain of interest. (This is the content of Chapter 7.) When a certain problem was solved, the values given in the

table below were obtained. This procedure does not give the temperatures at points other than the nodes of the grid; if they are desired, one can interpolate to find them. Use the data to estimate the values of the temperature at the points $(0.7, 1.2)$, $(1.6, 2.4)$, and $(0.65, 0.82)$.

x \ y	0.0	0.5	1.0	1.5	2.0	2.5
0.0	0.0	5.00	10.00	15.00	20.00	25.00
0.5	5.00	7.51	10.05	12.70	15.67	20.00
1.0	10.00	10.00	10.00	10.00	10.00	10.00
1.5	15.00	12.51	9.95	7.32	4.33	0.0
2.0	20.00	15.00	10.00	5.00	0.00	−5.00

Chapter Four

Numerical Differentiation and Numerical Integration

Differentiation and integration of functions are important mathematical operations, and the student of calculus spends much of his time learning techniques to do these operations. One learns the proofs of formulas that permit one to express the derivative of a function, given a statement of the function, as an expression in x. Suppose, however, that the function is not known as an explicit expression in terms of x, but only as a tabulation of values. In this chapter we study numerical techniques that allow the computation of values for the derivative of a tabulated function.

It is also frequently necessary to *integrate* a function when it is known only as a tabulation of data. Numerical methods permit us to do this. Furthermore, even when the function is explicitly known, computers cannot readily be programmed to analytically integrate an arbitrary function, and nearly all integrations performed on a computer must utilize numerical techniques. In addition to these two reasons for studying this topic, the integral of many functions cannot be expressed in terms of ordinary functions. In such cases, numerical methods permit us to evaluate the integral. These techniques will also be useful in solving differential equations.

4.1 FIRST DERIVATIVES FROM INTERPOLATING POLYNOMIALS

If a function is reasonably well approximated by an interpolating polynomial, we should expect that the slope of the function should also be approximated by the slope of the polynomial, although we shall discover that the error of estimating the slope is greater than the error of estimating the function.

We begin with a Newton–Gregory forward polynomial:

$$f(x_s) = P_n(x_s) + \text{error} = f_0 + s\,\Delta f_0 + \binom{s}{2}\Delta^2 f_0 + \ldots + \binom{s}{n}\Delta^n f_0 + \text{error}.$$

$$(4.1)$$

The error term of Eq. (4.1) is, by our rule for changing the next term,

$$\text{Error of } P_n(x_s) = \binom{s}{n+1} h^{n+1} f^{(n+1)}(\xi), \qquad x_0 < \xi < x_n. \tag{4.2}$$

Differentiating Eq. (4.1), remembering that f_0 and all the Δ-terms are constants (after all, they are just the numbers from the difference table), we have

$$f'(x_s) \doteq P'_n(x_s) = \frac{d}{dx}[P_n(x_s)] = \frac{d}{ds}[P_n(x_s)]\frac{ds}{dx} = \frac{d}{ds}[P_n(x_s)]\frac{1}{h}$$

$$= \frac{1}{h}\left(\Delta f_0 + \frac{1}{2}(s - 1 + s)\Delta^2 f_0\right.$$

$$\left. + \frac{1}{6}[(s-1)(s-2) + s(s-2) + s(s-1)]\Delta^3 f_0 + \ldots\right). \tag{4.3}$$

The derivatives of the factorials of s rapidly become algebraically complicated.* We get considerable simplification if we let $s = 0$, giving us the derivative corresponding to x_0 (where $s = 0$), however. If we let $s = 0$, Eq. (4.3) becomes

$$f'(x_0) \doteq \frac{1}{h}\left(\Delta f_0 - \frac{1}{2}\Delta^2 f_0 + \frac{1}{3}\Delta^3 f_0 - \frac{1}{4}\Delta^4 f_0 + \ldots \pm \frac{1}{n}\Delta^n f_0\right). \tag{4.4}$$

In Eq. (4.4), the derivative value, approximating to the derivative of the function, is the derivative of an nth-degree polynomial passing through the point (x_0, f_0) and n additional points to its right, evaluated at $x = x_0$. The error in Eq. (4.4) is obtained by differentiating the error term given in Eq. (4.2):

$$\text{Error of } P'_n(x_s) = h^{n+1} f^{(n+1)}(\xi)\left[\frac{d}{ds}\binom{s}{n+1}\right]\frac{1}{h} + \binom{s}{n+1}h^{n+1}\frac{d}{dx}[f^{(n+1)}(\xi)]. \tag{4.5}$$

The second term here cannot be evaluated, for the way that ξ varies with x is unknown but, when we let $s = 0$, this second term vanishes because

$$\binom{s}{n+1} = \frac{1}{(n+1)!}(s)(s-1)\cdots(s-n) = 0$$

at $s = 0$. Then we need only evaluate the first term of Eq. (4.5). Since

$$\frac{d}{ds}\binom{s}{n+1} =$$

$$\frac{(s-1)(s-2)\cdots(s-n) + s(s-2)(s-3)\cdots(s-n) + \ldots + s(s-1)\cdots(s-n+1)}{(n+1)!},$$

---

* In computing the derivatives, we have used the method:

$$\frac{d}{ds}(xyz) = \frac{dx}{ds}\cdot yz + x\frac{dy}{ds}z + xy\frac{dz}{ds}.$$

at $s = 0$, only the first term of the numerator remains, and Eq. (4.5) becomes

$$\text{Error of } P_n'(x_0) = h^{n+1}f^{(n+1)}(\xi)\left[(-1)^n \frac{n!}{(n+1)!}\right]\left(\frac{1}{h}\right) = \frac{(-1)^n}{n+1} h^n f^{(n+1)}(\xi). \qquad (4.6)$$

Note that even though the interpolating polynomial gives the function exactly at $s = 0$, our derivative formula is subject to an error of $O(h^n)$ at that point, unless $f^{(n+1)}(\xi) = 0$. In comparing Eqs. (4.6) and (4.4) we see that the error term of the derivative also can be obtained by changing the $\Delta^n f_0$ in the next term beyond the last one included, into $h^n f^{(n)}(\xi)$ exactly as with our interpolating formulas.

Example Use the data in Table 4.1 to estimate the derivative of y at $x = 1.7$. Use $h = 0.2$, and compute using one, two, three, or four terms of the formula.

Table 4.1

x	y	Δy	$\Delta^2 y$	$\Delta^3 y$	$\Delta^4 y$
1.3	3.669				
		0.813			
1.5	4.482		0.179		
		0.992		0.041	
1.7	5.474		0.220		0.007
		1.212		0.048	
1.9	6.686		0.268		0.012
		1.480		0.060	
2.1	8.166		0.328		0.012
		1.808		0.072	
2.3	9.974		0.400		
		2.208			
2.5	12.182				

With one term $y'(1.7) = \dfrac{1}{0.2}(1.212) = 6.060,$

With two terms $y'(1.7) = \dfrac{1}{0.2}\left(1.212 - \dfrac{1}{2}(0.268)\right) = 5.390,$

With three terms $y'(1.7) = \dfrac{1}{0.2}\left(1.212 - \dfrac{1}{2}(0.268) + \dfrac{1}{3}(0.060)\right) = 5.490,$

With four terms $y'(1.7) = \dfrac{1}{0.2}\left(1.212 - \dfrac{1}{2}(0.268) + \dfrac{1}{3}(0.060) - \dfrac{1}{4}(0.012)\right)$
     $= 5.475.$

The data tabulated in Table 4.1 are for $y = e^x$, rounded to three decimals.* Since the derivative of e^x is also e^x, we see that the error in the derivative is quite small with four terms, comparing 5.475 to 5.474. We can anticipate this; since the fourth differences in Table 4.1 do not vary greatly, the function is represented reasonably well by a fourth-degree polynomial.

Since we know that $f(x) = e^x$, the estimated errors for the above computations can be calculated by Eq. (4.6):

With one term,   error $= \dfrac{(-1)^1}{2}(0.2)f''(\xi),$ $1.7 \leqslant \xi \leqslant 1.9,$

$$= \frac{-0.2}{2} \begin{Bmatrix} e^{1.7} & (\text{min}) \\ e^{1.9} & (\text{max}) \end{Bmatrix} = \begin{Bmatrix} -0.547 & (\text{min}) \\ -0.669 & (\text{max}) \end{Bmatrix}.$$

(Actual error is −0.586.)

With two terms,   error $= \dfrac{(-1)^2}{3}(0.2)^2 f'''(\xi),$ $1.7 \leqslant \xi \leqslant 2.1,$

$$= \frac{0.04}{3} \begin{Bmatrix} e^{1.7} & (\text{min}) \\ e^{2.1} & (\text{max}) \end{Bmatrix} = \begin{Bmatrix} 0.073 & (\text{min}) \\ 0.109 & (\text{max}) \end{Bmatrix}.$$

(Actual error is 0.084.)

With three terms, error $= \dfrac{(-1)^3}{4}(0.2)^3 f^{iv}(\xi),$ $1.7 \leqslant \xi \leqslant 2.3,$

$$= \frac{-0.008}{4} \begin{Bmatrix} e^{1.7} & (\text{min}) \\ e^{2.3} & (\text{max}) \end{Bmatrix} = \begin{Bmatrix} -0.011 & (\text{min}) \\ -0.020 & (\text{max}) \end{Bmatrix}.$$

(Actual error is −0.016.)

With four terms, error $= \dfrac{(-1)^4}{5}(0.2)^4 f^{v}(\xi),$ $1.7 \leqslant \xi \leqslant 2.5,$

$$= \frac{0.0016}{5} \begin{Bmatrix} e^{1.7} & (\text{min}) \\ e^{2.5} & (\text{max}) \end{Bmatrix} = \begin{Bmatrix} 0.002 & (\text{min}) \\ 0.004 & (\text{max}) \end{Bmatrix}.$$

(Actual error is −0.001.)

In the case of four terms, the calculated and actual errors do not agree because of round-off errors; Eq. (4.6) accounts only for the truncation error. In the other cases, the round-off is not large enough relative to the truncation error to invalidate the estimates from Eq. (4.6).

The formulas for derivatives given by Eq. (4.4) are called forward-difference approximations because they involve only differences of function values forward

---

* The table incidentally illustrates the effect of rounding errors in the fourth differences. A more accurate table would show fourth differences of 0.0088, 0.0108, 0.0132.

from $f(x_0)$. For a forward-difference approximation, the interpolating polynomial that is used does not fit points that are symmetrical about x_0. We have seen in the previous chapter that interpolation is more accurate near the center of the range of fit. A similar effect is observed with derivatives. Consider the first two terms of Eq. (4.3):

$$f'(x) = \frac{1}{h}\left(\Delta f_0 + \frac{1}{2}(s - 1 + s)\Delta^2 f_0\right) + \text{error}. \tag{4.7}$$

The second-degree interpolating polynomial to which Eq. (4.7) corresponds fits at x_0, x_1, and x_2. Let $s = 1$ in Eq. (4.7), so that we can estimate $f'(x_1)$:

$$f'(x_1) = \frac{1}{h}\left(\Delta f_0 + \frac{1}{2}\Delta^2 f_0\right) + \text{error}.$$

Write out the differences in terms of $f's$:

$$f'(x_1) = \frac{1}{h}\left(f_1 - f_0 + \frac{1}{2}(f_2 - 2f_1 + f_0)\right) = \frac{f_2 - f_0}{2h} + \text{error}. \tag{4.8}$$

Equation (4.8) can be called a central-difference approximation. The value of x where it is applied is centered in the range of fit of the polynomial—differences of function values on either side of $f(x_1)$ are used.

The error term for Eq. (4.8) can be derived similarly to Eq. (1.6.) It is:

$$\text{Error in } P_2'(x_1) = -\tfrac{1}{6} h^2 f'''(\xi), \qquad x_0 \le \xi \le x_2. \tag{4.9}$$

Note particularly that the power of h in Eq. (4.9) is two, so the error is $O(h^2)$. Contrast this to the first power of h in the error term when only one term of Eq. (4.3) is used, although both computations involve only two functional values. The coefficient in the error term is also significantly smaller. Central-difference formulas are decidedly superior in calculating values for derivatives.

If we apply Eq. (4.8) to the data of Table 4.1 to estimate $f'(1.7)$, we get:

$$f'(1.7) \doteq \frac{6.686 - 4.482}{2(0.2)} = 5.510.$$

Estimate of error is $\begin{Bmatrix} -0.030 & (\text{min}) \\ -0.046 & (\text{max}) \end{Bmatrix}$.

Actual error is -0.036.

Central-difference formulas similar to Eq. (4.8) can be derived using higher-degree polynomials of even order. (The odd-degree polynomials do not have a range of fit that is symmetrical about any of the x-values.) For example, the formula corresponding to a fourth-degree polynomial, expressed in terms of function values rather than differences, and related to the point x_0, is:

$$f'(x_0) = \frac{1}{h}\frac{f_{-2} - 8f_{-1} + 8f_1 - f_2}{12}, \qquad \text{Error} = \frac{1}{30} h^4 f^v(\xi), \qquad x_{-2} \le \xi \le x_2.$$

In many applications it is preferable to use the simpler formula of Eq. (4.8) and control the error by making h small. Observe, however, that the error of the derivative from an nth-degree polynomial is $O(h^n)$, while the error of interpolation is $O(h^{n+1})$.

4.2 FORMULAS FOR HIGHER DERIVATIVES

We can get formulas for higher derivatives by further differentiating Eq. (4.3), and differentiating Eq. (4.5) will give the error terms. For example,

$$f''(x_s) \doteq P_n''(x_s) = \frac{1}{h^2} \frac{d^2}{ds^2} P_n(x_s)$$

$$\doteq \frac{1}{h^2} \left(\frac{1}{2} (2) \Delta^2 f_0 + \frac{1}{6} [(s-2) + (s-1) + (s-2) + s + (s-1) + s] \Delta^3 f_0 \right. $$
$$\left. + \ldots \right). \quad (4.10)$$

At $s = 0$,

$$f''(x_0) \doteq \frac{1}{h^2} (\Delta^2 f_0 - \Delta^3 f_0 + \ldots).$$

There is no simple pattern for the coefficients, and the error terms have complicated coefficients as well. The situation gets more complex with still higher derivatives, so we would like a simpler method. We have seen before that symbolic methods have given interpolation formulas with relative ease. It is true for derivative formulas also:

$$E = 1 + \Delta,$$

$$y_s = E^s y_0,$$

$$y_s' = \frac{d}{dx} (E^s y_0) = \frac{1}{h} \frac{d}{ds} (E^s y_0) = \frac{1}{h} (\ell n \, E) E^s y_0. \quad (4.11)$$

At $s = 0$,

$$y_0' = \frac{1}{h} (\ell n \, E) y_0 = \frac{1}{h} \ell n \, (1 + \Delta) y_0$$

$$= \frac{1}{h} \left(\Delta y_0 - \frac{1}{2} \Delta^2 y_0 + \frac{1}{3} \Delta^3 y_0 - \frac{1}{4} \Delta^4 y_0 + \ldots \right). \quad (4.12)$$

In writing Eq. (4.12), we have used the Maclaurin expansion for $\ell n \, (1 + \Delta)$. Note that Eq. (4.12) is identical to Eq. (4.4) that we derived before by nonsymbolic means.

If we use the symbol D for the derivative operator,

$$Dy_0 = \frac{1}{h} \ell n \, (1 + \Delta) y_0. \quad (4.13)$$

We abstract the equivalence between operators from Eq. (4.13):

$$D = \frac{1}{h} \ell n \, (1 + \Delta). \quad (4.14)$$

The value of the symbolic method is that algebraic operations on operator relations give valid results. Let us raise each side of Eq. (4.14) to powers:

$$D^2 = \frac{1}{h^2} \ln^2 (1 + \Delta),$$

$$D^3 = \frac{1}{h^3} \ln^3 (1 + \Delta), \tag{4.15}$$

$$D^n = \frac{1}{h^n} \ln^n (1 + \Delta).$$

Equations (4.15) show that we can get formulas for higher derivatives by multiplying the series in Eq. (4.12) by itself. The second derivative is given by:

$$D^2 y_0 = \frac{1}{h^2} \left(\Delta - \frac{1}{2} \Delta^2 + \frac{1}{3} \Delta^3 - \frac{1}{4} \Delta^4 + \dots \right)^2 y_0$$

$$= \frac{1}{h^2} \left(\Delta^2 - \Delta^3 + \frac{11}{12} \Delta^4 - \frac{5}{6} \Delta^5 + \dots \right) y_0$$

$$= \frac{1}{h^2} \left(\Delta^2 y_0 - \Delta^3 y_0 + \frac{11}{12} \Delta^4 y_0 - \frac{5}{6} \Delta^5 y_0 + \dots \right). \tag{4.16}$$

Formulas for derivatives of higher order can be similarly obtained by multiplication of the series in Eq. (4.12).

Equations (4.15) lead to an important generalization for the estimation of derivatives from forward-difference formulas. The first term of all the formulas in terms of differences has only Δ^n. Hence, as a first approximation,[*]

$$D^n y = \frac{1}{h^n} \Delta^n y + \text{error } O(h). \tag{4.17}$$

Example Use formula (4.16) to estimate $y''(1.7)$ from Table 4.1 using terms through Δ^3. Also estimate the error.

$$y''(1.7) = \frac{1}{(0.2)^2} (0.268 - 0.060)$$

$$= 5.200 \qquad \text{(compare to 5.474, exact answer)}.$$

$$\text{Error} = \frac{1}{h^2} \left(\frac{11}{12} h^4 y^{iv}(\xi) \right), \qquad 1.7 < \xi < 2.1,$$

by converting the next term in the formula. Since $y = e^x$, $\quad y^{iv} = e^x$.

$$\text{Max value of error} = \tfrac{11}{12}(0.2)^2(e^{2.1}) = 0.298,$$

$$\text{Min value of error} = \tfrac{11}{12}(0.2)^2(e^{1.7}) = 0.201.$$

Compare to actual error of 0.274.

---

[*] This was used in the previous chapter in developing the "next term rule" for the error of an interpolation formula.

In the absence of knowledge of the function, we could have used the fourth differences to estimate $h^4 f^{iv}$ as shown by Eq. (4.17). If the differences vary greatly, this is hazardous, however. In this instance we would estimate the error as about 0.23 if we used 0.010 (the average value of $\Delta^4 y$) as an estimate of $h^4 f^{iv}(\xi)$.

Central differences for higher derivatives will have more accuracy than the forward-difference formula represented by Eq. (4.16). We prefer to develop these through the lozenge diagrams of the next section.

4.3 LOZENGE DIAGRAM FOR DERIVATIVES

Since there is a variety of equivalent forms of the interpolating polynomial, we may obtain a variety of forms of derivative formulas by differentiating the various polynomials. They are interrelated, of course, as we have seen for interpolation formulas. It is convenient to exhibit this interrelation by a lozenge diagram as in Fig. 4.1.

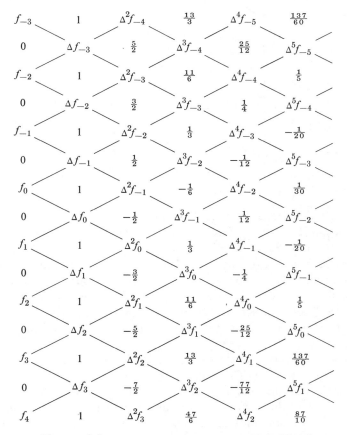

Figure 4.1   Note. All formulas must be multiplied by $1/h$.

There is a special property of lozenge diagrams that makes them easy to construct. The diagram can be considered to be of two parts, the differences of the function and the interspersed coefficients, one superimposed on the other. Consider the array of coefficients of Fig. 4.1, shown here as Fig. 4.2.

A	B	C	D	E	F
	1		$\frac{13}{3}$		$\frac{137}{60}$
0		$\frac{5}{2}$		$\frac{25}{12}$	
	1		$\frac{11}{6}$		$\frac{1}{5}$
0		$\frac{3}{2}$		$\frac{1}{4}$	
	1		$\frac{1}{3}$		$-\frac{1}{20}$
0		$\frac{1}{2}$		$-\frac{1}{12}$	
	1		$-\frac{1}{6}$		$\frac{1}{30}$
0		$-\frac{1}{2}$		$\frac{1}{12}$	
	1		$\frac{1}{3}$		$-\frac{1}{20}$
0		$-\frac{3}{2}$		$-\frac{1}{4}$	
	1		$\frac{11}{6}$		$\frac{1}{5}$
0		$-\frac{5}{2}$		$-\frac{25}{12}$	
	1		$\frac{13}{3}$		$\frac{137}{60}$
0		$-\frac{7}{2}$		$-\frac{77}{12}$	
	1		$\frac{47}{6}$		$\frac{87}{10}$
↑	↑	↑	↑	↑	↑

Figure 4.2   A   B   C   D   E   F

Compare the columns in pairs, beginning with columns *A* and *B*. Observe that the values in *A* are the differences of the values in *B*. Now compare columns *B* and *C*; column *B* is just the differences of the values in column *C*, with each value subtracted from the one above it. The same thing is true for columns *C* and *D*, columns *D* and *E*, and so on. In other words, the array of coefficients is a difference table, but written from right to left and upside down. This means that one line of coefficients, which we get from Eq. (4.4), suffices to create the entire array of coefficients! (This same property is true in Fig. 3.1 of Chapter 3, but is less obvious.)

Equation (4.16) serves to start a lozenge diagram for the second derivative at $x = x_0$ in a similar fashion. The result is Fig. 4.3.

We use the lozenge diagrams to write formulas for differentiation of a tabulated function in the same way we used similar diagrams to write interpolation formulas. Any path through the table may be chosen; the coefficients are then given by the rules discussed in Section 3.5. Suppose we use a horizontal path

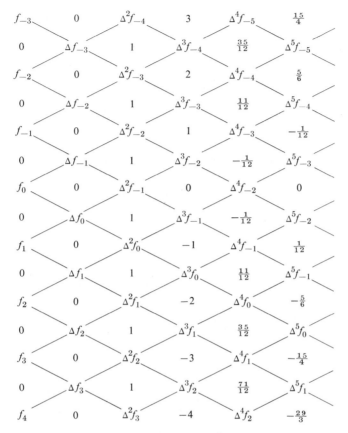

Figure 4.3   Note. All formulas must be multiplied by $1/h^2$.

beginning at f_0:

$$\frac{df}{dx}\bigg|_{x=x_0} = \frac{1}{h}\left[\frac{\Delta f_0 + \Delta f_{-1}}{2} + (0)\Delta^2 f_{-1} + \left(-\frac{1}{6}\right)\frac{\Delta^3 f_{-1} + \Delta^3 f_{-2}}{2} + \dots\right]$$

$$= \frac{1}{h}\left[\frac{(f_1 - f_0) + (f_0 - f_{-1})}{2} + O(h^3)\right] = \frac{f_1 - f_{-1}}{2h} + O(h^2). \quad (4.18)$$

Using Fig. 4.3 to find the second derivative, again following a horizontal path, we obtain

$$\frac{d^2 f}{dx^2}\bigg|_{x=x_0} = \frac{1}{h^2}\left[\Delta^2 f_{-1} + 0 + \left(-\frac{1}{12}\right)\Delta^4 f_{-2} + \dots\right]$$

$$= \frac{1}{h^2}\left[(f_1 - 2f_0 + f_{-1}) + O(h^4)\right]$$

$$= \frac{f_1 - 2f_0 + f_{-1}}{h^2} + O(h^2). \quad (4.19)$$

Equations (4.18) and (4.19) are formulas of special importance, Eq. (4.18) being the same as Eq. (4.8). Note that even though they are only one-term formulas, they have errors of $O(h^2)$, in contrast to the $O(h)$ error for Eq. (4.3) and Eq. (4.16), when only one term is used. As previously stated, these $O(h^2)$ formulas are known as *central-difference approximations*. Their good error terms occur because the point at $x = x_0$ is at the center of the range of x-values at which the interpolating polynomial matches the table. Central-difference formulas of higher order can be developed by continuing further on our horizontal path.

The efficiency of the central-difference approximations for second derivatives, as compared to the forward-difference approximation, is illustrated by the following example.

Using the data of Table 4.1, we compute $f''(1.7)$.

Using forward differences, Eq. (4.16),

$$f''(1.7) = \frac{0.268}{(0.2)^2} = 6.700,$$

$$\text{Error} = -1.226.$$

Using central differences, Eq. (4.19),

$$f''(1.7) = \frac{0.220}{(0.2)^2} = 5.500,$$

$$\text{Error} = -0.026.$$

The central-difference expression is much more accurate. The estimated errors for the above computations are -1.095 (min) to -1.633 (max) for forward differences, and -0.015 (min) to -0.022 (max) for central differences. (Note that, in the second case, round-off causes the actual error to fall outside these limits.)

Figure 4.3 gives a formula of $O(h^4)$ error for the second derivative if we extend the horizontal path until we reach the column of fourth differences:

$$\frac{d^2f}{dx^2}\bigg|_{x=x_0} = \frac{1}{h^2}\left[\Delta^2 f_{-1} - \frac{1}{12}\Delta^4 f_{-2}\right] + O(h^4).$$

Rewriting this in terms of the f-values, we get an equivalent expression:

$$\frac{d^2f}{dx^2}\bigg|_{x=x_0} = \frac{-f_2 + 16f_{-1} - 30f_0 + 16f_1 - f_2}{12h^2} + O(h^4).$$

Applying these formulas to the data of Table 4.1, we compute $f''(1.7)$ as:

$$f''(1.7) = \frac{1}{(0.2)^2}\left[0.220 - \frac{1}{12}(0.007)\right] = 5.485$$

$$= \frac{-8.166 + 16(6.686) - 30(5.474) + 16(4.482) - 3.669}{12(0.2)^2} = 5.485.$$

(Compare the result to the exact value: 5.474. The error, -0.011, is much larger than the truncation error (-0.0049 to -0.00108) due to round-off.)

The central-difference approximations extend directly to higher derivatives as well. Lozenge diagrams can be constructed for these, but if we need only the first nonzero term, a simple formula will suffice. For the even-order derivatives it is

$$f^{(n)}(x)\big|_{x=x_0} = \frac{1}{h^n}\Delta^n f_{-n/2} + O(h^2), \qquad n \text{ even.}$$

The difference required is always found in the table on the horizontal line through x_0 and f_0. If the difference table has not been constructed, we would express in terms of the f-values.

For odd-order derivatives, the expression involves the average of differences found just above and just below the horizontal line through x_0:

$$f^{(n)}(x)\big|_{x=x_0} = \frac{\Delta^n f_{(-n+1)/2} + \Delta^n f_{(-n-1)/2}}{2h^n} + O(h^2), \qquad n \text{ odd.} \qquad (4.20)$$

4.4 EXTRAPOLATION TECHNIQUES

There is an alternative way of successively improving the accuracy of our estimates of the derivative which is equivalent to adding terms as given by the lozenge diagram. This is the technique of *successive extrapolations*. It has special utility in computer programs for differentiation of arbitrary functions.

We illustrate the method first for a function known only from a table, such as Table 4.2. We desire the derivative $f'(x)$ at $x = 2.5$. The central-difference formula, Eq. (4.8) or Eq. (4.18), gives

$$f'(x)\big|_{x=2.5} \doteq \frac{f_1 - f_{-1}}{2h} = \frac{0.25337 - 0.31729}{2(0.1)} = -0.3196.$$

This estimate has an error $O(h^2)$.

Table 4.2

x	$f(x)$	x	$f(x)$
2.0	0.42298	2.6	0.25337
2.1	0.40051	2.7	0.22008
2.2	0.37507	2.8	0.18649
2.3	0.34718	2.9	0.15290
2.4	0.31729	3.0	0.11963
2.5	0.28587		

We can also compute the derivative by using values at $x = 2.7$ and $x = 2.3$, for which the spacing is 0.2, and get

$$f'(x)|_{x=2.5} \doteq \frac{0.22008 - 0.34718}{2(0.2)} = -0.3178.$$

The error here is also $O(h^2)$, but now h is twice as large as in the previous calculation.

In order to combine these two estimates and extrapolate to a more accurate estimate, we examine in greater detail the nature of the error term of Eq. (4.18). Using the lozenge diagram, Fig. 4.1, we have

$$f_0' = \frac{1}{h}\left(\frac{\Delta f_{-1} + \Delta f_0}{2} - \frac{1}{6}\frac{\Delta^3 f_{-2} + \Delta^3 f_{-1}}{2} + \frac{1}{30}\frac{\Delta^5 f_{-3} + \Delta^5 f_{-2}}{2} - \ldots\right). \quad (4.21)$$

If we truncate after the first term, we get an error of $O(h^2)$, but Eq. (4.20) shows that

$$f_0''' = \frac{\Delta^3 f_{-2} + \Delta^3 f_{-1}}{2h^3} + O(h^2).$$

We may write equivalently,

$$-\frac{1}{6}\frac{\Delta^3 f_{-2} + \Delta^3 f_{-1}}{2} = -\frac{1}{6}h^3 f_0''' + O(h^5) = Ch^3 + O(h^5).$$

In this expression C is a constant; we merely recognize that the third derivative of f evaluated at x_0 has a fixed value. Equation (4.21) can hence be written as:

$$f_0' = \frac{1}{h}\frac{\Delta f_{-1} + \Delta f_0}{2} + Ch^2 + O(h^4) + \frac{1}{30}h^4 f^v(\xi)$$

$$= \frac{1}{h}\frac{\Delta f_{-1} + \Delta f_0}{2} + Ch^2 + O(h^4).$$

We see that we make only an order h^4 error in assuming that the error of Eq. (4.18) is proportional to h^2. We then combine the two estimates of the derivative, -0.3196 with $h = 0.1$ and -0.3178 with $h = 0.2$, as follows:

$$f_0' \text{ (exact)} = -0.3178 + Ch^2 + O(h^4), \quad \text{based on less accurate value,}$$
$$\doteq -0.3178 + C(0.2)^2, \quad \text{if we neglect the } O(h^4) \text{ term.}$$

$$f_0' \text{ (exact)} = -0.3196 + Ch^2 + O(h^4), \quad \text{based on more accurate value,}$$
$$\doteq -0.3196 + C(0.1)^2, \quad \text{if we neglect the } O(h^4) \text{ term.}$$

We solve for the unknown value of f_0 (exact) by eliminating the terms in C between the two expressions,

$$f_0' \text{ (exact)} \doteq -0.3196 + \tfrac{1}{3}(-0.3196 - (-0.3178)]$$
$$\doteq -0.3203.$$

The method can be applied to any two estimates, each with $O(h^n)$ errors and with values of h varying by 2 to 1. The rule is:

$$\text{Exact value} \doteq \text{more accurate} + \frac{1}{2^n - 1}(\text{more accurate} - \text{less accurate}). \qquad (4.22)$$

We show that the relationship is not perfect by using the $\doteq$ sign; this lack of exactness is the result of our neglect of the $O(h^{n+2})$ terms in the two expressions.

This extrapolation method can be extended since the first-order extrapolations with error $O(h^4)$ can be shown to have errors of the form $Ch^4 + O(h^6)$. (Observe that this follows from the alternately zero coefficients on the horizontal path through Fig. 4.1, so this principle extends to all orders of extrapolation. Errors due to rounding of the original data eventually dominate, however, so the extrapolation technique must be abandoned at some point.)

For the data in Table 4.2 we can extend to one more order of extrapolation. We arrange the data in tabular form.

h	Initial estimate	First-order extrapolation	Second-order extrapolation
0.1	−0.31960		
		−0.32022	
0.2	−0.31775		−0.32020
		−0.32050	
0.4	−0.30951		

The last entry was calculated as

$$-0.32022 + \frac{1}{2^4 - 1}(-0.32022 + 0.32050) = -0.32020.$$

An extra guard figure was carried to minimize round-off in the computations.

Since the lozenge for second derivatives. Fig. 4.3, also shows alternate coefficients to be zero, approximations for the second derivative can be extrapolated in exactly the same way, the order of the error increasing to h^4, h^6, h^8, etc., at each step.

It is left as an exercise for the student to show that this extrapolation method is the equivalent to passing a higher-degree interpolating polynomial through the data (with the point $x = x_0$ at the center of the region of fit for each of these), and then finding the derivative by central-difference formulas.

The application of this extrapolation method in a computer program is important because successive estimates, with h decreasing, have smaller and smaller truncation errors, and hence provide a built-in measure of accuracy. The method is generally employed to differentiate a function of known form, rather than a tabulated function. The basic scheme is to compute the derivative at two values of h, one being one-half the value of the other, and to extrapolate as in the above example. If a significant change is made by the extrapolation, compared to the more accurate of the original estimates, h is halved again and a second-order extrapolation is made. Again, if a significant improvement results, another computation is made with h halved once more and a third-order extrapolation is made. The process is continued until the improvement is less than some criterion.

Since round-off affects calculation by computers just as it does hand computations, eventually these errors will control the accuracy of successive values, so the student is cautioned against indiscriminate use of extrapolation. It is generally good practice to put a limit to the number of repetitions of the process. This is discussed in the next section.

4.5 ROUND-OFF AND ACCURACY OF DERIVATIVES

It is important to realize the effect of round-off error on the accuracy of derivatives computed by finite-difference formulas. As we have seen, decreasing the value of h or increasing the degree of the interpolating polynomial increases the accuracy of our derivative formulas. (These procedures reduce the truncation error.)

As h is reduced, however, we are required to subtract function values that are more nearly the same value, and this, we have seen, incurs a large error due to round-off. The increase in round-off error as h gets smaller causes the best accuracy to occur at some intermediate point. The data in Fig. 4.4(a) and (b) which were obtained with an IBM Model 360 computer, illustrate this phenomenon. The test was made with both single-precision arithmetic (giving 6–7 significant figure accuracy) and using double-precision (14–15 figures). The function tested was $f(x) = e^x$, at $x = 0$. The true value of $f'(0)$ is 1.0, of course. Central-difference approximations were employed.

In the single-precision test, the optimum value of h was only 0.01 for the first derivative, and the best accuracy of the second derivative was at $h = 0.1$, a surprisingly large value. When the round-off errors were reduced by using double-precision, essentially exact results were obtained for the first derivative throughout the range of h values. For the second derivative, $h = 0.1 \times 10^{-2}$ or $h = 0.1 \times 10^{-3}$ gives the best accuracy. Note how large the error grows; it is especially marked for second derivatives when h becomes much smaller than the optimum. Higher derivatives than the second will be expected to show an even more exaggerated influence of round-off. It is not too surprising, then, that IBM has provided for Extended Precision in their Model 370 computers, giving about twice the number of accurate digits as double precision does.

Different models of computers vary widely in the number of digits of accuracy they provide, due to the different number of bits provided for the storage of floating-point values and also, to a lesser degree, due to the format of their floating-point representations. For example, a CDC 6400 provides 60 bits for floating-point values, and gives the equivalent of about 14 decimal digits of accuracy. The Control Data Corporation did not provide for double precision beyond this accuracy in this model. Another factor that affects the accuracy of functional values is the algorithm used in their evaluation. Chapter 10 gives some insight into this. Different software packages that are used with computer operating systems employ different algorithms, which are of varying accuracy. Further, as Chapter 10 will show, the error of approximating the function differs with the value of the argument.

a) RESULTS WITH SINGLE PRECISION

H	DX	DDX
0.1E 00	0.1001663E 01	0.1000761E 01
0.1E−01	0.1000001E 01	0.9959934E 00
0.1E−02	0.9999569E 00	0.8940693E 00
0.1E−03	0.9959930E 00	−0.8344640E 02
0.1E−04	0.9775158E 00	−0.4768363E 04
0.1E−05	0.9834763E 00	−0.5960462E 05
0.1E−06	0.5960463E 00	−0.1192093E 08
0.1E−07	0.2980230E 01	−0.5960458E 09
0.1E−08	0.2980229E 02	−0.5960459E 11
0.1E−09	0.2980229E 03	−0.5960456E 13

b) RESULTS WITH DOUBLE PRECISION

H	DX	DDX
0.1E 00	0.1001667E 01	0.1000834E 01
0.1E−01	0.1000016E 01	0.1000008E 01
0.1E−02	0.1000000E 01	0.1000000E 01
0.1E−03	0.1000000E 01	0.9999999E 00
0.1E−04	0.1000000E 01	0.9999989E 00
0.1E−05	0.1000000E 01	0.1000031E 01
0-1E−06	0.9999999E 01	0.9894834E 00
0.1E−07	0.9999999E 00	−0.4163322E 00
0.1E−08	0.9999999E 00	−0.4163319E 02
0.1E−09	0.1000000E 01	0.2775545E 04

Figure 4.4 Derivatives computed by central-difference formulas $f(x) = e^x$, at $x = 0$.

No analytical way of predicting the effects of round-off error seems to exist. The best we can do is to warn the budding numerical analyst so he will be properly wary when computing derivatives numerically. In the words of many authorities, *the process of differentiating is basically an unstable process*—meaning that small errors made during the process cause greatly magnified errors in the final result.

Differentiation of "noisy" data encounters a similar problem. If the data being differentiated are from experimental tests, or are observations subject to errors of measurement, the errors so influence the derivative values calculated by numerical procedures that they may be meaningless. The usual recommendation is to smooth the data first, using methods that are discussed in Chapter 10. Passing a cubic spline through the points, and then getting the derivative of this approximation to the data has become quite popular. A least-squares curve may also be used. The strategy involved is straightforward—we don't try to represent the function by one that fits exactly to the data points, for this fits to the errors as well as to the trend of the information. Rather, we approximate with a smoother curve that we hope is closer to the truth than the data themselves. The problem, of course, is how much smoothing should be done. One can go too far and "smooth" beyond the point where only errors are eliminated.

A final situation should be mentioned. Some functions, or data from a series of tests, are inherently "rough." By this we mean that the function values change rapidly; a graph would show sharp local variations. When the derivative values of the function incur rapid changes, a sampling of the information may not reflect them. In this instance, the data indicate a smoother function than actually exists. Unless enough data is at hand to show the local variations, valid values of the derivatives just cannot be obtained. The only solution is more data, especially near the "rough" spots. And then we are beset by problems of accuracy of the data!

For convenience, we collect here formulas for computing derivatives.

Formulas for Computing Derivatives

Formulas for the first derivative:

$$f'(x_0) = \frac{f_1 - f_0}{h} + O(h).$$

$$f'(x_0) = \frac{f_1 - f_{-1}}{2h} + O(h^2).$$ (Central difference)

$$f'(x_0) = \frac{-f_2 + 4f_1 - 3f_0}{2h} + O(h^2).$$

$$f'(x_0) = \frac{-f_2 + 8f_1 - 8f_{-1} + f_{-2}}{12h} + O(h^4).$$ (Central difference)

Formulas for the second derivative:

$$f''(x_0) = \frac{f_2 - 2f_1 + f_0}{h^2} + O(h).$$

$$f''(x_0) = \frac{f_1 - 2f_0 + f_{-1}}{h^2} + O(h^2).$$ (Central difference)

$$f''(x_0) = \frac{-f_3 + 4f_2 - 5f_1 + 2f_0}{h^2} + O(h^2).$$

$$f''(x_0) = \frac{-f_2 + 16f_1 - 30f_0 + 16f_{-1} - f_{-2}}{12h^2} + O(h^4).$$ (Central difference)

Formulas for the third derivative:

$$f'''(x_0) = \frac{f_3 - 3f_2 + 3f_1 - f_0}{h^3} + O(h).$$

$$f'''(x_0) = \frac{f_2 - 2f_1 + 2f_{-1} - f_{-2}}{2h^3} + O(h^2). \qquad \text{(Averaged differences)}$$

Formulas for the fourth derivative:

$$f^{iv}(x_0) = \frac{f_4 - 4f_3 + 6f_2 - 4f_1 + f_0}{h^4} + O(h).$$

$$f^{iv}(x_0) = \frac{f_2 - 4f_1 + 6f_0 - 4f_{-1} + f_{-2}}{h^4} + O(h^2). \qquad \text{(Central differences)}$$

4.6 NEWTON–COTES INTEGRATION FORMULAS

The usual strategy in developing formulas for numerical integration is similar to that for numerical differentiation. We pass a polynomial through points of the function, and then integrate this polynomial approximation to the function. This permits us to integrate a function known only as a table of values. When the values are equispaced, our familiar Newton–Gregory forward polynomial is a convenient starting point, so

$$\int_a^b f(x)\, dx \doteq \int_a^b P_n(x_s)\, dx. \qquad (4.23)$$

The formula we get from Eq. (4.23) will not be exact because the polynomial is not identical with $f(x)$. We get an expression for the error by integrating the error term of $P_n(x_s)$:

$$\text{Error} = \int_a^b \binom{s}{n+1} h^{n+1} f^{(n+1)}(\xi)\, dx.$$

There are various way that we can employ Eq. (4.23). The interval of integration (a, b) can match the range of fit of the polynomial, (x_0, x_n). In this case, we get the Newton–Cotes formulas; these are a set of integration rules corresponding to the varying degrees of the interpolating polynomial. The first three, with the degree of the polynomial one, two, or three, are particularly important, and we discuss them at length is the following sections.

If the degree of the polynomial is too high, errors due to round-off and local irregularities can cause a problem. This explains why it is only the lower-degree Newton–Cotes formulas that are often used.

The range of the polynomial and the interval of integration do not have to be the same. If the interval of integration extends outside the range of fit, however, we are extrapolating, and this incurs larger errors. If we only desire to get the integral of a function whose values are known, we will normally avoid extrapolation. As we shall see in the next chapter, however, integrating the polynomial outside its range of fit leads to some important methods for solving differential equations. Using an interval of integration that is a subset of the points at which the polynomial agrees with the function also has special application in solving differential equations numerically.

The utility of numerical integration extends beyond the need to integrate a function known only as a table of values. Most computer programs for integration of functions whose form is known use these numerical techniques rather than the analytical methods of the calculus. While it is possible to write a program to do symbol manipulation and hence to perform analytical integration for simple forms of elementary functions, these routines lack generality. Furthermore, in a large number of cases no closed form for the integral exists. Numerical integration applies regardless of the complexity of the integrand or the existence of a closed form for the integral.

Let us now develop our three important Newton–Cotes formulas. During the integration, we will need to change the variable of integration from x to s since our polynomials are expressed in terms of s. Observe that $dx = h\,ds$.

For $n = 1$,

$$\int_{x_0}^{x_1} f(x)\,dx \doteq \int_{x_0}^{x_1} (f_0 + s\Delta f_0)\,dx = h\int_{s=0}^{s=1} (f_s + s\Delta f_0)\,ds$$

$$= hf_0 s\Big]_0^1 + h\Delta f_0 \frac{s^2}{2}\Big]_0^1 = h\left(f_0 + \frac{1}{2}\Delta f_0\right)$$

$$= \frac{h}{2}(2f_0 + (f_1 - f_0)] = \frac{h}{2}(f_0 + f_1). \tag{4.24}$$

$$\text{Error} = \int_{x_0}^{x_1} \frac{s(s-1)}{2} h^2 f''(\xi)\,dx = h^3 f''(\xi_1)\int_0^1 \frac{s^2 - s}{2}\,ds$$

$$= h^3 f''(\xi_1)\left(\frac{s^3}{6} - \frac{s^2}{4}\right)\Big]_0^1 = -\frac{1}{12} h^3 f''(\xi_1), \qquad x_0 \le \xi_1 \le x_1. \tag{4.25}$$

In getting the error term of Eq. (4.25), we have used the mean-value theorem for integrals, which is that

$$\int_a^b f(x)g(x)\,dx = f(\xi)\int_a^b g(x)\,dx, \qquad a \le \xi \le b,$$

provided that $g(x)$ does not change sign in the interval (a, b). This applies to Eq. (4.25) because $s(s - 1)$ is always less than or equal to zero on $(0, 1)$. The value of ξ_1 is not necessarily the same as ξ, though both fall in the interval (x_0, x_1).

For $n = 2$,

$$\int_{x_0}^{x_2} f(x)\, dx \doteq \int_{x_0}^{x_2} \left(f_0 + s\Delta f_0 + \frac{s(s-1)}{2}\Delta^2 f_0 \right) dx$$

$$= h\int_0^2 \left(f_0 + s\Delta f_0 + \frac{s(s-1)}{2}\Delta^2 f_0 \right) ds$$

$$= hf_0 s\Big]_0^2 + h\Delta f_0 \frac{s^2}{2}\Big]_0^2 + h\Delta^2 f_0\left(\frac{s^3}{6} - \frac{s^2}{4}\right)\Big]_0^2$$

$$= h\left(2f_0 + 2\Delta f_0 + \frac{1}{3}\Delta^2 f_0 \right) = \frac{h}{3}(f_0 + 4f_1 + f_2). \tag{4.26}$$

When we integrate the next term to get the error, we find

$$\int_{x_0}^{x_2} \frac{s(s-1)(s-2)}{6}\Delta^3 f_0\, dx = 0,$$

so, to get the error, we must integrate the succeeding term:

$$\text{Error} = \int_{x_0}^{x_2} \frac{s(s-1)(s-2)(s-3)}{24}h^4 f^{iv}(\xi)\, dx$$

$$= -\tfrac{1}{90}h^5 f^{iv}(\xi_1), \qquad x_0 \le \xi_1 \le x_2. \tag{4.27}$$

We omit the details of the integration in Eq. (4.27).

Similarly, for $n = 3$, we find

$$\int_{x_0}^{x_3} f(x)\, dx = \int_{x_0}^{x_3} P_3(x_s)\, dx = \frac{3h}{8}(f_0 + 3f_1 + 3f_2 + f_3). \tag{4.28}$$

$$\text{Error} = -\tfrac{3}{80}h^5 f^{v}(\xi_1), \qquad x_0 \le \xi_1 \le x_3. \tag{4.29}$$

An important item to observe is that the error terms for both $n = 2$ and $n = 3$ are $O(h^5)$. This means that the error of integration using a quadratic is similar to the integral using a cubic; it is a consequence of finding the integral of the next term equal to zero in deriving the error term of Eq. (4.27). Note also that the coefficient in Eq. (4.27)—($-\tfrac{1}{90}$)—is smaller than that in Eq. (4.29)—($-\tfrac{3}{80}$). The formula based on a quadratic is unexpectedly accurate.

This phenomenon is true of all the even-order Newton–Cotes formulas; each has an order of h in its error term the same as for the formula of next higher order. This suggests that the even-order rules are especially useful.

4.7 THE TRAPEZOIDAL RULE

The first of the Newton–Cotes formulas, based on approximating $f(x)$ on (x_0, x_1) by a straight line, is also called the *trapezoidal rule*. We have derived it by integrating $P_1(x_s)$, but the familiar and simple trapezoidal rule also can be considered to be an adaptation of the definition of the definite integral as a sum.

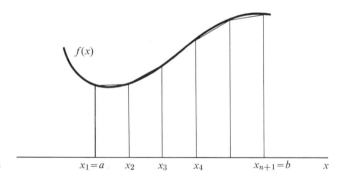

Figure 4.5 $x_1=a$ x_2 x_3 x_4 $x_{n+1}=b$ x

To evaluate $\int_a^b f(x)\,dx$, we subdivide the interval from a to b into n subintervals, as in Fig. 4.5. The area under the curve in each subinterval is approximated by the trapezoid formed by replacing the curve by its secant line drawn between the endpoints of the curve. The integral is then approximated by the sum of all the trapezoidal areas. There is no necessity to make the subintervals equal in width, but our formula is simpler if this is done. Let h be the constant Δx. Since the area of a trapezoid is its average height times the base, for each subinterval,

$$\int_{x_i}^{x_{i+1}} f(x)\,dx \doteq \frac{f(x_i) + f(x_{i+1})}{2}(\Delta x) = \frac{h}{2}(f_i + f_{i+1}), \tag{4.30}$$

and for $[a, b]$ subdivided into subintervals of size h,

$$\int_a^b f(x)\,dx \doteq \sum_{i=1}^{n} \frac{h}{2}(f_i + f_{i+1}) = \frac{h}{2}(f_1 + f_2 + f_2 + f_3 + \ldots + f_n + f_{n+1})$$

$$= \frac{h}{2}(f_1 + 2f_2 + 2f_3 + \ldots + 2f_n + f_{n+1}). \tag{4.31}$$

Equation (4.30) is identical to Eq. (4.24). Equation (4.31) is called the extended trapezoidal rule; it lets us apply the formula over an extended region where $f(x)$ is far from linear by applying the procedure to subintervals in which it can be approximated by linear segments. The formula is beautifully simple, and its applicability to unequally spaced values is useful in finding the integral of an experimentally determined function. It is obvious from Fig. 4.5 that the method is subject to large errors unless the subintervals are small, for replacing a curve by a straight line is hardly accurate.

Example Suppose we wished to integrate the function tabulated in Table 4.3 over the interval from $x = 1.8$ to $x = 3.4$. The extended trapezoidal rule gives

$$\int_{1.8}^{3.4} f(x)\,dx \doteq \frac{0.2}{2}[6.050 + 2(7.389) + 2(9.025) + 2(11.023) + 2(13.464)$$

$$+ 2(16.445) + 2(20.086) + 2(24.533) + 29.964] = 23.9944.$$

The data in Table 4.3 are for $f(x) = e^x$, so the true value of the integral is $e^{3.4} - e^{1.8} = 23.9144$. We are off in the second decimal.

We have previously derived the error of the trapezoidal rule as Eq. (4.25). We repeat it here:

$$\text{Local error of trapezoidal rule} = -\tfrac{1}{12}h^3 f''(\xi_1), \qquad x_0 \leqslant \xi_1 \leqslant x_1.$$

Table 4.3

x	$f(x)$	x	$f(x)$
1.6	4.953	2.8	16.445
1.8	6.050	3.0	20.086
2.0	7.389	3.2	24.533
2.2	9.025	3.4	29.964
2.4	11.023	3.6	36.598
2.6	13.464	3.8	44.701

This error, it should be emphasized, is the error of only a single step, and is hence called the *local error*. We normally apply the trapezoidal formula to a series of subintervals to get the integral over a large interval from $x = a$ to $x = b$. We are interested in the total error, which is called the *global error*.

To develop the formula for global error of the trapezoidal rule, we note that it is the sum of the local errors:

$$\text{Global error} = -\tfrac{1}{12}h^3[f''(\xi_1) + f''(\xi_2) + \ldots + f''(\xi_n)]. \qquad (4.32)$$

In Eq. (4.32) each of the values of ξ_i is found in the n successive subintervals. If we assume that $f''(x)$ is continuous on (a, b), there is some value of x in (a, b), say $x = \xi$, at which the value of the sum in Eq. (4.23) is equal to $n \cdot f''(\xi)$. Since $nh = b - a$, the global error becomes

$$\text{Global error of trapezoidal rule} = -\frac{1}{12}h^3 n f''(\xi) = \frac{-(b-a)}{12}h^2 f''(\xi) = O(h^2). \qquad (4.33)$$

The fact that the global error is $O(h^2)$ while the local error is $O(h^3)$ is reasonable since, for example, if h is halved the number of subintervals is doubled, so we add together twice as many errors.

When the function $f(x)$ is known, Eq. (4.33) permits us to estimate the error of numerical integration by the trapezoidal rule. In applying this equation we bracket the error by calculating with the maximum and the minimum values of $f''(x)$ on the interval $[a, b]$.

For the example above, our error expression gives these estimates:

$$\text{Error} = -\tfrac{1}{12}h^3 n f''(\xi), \qquad 1.8 \leq \xi \leq 3.4,$$

$$= -\tfrac{1}{12}(0.2)^3(8)\begin{Bmatrix} e^{1.8} & \text{(min)} \\ e^{3.4} & \text{(max)} \end{Bmatrix} = \begin{Bmatrix} -0.0323 & \text{(min)} \\ -0.1598 & \text{(max)} \end{Bmatrix}.$$

Alternatively,

$$\text{Error} = -\tfrac{1}{12}(0.2)^2(3.4 - 1.8)\begin{Bmatrix} e^{1.8} & \text{(min)} \\ e^{3.4} & \text{(max)} \end{Bmatrix} = \begin{Bmatrix} -0.0323 \\ -0.1598 \end{Bmatrix}.$$

The actual error was -0.080.

If we did not know the form of the function for which we have tabulated values, we would estimate $h^2 f''(\xi)$ from the second differences.

4.8 ROMBERG INTEGRATION

There is a way to improve on the accuracy of the simple trapezoidal rule when the function is known at equispaced intervals, or when it is explicitly known so that it can be computed as desired. This method is known as Romberg integration, or as extrapolation to the limit. It is based on the same principle as that used in Section 4.4 for the extrapolation of derivative values. We illustrate the method by a reconsideration of the data in Table 4.3. We saw that the trapezoidal rule gave a value of 23.9944 for $\int_{1.8}^{3.4} f(x)\,dx$, which differs from the true value of 23.9144. Purely for illustrative purposes, we do not desire to take advantage of our knowledge of the function, so we do not make h smaller to improve our estimate of the integral.

Even if we should not make the interval smaller, we can certainly make it larger, by using only every other value, taking $h = 0.4$. If we do this, we get:

$$I(h = 0.4) \doteq \frac{0.4}{2}[6.050 + 2(9.025) + 2(13.464) + 2(20.086) + 29.964]$$

$$= 24.2328.$$

This value is even more in error than before, which, of course, we expect with h larger; but we can take advantage of this to nearly eliminate the error of our former computation if we assume that errors of $O(h^2)$ can be interpreted as proportional to h^2, say error $= Ch^2$. Taking C to be a constant applies strictly only in the limit, of course, and at $h = 0.2$ and $h = 0.4$ we are not very near zero. It can be shown* however, that we make an error of only $O(h^4)$ by assuming

---

* See, for example, Davis and Rabinowitz (1967), who show that the error can be expressed as $C_1 h^2 + C_2 h^4 + C_3 h^6 + \ldots$.

that the errors are proportional to h^2. We then have, for the two computations above,

$$\text{True value} = \text{computed value} + \text{error,}$$
$$\text{True value} = 23.9944 + C(0.2)^2, \qquad (4.34)$$
$$\text{True value} = 24.2328 + C(0.4)^2.$$

In Eq. (4.34) there are two unknowns, the "true value" and the proportionality constant C, but we can certainly solve for these with two equations. We get, subtracting the second equation from four times the first,

$$\text{True value} = 23.9944 + \tfrac{1}{3}(23.9944 - 24.2328)$$
$$= 23.9149 \qquad \text{(versus 23.9144, exact value).}$$

We have extrapolated from two inexact values to an improved one. (What we have called "true value" is not exact because it used the erroneous assumption that $O(h^2)$ was the same as Ch^2.)

We can make a further improvement by extrapolation again, if we have two such estimates of $O(h^4)$ accuracy, by assuming that their errors can be stated as Ch^4. Such a second estimate can be obtained from the data in Table 4.3 by backing off once more, and computing the area with $h = 0.8$. If we do so we get $I(h = 0.8) = 25.1768$. We use this with $I(h = 0.4) = 24.2328$, as shown below.

h	Trapezoidal rule estimate $O(h^2)$ errors	Extrapolated value $O(h^4)$ errors
0.2	23.9944	
		23.9149
0.4	24.2328	
		23.9181
0.8	25.1768	

As before, we set up relationships:

$$\text{True value} = 23.9149 + C(0.2)^4,$$
$$\text{True value} = 23.9181 + C(0.4)^4.$$

Solving, we get

$$\text{True value} = 23.9149 + \tfrac{1}{15}(23.9149 - 23.9181)$$
$$= 23.9147.$$

This is now as close as can be determined to the exact value, 23.9144. One further order of extrapolation could be made, for we could start with $h = 1.6$ and, by combining with $h = 0.8$ and $h = 0.4$, get another second-order extrapolation. These second-order extrapolations can be shown to be of $O(h^6)$ accuracy and can be combined to an estimate of $O(h^8)$. No further improvement results from this, however, due to the influence of round-off errors in the original data. With only three-decimal accuracy there, it is not surprising to find errors in the fourth decimal of the integrals.

The number of successive improvements that can be made, and hence the order of the error that can be attained, obviously depends on the number of functional values that are available, or that are computed when $f(x)$ is a known function. The number of successive improvements that *should* be made may be less than the number that can, in principle, be made, because of the effects of round-off error.

We summarize our extrapolation rule, to be used for improving two values for which h varies 2:1 (it is identical to Eq. (4.22) that was derived for derivatives):

$$\text{Improved value} = \text{more accurate} + \left(\frac{1}{2^n - 1}\right)(\text{more accurate} - \text{less accurate}).$$
$$(4.35)$$

The exponent n in the coefficient is the exponent on h in the error term $O(h^n)$ that applies to the two values used for extrapolation. One always knows that the more accurate value is the one computed with smaller h.

When this technique of extrapolation is applied to integrals of known functions so that values can be computed for smaller values of h, it is known as *Romberg integration.* It is quite popular in computer programs,[*] often being available as a stock subroutine in the computer library. The technique is to compute the integral for an arbitrarily chosen h, and then to compute it again with h halved. If the two values for the integral differ by more than some tolerance value, the value is improved by extrapolation, and a second integration is computed with h halved again. This is combined with the previous value of the integral to give a second extrapolated value which is compared to the first. If necessary, these two are combined to give a higher-order extrapolation. The method is continued until a pair of extrapolated values of the integral agree satisfactorily.

The Romberg method is applicable to a wide class of functions—to any Reimann-integrable function, in fact. Smoothness and continuity are not required.

4.9 SIMPSON'S $\frac{1}{3}$ RULE

The second Newton–Cotes formula in Section 4.6 is for a quadratic integrated over two Δx intervals that are of uniform width. (Such intervals are often called "panels.") We repeat it from Section 4.6:

$$\int_{x_0}^{x_1} f(x)\, dx = \frac{h}{3}(f_0 + 4f_1 + f_2) - \frac{h^5}{90} f^{\text{iv}}(\xi), \qquad x_0 \le \xi \le x_2. \qquad (4.36)$$

---

[*] A subroutine that performs Romberg integration is given at the end of this chapter.

This is the very popular Simpson's $\frac{1}{3}$ rule, which has a local error of $O(h^5)$. If we apply this to a succession of pairs of panels to evaluate $\int_a^b f(x)\,dx$, we get

$$\int_a^b f(x)\,dx = \frac{h}{3}(f_1 + 4f_2 + 2f_3 + 4f_4 + 2f_5 + \ldots + 2f_{n+1} + 4f_n + f_{n+1})$$

$$-\frac{(b-a)}{180}h^4 f^{iv}(\xi), \qquad x_1 < \xi < x_{n+1}. \tag{4.37}$$

In Eq. (4.37) we have subscripted the x-values so that $x_1 = a$ and $x_{n+1} = b$. In getting the global error we have used the fact that $\sum f^{iv}(\xi_i) = nf^{iv}(\xi)$, which imposes a continuity condition on the fourth derivative. Here n, the number of times that the local rule was used, is only one-half the number of Δx steps, because we integrate over two panels each time; this causes the denominator to change to 180 from 90. Note that Simpson's $\frac{1}{3}$ rule has a global error of $O(h^4)$. Its use requires that we subdivide into an even number of panels.

Example   Apply Simpson's $\frac{1}{3}$ rule to the data of Table 4.3 to evaluate $\int_{1.8}^{3.4} f(x)\,dx$. Using $h = 0.2$, we get

$$\int_{1.8}^{3.4} f(x)\,dx \doteq \frac{0.2}{3}[6.050 + 4(7.389) + 2(9.025) + 4(11.023)$$

$$+ 2(13.464) + 4(16.445) + 2(20.086)$$

$$+ 4(24.533) + 29.964]$$

$$= 23.9149.$$

It is no coincidence that this value is the same as the extrapolated value from the trapezoidal-rule estimates at $h = 0.2$ and $h = 0.4$. The student should demonstrate that these two procedures always give identical results.

The estimated error for our example is

$$\text{Error} = \frac{-(3.4 - 1.8)}{180}(0.2)^4 \begin{Bmatrix} e^{1.8} & (\text{min}) \\ e^{3.4} & (\text{max}) \end{Bmatrix} = \begin{Bmatrix} -8.6 \times 10^{-5} \\ -4.3 \times 10^{-4} \end{Bmatrix}.$$

The actual error is -5×10^{-4}; round-off has magnified it.

An extrapolation can also be made for this example. We can integrate $\int_{1.8}^{3.4} f(x)\,dx$ using $h = 0.4$ and still apply the $\frac{1}{3}$ rule, getting a second estimate of $O(h^4)$ with h doubled (the value is 23.9181). Extrapolation from these two estimates gives an estimate of $O(h^6)$:

$$[23.9149 + \tfrac{1}{15}(23.9149 - 23.9181) = 23.9147].$$

Note that the coefficient of the "correction term," which involves the difference between the more accurate and less accurate estimate, is $1/(2^n - 1)$, the same as in Eq. (4.35). Since the error term for Simpson's $\frac{1}{3}$ rule can be written as $C_1 h^4 + C_2 h^6 + C_3 h^8 + \ldots$, we can do a Romberg-like set of successive extrapolations.

4.10 SIMPSON'S $\frac{3}{8}$ RULE

The third Newton–Cotes formula that finds frequent use is obtained by integrating a cubic interpolating polynomial over its range of fit. We have already derived it as Eq. (4.28) and the error term as Eq. (4.29). The local rule is

$$\int_{x_0}^{x_3} f(x)\,dx = \frac{3h}{8}(f_0 + 3f_1 + 3f_2 + f_3) - \frac{3}{80}h^5 f^{iv}(\xi), \qquad x_0 \leqslant \xi \leqslant x_3.$$

The value of the coefficient $3h/8$ gives it its common name, *Simpson's $\frac{3}{8}$ Rule*. When applied over the interval from $x_1 = a$ to $x_{n+1} = b$ (the interval must be divided into a multiple of three panels), we get

$$\int_a^b f(x)\,dx = \frac{3h}{8}(f_1 + 3f_2 + 3f_3 + 2f_4 + 3f_5 + 3f_6 + \ldots + 2f_{n-2} + 3f_{n-1}$$

$$+ 3f_n + f_{n+1}) - \frac{(b-a)}{80}h^4 f^{iv}(\xi), \qquad x_1 \leqslant \xi \leqslant x_{n+1}. \tag{4.38}$$

The global rule has a coefficient in the error term of $\frac{1}{80}$, reduced from $\frac{3}{80}$ in the local error, because we apply the local rule $n/3$ times when there are n panels.

As we have previously seen, the order of the global error, $O(h^4)$, is no higher than for Simpson's $\frac{1}{3}$ rule and the coefficient is larger. One might wonder whether the $\frac{3}{8}$ rule should ever be used.

As we have seen, Simpson's $\frac{1}{3}$ rule requires that the interval be subdivided into an even number of panels. When applied to tabulated data, this is often impossible. In such cases, one can combine the two rules, applying the $\frac{1}{3}$ rule as far as possible and completing the integration by picking up three panels with the $\frac{3}{8}$ rule. An alternative method for an odd number of panels is to integrate over one panel by the trapezoidal rule. This involves a penalty of an $O(h^2)$ error over that subinterval, however, and is somewhat less accurate.

Example   Use the data of Table 4.3 to find the integral from $x = 1.6$ to $x = 3.4$. There are nine panels, so we cannot apply the $\frac{1}{3}$ rule directly. We have several choices to handle this problem: Apply the trapezoidal rule; apply Simpson's $\frac{3}{8}$ rule, or combine Simpson's $\frac{1}{3}$ rule with either one of the others. We investigate each choice.

With the trapezoidal rule:

$$\int_{1.6}^{3.4} f(x)\,dx = 25.0947, \qquad \text{error} = -0.0836.$$

With Simpson's $\frac{3}{8}$ rule:

$$\int_{1.6}^{3.4} f(x)\,dx = 25.0119, \qquad \text{error} = -0.0008.$$

With Simpson's $\frac{1}{3}$ rule from 1.8 to 3.4, and the trapezoidal rule from 1.6 to 1.8:

$$\int_{1.6}^{3.4} f(x)\, dx = 25.0152, \qquad \text{error} = -0.0041.$$

With Simpson's $\frac{1}{3}$ rule from 2.2 to 3.4, and the $\frac{3}{8}$ rule from 1.6 to 2.2:

$$\int_{1.6}^{3.4} f(x)\, dx = 25.0115, \qquad \text{error} = -0.0004.$$

The results are as we would anticipate: The best accuracy is obtained when the $\frac{1}{3}$ and $\frac{3}{8}$ rules are combined. We still have a choice to make when we combine the rules: At which three panels should we use the $\frac{3}{8}$ rule? A consideration of the error term of the $\frac{3}{8}$ rule gives the answer: Apply it where its error is least. This means that we should choose the three panels where the fourth derivative of the function is smallest. For the data of this example, we know the function $[f(x) = e^x]$ so the choice we made is best. The choice will not always be obvious when we know $f(x)$ only as the table of values. The differences will be minor, however, and round-off errors distort the picture, so this point is of more theoretical than practical importance.

For convenience, we collect here the Newton–Cotes integration formulas.

Formulas for integration (Uniform spacing, $\Delta x = h$)

Trapezoidal Rule:

$$\int_a^b f(x)\, dx = \frac{h}{2}(f_1 + 2f_2 + 2f_3 + \ldots + 2f_n + f_{n+1})$$

$$- \frac{(b-a)}{12} h^2 f''(\xi), \qquad a \leqslant \xi \leqslant b.$$

$\left(h^2 \right)$

Simpson's $\frac{1}{3}$ Rule:

$$\int_a^b f(x)\, dx = \frac{h}{3}(f_1 + 4f_2 + 2f_3 + 4f_4 + 2f_5 + \ldots + 4f_n + f_{n+1})$$

$$- \frac{(b-a)}{180} h^4 f^{iv}(\xi), \qquad a \leqslant \xi \leqslant b.$$

$\left(h^4 \right)$

(Requires an even number of panels.)

Simpson's $\frac{3}{8}$ Rule:

$$\int_a^b f(x)\, dx = \frac{3h}{8}(f_1 + 3f_2 + 3f_3 + 2f_4 + 3f_5 + 3f_6 + \ldots + 3f_n + f_{n+1})$$

$$- \frac{(b-a)}{80} h^4 f^{iv}(\xi), \qquad a \leqslant \xi \leqslant b.$$

$\left(h^4 \right)$

(Requires a number of panels divisible by 3.)

4.11 OTHER WAYS TO DERIVE INTEGRATION FORMULAS

It is interesting to examine other ways of deriving integration formulas than by integrating the interpolating polynomial. One way is to use symbolic methods, similar to those of Section 4.2 and Section 3.7.

In terms of the stepping operator E,

$$f(x_s) = f_s = E^s f_0.$$

Multiplying by $dx = h\,ds$ and integrating from x_0 to x_1 ($s = 0$ to $s = 1$), we obtain

$$\int_{x_0}^{x_1} f(x)\,dx = h \int_0^1 E^s f_0\,ds = \left[\frac{hE^s}{\ell n\,E} f_0\right]_0^1 = \frac{h(E-1)}{\ell n\,E} f_0.$$

Let $E = 1 + \Delta$ and expand $\ell n\,(1 + \Delta)$ as a power series:

$$\ell n\,(1 + \Delta) = \Delta - \tfrac{1}{2}\Delta^2 + \tfrac{1}{3}\Delta^3 - \tfrac{1}{4}\Delta^4 + \ldots.$$

On dividing Δ by this series, we get

$$\int_{x_0}^{x_1} f(x)\,dx = \frac{h\Delta}{\Delta - \tfrac{1}{2}\Delta^2 + \tfrac{1}{3}\Delta^3 - \tfrac{1}{4}\Delta^4 + \ldots} f_0$$

$$= h(f_0 + \tfrac{1}{2}\Delta f_0 - \tfrac{1}{12}\Delta^2 f_0 + \tfrac{1}{24}\Delta^3 f_0 - \ldots). \qquad (4.39)$$

The coefficients are considerably easier to get by this technique than by the term-by-term integration of Section 4.6.

Equation (4.39) is not a Newton–Cotes formula unless we use only the first two terms, making the interval of integration and the range of fit of the polynomial agree. When n terms are used, it represents a polynomial of degree n, fitting from x_0 to x_n, but integrated only from x_0 to x_1. This is especially useful in connection with differential-equation methods.

We can develop the formula for an nth-degree interpolating polynomial integrated over two panels, from x_0 to x_2, in a similar fashion:

$$\int_{x_0}^{x_2} f(x)\,dx = \left[\frac{hE^s}{\ell n\,E} f_0\right]_0^2 = \frac{h(E^2-1)}{\ell n\,E} f_0 = \frac{h(E+1)(E-1)}{\ell n\,E} f_0.$$

Again letting $E = 1 + \Delta$ so $E - 1 = \Delta$ and $E + 1 = 2 + \Delta$, and dividing Δ by the series for $\ell n\,(1 + \Delta)$, we get

$$\int_{x_0}^{x_2} f(x)\,dx = h(2 + \Delta)(f_0 + \tfrac{1}{2}\Delta f_0 - \tfrac{1}{12}\Delta^2 f_0 + \tfrac{1}{24}\Delta^3 f_0 - \ldots)$$

$$= h(2f_0 + \Delta f_0 - \tfrac{1}{6}\Delta^2 f_0 + \tfrac{1}{12}\Delta^3 f_0 - \ldots$$

$$+ \Delta f_0 + \tfrac{1}{2}\Delta^2 f_0 - \tfrac{1}{12}\Delta^3 f_0 + \ldots)$$

$$= h(2f_0 + 2\Delta f_0 + \tfrac{1}{3}\Delta^2 f_0 + 0 - \ldots). \qquad (4.40)$$

The method can obviously be extended. One reason why we might desire formulas such as Eqs. (4.39) and (4.40) is to use them to construct lozenge diagrams similar to Figs. 4.1 and 4.3. These would permit integration over m panels based on polynomials of degree n. The diagrams are easy to construct if we remember that the coefficients themselves form a difference table that is interlaced with the function differences. Since they are of rather special application, we do not exhibit them here.

We now present still another interesting method of deriving formulas that can be applied to a variety of situations including the development of integration formulas. It may be called the method of *undetermined coefficients*.* We express the formula as a sum of $n + 1$ terms with unknown coefficients, and then evaluate the coefficients by requiring that the formula be exact for all polynomials of degree n or less. We illustrate it here by finding Simpson's $\frac{1}{3}$ rule by this other technique. Express the integral as a weighted sum of three equispaced function values:

$$\int_{-1}^{1} f(x)\, dx = af(-1) + bf(0) + cf(+1). \tag{4.41}$$

The symmetrical interval of integration simplifies the arithmetic. We stipulate that the function is to be evaluated at three equally spaced intervals, the two end values and the midpoint. Since the formula contains three terms, we can require it to be correct for all polynomials of degree 2 or less. If that is true, it certainly must be true for the three special cases of $f(x) = x^2$, $f(x) = x$, and $f(x) = 1$. We rewrite Eq. (4.41) three times, applying each definition of $f(x)$ in turn:

$$f(x) = 1: \qquad \int_{-1}^{1} dx = 2 = a(1) + b(1) + c(1) = a + b + c;$$

$$f(x) = x: \qquad \int_{-1}^{1} x\, dx = 0 = a(-1) + b(0) + c(1) = -a + c;$$

$$f(x) = x^2: \qquad \int_{-1}^{1} x^2\, dx = \tfrac{2}{3} = a(1) + b(0) + c(1) = a + c.$$

Solving the three equations simultaneously gives $a = \frac{1}{3}$, $b = \frac{4}{3}$, $c = \frac{1}{3}$. Here the spacing between points was unity; obviously the integral is proportional to $\Delta x = h$. We then get Simpson's $\frac{1}{3}$ rule:

$$\int_{-h}^{h} f(x)\, dx = h[\tfrac{1}{3}f(-h) + \tfrac{4}{3}f(0) + \tfrac{1}{3}f(h)].$$

---

* This method is described in more detail in the Appendix, and several examples are presented.

4.12 GAUSSIAN QUADRATURE

Our previous formulas for numerical integration were all predicated on evenly spaced x-values; this means the x-values were predetermined. With a formula of three terms, then, there were three parameters, the coefficients (weighting factors) applied to each of the functional values. A formula with three parameters corresponds to a polynomial of the second degree, one less than the number of parameters. Gauss observed that if we remove the requirement that the function be evaluated at predetermined x-values, a three-term formula will contain six parameters (the three x-values are now unknowns, plus the three weights) and should correspond to an interpolating polynomial of degree 5. Formulas based on this principle are called *Gaussian quadrature formulas*. They can be applied only when $f(x)$ is explicitly known, so that it can be evaluated at any desired value of x.

We shall determine the parameters in the simple case of a two-term formula containing four unknown parameters:

$$\int_{-1}^{1} f(t) \doteq af(t_1) + bf(t_2).$$

The method is the same as that illustrated in the previous section, by determining unknown parameters. We use a symmetrical interval of integration to simplify the arithmetic, and call our variable t. (This agrees with the notation of most authors. Since the variable of integration is only a dummy variable, its name is unimportant.) Our formula is to be valid for any polynomial of degree 3; hence it will hold if $f(t) = t^3$, $f(t) = t^2$, $f(t) = t$, and $f(t) = 1$:

$$
\begin{aligned}
f(t) = t^3: \quad & \int_{-1}^{1} t^3\, dt = 0 = at_1^3 + bt_2^3; \\
f(t) = t^2: \quad & \int_{-1}^{1} t^2\, dt = \tfrac{2}{3} = at_1^2 + bt_2^2; \\
f(t) = t; \quad & \int_{-1}^{1} t\, dt = 0 = at_1 + bt_2; \\
f(t) = 1: \quad & \int_{-1}^{1} dt = 2 = a + b.
\end{aligned}
\tag{4.42}
$$

Multiplying the third equation by t_1^2, and subtracting from the first, we have

$$0 = 0 + b(t_2^3 - t_2 t_1^2) = b(t_2)(t_2 - t_1)(t_2 + t_1). \tag{4.43}$$

We can satisfy Eq. (4.43) by either $b = 0$, $t_2 = 0$, $t_1 = t_2$ or $t_2 = -t_1$. Only the last of these possibilities is satisfactory, the others being invalid, or else reduce our formula to only a single term, so we choose $t_1 = -t_2$. We then find that

$$a = b = 1,$$
$$t_2 = -t_1 = \sqrt{\tfrac{1}{3}} = 0.5773,$$
$$\int_{-1}^{1} f(t)\, dx \doteq f(-0.5773) + f(0.5773).$$

It is remarkable that adding these two values of the function gives the exact value for the integral of any cubic polynomial over the interval from -1 to 1.

Suppose our limits of integration are from a to b, and not -1 to 1 for which we derived this formula. To use the tabulated Gaussian quadrature parameters, we must change the interval of integration to $(-1, 1)$ by a change of variable. We replace the given variable by another to which it is linearly related according to the following scheme:

If we let

$$x = \frac{(b - a)t + b + a}{2}, \qquad \text{so that } dx = \left(\frac{b - a}{2}\right) dt,$$

then

$$\int_a^b f(x)\, dx = \frac{b - a}{2} \int_{-1}^1 f\left(\frac{(b - a)t + b + a}{2}\right) dt.$$

Example Evaluate $I = \int_0^{\pi/2} \sin x\, dx$. (Obviously, $I = 1.0$, so we can readily see the error of our estimate.)

To use the two-term Gaussian formula, we must change the variable of integration to make the limits of integration from -1 to 1.

Let

$$x = \frac{(\pi/2)t + \pi/2}{2}, \qquad \text{so } dx = (\pi/4)\, dt.$$

Observe that when $t = -1$, $x = 0$; when $t = 1$, $x = \pi/2$.) Then

$$I = \frac{\pi}{4} \int_{-1}^1 \sin\left(\frac{\pi t + \pi}{4}\right) dt.$$

The Gaussian formula calculates the value of the new integral as a weighted sum of two values of the integrand, at $t = -0.5773$ and at $t = 0.5773$. Hence,

$$I = \frac{\pi}{4}[(1.0)(\sin(0.10566\pi)) + (1.0)(\sin(0.39434\pi))]$$

$$= 0.99847.$$

The error is 1.53×10^{-3}.

The power of the Gaussian method derives from the fact that we need only two functional evaluations. If we had used the trapezoidal rule, which also requires only two evaluations, our estimate would have been $(\pi/4)(0.0 + 1.0) = 0.7854$, an answer quite far from the mark. Simpson's $\frac{1}{3}$ rule requires three functional evaluations and gives $I = 1.0023$, with an error of -2.3×10^{-3}, somewhat greater than for Gaussian quadrature.

To obtain the parameters for Gaussian formulas of higher order than the simple two-term formula we have derived, we must solve a set of simultaneous equations in exactly the same way. The set of equations that results from writing the definition of $f(t)$ as a succession of polynomials is not easily solved, however, because of the nonlinearity with respect to t. The theory of orthogonal polynomials is employed; the values of t that satisfy the equations are the roots of the Legendre polynomials. This is easier than the methods for a nonlinear system that we used in Chapter 2.

The Legendre polynomials are defined by recursion:

$$(n + 1)L_{n+1}(x) - (2n + 1)xL_n(x) + nL_{n+1}(x) = 0,$$

with $L_0(x) = 1$, $L_1(x) = x$.

Then $L_2(x)$ is

$$L_2(x) = \frac{3xL_1(x) - (1)L_0(x)}{2} = \frac{3}{2}x^2 - \frac{1}{2};$$

here roots are $\pm\sqrt{\frac{1}{3}} = \pm0.5773$, precisely the t-values for the two-term formula.

By using the recursion relation, we find

$$L_3(x) = \frac{5x^3 - 3x}{2},$$

$$L_4(x) = \frac{35x^4 - 30x^2 + 3}{8}, \quad \text{etc.}$$

The methods of Chapter 1 allow us to find the roots of these polynomials. After these have been determined, the set of equations analogous to Eqs. (4.42) can easily be solved for the weighting factors because the equations are linear with respect to these unknowns.

The weighting factors and t-values for Gaussian quadrature have been tabulated. (Love (1966) gives values for up to 200-term formulas.) We are content to give a few of the values in Table 4.4.

Table 4.4 Values for Gaussian Quadrature.

Number of terms	Values of t	Weighting factor	Valid up to degree
2	−0.57735027	1.0	3
	0.57735027	1.0	
3	−0.77459667	0.55555555	5
	0.0	0.88888889	
	0.77459667	0.55555555	
4	−0.86113631	0.34785485	7
	−0.33998104	0.65214515	
	0.33998104	0.65214515	
	0.86113631	0.34785485	

We illustrate the three-term formula with an example.

Example Evaluate $I = \int_{0.2}^{1.5} e^{-x^2} dx$ using the three-term Gaussian formula.
Change variable: Let

$$x = \frac{(1.5 - 0.2)t + 1.5 + 0.2}{2} = 0.65t + 0.85.$$

Then

$$I = \frac{1.5 - 0.2}{2} \int_{-1}^{1} e^{-(0.65t+0.85)^2} dt$$

$$= 0.65[0.555\ldots e^{-[0.65(-0.774\ldots)+0.85]^2} + 0.888\ldots e^{-[0.65(0.0)+0.85]^2}$$

$$+ 0.555\ldots e^{-[0.65(0.774\ldots)+0.85]^2}]$$

$$= 0.6586 \qquad \text{(cf. to correct value 0.65882).}$$

4.13 IMPROPER INTEGRALS AND INDEFINITE INTEGRALS

Our previous examples have been for integrals with a finite upper and lower limit and with bounded integrands. Sometimes we want to evaluate integrals such as:

$$I_1 = \int_0^\infty xe^{-x} dx; \tag{4.44}$$

$$I_2 = \int_0^2 \frac{dx}{\sqrt{x}}; \tag{4.45}$$

$$I_3 = \int_0^t x\sqrt{x^2 + 1} \, dx. \tag{4.46}$$

Equation (4.44) has an infinite upper limit; Eq. (4.45) has an infinite value for the integrand at the lower limit, $x = 0$; Eq. (4.46) is an indefinite integral whose value is a function of t.

Sometimes we can transform such integrals into the simpler forms that we have discussed above, by a change of variable. For example, we can break I_1 into the sum of two integrals:

$$I_1 = \int_0^1 xe^{-x} dx + \int_1^\infty xe^{-x} dx,$$

and then change the variable in the second term by substituting $y = 1/x$, so that

$$\int_1^\infty xe^{-x} dx = \int_1^0 \frac{1}{y} e^{-1/y} \left(-\frac{dy}{y^2}\right) = \int_0^1 \frac{e^{-1/y}}{y^3} \, dy.$$

In the transformed integral, the value of the integrand at $y = 0$ is indeterminate (0/0) but this causes us no trouble because the limit of the integrand is zero as $y \to 0$.

Where the integral exists, as in Eq. (4.44), we can apply our numerical methods directly, however. We just evaluate

$$\int_0^A xe^{-x}\,dx$$

for increasing values of A, and use the limiting value of these results as the value of the integral with infinite upper limit. Observe that this is just an adaptation of the usual calculus definition of the value of an integral with an infinite upper limit. One must be on guard in doing this when the convergence of the integral is unknown, for round-off error may give an upper limit to the sequence of results even though the integral actually diverges.

Example   Table 4.5 below gives results for the integral in Eq. (4.44). They show how the value of the integral reaches a limiting value as the value of A increases. (A computer program that implements Romberg integration was used.)

Table 4.5   Values for $I = \int_0^A xe^{-x}\,dx,\ \ A \to \infty$.

A	I	
1	0.26424	
10	0.99950	
100	1.00001	
1000	1.00001	
10000	1.00001	
(∞)	(1.00000)	(analytical result)

The value of the integral in Eq. (4.45) can be found similarly by finding the value of I_2 with a variable lower limit that approaches closer and closer to zero, the point where the integrand is undefined. Equation (4.46) is also amenable to this same approach; we let t take on a series of values and produce a table that relates the value of I_3 to values of t.

Equation (4.46) is a simple special case of a situation we shall treat at length in the next chapter, which deals with differential equations.

4.14   MULTIPLE INTEGRALS

We consider first the case when the limits of integration are constants. In the calculus we learned that a double integral may be evaluated as an iterated integral; in other words, we may write:

$$\iint_A f(x, y)\,dA = \int_a^b \left(\int_c^d f(x, y)\,dy \right) dx = \int_c^d \left(\int_d^b f(x, y)\,dx \right) dy. \qquad (4.47)$$

In Eq. (4.47) the rectangular region A is bounded by the lines

$$x = a, \qquad x = b, \qquad y = c, \qquad y = d.$$

In computing the iterated integrals, we hold x constant while integrating with respect to y (vice versa in the second case).

Adapting the numerical integration formulas developed earlier in this chapter for integration with respect to just one independent variable is quite straightforward. Recall that any of the integration formulas is just a linear combination of values of the function, evaluated at varying values of the independent variable. In other words, a quadrature formula is just a weighted sum of certain functional values. The inner integral is written then as a weighted sum of function values with one variable held constant. We then add together a weighted sum of these sums. If the function is known only at the nodes of a rectangular grid through the region, we are constrained to use these values. The Newton–Cotes formulas are a convenient set to employ. There is no reason why the same formula must be used in each direction, although it is often particularly convenient to do so.

We illustrate this technique by evaluating the integral of the function of Table 4.6 over the rectangular region bounded by

$$x = 1.5, \qquad x = 3.0, \qquad y = 0.2, \qquad y = 0.6.$$

Let us use the trapezoidal rule in the x-direction and Simpson's $\frac{1}{3}$ rule in the y-direction. (Since the number of panels in the x-direction is not even, Simpson's $\frac{1}{3}$ rule does not apply readily.) It is immaterial which integral we evaluate first. Suppose we start with y constant:

$$y = 0.2: \qquad \int_{1.5}^{3.0} f(x, y)\, dx = \int_{1.5}^{3.0} f(x, 0.2)\, dx = \frac{h}{2}(f_1 + 2f_2 + 2f_3 + f_4)$$

$$= \frac{0.5}{2}[0.990 + 2(1.568) + 2(2.520) + 4.090]$$

$$= 3.3140;$$

$$y = 0.3: \qquad \int_{1.5}^{3.0} f(x, 0.3)\, dx = \frac{0.5}{2}[1.524 + 2(2.384) + 2(3.800) + 6.136]$$

$$= 5.0070.$$

Similarly, at

$$y = 0.4, \qquad I = 6.6522;$$
$$y = 0.5, \qquad I = 8.2368;$$
$$y = 0.6, \qquad I = 9.7435.$$

We now sum these in the y-direction according to Simpson's rule:

$$f(x, y)\, dx = \frac{0.1}{3}[3.3140 + 4(5.0070) + 2(6.6522) + 4(8.2368) + 9.7435]$$

$$= 2.6446.$$

Table 4.6 Tabulation of a function of two variables, $u = f(x, y)$.

x \ y	0.1	0.2	0.3	0.4	0.5	0.6
0.5	0.165	0.428	0.687	0.942	1.190	1.431
1.0	0.271	0.640	1.003	1.359	1.703	2.035
1.5	0.447	0.990	1.524	2.045	2.549	3.031
2.0	0.738	1.568	2.384	3.177	3.943	4.672
2.5	1.216	2.520	3.800	5.044	6.241	7.379
3.0	2.005	4.090	6.136	8.122	10.030	11.841
3.5	3.306	6.679	9.986	13.196	16.277	19.198

(In this example our answer does not check well with the analytical value of 2.5944 because the x-intervals are large. We could improve our estimate somewhat by fitting a higher-degree polynomial than the first to provide the integration formula. We can even use values outside the range of integration for this, using the techniques of Section 4.11 to derive the formulas.)

The previous example shows that double integration by numerical means reduces to a double summation of weighted function values. The calculations we have just made could be written in the form:

$$
\int f(x, y) \, dx \, dy = \sum_{j=1}^{m} v_j \sum_{i=1}^{n} w_i f_{ij}
$$

$$
= \frac{\Delta y}{3} \frac{\Delta x}{2} [(f_{1,1} + 2f_{2,1} + 2f_{3,1} + f_{4,1})
$$

$$
+ 4(f_{1,2} + 2f_{2,2} + 2f_{3,2} + f_{4,2})
$$

$$
+ \ldots + (f_{1,5} + 2f_{2,5} + 2f_{3,5} + f_{4,5})].
$$

It is convenient to write this in a pictorial operator form, in which the weighting factors are displayed in an array that is a map to the location of the functional values to which they are applied.

$$
\int f(x, y) \, dx \, dy = \frac{\Delta y}{3} \frac{\Delta x}{2}
\begin{bmatrix}
1 & 4 & 2 & 4 & 1 \\
2 & 8 & 4 & 8 & 2 \\
2 & 8 & 4 & 8 & 2 \\
1 & 4 & 2 & 4 & 1
\end{bmatrix} f_{i,j}.
\tag{4.48}
$$

We interpret the numbers in the array of Eq. (4.48) in this manner: We use the values 1, 4, 2, 4, and 1 as weighting factors for functional values in the top row of the portion of Table 4.6 that we integrate over (values where $x = 1.5$ and y varies from 0.2 to 0.6.) Similarly, the second column of the array in Eq. (4.48) represents weighting factors that are applied to a column of function values where $y = 0.4$ and x varies from 1.5 to 3.0. Observe that the values in the pictorial operator of Eq. (4.48) follow immediately from the Newton–Cotes coefficients for single variable integration.

Other combinations of Newton–Cotes formulas give similar results. It is probably easiest for hand calculation to use these pictorial integration operators. Pictorial integration is readily adapted to any desired combination of integration formulas. Except for the difficulty of representation beyond two dimensions, this operator technique also applies to triple and quadruple integrals.

There is an alternative representation to such pictorial operators that is easier to translate into a computer program. We also derive it somewhat differently. Consider the numerical integration formula for one variable,

$$\int_{-1}^{1} f(x)\, dx \doteq \sum_{i=1}^{n} a_i x_i. \tag{4.49}$$

We have seen in Section 4.11 that such formulas can be made exact if $f(x)$ is any polynomial of a certain degree. Assume that Eq. (4.49) holds for polynomials up to degree s.*

We now consider the multiple integral formula

$$\int_{-1}^{1} \int_{-1}^{1} \int_{-1}^{1} f(x, y, z)\, dx\, dy\, dz \stackrel{?}{=} \sum_{i=1}^{n} \sum_{j=1}^{n} \sum_{k=1}^{n} a_i a_j a_k f(x_i, y_j, z_k). \tag{4.50}$$

We wish to show that Eq. (4.50) is exact for all polynomials in x, y, and z up to degree s. Such a polynomial is a linear combination of terms of the form $x^\alpha y^\beta z^\gamma$, where α, β, and γ are nonnegative integers whose sum is equal to s or less. If we can prove that Eq. (4.50) holds for the general term of this form, it will then hold for the polynomial.

To do this we assume that

$$f(x, y, z) = x^\alpha y^\beta z^\gamma.$$

Then, since the limits are constants and the integrand is factorable,

$$I = \int_{-1}^{1} \int_{-1}^{1} \int_{-1}^{1} x^\alpha y^\beta z^\gamma\, dx\, dy\, dz$$

$$= \left(\int_{-1}^{1} x^\alpha\, dx \right) \left(\int_{-1}^{1} y^\beta\, dy \right) \left(\int_{-1}^{1} z^\gamma\, dz \right).$$

Replacing each term according to Eq. (4.49), we get,

$$I = \left(\sum_{i=1}^{n} a_i x_i \right) \left(\sum_{j=1}^{n} a_j y_j \right) \left(\sum_{k=1}^{n} a_k z_k \right) = \sum_{i=1}^{n} a_i x_i \sum_{j=1}^{n} a_j y_j \sum_{k=1}^{n} a_k z_k. \tag{4.51}$$

---

* For Newton–Cotes formulas, $s = n - 1$ for n even and $s = n$ for n odd. For Gaussian quadrature formulas, $s = 2n - 1$, and the x_i will be unevenly spaced.

We need now an elementary rule about the product of summations. We illustrate it for a simple case. We assert that

$$\left(\sum_{i=1}^{3} u_i\right)\left(\sum_{j=1}^{2} v_j\right) = \sum_{i=1}^{3}\left(\sum_{j=1}^{2} u_i v_j\right)$$

$$= \sum_{i=1}^{3}\sum_{j=1}^{2} u_i v_j.$$

The last equality is purely notational. We prove the first by expanding both sides:

$$\left(\sum_{i=1}^{3} u_i\right)\left(\sum_{j=1}^{2} v_j\right) = \sum_{i=1}^{3} u_i \sum_{j=1}^{2} v_i$$

$$= (u_1 + u_2 + u_3)(v_1 + v_2)$$

$$= u_1 v_1 + u_1 v_2 + u_2 v_1 + u_2 v_2 + u_3 v_1 + u_3 v_2;$$

$$\sum_{i=1}^{3}\sum_{j=1}^{2} u_i v_j = (u_1 v_1 + u_1 v_2) + (u_2 v_1 + u_2 v_2) + (u_3 v_1 + u_3 v_2).$$

On removing parentheses we see the two sides are the same. Using this principle, we can write Eq. (4.51) in the form

$$I = \sum_{i=1}^{n}\sum_{j=1}^{n}\sum_{k=1}^{n} a_i a_j a_k x_i y_j z_k, \tag{4.52}$$

which shows that the questioned equality of Eq. (4.50) is valid, and we can write a program for a triple integral by three nested DO loops. The coefficients a_i are chosen from any numerical integration formula. If the three one-variable formulas corresponding to Eq. (4.50) are not identical, an obvious modification of Eq. (4.52) applies. In some cases a change of variable is needed to correspond to Eq. (4.49).

　　If we are evaluating a multiple integral numerically where the integrand is a known function, our choice of the form of Eq. (4.49) is wider. Of higher efficiency than the Newton–Cotes formulas is Gaussian quadrature. Since it also fits the pattern of Eq. (4.49), the formula of Eq. (4.52) applies. We illustrate with a simple example.

Example Evaluate

$$I = \int_{0}^{1}\int_{-1}^{0}\int_{-1}^{1} yze^x \, dx \, dy \, dz$$

by Gaussian quadrature using a three-term formula for x and two-term formulas for y and z. We first make the changes of variables to adjust the limits for y and z to $(-1, 1)$:

$$y = \tfrac{1}{2}(u + 1), \qquad dy = \tfrac{1}{2}du;$$
$$z = \tfrac{1}{2}(v - 1), \qquad dz = \tfrac{1}{2}dv.$$

Our integral becomes

$$I = \tfrac{1}{16} \int_{-1}^{1} \int_{-1}^{1} \int_{-1}^{1} (u + 1)(v - 1)e^x \, dx \, du \, dv.$$

The two- and three-point Gaussian formulas are, from Section 4.12:

$$\int_{-1}^{1} f(x) \, dx = (1)f(-0.5774) + (1)f(0.5774),$$

$$\int_{-1}^{1} f(x) \, dx = (\tfrac{5}{9})f(-0.7746) + (\tfrac{8}{9})f(0) + (\tfrac{5}{9})f(0.7746).$$

The integral is then

$$I = \tfrac{1}{16} \sum_{i=1}^{2} \sum_{j=1}^{2} \sum_{k=1}^{3} a_i a_j b_k (u_i + 1)(v_j - 1)e^{x_k},$$

$$a_1 = 1, \qquad a_2 = 1,$$
$$b_1 = \tfrac{5}{9}, \qquad b_2 = \tfrac{8}{9}, \qquad b_3 = \tfrac{5}{9},$$

and values of u, v, and x as above.

A few representative terms of the sum are

$$I = \tfrac{1}{16}[(1)(1)(\tfrac{5}{9})(-0.5774 + 1)(-0.5774 - 1)e^{-0.7446}$$
$$+ (1)(1)(\tfrac{8}{9})(-0.5774 + 1)(-0.5774 - 1)e^{0}$$
$$+ (1)(1)(\tfrac{5}{9})(-0.5774 + 1)(-0.5774 - 1)e^{0.7746}$$
$$+ (1)(1)(\tfrac{5}{9})(0.5774 + 1)(-0.5774 - 1)e^{-0.7746}$$
$$+ \ldots].$$

On evaluating, we get $I = -0.58758$. The analytical value is

$$-\tfrac{1}{4}(e - e^{-1}) = -0.58760.$$

4.15 ERRORS IN MULTIPLE INTEGRATION AND EXTRAPOLATIONS

The error term of a one-variable quadrature formula is an additive one just like the other terms in the linear combination (although of special form). It would seem reasonable that it would go through the multiple summations in a similar fashion, so we should expect error terms for multiple integration that are analogous to the one-dimensional case. We illustrate that this is true for double integration using the trapezoidal rule in both directions, with uniform spacings, choosing n intervals in the x-direction and m in the y-direction.

From Section 4.7 we have

$$\text{Error of} \int_{a}^{b} f(x) \, dx = -\frac{b - a}{12} h^2 f''(\xi) = O(h^2),$$

$$h = \Delta x = \frac{b - a}{n}.$$

In developing Romberg integration, we found that the error term could be written as:

$$\text{Error} = O(h^2) = Ah^2 + O(h^4) \doteq Ah^2 + Bh^4,$$

where A is a constant and the value of B depends on a fourth derivative of the function. Appending this error term to the trapezoidal rule, we get (equivalent to Eq. (4.49)):

$$\int_a^b f(x)\, dx \bigg|_{y=y_i} = \frac{h}{2}(f_{0,j} + 2f_{1,j} + 2f_{2,j} + \ldots + f_{n,j}] + A_j h^2 + B_j h^4.$$

Summing these in the y-direction and retaining only the error terms, we have:

$$\int_c^d \int_a^b f(x, y)\, dx\, dy = \frac{k}{2}\frac{h}{2}\sum_{i=0}^n \sum_{j=0}^m a_i a_j f_{i,j}$$

$$+ \frac{k}{2}(A_0 + 2A_1 + 2A_2 + \ldots + A_m)h^2$$

$$+ \frac{k}{2}(B_0 + 2B_1 + 2B_2 + \ldots + B_m)h^4 \qquad (4.53)$$

$$+ \bar{A}k^2 + \bar{B}k^4,$$

$$k = \Delta y = \frac{d - c}{m}.$$

In Eq. (4.53), $\bar{A}$ and $\bar{B}$ are the coefficients of the error term for y. The coefficients A and B for the error terms in the x-direction may be different for each of the $(m + 1)$ y-values, but each of the sums in parentheses in Eq. (4.53) is $2n$ times some average value of A or B, so the error terms become:

$$\text{Error} = \frac{k}{2}(nA_{av})h^2 + \frac{k}{2}(nB_{av})h^4 + \bar{A}k^2 + \bar{B}k^4. \qquad (4.54)$$

Since both Δx and Δy are constant, we may take $\Delta y = k = \alpha \Delta x = \alpha h$, where $\alpha = \Delta y/\Delta x$, and Eq. (4.54) can be written, with $nh = (b - a)$:

$$\text{Error} = \left(\frac{b - a}{2}A_{av}\alpha\right)h^2 + \left(\frac{b - a}{2}B_{av}\alpha\right)h^4 + A\alpha^2 h^2 + B\alpha^4 h^4$$

$$= K_1 h^2 + K_2 h^4.$$

Here K_2 will depend on fourth-order partial derivatives. This confirms our expectation that the error term of double integration by numerical means is of the same form as for single integration.

Since this is true, a Romberg integration may be applied to multiple integration, whereby we extrapolate to an $O(h^4)$ estimate from two trapezoidal computations at a 2-to-1 interval ratio. From two such $O(h^4)$ computations we may extrapolate to one of $O(h^6)$ error.

4.16 MULTIPLE INTEGRATION WITH VARIABLE LIMITS

If the limits of integration are not constant, so that the region in the x,y-plane upon which the integrand $f(x, y)$ is to be summed is not rectangular, we must modify the above procedure. Let us consider a simple example to illustrate the method. Evaluate

$$\int\int f(x, y) \, dy \, dx$$

over the region bounded by the lines $x = 0$, $x = 1$, $y = 0$, and the curve $y = x^2 + 1$. The region is sketched in Fig. 4.6. If we draw vertical lines spaced at $\Delta x = 0.2$ apart, shown as dashed lines in Fig. 4.6, it is obvious that we can approximate the inner integral at constant x-values along any one of the vertical lines (including $x = 0$ and $x = 1$). If we use the trapezoidal rule with five panels

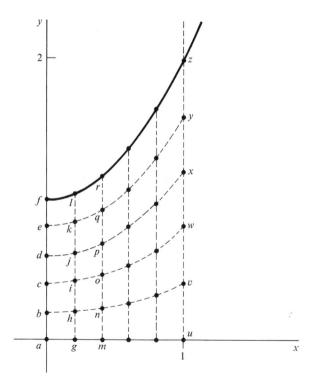

Figure 4.6

for each of these, we get the series of sums

$$S_1 = \frac{h_1}{2}(f_a + 2f_b + 2f_c + 2f_d + 2f_e + f_f),$$

$$S_2 = \frac{h_2}{2}(f_g + 2f_h + 2f_i + 2f_j + 2f_k + f_l),$$

$$S_3 = \frac{h_3}{2}(f_m + 2f_n + \ldots),$$

.

.

.

$$S_6 = \frac{h_6}{2}(f_u + 2f_v + 2f_w + 2f_x + 2f_y + f_z).$$

The subscripts here indicate the values of the function at the points so labeled in Fig. 4.6. The values of the h_i are not the same in the above equations, but in each they are the vertical distances divided by 5. The combination of these sums to give an estimate of the double integral will then be

$$\text{Integral} = \frac{0.2}{2}(S_1 + 2S_2 + 2S_3 + 2S_4 + 2S_5 + S_6).$$

To be even more specific, suppose $f(x, y) = xy$. Then

$$S_1 = \frac{1.0/5}{2}(0 + 0 + 0 + 0 + 0 + 0) = 0,$$

$$S_2 = \frac{1.04/5}{2}(0 + 0.0832 + 0.1664 + 0.2496 + 0.3328 + 0.208) = 0.1082,$$

$$S_3 = \frac{1.16/5}{2}(0 + 0.1856 + 0.3712 + 0.5568 + 0.7428 + 0.464) = 0.2692,$$

$$S_4 = \frac{1.36/5}{2}(0 + 0.3264 + 0.6528 + 0.9792 + 1.3056 + 0.816) = 0.5549,$$

$$S_5 = \frac{1.64/5}{2}(0 + 0.5248 + 1.0496 + 1.5744 + 2.0992 + 1.312) = 1.0758,$$

$$S_6 = \frac{2.0/5}{2}(0 + 0.8 + 1.6 + 2.4 + 3.2 + 2.0) = 2.0;$$

$$\text{Integral} = \frac{0.2}{2}(0 + 0.2164 + 0.5384 + 1.1098 + 2.1516 + 2.0)$$

$$= 0.6016 \quad \text{versus analytical value of } 0.6666.$$

The extension of this to more complicated regions and the adaptation to the use of Simpson's rule should be obvious. If the functions that define the region are not single-valued, one must divide the region into subregions to avoid the problem, but this must also be done to integrate analytically.

4.17. PROGRAMS FOR DIFFERENTIATION AND INTEGRATION

We conclude this chapter with a series of computer programs. Programs 1 (Fig. 4.7) and 2 (Fig. 4.9) perform differentiation. The first is a subroutine named DERIV that finds the first derivative of tabular data with equispaced x-values. Parabolas are passed through groups of three points and the derivative is computed as the slope of the parabola at the point in question. For each interior point, this amounts to using a central-difference formula for the derivative of $f(x)$. The endpoints use three-point formulas for the derivative, but the point at which the derivative is estimated is not a central point. The lozenge diagram of Fig. 4.1 is the source of these formulas; they have been rewritten in terms of function values in the routine. The subroutine is passed the values of the function in an array, the number of points, and the uniform x-interval. An array holding the derivative values is returned.

When the tabular data given below is passed to the subroutine, the results in Fig. 4.8 were obtained. The data are recordings of the speed of a car at 10-second intervals:

Time, sec	0	10	20	30	40	50	60	70	80	90	100
Speed, mph	48	38	31	33	36	41	51	43	35	29	28

In view of the discussion of "noisy data" in Section 4.5, what can you say about the accuracy of these results? How could the accuracy of the acceleration determination be improved?

```
      SUBROUTINE DERIV (F, DF, N, H)
C   THIS SUBROUTINE FINDS THE FIRST DERIVATIVE OF EQUISPACED DATA.
C   CENTRAL DIFFERENCE FORMULAS ARE USED FOR ALL POINTS EXCEPT THE
C   FIRST AND THE LAST.  FOR THESE, A QUADRATIC IS PASSED THROUGH
C   THREE SUCCESSIVE POINTS TO OBTAIN THE DERIVATIVE.
C   PARAMETERS ARE -
C      F      THE ARRAY OF FUNCTION VALUES
C      DF     AN ARRAY THAT HOLDS THE DERIVATIVE VALUES, RETURNED TO CALLER
C      N      THE NUMBER OF POINTS
C      H      THE UNIFORM SPACING BETWEEN X VALUES
         DIMENSION F(N), DF(N)
C   COMPUTE THE DERIVATIVES AT X(2) THROUGH X(N-1)
         NM1 = N - 1
         DO 10 I = 2,NM1
            DF(I) = (F(I+1) - F(I-1))/2./H
10       CONTINUE
C   NOW COMPUTE DERIVATIVE AT X(1)
         DF(1) = (2.*F(2) - 1.5*F(1) - 0.5*F(3))/H
C   AND GET IT AT X(N)
         DF(N) = (1.5*F(N) - 2.*F(N-1) + 0.5*F(N-2))/H
         RETURN
         END
```

Figure 4.7 Program 1.

The second program (DER2), in Fig. 4.9, is a subroutine that finds the first and second derivatives of a function whose form is known. The name of the function is passed as a parameter; it must therefore be declared EXTERNAL in the calling program. Central-difference formulas are used, values of $f(x)$ being used at the distance H on either side of the point where the derivatives are desired. This routine and a modified version that employed double-precision arithmetic were used to create the tables in Section 4.5 that illustrate how the best accuracy in computing derivatives occurs at a step size where the sum of round-off and truncation errors are minimized.

Two integration programs are presented. Program 3 (Fig. 4.10) is a subroutine named SIMPS designed to integrate uniformly spaced data that is known from a table. A combination of Simpson's $\frac{1}{3}$ and $\frac{3}{8}$ rules are used when there is an odd number of intervals. If there is an even number, only the $\frac{1}{3}$ rule is used, which normally gives better accuracy. An array containing the function values is passed, together with $H = \Delta x$ and the number of points. The program first checks to see whether the number of panels is odd or even. If it is odd, one application of the $\frac{3}{8}$

PROGRAM 1 OUTPUT

FUNCTION VALUES AND DERIVATIVE VALUES

POINT #	F	DF
1	0.480000E 02	−0.115000E 01
2	0.380000E 02	−0.850000E 00
3	0.310000E 02	−0.250000E 00
4	0.330000E 02	0.250000E 00
5	0.360000E 02	0.400000E 00
6	0.410000E 02	0.750000E 00
7	0.510000E 02	0.100000E 00
8	0.430000E 02	−0.800000E 00
9	0.350000E 02	−0.700000E 00
10	0.290000E 02	−0.350000E 00
11	0.280000E 02	0.150000E 00

Figure 4.8

```
      SUBROUTINE DER2 (FCT, X, H, DX, DDX)
C  THIS ROUTINE COMPUTES THE FIRST AND SECOND DERIVATIVES
C  OF THE FUNCTION FCT AT THE POINT X.  CENTRAL DIFFERENCE
C  FORMULAS ARE USED, EMPLOYING A STEP SIZE EQUAL TO H.
C  PARAMETERS ARE -
C    FCT    FUNCTION WHOSE DERIVATIVES ARE DESIRED.  MUST BE DECLARED
C           EXTERNAL IN CALLING PROGRAM.
C    X      POINT AT WHICH DERIVATIVES ARE TO BE CALCULATED.
C    H      STEP SIZE TO BE USED.
C    DX     RETURNS THE FIRST DERIVATIVE TO CALLER.
C    DDX    RETURNS THE SECOND DERIVATIVE.
      FO = FCT(X)
      FR = FCT(X + H)
      FL = FCT (X - H)
      DX = (FR - FL)/2./H
      DDX = (FR - 2.*FO + FL)/H/H
      RETURN
      END
```

Figure 4.9 Program 2.

rule is performed to integrate over the first three panels, and the $\frac{1}{3}$ rule is applied over all the others. The parameter RESULT returns the value of the integral. When this subroutine is applied to the velocity data tabulated with Program 1, the result obtained is 3726.67 mile-sec/hr. If we unscramble the mixed time units by dividing by 3600, we find that the distance travelled, as indicated by the speedometer, is 1.035 mi.

Program 4 (Fig. 4.11) is a subroutine that applies the Romberg method to integrate a function. The name of the function is passed as a parameter (thus it must be declared EXTERNAL), along with the limits of integration and a tolerance value to control the error. RESULT returns the final value to the caller, but intermediate results are printed out by the subroutine. The value of H is halved at each stage of the integration until TOL is met on that stage. The program avoids recomputing function values that have been used in prior stages.

A main program and a function are also shown with Program 4, illustrating how the subroutine can be used to find an integral with infinite upper limit. The integral being evaluated is

$$\int_0^\infty xe^{-x}\,dx,$$

whose analytical result is unity. The routine, with TOL $= 0.0001$, finds a result of almost perfect accuracy. In the program, the upper limit is increased progressively until a negligibly small increment is computed.

```
      SUBROUTINE SIMPS (F, N, H, RESULT)
C   THIS ROUTINE PERFORMS SIMPSON'S RULE INTEGRATION OF A FUNCTION
C   DEFINED BY A TABLE OF EQUISPACED VALUES.
C   PARAMETERS ARE -
C     F      ARRAY OF VALUES OF THE FUNCTION
C     N      NUMBER OF POINTS
C     H      THE UNIFORM SPACING BETWEEN X VALUES
C     RESULT  ESTIMATE OF THE INTEGRAL THAT IS RETURNED TO CALLER
          DIMENSION F(100)
C   CHECK TO SEE IF NUMBER OF PANELS IS EVEN.  NUMBER OF PANELS
C   IS N - 1
          NPANEL = N - 1
          NHALF = NPANEL/2
          NBEGIN = 1
          RESULT = 0.
          IF ((NPANEL - 2*NHALF) .EQ. 0)  GO TO 5
C   NUMBER OF PANELS IS ODD.  USE 3/8 RULE ON FIRST THREE
C   PANELS, 1/3 RULE ON REST OF THEM.
          RESULT = 3.*H/8.*(F(1) + 3.*F(2) + 3.*F(3) + F(4))
          NBEGIN = 4
C   APPLY 1/3 RULE - ADD IN FIRST, SECOND, LAST VALUES.
5         RESULT = RESULT + H/3.*(F(NBEGIN) + 4.*F(NBEGIN+1) + F(N))
          NBEGIN = NBEGIN + 2
          IF (NBEGIN .EQ. N)  RETURN
C   THE PATTERN AFTER NBEGIN+2 IS REPETITIVE. GET NEND, THE PLACE TO STOP.
          NEND = N - 2
          DO 10 I = NBEGIN, NEND, 2
10        RESULT = RESULT + H/3.*(2.*F(I) + 4.*F(I+1))
          RETURN
          END
```

Figure 4.10 Program 3.

```
            SUBROUTINE ROMB (FCN, A, B, TOL, RESULT)
C    SUBROUTINE FOR ROMBERG INTEGRATION.  PROGRAM BEGINS WITH TRAPEZOIDAL
C    INTEGRATION WITH 10 SUBINTERVALS.  INTERVALS ARE THEN HALVED AND
C    RESULTS ARE EXTRAPOLATED UP TO EIGHTH ORDER.
C    MAXIMUM NUMBER OF SUBINTERVALS USED IN PROGRAM IS 2560.
C    PARAMETERS ARE -
C       FCN    FUNCTION THAT COMPUTES F(X), DECLARED EXTERNAL IN MAIN
C       A,B    INITIAL AND FINAL X VALUES
C       TOL    TOLERANCE VALUE USED TO TERMINATE ITERATIONS
C       RESULT RETURNS VALUE OF ITEGRAL TO CALLER
C       TRAP   DOUBLY SUBSCRIPTED ARRAY THAT HOLDS INTERMEDIATE VALUES
C              FOR COMPARISONS AND EXTRAPOLATION
C       KFLG   FLAG USED INTERNALLY TO SIGNAL NON-CONVERGENCE
C              WHEN KFLG = 0, MEANS NON-CONVERGENT, = 1 MEANS ALL OK.
            DIMENSION TRAP(9,9)
C    SET FLAG AT 1 INITIALLY
            KFLG = 1
C    COMPUTE FIRST INTEGRAL WITH 10 SUBINTERVALS AND USING TRAP RULE
            H = (B - A)/10.
            SUM = FCN(A) + FCN(B)
            X = A
            DO 10 I = 2,10
            X = X + H
10          SUM = SUM + FCN(X)*2.
            TRAP(1,1) = H/2.*SUM
C    RECOMPUTE INTEGRAL WITH H HALVED, EXTRAPOLATE AND TEST.  REPEAT UP TO
C    8 TIMES.
            DO 20 I = 1,8
               H = H/2.
               X = A + H
               K = 10*2**I
               DO 30 J = 2,K,2
                  SUM = SUM + FCN(X)*2.
                  X = X + H + H
30             CONTINUE
               TRAP(1,I+1) = H/2.*SUM
               DO 40 L = 1,I
                  TRAP(L+1,I+1) = TRAP(L,I+1) + 1./(4.**L - 1.)*(TRAP(L,I+1) -
     +            TRAP(L,I))
40             CONTINUE
               IF (ABS(TRAP(I+1,I+1) - TRAP(I,I+1)) - TOL) 50,50,20
20          CONTINUE
C    IF TOLERANCE NOT MET AFTER 8 EXTRAPOLATIONS, PRINT NOTE.  ALSO
C    SET KFLG = 0
            KFLG = 0
            WRITE (6,200)
200         FORMAT (1H0, 42HTOLERANCE MOT MET.  CALCULATED VALUES WERE )
C    PRINT INTERMEDIATE RESULTS
50          I = I + 1
            DO 70 L = 1,I
               WRITE (6,203) (TRAP(J,L), J = 1,L)
203         FORMAT (1H , 8F12.6)
70          CONTINUE
            IF (KFLG .EQ. 0)  STOP
            RESULT = TRAP (I,I)
            RETURN
            END
```

Figure 4.11 Program 4.

The fifth program (DBLINT) (see Fig. 4.12) integrates a function of two independent variables over a rectangular region. The Romberg method again is used in the program, and the name of the function is passed as a parameter to

```
EXTERNAL FCT
TOL = 0.0001
TOTAL = 0.0
A = 0.0
B = 1.0
DO 10 I = 1,5
   CALL ROMB (FCT, A, B, TOL, RESULT)
   TOTAL = TOTAL + RESULT
   WRITE (6,200) B, RESULT, TOTAL
200   FORMAT ('0', ' B = ', F7.0, '   INCREMENT IS ', E15.7,
   +   ' INTEGRAL IS ', E15.7)
   A = B
   B = B*10.
10   CONTINUE
STOP
END


FUNCTION FCT (X)
FCT = X*EXP(-X)
RETURN
END


RESULTS OF PROGRAM 4

   0.263408
   0.264032      0.264241
   0.264188      0.264240      0.264240

B =      1.   INCREMENT IS   0.2642395E 00 INTEGRAL IS   0.2642395E 00
   0.735877
   0.735292      0.735097
   0.735259      0.735248      0.735258

B =     10.   INCREMENT IS   0.7352578E 00 INTEGRAL IS   0.9994974E 00
   0.002044
   0.001055      0.000725
   0.000661      0.000530      0.000517

B =    100.   INCREMENT IS   0.5171911E-03 INTEGRAL IS   0.1000014E 01
   0.000000
   0.000000      0.000000

B =   1000.   INCREMENT IS   0.5580114E-40 INTEGRAL IS   0.1000014E 01
   0.000000
   0.000000      0.000000

B =  10000.   INCREMENT IS   0.0000000E 00 INTEGRAL IS   0.1000014E 01
```

Figure 4.11 (continued)

provide greater flexibility. This program is similar to Program 4, except that a series of integrals in the x-direction is combined to integrate in the y-direction; a function subprogram is called to do this. Program 5 is less efficient than Program 4 in that some function values are recomputed as the step size is reduced.

```
      SUBROUTINE DBLINT (FCT, XA, XB, YA, YB, TOL, RESULT)
C THIS ROUTINE COMPUTES THE INTEGRAL OF A FUNCTION OF TWO
C VARIABLES.  THE ROMBERG METHOD IS USED.  INTIIALLY, FOUR
C SUBDIVISIONS ARE USED.  THESE ARE HALVED UNTIL THE TOLERANCE
C IS MET, WITH A MAXIMUM NUMBER OF SUBDIVIDINGS OF 5.
C PARAMETERS ARE -
C   FCT    FUNCTION SUBPROGRAM TO COMPUTE F(X, Y).  DECLARED
C          EXTERNAL IN CALLING PROGRAM.
C   XA,XB LOWER AND UPPER LIMITS FOR X.
C   YA,YB LOWER AND UPPER LIMITS FOR Y.
C   TOL   TOLERANCE TO TERMINATE INTEGRATION.  WHEN NOT MET,
C          A MESSAGE IS PRINTED AND LAST VALUE RETURNED.
C   RESULT  RETURNS VALUE OF INTEGRAL TO CALLER.
C   ARRAY DOUBLY SUBSCRIPTED ARRAY TO HOLD INTERMEDIATE VALUES FOR
C          COMPARISON AND EXTRAPOLATION.
C A FUNCTION SUBPROGRAM NAMED SUMROW IS CALLED TO COMPUTE SUMS
C ACROSS ONE ROW OF THE REGION.
      DIMENSION ARRAY(6,6)
      EXTERNAL FCT
      DELX = (XB - XA)/4.
      DELY = (YB - YA)/4.
      N = 4
C GET THE SUM FOR TOP AND BOTTOM ROWS
      ARRAY(1,1) = SUMROW(FCT, XA, XB, YA, DELX, N) +
     +   SUMROW(FCT, XA, XB, YB, DELX, N)
C GET THE SUMS FOR INTERMEDIATE ROWS
      Y = YA
      DO 10 I = 2,N
        Y = Y + DELY
        ARRAY(1,1) = ARRAY(1,1) + 2.*SUMROW(FCT, XA, XB, Y, DELX, N)
10    CONTINUE
      ARRAY(1,1) = ARRAY(1,1) *DELX*DELY/4.
C NOW HALVE THE VALUES OF DELX AND DELY, RECOMPUTE THE INTEGRAL,
C AND EXTRAPOLATE, THEN TEST TO SEE IF TOLERANCE IS MET.  TEPEAT
C UP TO 5 TIMES.
      DO 40 J = 1,5
        DELX = DELX/2.
        DELY = DELY/2.
        N = 2*N
C DO TOP AND BOTTOM ROWS FIRST
        ARRAY(J+1,1) = SUMROW(FCT, XA, XB, YA, DELX, N) +
     +     SUMROW(FCT, XA, XB, YB, DELX, N)
C THEN THE INTERMEDIATE ROWS
        Y = YA
        DO 20 I = 2,N
          Y = Y + DELY
          ARRAY(J+1,1) = ARRAY(J+1,1) +
     +       2.*SUMROW(FCT, XA, XB, Y, DELX, N)
20      CONTINUE
        ARRAY(J+1,1) = ARRAY(J+1,1)*DELX*DELY/4.
C NOW WE EXTRAPOLATE
        DO 30 K = 1,J
          ARRAY(J+1,K+1) = ARRAY(J+1,K) +
     +     1./(4.**K - 1.)*(ARRAY(J+1,K) - ARRAY(J,K))
30      CONTINUE
        IF (ABS(ARRAY(J+1,J+1) - ARRAY(J+1,J)) - TOL)  50, 50, 40
40      CONTINUE
```

Figure 4.12 Program 5.

```
C  WE HAVE A NORMAL TERMINATION OF LOOP 40 ONLY WHEN THE
C  TOLERANCE IS NOT MET, SO PRINT MESSAGE AND RETURN
       WRITE (6,201) TOL
201    FORMAT (1H0, 14H TOLERANCE OF ,E14.7,
      +  31H NOT MET AFTER 5 EXTRAPOLATIONS /)
       RESULT = ARRAY(6,6)
       RETURN
50     RESULT = ARRAY(J+1,J+1)
       RETURN
       END



       FUNCTION SUMROW (FCT, XA, XB, Y, DELX, N)
C  THIS FUNCTION COMPUTES THE WEIGHTED SUM FOR TRAPEZOIDAL
C  RULE INTEGRATION ACROSS ONE ROW OF A REGION, FROM XA TO XB WITH
C  INTERVALS OF DELX, WHERE THE VALUE OF Y IS Y.
C  PARAMETERS ARE -
C    FCT    FUNCTION THAT COMPUTES F(X, Y), DECLARED EXTERNAL
C             BY CALLER.
C    XA,XB  LIMITS FOR X-VALUES.
C    Y      VALUE OF Y
C    DELX   STEP SIZE FOR X.
C    N      NUMBER OF INTERVALS
C  GET FIRST AND LAST VALUES TO START.
       SUMROW = FCT(XA, Y) + FCT(XB, Y)
C  NOW ADD IN THE INTERMEDIATE VALUES.
       X = XA
       DO 10 I = 2,N
        X = X + DELX
        SUMROW = SUMROW + 2.*FCT(X, Y)
10     CONTINUE
       RETURN
       END
```

Figure 4.12 (continued)

EXERCISES

Section 4.1

▶1. The following table is for $(1 + \log x)$. Determine estimates of $d(1 + \log x)/dx$ at $x = 0.15, 0.19$, and 0.23, using (a) one term, (b) two terms, and (c) three terms of Eq. (4.4). By comparing to the analytical values, determine the errors of each estimate.

x	$1 + \log x$	Δ	Δ^2	Δ^3
0.15	0.1761			
		0.0543		
0.17	0.2304		−0.0059	
		0.0484		0.0009
0.19	0.2788		−0.0050	
		0.0434		0.0011
0.21	0.3222		−0.0039	
		0.0395		0.0006
0.23	0.3617		−0.0033	
		0.0362		0.0005
0.25	0.3979		−0.0027	
		0.0335		0.0002
0.27	0.4314		−0.0025	
		0.0310		0.0005
0.29	0.4624		−0.0020	
		0.0290		
0.31	0.4914			

▶2. Write expressions for the errors in each computation of Exercise 1, by properly interpreting Eq. (4.6). From these expressions find upper and lower bounds for each of the computations.

3. If one wished the derivative of $(1 + \log x)$ at $x = 0.31$, the table in Exercise 1 would be inadequate for this, using Eq. (3.4), because the necessary forward differences cannot be computed. A formula in terms of backward differences can be derived, however, by differentiating the Newton–Gregory backward-interpolation polynomial. Do this, and use three terms to obtain $d(1 + \log x)/dx$ at $x = 0.31$ from the data given in the table of Exercise 1.

4. Derive the error term for the formula of Exercise 3, and show that the "next-term rule" applies.

5. A central-difference approximation for derivatives is more accurate than a forward- or backward-difference approximation, as discussed in Section 4.1. Repeat Exercise 1(b), but use Eq. (4.8) and compare the errors.

6. Find bounds on the errors of Exericse 5 and compare the actual errors to the bounds.

7. Equation (4.8) is based on a quadratic interpolating polynomial. Derive the central-difference formula for the derivative based on a fourth-degree interpolation. Also derive the error term.

Section 4.2

8. The Newton–Gregory backward-interpolating formula can be developed by expanding the symbolic relation

$$f_s = E^s f_0 = (1 - \nabla)^s f_0.$$

Using this representation of the interpolation formula, derive the derivative formula required in Exercise 3, by the symbolic technique.

▶9. Suppose the nature of the function tabulated in Exercise 1 were unknown. How could estimates of the errors of each computation in that problem be determined? Compare with the error bounds determined in Exercise 2.

Section 4.3

10. If one begins with Stirling's interpolation formula (page 162), differentiates, and then evaluates at $x = x_0$, one obtains a derivative formula in terms of central differences. Show that, when this is done, the coefficients are the same as those given by a horizontal path through the lozenge diagram, Fig. 4.1, starting at f_0.

11. If one differentiates Stirling's formula twice and then sets $s = 0$, a central-difference formula for the second derivative results. Show that the coefficients so obtained match those on a horizontal path from f_0 in Fig. 4.3.

12. Make a lozenge diagram, similar to Figs. 4.1 and 4.3, for the third derivative.

13. If the difference table of Exercise 1 is extended to the right, differences up to the eighth order can be calculated. The lozenge diagram, Fig. 4.1, can also be extended to a similar order of differences. Discuss whether the accuracy of estimates of the derivative of the logarithm function will be improved by such extensions.

▶14. The following table is for the first-order Bessel function $J_1(x)$. Prepare tables for $J_1'(x)$ and $J_1''(x)$ for $x = 0(0.1)1$.

x	$J_1(x)$	x	$J_1(x)$
0	0.0000	0.6	0.2867
0.1	0.0499	0.7	0.3290
0.2	0.0995	0.8	0.3688
0.3	0.1483	0.9	0.4059
0.4	0.1960	1.0	0.4401
0.5	0.2423		

15. An alternative way of deriving error terms for derivative formulas is through Taylor-series expansions. For example, the error term of Eq. (4.18) can be obtained by expanding $f_1 = f(x_0 + h)$ and $f_{-1} = f(x_0 - h)$ each about the point $x = x_0$, and then combining the two series. Show that the error term of Eq. (4.18) is $-\frac{1}{6} h^2 f'''(\xi)$.

16. For unevenly spaced data, a derivative expression analogous to Eq. (4.18) is:

$$f'(x)\bigg|_{x=x_0} \doteq \frac{f(x_1) - f(x_{-1})}{x_1 - x_{-1}}.$$

In this case $x_1 - x_0 \neq x_0 - x_{-1}$. Take $x_1 - x_0 = h$ and $x_0 - x_{-1} = \alpha h$, $\alpha \neq 1$. Show by the Taylor-series method that the error is now $O(h)$ and not $O(h^2)$ unless $\alpha = 1$.

Section 4.4

▶17. Use the extrapolation method to determine $f'(0.23)$ for the data of Exercise 1, to an $O(h^6)$ extrapolation.

18. Show that the first-order extrapolation for $f'(x_0)$ with Δx-values differing by $2 : 1$ is the same as the formula

$$f'_0 = \frac{1}{h}\left(\frac{\Delta f_{-1} + \Delta f_0}{2} - \frac{1}{6}\frac{\Delta^3 f_{-2} + \Delta^3 f_{-1}}{2}\right),$$

where h is the smaller of the two Δx-values. Note that this comes directly from the lozenge diagram.

19. Show that a second-order extrapolation for $f'(x_0)$ is the equivalent of using central differences through $\Delta^5 f$.

20. Repeat Exercises 18 and 19 for $f''(x_0)$, deriving analogous relations.

Section 4.5

21. The data below is for $\tan \theta$ in the neighborhood of $\theta = 1$ radian.
   ▶a) Truncating the data to four significant digits, compute the derivative at $\theta = 1.0$ using a central-difference formula (Eq. (4.8)) but with varying step size. At what point is the greatest accuracy obtained?
   ▶b) Repeat part (a) but with the data rounded to four digits.
   c) Repeat with the data as given below, which are rounded to six digits.

θ	$\tan \theta$	θ	$\tan \theta$
0.9000	1.26016	1.0001	1.55775
0.9900	1.52368	1.0010	1.57084
0.9990	1.55399	1.0100	1.59221
0.9999	1.55706	1.1000	1.96476
1.0000	1.55741		

22. Repeat Exercise 21, but use forward- and backward-difference formulas.

23. Repeat Exercise 21, but use a central-difference formula for second derivatives.

24. Use Program 2 (Fig. 4.9) to determine the optimum step size for most accurate computation of first and second derivatives for a number of functions, including $\sin x$, $\cos x$, $\sinh x$, $\cosh x$, $\ln x$. If your computer has double precision in FORTRAN (also if extended precision is available), modify the program to investigate for these cases.

Section 4.6

25. Equations (4.24) through (4.29) are Newton–Cotes formulas and their error terms are derived from forward-difference interpolating polynomials. Rederive these, beginning with backward-difference formulas.

▶26. Prove the assertion made at the end of Section 4.6 that all Newton–Cotes integration formulas of even order have an order of h in the error term one greater than would normally be expected.

27. Derive the Newton–Cotes formulas of orders four and five.

Section 4.7

▶28. The following values of a function are given.

x	$f(x)$	x	$f(x)$
1.0	1.543	1.5	2.352
1.1	1.668	1.6	2.577
1.2	1.811	1.7	2.828
1.3	1.971	1.8	3.107
1.4	2.151		

Find $\int_{1.0}^{1.8} f(x)\, dx$, using the trapezoidal rule with

a) $h = 0.1$ b) $h = 0.2$ c) $h = 0.4$.

29. The function tabulated in Exercise 28 is $\cosh x$. What are the errors of the computations in parts (a), (b), (c)? How closely are the errors proportional to h^2? What other errors besides that of Eq. (4.33) are present?

30. In Section 4.7, the value of $\int_{1.8}^{3.4} e^x\, dx$ is computed as 23.9944 using the trapezoidal rule with $h = 0.2$, and is in error by -0.08. Equation (4.33) shows this to fall within the predicted bounds for the error. Suppose we did not know what $f(x)$ was; estimate the error using second-order differences of the data in Table 4.3.

▶31. If one wished to compute $\int_{1.8}^{3.4} e^x\, dx$ using the trapezoidal rule, and wished to be certain of five-decimal accuracy (error $\leqslant 0.000005$), how small must h be?

32. Find $\int_0^2 f(x)\, dx$:

x	$f(x)$	x	$f(x)$
0	1.0000	1.08	0.3396
0.12	0.8869	1.43	0.2393
0.53	0.5886	2.00	0.1353
0.87	0.4190		

Section 4.8

33. Extrapolate the individual answers of Exercise 28 to get estimates of improved accuracy. What are the orders of the errors of these extrapolations?

▶34. Use extrapolation to the limit to evaluate $\int_0^1 f(x)\, dx$, getting a result with error $O(h^6)$:

x	$f(x)$	x	$f(x)$
0	0.3989	0.50	0.3521
0.25	0.3867	0.75	0.3011
		1.00	0.2420

35. Use Romberg integration (successive extrapolations with h halved each time) to evaluate $\int_1^2 dx/x$. Carry six decimals and continue until no change in the fifth place occurs. Compare to the analytical value $\ell n\, 2 = 0.69315$.

Section 4.9

36. Repeat Exercise 28 using Simpson's $\frac{1}{3}$ rule with (a) $h = 0.1$, (b) $h = 0.2$, (c) $h = 0.4$.

37. Use the error expression of Eq. (4.37) to estimate the maximum and minimum errors to be expected in Exercise 36. [$f(x)$ in Exercise 36 is cosh x.]

▶38. Evaluate $\int_0^{1.0} e^x\, dx$ by Simpson's $\frac{1}{3}$ rule choosing h small enough to guarantee five-decimal accuracy. How large can h be?

39. In the text it is stated that the results of using Simpson's $\frac{1}{3}$ rule and the extrapolated result from two applications of the trapezoidal rule (with h changed by $2:1$) are identical. Show that this will always be true.

40. Compute

$$\int_0^1 \frac{\sin x}{x}\, dx$$

using Simpson's $\frac{1}{3}$ rule with $h = 0.5$ and with $h = 0.25$. Extrapolate the results. What is the order of the error for the extrapolated result?

Section 4.10

41. Repeat Exercise 38, but use Simpson's $\frac{3}{8}$ rule.

▶42. Apply a combination of the two Simpson rules to evaluate the integral from $x = 3$ to $x = 6.5$. Note that this is a technique to handle an odd number of panels:

x	$f(x)$	x	$f(x)$
3.0	0.33906	5.0	-0.32758
3.5	0.13738	5.5	-0.34144
4.0	-0.06604	6.0	-0.27668
4.5	-0.23106	6.5	-0.15384

43. The fact that Simpson's $\frac{1}{3}$ rule has an error term dependent on the third derivative and not the second (as is also true for the $\frac{3}{8}$ rule) means that these integrations are exact for any cubic polynomial. However, the $\frac{1}{3}$ rule is based on fitting a quadratic through three equispaced points. The implication of this is that the area under any cubic, from $x = a$ to $x = b$, is exactly the same as the area under the parabola that intersects the cubic at the two endpoints, and also at $x = \frac{1}{2}(a + b)$. Prove this.

Section 4.11

▶44. Use the symbolic method to determine a formula for three-panel integration, using a polynomial of degree m.

45. Integration from x_0 to x_1, as in Eq. (4.39), is an arbitrary choice taken for convenience only. If the limits were taken from x_{-1} to x_0, a similar formula for one-panel integration would result, but different coefficients would be obtained. Carry out this computation.

46. Perform a computation similar to Exercise 45, but get a two-panel integration formula from x_{-1} to x_1.

47. Use the method of undertermined coefficients to derive (a) the trapezoidal rule, (b) Simpson's $\frac{3}{8}$ rule.

48. Use the method of undetermined coefficients to derive the central-difference formulas, Eqs. (4.18) and (4.19) for $f'(x_0)$ and $f''(x_0)$.

Section 4.12

▶49. Evaluate

$$\int_0^1 \frac{\sin x}{x}\, dx$$

by a three-term Gaussian quadrature formula.

50. By computing with Gaussian quadrature formulas of increasing complexity, determine how many terms are needed to evaluate $\int_{1.8}^{3.4} e^x\, dx$ to five-decimal accuracy. (The exact value of the integral is 23.9144526.)

51. An n-term Gaussian quadrature formula assumes that a polynomial of degree $2n - 1$ is used to fit the function between $x = a$ and $x = b$. Does this mean that the error term is the same as for the Newton–Cotes integration formulas based on polynomials of $(2n - 1)$ degree?

Section 4.13

52. As discussed in Section 4.13, an integral with infinite limits can be approximated numerically if it is convergent. For example, the exponential integral $Ei(x)$ can be evaluated by taking the upper limit, U, sufficiently large in

$$Ei(x) = \int_x^\infty \frac{e^{-v}}{v}\, dv \doteq \int_x^U \frac{e^{-v}}{v}\, dv.$$

We know that U is "sufficiently large" when the additional contributions of making U larger are negligible. Estimate $Ei\ (0.5)$ using the trapezoidal rule. Note that one can use larger subintervals as v increases. Compare to the tabular value of 0.5598.

53. Apply the method of Section 4.13 to show that

$$\int_{-\infty}^{\infty} \frac{1}{\sqrt{2\pi}} e^{-x^2/2} dx = 1.$$

(The values of the integrand are available as ordinates of the standard normal curve.)

▶54. Evaluate these integrals

a) $\displaystyle\int_0^2 \frac{dx}{\sqrt{x}}$ b) $\displaystyle\int_0^1 \frac{dx}{\sqrt{1-x}}$ c) $\displaystyle\int_0^{\infty} \frac{x\,dx}{e^x + 1}$ (Analytical value is $\pi^2/12$)

55. In integrating $\int_a^b f(x)\,dx$, Gaussian quadrature does not require the value of $f(x)$ at the endpoints. Therefore that technique has special utility when there are singularities at the endpoints of the interval of integration. Apply this to Exercise 54 (a) and (b).

▶56. It often helps to break up an integral into parts. Consider

$$\int_0^{\infty} \frac{dx}{e^x + e^{-x}}.$$

At $x = 10$, e^{-x} is only 4×10^{-5} while e^x is 22026.5; for $x \geq 10$, the integral can be considered to be just $\int dx/e^x$. Evaluate the integral numerically as

$$\int_0^{\infty} \frac{dx}{e^x + e^{-x}} = \int_0^{10} \frac{dx}{e^x + e^{-x}} + \int_{10}^{\infty} \frac{dx}{e^x}.$$

57. One must be cautious in integrating on a computer. $\int_0^1 dx/x$ surely diverges, but you will probably not find this to be true when calculating numerically because the small values of delta x near $x = 0$ may cancel the contribution to the total. It will depend on how one defines the step size as x gets small. See what you get as a result of numerically evaluating this integral.

Section 4.14

58. In connection with the first example of Section 4.14 it was stated that it is immaterial which integral of a double integral (with constant limits) we integrate first. Confirm this by evaluating $\int_{1.5}^{3.0}\int_{0.2}^{0.6} f(x, y)\,dy\,dx$, with $f(x, y)$ given by Table 4.6, performing the integration first with respect to y.

▶59. Write the pictorial operators similar to that portrayed in Eq. (4.48) that result by using:

a) Simpson's $\frac{1}{3}$ rule in the y-direction and the trapezoidal rule in the x-direction;

b) Simpson's $\frac{1}{3}$ rule in both directions;

c) Simpson's $\frac{3}{8}$ rule in both directions.

d) What conditions are placed on the number of panels in each direction by the methods employed in parts (a), (b), and (c)?

60. Since Simpson's $\frac{1}{3}$ rule is accurate for a cubic, evaluation of the triple integral below using this rule should be exact. Confirm this by evaluating both numerically and analytically:

$$\int_0^1 \int_0^1 \int_0^1 x^3 y z^2 \, dx \, dy \, dz.$$

Use Eq. (4.52) adapted to this case.

61. Draw a pictorial operator (three dimensions) to represent the formula used in Exercise 60. It is perhaps easiest to do this with three widely separated planes on which one indicates the coefficients (see illustration).

▶62. Evaluate

$$\int_{0.1}^{0.7} \int_{-.2}^{0.6} e^x \sin y \, dy \, dx$$

a) Using the trapezoidal rule in both directions, $\Delta x = \Delta y = 0.1$.
b) Using Simpson's $\frac{1}{3}$ rule in both directions, $\Delta x = \Delta y = 0.1$.
c) Using Gaussian quadrature, three-term formulas in both directions.

Section 4.15

63. Solve Exercise 62 by extrapolating from the trapezoidal-rule evaluations using $\Delta x = \Delta y = 0.2$, and $\Delta x = \Delta y = 0.1$. This should give the same result as part (b) of Exercise 62. Does it?

64. Integrate $\int_0^1 \int_0^1 (x^2 + y^2) \, dx \, dy$ using the trapezoidal rule with varying Δx and Δy. Show that the errors decrease proportionately to h^2.

Section 4.16
▶65. Integrate $\iint \sin x \sin y \, dx \, dy$ over the region defined by that portion of the unit circle that lies in the first quadrant. Integrate first with respect to x holding y constant with $\Delta x = 0.25$. Subdivide the vertical lines into four panels.
a) Use the trapezoidal rule.   b) Use Simpson's $\frac{1}{3}$ rule.

66. The order of integration in multiple integration may usually be changed. Evaluate the integral of Exercise 65. by integrating first with respect to y, holding x constant. The integral then becomes

$$\int_0^1 \int_0^{\sqrt{1-x^2}} \sin x \sin y \, dy \, dx.$$

67. Reverse the order of integration to solve the example problem of Section 4.16.

▶68. Integrate the function $e^{-x^2y^2}$ over the region bounded by the two parabolas $y = x^2$ and $y = 2x^2 - 1$. Note that the integrand is an even function and that the region is symmetrical about the y-axis, so that the integral over half the region may be evaluated and then doubled. Choose reasonable values for Δx and Δy.

APPLIED PROBLEMS AND PROJECTS

69. Differential thermal analysis is a specialized technique that can be used to determine transition temperatures and the thermodynamics of chemical reactions. It has special application in the study of minerals and clays. Vold (*Anal. Chem.* **21,** 683 (1949)) describes the technique. In this method, the temperature of a sample of the material being studied is compared to the temperature of an inert reference material when both are heated simultaneously under identical conditions. The furnace housing the two materials is normally heated so that its temperature T_f increases (approximately) linearly with time (t), and the difference in temperatures (ΔT) between the sample and the reference is recorded. Some typical data are:

t, min	0	1	2	3	4	5	6	7
ΔT, °F	0.00	0.34	1.86	4.32	8.07	13.12	16.80	18.95
T_f, °F	86.2	87.8	89.4	91.0	92.7	94.3	95.9	97.5

t	8	9	10	11	12	13	14	15	16
ΔT	18.07	16.69	15.26	13.86	12.58	11.40	10.33	8.95	6.46
T_f	99.2	100.8	102.3	103.9	105.5	107.1	108.6	110.2	111.8

t	17	18	19	20	21	22	23	24	25
ΔT	4.65	3.37	2.40	1.76	1.26	0.88	0.63	0.42	0.30
T_f	113.5	115.1	116.8	118.4	120.0	121.6	123.2	124.9	126.5

The ΔT values increase to a maximum, then decrease. This is due to the heat evolved in an exothermic reaction. One item of interest is the time (and furnace temperature) when the reaction is complete. Vold shows that the logarithm of ΔT should decrease linearly after the reaction is over; while the chemical reaction is occurring, the data depart from this linear relation. She used a graphical method to find this point. Perform numerical computations to find, from the above data, the time and the furnace temperature when the reaction terminates. Compare the merits of doing it graphically or numerically.

70. The temperature difference data in Problem 69 can be used to compute the heat of reaction. To do this, the integral of the values of ΔT are required, from the point where the reaction begins (which is at the point where ΔT becomes nonzero) to the time when the reaction ceases, as found in Exercise 69. Determine the value of the required integral. Which of the methods of this chapter should give the best value for the integral?

71. Fugacity is a term used by engineers to describe the available work from an isothermal

process. For an ideal gas, the fugacity, f, is equal to its pressure, P, but for real gases,

$$\ln\frac{f}{P} = \int_0^P \frac{C-1}{P}\, dp,$$

where C is the experimentally determined *compressibility factor*. For methane, values of C are:

P (atm)	C	P (atm)	C
1	0.9940	80	0.3429
10	0.9370	120	0.4259
20	0.8683	160	0.5252
40	0.7043	250	0.7468
60	0.4515	400	1.0980

Write a program that reads in the P and C values and uses them to compute and print f corresponding to each pressure given in the table. Assume that the value of C varies linearly between the tabulated values (a more precise assumption would fit a polynomial to the tabulated C values). The value of C approaches 1.0 as P approaches 0.

72. The stress developed in a rectangular bar when it is twisted can be computed if one knows the values of a torsion function, U, that satisfies a certain partial-differential equation. Chapter 7 describes a numerical method that can determine values of U. To compute the stress, it is necessary to integrate $\iint U\, dx\, dy$ over the rectangular region for which the data given below applies. (You may be able to simplify the integration because of the symmetry in the data.)

x \ y	0.0	0.2	0.4	0.6	0.8	1.0	1.2
0.0	0	0	0	0	0	0	0
0.2	0	2.043	3.048	3.354	3.048	2.043	0
0.4	0	3.123	4.794	5.319	4.794	3.123	0
0.6	0	3.657	5.686	6.335	5.686	3.657	0
0.8	0	3.818	5.960	6.647	5.960	3.818	0
1.0	0	3.657	5.686	6.336	5.686	3.657	0
1.2	0	3.123	4.794	5.319	4.794	3.123	0
1.4	0	2.043	3.048	3.354	3.048	2.043	0
1.6	0	0	0	0	0	0	0

73. Make a critical comparison of the accuracy of Newton–Cotes integration formulas compared to Gaussian quadrature. Test the formulas for a variety of functions for which you can calculate the integrals analytically. Select some functions that are smooth, some that have sharp changes in value, and some with periodic behavior.

74. Write a general-purpose subroutine that performs Gaussian quadrature. You should have the subroutine change the limits of the integration appropriately and call a function subprogram to compute function values, with the name of the function subprogram passed as an argument. It should also receive as an argument the degree of the formula to be employed.

Chapter Five

Numerical Solution of Ordinary Differential Equations

5.1 POPULATION CHARACTERISTICS OF FIELD MICE

An ecologist has been studying the effects of the environment on the population of field mice. His research shows that the number of mice born each month is proportional to the number of females in the group, and that the fraction of females is normally constant in any group. This implies that the number of births per month is proportional to the total population.

He has located a test plot for further research, which is a restricted area of semi-arid land. He has constructed barriers around the plot so mice cannot enter or leave. Under the conditions of the experiment, the food supply is limited, and it is found that the death rate is affected thereby, with mice dying of starvation at a rate proportional to some power of the population. (He also hypothesizes that when the mother is undernourished, the babies have less chance for survival, and that starving males tend to attack each other, but these factors are only speculation.)

The net result of this scientific analysis is the following equation, with N being the number of mice at time t (with t expressed in months.) He has come to you for help in solving the equation; his calculus doesn't seem to apply.

$$\frac{dN}{dt} = aN - BN^{1.7}, \qquad \text{with } B \text{ given by Table 5.1.}$$

As the season progresses, the amount of vegetation varies. He accounts for this change in the food supply by using a "constant," B, that varies with the season.

If 100 mice were initially released into the test plot and if $a = 0.9$, estimate the number of mice as a function of t, for $t = 0$ to $t = 8$.

In this chapter we will find how such a problem can be solved by numerical methods.

Table 5.1

t	B	t	B
0	0.0070	5	0.0013
1	0.0036	6	0.0028
2	0.0011	7	0.0043
3	0.0001	8	0.0056
4	0.0004		

Because the rate of change of a quantity is generally more easily determined from the various factors that influence a given physical system than from the values of the quantity itself, most scientific and engineering laws are expressed as differential equations. We mean by the term "differential equation" a relationship involving derivatives of the function as well as the function and the independent variable. For example,

$$\frac{du}{dt} = -0.27(u - 60)^{5/4}$$

is an equation describing (approximately) the rate of change of temperature u of a body losing heat by natural convection with constant-temperature surroundings. This is termed a first-order differential equation because the highest-order derivative is the first.

If the equation contains derivatives of nth order, it is said to be an nth-order differential equation. For example, a second-order equation describing the oscillation of a weight acted upon by a spring, with resistance to motion proportional to the square of the velocity, might be

$$\frac{d^2x}{dt^2} + 4\left(\frac{dx}{dt}\right)^2 + 0.6x = 0,$$

where x is the displacement and t is time.

The solution to a differential equation is the function which satisfies the differential equation and which also satisfies certain initial conditions on the function. In solving a differential equation analytically, one usually finds a general solution containing arbitrary constants and then evaluates the arbitrary constants so that the expression agrees with the initial conditions. For an nth-order equation, n independent initial conditions must usually be known. The analytical methods are limited to certain special forms of equations; elementary courses normally treat only linear equations with constant coefficients when the degree of the equation is higher than the first. Neither of the above examples is linear.

Numerical methods have no such limitations to only standard forms. We obtain the solution as a tabulation of the values of the function at various values of the independent variable, however, and not as a functional relationship. We also must pay a price for our ability to solve practically any equation in that we must recompute the entire table if the initial conditions are changed.

Our procedure will be to explore several methods of solving first-order equations, and then show how these same methods can be applied to systems of simultaneous first-order equations and to higher-order differential equations. We shall use for our typical first-order equation the form

$$\frac{dy}{dx} = f(x, y),$$

$$y(x_0) = y_0.$$

5.2 TAYLOR-SERIES METHOD

The first method we discuss is not strictly a numerical method, but it is sometimes used in conjunction with the numerical schemes, is of general applicability, and serves as an introduction to the other techniques we shall study. Consider the example problem

$$\frac{dy}{dx} = x + y, \qquad y(0) = 1. \tag{5.1}$$

(This particularly simple example is chosen to illustrate the method so that the student can readily check the computational work. The analytical solution, $y = 2e^x - x - 1$, is obtained immediately by application of the standard methods, and will be compared with our numerical results to show exactly the error at any step.)

We develop the relation between y and x by finding the coefficients of the Taylor series in which we expand y about the point $x = x_0$:

$$y(x) = y(x_0) + y'(x_0)(x - x_0) + \frac{y''(x_0)}{2!}(x - x_0)^2 + \frac{y'''(x_0)}{3!}(x - x_0)^3 + \dots .$$

If we let $x - x_0 = h$, we can write the series as

$$y(x) = y(x_0) + y'(x_0)h + \frac{y''(x_0)}{2}h^2 + \frac{y'''(x_0)}{6}h^3 + \dots .$$

Since $y(x_0)$ is our initial condition, the first term is known from the initial condition $y(0) = 1$. (Since the expansion is about the point $x = 0$, our Taylor series is actually a Maclaurin series in this example.)

We get the coefficient of the second term by substituting $x = 0$, $y = 1$ into the equation for the first derivative, Eq. (5.1):

$$y'(x_0) = y'(0) = 0 + 1 = 1.$$

We get equations for the second- and higher-order derivatives by successively differentiating the equation for the first derivative. Each of these is evaluated corresponding to $x = 0$ to get the various coefficients:

$$y''(x) = 1 + y', \qquad y''(0) = 1 + 1 = 2,$$
$$y'''(x) = y'', \qquad y'''(0) = 2,$$
$$y^{iv}(x) = y''', \qquad y^{iv}(0) = 2,$$
$$\text{etc.} \qquad y^{(n)}(0) = 2.$$

[Getting the derivatives is deceptively easy in this example. The student should compare this to the function $f(x, y) = x/y$, with $y(1) = 1$.]

We then write our series solution for y, letting $x = h$ be the value at which we wish to determine y:

$$y(h) = 1 + h + h^2 + \tfrac{1}{3}h^3 + \tfrac{1}{12}h^4 + \text{error}.$$

The solution to our differential equation, $dy/dx = x + y$, $y(0) = 1$, is then given by Table 5.2.

Table 5.2

x	y	y, analytical
0	1.000	1.0000
0.1	1.1103	1.1103
0.2	1.2428	1.2428
0.3	1.3997	1.3997
0.4	1.5835 (?)	1.5836
0.5	1.7969 (?)	1.7974

As we computed the last two entries, there was some doubt in our minds as to their accuracy without using more terms of the Taylor series, because the successive terms were decreasing less and less rapidly. We need more terms than we have calculated to get four-decimal-place accuracy.

The error when a convergent Taylor series is truncated is simple to express. We merely take the next term and evaluate the derivative at the point $x = \xi$, $0 < \xi < h$, instead of at the point $x = x_0$. This is exactly what we did to write error terms in the previous chapters. The error term of the Taylor series after the h^4 term is:

$$\text{error} = \frac{y^{(v)}(\xi)}{5!} h^5, \qquad 0 < \xi < h.$$

However, this cannot be computed because evaluating the derivative at $x = \xi$ is impossible with ξ unknown, and even bounding it in the interval $[0, h]$ is impossible because the derivatives are known only at $x = 0$ and not at $x = h$.

Numerical analysis is sometimes termed an art instead of a science because, in situations like this, the number of Taylor-series terms to be included is a matter of judgment and experience. We normally truncate the Taylor series when the contribution of the last term is negligible to the number of decimal places to which we are working. However, this is correct only when the succeeding terms become small rapidly enough—in some cases the sum of the many neglected small terms is significant.

The Taylor series is easily applied to a higher-order equation. For example, if we are given

$$y'' = 3 + x - y^2, \qquad y(0) = 1, \qquad y'(0) = -2,$$

we can find the derivative terms in the Taylor series as follows:

$y(0)$, $y'(0)$ are given by the initial conditions.

$y''(0)$, results by substitution into the differential equation from $y(0)$ and $y'(0)$.

$y'''(0)$, $y^{iv}(0)$, ... are found by differentiating the equation for the previous order of derivative and substituting previously computed values.

5.3 EULER AND MODIFIED EULER METHODS

As we have seen, the Taylor-series method is awkward to apply if the various derivatives are complicated, and the error is difficult to determine. An even more significant criticism in this computer age is that taking derivatives of arbitrary functions cannot easily be written into a computer program. We look for an approach that is not subject to these disadvantages.

One thing we do know about the Taylor series. The error will be small if the step size h (the interval beyond x_0 where we evaluate the series) is small. In fact, if it is small enough, only a few terms are needed for good accuracy. The Euler method may be thought of as following this idea to the extreme for first-order differential equations. Suppose we choose h small enough that we may truncate after the first derivative term. Then

$$y(x_0 + h) = y(x_0) + hy'(x_0) + \frac{y''(\xi)}{2} h^2, \qquad x_0 < \xi < x_0 + h.$$

We have written the usual form of the error term for the truncated Taylor series.

In using this equation, the value of $y(x_0)$ is given by the initial condition and $y'(x_0)$ is evaluated from $f(x_0, y_0)$, given by the differential equation, $dy/dx = f(x, y)$. It will of course be necessary to use this method iteratively, advancing the solution to $x = x_0 + 2h$ after $y(x_0 + h)$ has been found, then to $x = x_0 + 3h$, etc. Adopting a subscript notation for the successive y-values and representing the error by the order relation, we may write the algorithm for the Euler method as:

$$y_{n+1} = y_n + hy'_n + O(h^2)\text{error}.^* \tag{5.2}$$

---

* This is the order of error for one step only, the "local error." As detailed in a later section, over many steps the error becomes $O(h)$. Such accumulated error is termed the *global error.*

Table 5.3

x_n	y_n	y'_n	hy'_n
0	1.0000	1.0000	0.0200
0.02	1.0200	1.0400	0.0208
0.04	1.0408	1.0808	0.0216
0.06	1.0624	1.1224	0.0224
0.08	1.0848	1.1648	0.0233
0.10	1.1081		

(1.1103 analytical, error is 0.0022)

As an example, we apply this to the simple equation

$$\frac{dy}{dx} = x + y, \qquad y(0) = 1,$$

where the computations can be done mentally. It is convenient to arrange the work as in Table 5.3. Take $h = 0.02$.

Each of the y_n values is computed using Eq. (5.2), adding hy'_n and y_n of the previous line. Comparing the last result to the analytical answer $y(0.10) = 1.1103$, we see that there is only two-decimal-place accuracy, even though we have advanced the solution only five steps. To gain four-decimal accuracy, we must reduce the error at least 22-fold. Since the global error is about proportional to h, we will need to reduce the step about 22-fold to < 0.004.

The trouble with this most simple method is its lack of accuracy, requiring an extremely small step size. Figure 5.1 suggests how we might improve this method with little additional effort.

In the simple Euler method, we use the slope at the beginning of the interval, y'_n, to determine the increment to the function, but this is always wrong. After all, if the slope of the function were constant, the solution is an obvious linear relation. We need to use an average slope over the interval if we hope to estimate the change in y with precision.

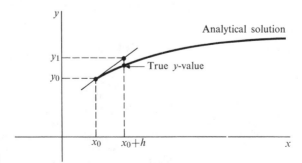

Figure 5.1

Suppose we do this, using the arithmetic average of the slopes at the beginning and end of the interval:

$$y_{n+1} = y_n + h\frac{y'_n + y'_{n+1}}{2}.$$ (5.3)

This will surely give an improved estimate for y at x_{n+1}. However, we are unable to employ Eq. (5.3) immediately, because, since the derivative is a function of both x and y, we cannot evaluate y'_{n+1} with y_{n+1} unknown. The modified Euler method surmounts the difficulty by estimating, or "predicting." a value of y_{n+1} by the simple Euler relation, Eq. (5.2), and uses this value to compute y'_{n+1}, giving an improved estimate ("corrected" value) for y_{n+1}. Since the value of y'_{n+1} was computed using the predicted value, of less than perfect accuracy, one is tempted to recorrect the y_{n+1} value as many times as will make a significant difference. (If more than two or three recorrections are required, it is more efficient to reduce the step size.)

We will illustrate the modified Euler method which we also call the Euler predictor–corrector method, on the same problem previously treated. Table 5.4 is convenient.

$$\frac{dy}{dx} = x + y, \qquad y(0) = 1, \qquad h = 0.02.$$

Table 5.4

x_n	y_n	y'_n	hy'_n	y_{n+1}	y'_{n+1}	y'_{av}	hy'_{av}
0	1.0000	1.0000	0.0200	1.0200	1.0400	1.0200	0.0204
				1.0204*	1.0404	1.0202	0.0204
0.02	1.0204	1.0404	0.0208	1.0412	1.0812	1.0608	0.0212
				1.0416	1.0816	1.0610	0.0212
0.04	1.0416	1.0816	0.0216	1.0632	1.1232	1.1024	0.0220
				1.0636	1.1236	1.1026	0.0221
				1.0637	1.1237	1.1027	0.0221
0.06	1.0637	1.1237	0.0225	1.0862	1.1662	1.1449	0.0229
				1.0866	1.1666	1.1451	0.0229
0.08	1.0866	1.1666	0.0233	1.1099	1.2099	1.1883	0.0238
				1.1104	1.2104	1.1885	0.0238
0.10	1.1104						

(1.1103 analytical value)

* It is convenient to use this column for both the predicted and corrected values of y_{n+1}. The first entry in any x-row is the predicted value. Corrected and recorrected values follow. In this example, it was necessary to recorrect only where $x = 0.04$. All the other calculations indicate no need to recorrect the values of y_{n+1}. We shall discuss in a later section whether one should recorrect the first corrected value of y_{n+1} in this type of method. In general, it is better to make h small and not recorrect.

In this table, we tabulate the corrected values of y_{n+1} in the same column as the predicted ones. y'_{av} is the mean of y'_n and the last value of y'_{n+1}. The answer agrees within 1 in the fourth decimal place. We have done more work than in the simple Euler method, but certainly not the 22 times more that would have been needed with that method to attain four-decimal-place accuracy. Hence the modified method is more efficient.

We can find the error of the modified Euler method by comparing with the Taylor series.

$$y_{n+1} = y_n + y'_n h + \tfrac{1}{2} y''_n h^2 + \frac{y'''(\xi)}{6} h^3, \qquad x_n < \xi < x_n + h.$$

Replace the second derivative by the forward-difference approximation for y'', $(y'_{n+1} - y'_n)/h$, which has error of $O(h)$, and write the error term as $O(h^3)$:

$$y_{n+1} = y_n + h\left(y'_n + \frac{1}{2}\left[\frac{y'_{n+1} - y'_n}{h} + O(h)\right]h\right) + O(h^3),$$

$$y_{n+1} = y_n + h(y'_n + \tfrac{1}{2}y'_{n+1} - \tfrac{1}{2}y'_n) + O(h^3),$$

$$y_{n+1} = y_n + h\left(\frac{y'_n + y'_{n+1}}{2}\right) + O(h^3).$$

This shows that the error of one step of the modified Euler method is $O(h^3)$. This is the "local error." There is an accumulation of errors from step to step, so that the error over the whole range of application, the so-called "global error," is $O(h^2)$. This seems intuitively reasonable since the number of steps into which the interval is subdivided is proportional to $1/h$; hence the order of error is reduced to $O(h^2)$ on the continuing application. We treat the accumulation of errors more fully in a later section.

5.4 RUNGE–KUTTA METHODS

A further advance in efficiency (i.e., obtaining the most accuracy per unit of computational effort) can be secured with a group of methods due to the German mathematicians Runge and Kutta. The fourth-order Runge–Kutta methods are widely used in computer solutions to differential equations. The development of this technique is algebraically complicated.*

To convey some idea of how the Runge–Kutta methods are developed, we show the derivation of a second-order method. We write the increment to y as a weighted average of two estimates of Δy, k_1 and k_2. For the equation $dy/dx = f(x, y)$,

$$y_{n+1} = y_n + ak_1 + bk_2,$$
$$k_1 = hf(x_n, y_n), \qquad\qquad (5.4)$$
$$k_2 = hf(x_n + \alpha h, y_n + \beta k_1).$$

---

* The development is given rather completely in Kopal (1955), pp. 195–213.

We can think of the values k_1 and k_2 as estimates of the change in y when x advances by h because they are the product of the change in x and a value for the slope of the curve, dy/dx. The Runge–Kutta methods always use as the first estimate of Δy the simple Euler estimate; the other estimates are taken with x and y stepped up the fractions α and β of h and of the earlier estimate of Δy, k_1. Our problem is to devise a scheme of choosing the four parameters, a, b, α, β. We do this by making Eq. (5.4) agree as well as possible with the Taylor-series expansion, in which the y-derivatives are written in terms of f, from $dy/dx = f(x, y)$,

$$y_{n+1} = y_n + hf(x_n, y_n) + (h^2/2)f'(x_n, y_n) + \ldots.$$

An equivalent form, since $df/dx = f_x + f_y\,dy/dx = f_x + f_yf$, is

$$y_{n+1} = y_n + hf_n + h^2(\tfrac{1}{2}f_x + \tfrac{1}{2}f_yf)_n. \tag{5.5}$$

(All the derivatives in Eq. (5.5) are evaluated at the point (x_n, y_n).) We now rewrite Eq. (5.4) by substituting the definitions of k_1 and k_2:

$$y_{n+1} = y_n + ahf(x_n, y_n) + bhf[x_n + \alpha h, y_n + \beta hf(x_n, y_n)]. \tag{5.6}$$

To make the last term of Eq. (5.6) comparable to Eq. (5.5) we expand $f(x, y)$ in a Taylor series in terms of x_n, y_n, remembering that f is a function of two variables,* retaining only first derivative terms,

$$f[x_n + \alpha h, y_n + \beta hf(x_n, y_n)] \doteq (f + f_x\alpha h + f_y\beta hf)_n. \tag{5.7}$$

On the right side of both Eqs. (5.5) and (5.7), f and its partial derivatives are all to be evaluated at (x_n, y_n).

Substituting from Eq. (5.7) into Eq. (5.6), we have:

$$y_{n+1} = y_n + ahf_n + bh(f + f_x\alpha h + f_y\beta hf)_n,$$

or, rearranging,

$$y_{n+1} = y_n + (a + b)hf_n + h^2(\alpha bf_x + \beta bf_yf)_n. \tag{5.8}$$

Equation (5.8) will be identical to Eq. (5.5) if

$$a + b = 1, \qquad \alpha b = \tfrac{1}{2}, \qquad \beta b = \tfrac{1}{2}.$$

Note that there are only three equations that need to be satisfied by the four unknowns. We can choose one value arbitrarily (with minor restrictions); hence we have a set of second-order methods. For example, taking $a = \tfrac{2}{3}$, we have $b = \tfrac{1}{3}$, $\alpha = \tfrac{3}{2}$, $\beta = \tfrac{3}{2}$. Other choices give other sets of parameters that agree with the Taylor-series expansion. If one takes $a = \tfrac{1}{2}$, the other variables are $b = \tfrac{1}{2}$, $\alpha = 1$, $\beta = 1$. This last set of parameters gives the modified Euler algorithm that we have previously discussed; the modified Euler method is a special case of a second-order Runge–Kutta method.

---

* Appendix A will remind the student of this expansion.

Fourth-order Runge–Kutta methods are most widely used and are derived in similar fashion. Greater complexity results from having to compare terms through h^4, and gives a set of eleven equations in thirteen unknowns. The set of eleven equations can be solved with two unknowns being chosen arbitrarily. The most commonly used set of values leads to the algorithm:

$$y_{n+1} = y_n + \tfrac{1}{6}(k_1 + 2k_2 + 2k_3 + k_4),$$
$$k_1 = hf(x_n, y_n),$$
$$k_2 = hf(x_n + \tfrac{1}{2}h, y_n + \tfrac{1}{2}k_1),$$
$$k_3 = hf(x_n + \tfrac{1}{2}h, y_n + \tfrac{1}{2}k_2),$$
$$k_4 = hf(x_n + h, y_n + k_3).$$

As an example, we again solve $dy/dx = x + y$, $y(0) = 1$, this time taking $h = 0.1$:

$$k_1 = 0.1(0 + 1) = 0.10000,$$
$$k_2 = 0.1(0.05 + 1.05) = 0.11000,$$
$$k_3 = 0.1(0.05 + 1.055) = 0.11050,$$
$$k_4 = 0.1(0.10 + 1.1105) = 0.12105,$$
$$y(0.1) = 1.0000 + \tfrac{1}{6}(0.10000 + 0.2200 + 0.22100 + 0.12105)$$
$$= 1.11034.$$

This agrees to five decimals with the analytical result, and illustrates a further gain in accuracy with less effort than required by our example of Section 2.

The local error term for the fourth-order Runge–Kutta is $O(h^5)$; the global error would be about $O(h^4)$. It is computationally more efficient than the modified Euler method because, while four evaluations of the function are required per step rather than two, the steps can be many-fold larger for the same accuracy.

Let us summarize and compare (see Table 5.5) the three numerical methods we have studied for solving $y' = f(x, y)$.

Table 5.5

Method	Estimate of slope over x-interval	Local error	Global error	Evaluations of $f(x, y)$ per step
Euler	Initial value	$O(h^2)$	$O(h)$	1
Modified Euler	Arith. average of initial and final value*	$O(h^3)$	$O(h^2)$	2*
Runge–Kutta fourth-order	Weighted average of four values	$O(h^5)$	$O(h^4)$	4

*Final slope calculated using "predicted" value of y at end of interval. More than two function evaluations may be used if recorrections are made.

In the example problem, $y' = x + y$, $y(0) = 1$, we obtained the comparative results shown in Table 5.6 for the y-value at $x = 0.1$.

It is easy to see why the Runge–Kutta technique is so popular. Since going from second to fourth order was so beneficial, we may wonder whether we should use a still higher order of formula. Higher-order (fifth, sixth) Runge–Kutta formulas have been developed, but the general conclusion of most authorities is that the additional accuracy is about offset by the additional computational effort. In other words, there is no significant improvement in efficiency over fourth-order schemes.

Table 5.6

Method	Step size	Result	Error	Number of function evaluations
Euler	0.02	1.1081	0.0022	5
Modified Euler	0.02	1.1104	0.0001	12
Runge–Kutta fourth-order	0.1	1.11034	0.00000	4

5.5 MULTISTEP METHODS

Runge–Kutta type methods (which include Euler and Modified Euler as special cases) are called single-step methods because they use only the information from the last step computed. In this they have the ability to perform the next step with a different step size and are ideal for beginning the solution where only the initial conditions are available. After the solution has begun, however, there is additional information available about the function (and its derivative) if we are wise enough to retain it in the memory of the computer. A multistep method is one that takes advantage of this fact.

The principle behind a multistep method is to utilize the past values of y and/or y' to construct a polynomial that approximates the derivative function, and extrapolate this into the next interval. Most methods use equispaced past values to make the construction of the polynomial easy. The Adams method is typical. The number of past points that are used sets the degree of the polynomial and thereby is responsible for the truncation error. The order of the method is equal to the power of h in the global error term of the formula which is also equal to one more than the degree of the polynomial.

To derive the relations for the Adams method, we write the differential equation $dy/dx = f(x, y)$ in the form

$$dy = f(x, y) \, dx,$$

and we integrate between x_n and x_{n+1}:

$$\int_{x_n}^{x_{n+1}} dy = y_{n+1} - y_n = \int_{x_n}^{x_{n+1}} f(x, y) \, dx.$$

In order to integrate the term on the right, we approximate $f(x, y)$ as a polynomial in x, deriving this by making it fit at several past points. If we use three past points, the approximating polynomial will be a quadratic. If we use four points, it will be a cubic. The more points we use, the better the accuracy (until round-off interferes, of course).

Suppose we fit a second-degree polynomial through three past points, writing this as a Newton–Gregory backward polynomial:

$$\int_{x_n}^{x_{n+1}} dy = y_{n+1} - y_n = \int_{x_n}^{x_{n+1}} \left(f_n + s\Delta f_{n-1} + \frac{(s+1)s}{2} \Delta^2 f_{n-2} + \text{error} \right) dx$$

$$= \int_{s=0}^{s=1} \left(f_n + s\Delta f_{n-1} + \frac{(s+1)s}{2} \Delta^2 f_{n-2} \right) h \, ds + \int_{s=0}^{s=1} \frac{s(s-1)(s-2)}{6} h^3 f'''(\xi) h \, ds.$$

In the above, we have changed the variable to s and identified x_n as x_0. The interval of integration becomes $s = 0$ to $s = 1$. Performing the integration, we get

$$y_{n+1} - y_n = h(f_n + \tfrac{1}{2}\Delta f_{n-1} + \tfrac{5}{12}\Delta^2 f_{n-2}) + O(h^4). \tag{5.9}$$

While it is perfectly possible to use Eq. (5.9) as it stands, constructing a difference table is not necessary. If we expand the differences of f in terms of the values of f_n, f_{n-1}, and f_{n-2}, we get a more useful formula:

$$y_{n+1} = y_n + h\left[f_n + \frac{f_n - f_{n-1}}{2} + \frac{5(f_n - 2f_{n-1} + f_{n-2})}{12} \right]$$

$$= y_n + \frac{h}{12} [23f_n - 16f_{n-1} + 5f_{n-2}] + O(h^4). \tag{5.10}$$

Observe that Eq. (5.10) resembles the single-step formulas of the previous sections in that the increment to y is a weighted sum of the derivatives times the step size, but differs in that past values are used rather than estimates in the forward direction.

Example We illustrate the use of Eq. (5.10) to calculate $y(0.6)$ for $dy/dx = x + y$, $y(0) = 1$, assuming that we already have good values of $y(0.2)$ and $y(0.4)$, with $h = 0.2$. We first tabulate the past values (Table 5.7).

Table 5.7

x	y	$y' = f(x, y) = x + y$
0	1	1.0
0.2	1.2428	1.4428
0.4	1.5836	1.9836

Table 5.8

x	y	$f(x, y)$
0.3	1.3997	1.6997
0.4	1.5836	1.9836
0.5	1.7974	2.2974

Then, from Eq. (5.10):

$$y(0.6) = 1.5836 + \frac{0.2}{12}[23(1.9836) - 16(1.4428) + 5(1.0)] = 2.0426.$$

Comparing our result with the exact solution (2.04424), we find that the computed value has an error of 0.0017. We can reduce the size of the error by doing the calculations with a smaller step size. Let us begin by establishing a step size of 0.1, requiring values at $x = 0.3$, $x = 0.4$, and $x = 0.5$ (see Table 5.8). Then we repeat the computations:

$$y(0.6) = 1.7974 + \frac{0.1}{12}[23(2.2974) - 16(1.9836) + 5(1.6997)] = 2.0441,$$

which has an error of 0.0002.

We can also reduce the error by using more past points for fitting the polynomial. When the derivation is done for this case, we get these results:

$$y_{n+1} = y_n + h(f_n + \tfrac{1}{2}\Delta f_{n-1} + \tfrac{5}{12}\Delta^2 f_{n-2} + \tfrac{3}{8}\Delta^3 f_{n-3}) + O(h^5),$$

or

$$y_{n+1} = y_n + \frac{h}{24}[55f_n - 59f_{n-1} + 37f_{n-2} - 9f_{n-3}] + O(h^5).$$

We repeat the above example with $h = 0.1$ to compute $y(0.6)$, using the past values shown in Table 5.9 and computing:

$$y(0.6) = 1.7974 + \frac{0.1}{24}[55(2.2974) - 59(1.9836) + 37(1.6997) - 9(1.4428)]$$

$$= 2.0442.$$

The error of this computation has been reduced to 0.5×10^{-4}.

Table 5.9

x	y	$f(x, y)$
0.2	1.2428	1.4428
0.3	1.3997	1.6997
0.4	1.5836	1.9836
0.5	1.7974	2.2974

We can continue this attack in the direction of higher-order Adams formulas, but there is a better alternative. The errors of an extrapolated polynomial are greater than when one interpolates. Suppose one uses the first calculated value for y_{n+1} only as a predicted value and then corrects it before going on to the next interval. This technique is similar to the strategy used in the Euler predictor–corrector method. We explore this and the Milne method in the next sections.

5.6 MILNE'S METHOD

The method of Milne is a multistep method that first predicts a value for y_{n+1} by extrapolating values for the derivative. It differs from the Adams method in that it corrects this predicted value before going on to the next step. The past values that we require may have been computed by the Runge–Kutta method, or possibly by the Taylor-series method. In the Milne method, we suppose that four equispaced starting values of y are known, at the points x_n, x_{n-1}, x_{n-2}, and x_{n-3}. We can employ quadrature formulas to integrate as follows:

$$\frac{dy}{dx} = f(x, y),$$

$$\int_{x_{n-3}}^{x_{n+1}} \left(\frac{dy}{dx}\right) dx = \int_{x_{n-3}}^{x_{n+1}} f(x, y)\, dx = \int_{x_{n-3}}^{x_{n+1}} P_2(x)\, dx; \tag{5.11}$$

so

$$y_{n+1} - y_{n-3} = \frac{4h}{3}(2f_n - f_{n-1} + 2f_{n-2}) + \frac{28}{90} h^5 y^v(\xi_1), \qquad x_{n-3} < \xi_1 < x_{n+1}.$$

We integrate the function $f(x, y)$ by replacing it with a quadratic interpolating polynomial that fits at the three points, where $x = x_n$, x_{n-1}, and x_{n-2}, and integrating according to the methods of Chapter 4.* Note that we extrapolate in the integration by one panel both to the left and to the right of the region of fit. Hence the error is larger, because of the extrapolation, than it would be with only interpolation.

With this value of y_{n+1} one can calculate f_{n+1} reasonably accurately. In Milne's method, we use Eq. (5.11) as a predictor formula and then correct with

$$\int_{x_{n-1}}^{x_{n+1}} \left(\frac{dy}{dx}\right) dx = \int_{x_{n-1}}^{x_{n+1}} f(x, y)\, dx = \int_{x_{n-1}}^{x_{n+1}} P_2(x)\, dx, \tag{5.12}$$

$$y_{n+1,c} - y_{n-1} = \frac{h}{3}(f_{n+1} + 4f_n + f_{n-1}) - \frac{h^5}{90} y^v(\xi_2), \qquad x_{n-1} < \xi_2 < x_{n+1}.$$

---

* The formula can also be derived by the method of undetermined coefficients, as shown in the Appendix.

Table 5.10

	x	y	$f(x, y) = x + y$	
From	0	1.00000	1.00000	
Taylor	0.1	1.11034	1.21034	
series	0.2	1.24280	1.44280	
	0.3	1.39971	1.69971	
	0.4	(1.58364)	(1.98364)	Predictor value
		1.58364	1.98364	Corrector value
	0.5	(1.79742)	(2.29742)	Predictor value
		1.79743	2.29743	Corrector value

Analytical value at $x = 0.5$ is 1.79744

In Eq. (5.12), the polynomial P_2 is not identical to that in Eq. (5.11) because they do not fit the function at the same three points. In Eq. (5.12) the polynomial fits at x_{n+1}, x_n, and x_{n-1}. Note the changed range of integration and the smaller coefficient in the error term of Eq. (5.12) because the polynomial is not extrapolated; f_{n+1} is calculated using y_{n+1} from the predictor formula. The integration formula is the familiar Simpson's $\frac{1}{3}$ rule, since we integrate a quadratic over two panels within the region of fit. In this method we do not try to recorrect y_{n+1}.

We illustrate with our familiar simple problem, $dy/dx = x + y$, $y(0) = 1$. From Section 1, we take the four values calculated by Taylor series and carry five decimals (see Table 5.10).

For this example, the first predictor and corrector values agree, and the next set differ by only one in the fifth decimal. The corrected value at $x = 0.5$ is in error by only one in the fifth decimal. This discrepancy is mostly due to round-off error in the computed values. In this case, the value of h could have been chosen larger. With the set of values available, h cannot be increased without additional computations, but if we had seven equally spaced values, we could double h by taking only every other one and still have four values to move ahead from.

Normally, the value of y_{n+1} from the predictor and the corrector do not agree. Consideration of the error terms of Eqs. (5.11) and (5.12) suggests that the true value should usually lie between the two values and closer to the corrector value. While ξ_1 and ξ_2 are not necessarily the same value, they lie in similar intervals. If one assumes that the values of $y^v(\xi_1)$ and $y^v(\xi_2)$ are equal, the error in the corrector formula is $\frac{1}{28}$ times the error in the predictor formula. Hence the difference between the predictor and corrector formula is about 29 times the error in the corrected value. This is frequently used as a criterion of accuracy for Milne's method.* The ease with which we can monitor the error is a particular

---

* A further criterion as to whether or not the single correction that is normally used in Milne's method is adequate is discussed in Section 5.8.

advantage of predictor–corrector methods. We are able to know immediately whether the step size is too big to give the desired degree of accuracy. This is in strong contrast to the Runge–Kutta methods of the previous section.

Milne's method is simple and has a good error term, $O(h^5)$ for local error. It is subject to an instability problem in certain cases, however, wherein the errors do not tend to zero as h is made smaller. This unexpected phenomenon is discussed below. Because of possible instability, another method, a modification of that of Adams, is more widely used than Milne's method.

It is sufficient to show the instability of Milne's method for one simple case. We shall be able to draw the necessary conclusions from this. Consider the differential equation, $dy/dx = Ay$, where A is a constant. The general solution is $y = ce^{Ax}$. If $y(x_0) = y_0$ is the initial condition which the solution must satisfy, $c = y_0 e^{-Ax_0}$. Hence, letting y_n be the value of the function when $x = x_n$, the analytical solution is

$$y_n = y_0 e^{A(x_n - x_0)}. \tag{5.13}$$

If we solve the differential equation by the method of Milne, we have, from the corrector formula,

$$y_{n+1} = y_{n-1} + \frac{h}{3}(y'_{n+1} + 4y'_n + y'_{n-1}).$$

Letting $y'_n = Ay_n$, from the original differential equation, and rearranging, we get

$$y_{n+1} = y_{n-1} + \frac{h}{3}(Ay_{n+1} + 4Ay_n + Ay_{n-1}),$$

$$\left(1 - \frac{hA}{3}\right)y_{n+1} - \frac{4hA}{3}y_n - \left(1 + \frac{hA}{3}\right)y_{n-1} = 0. \tag{5.14}$$

We would like to solve this equation for y_n in terms of y_0 to compare to Eq. (5.13). Equation (5.14) is a second-order linear difference equation which can be solved in a manner analogous to that for differential equations. The solution is:

$$y_n = C_1 Z_1^n + C_2 Z_2^n, \tag{5.15}$$

where Z_1, Z_2 are the roots of the quadratic

$$\left(1 - \frac{hA}{3}\right)Z^2 - \frac{4hA}{3}Z - \left(1 + \frac{hA}{3}\right) = 0. \tag{5.16}$$

(The student should check that Eq. (5.15) is a solution of Eq. (5.14) by direct substitution.) For simplification, let $hA/3 = r$; the roots of Eq. (5.16) are then:

$$Z_1 = \frac{2r + \sqrt{3r^2 + 1}}{1 - r},$$

$$Z_2 = \frac{2r - \sqrt{3r^2 + 1}}{1 - r}. \tag{5.17}$$

We are interested in comparing the behavior of Eqs. (5.15) and (5.13) as the step size h becomes small. As $h \to 0$, $r \to 0$, and $r^2 \to 0$ even faster. Neglecting the $3r^2$ terms in comparison to the constant 1 under the radical in Eq. (5.17) gives:

$$Z_1 \doteq \frac{2r + 1}{1 - r} = 1 + 3r + O(r^2) = 1 + Ah + O(h^2),$$

$$Z_2 \doteq \frac{2r - 1}{1 - r} = -1 + r + O(r^2) = -\left(1 - \frac{Ah}{3}\right) + O(h^2).$$

(5.18)

The last results are obtained by dividing the fractions. We now compare Eq. (5.18) with the Maclaurin series of the exponential function,

$$e^{hA} = 1 + hA + O(h^2),$$

$$e^{-hA/3} = 1 - \frac{hA}{3} + O(h^2).$$

we see that, for $h \to 0$,

$$Z_1 \doteq e^{-hA}, \qquad Z_2 \doteq e^{-hA/3}.$$

Hence the Milne solution is represented by

$$y_n = C_1(e^{hA})^n + C_2(e^{-hA/3})^n = C_1 e^{A(x_n - x_0)} + C_2 e^{-A(x_n - x_0)/3}. \qquad (5.19)$$

In Eq. (5.19), we have used $x_n - x_0 = nh$. The solution consists of two parts. The first term obviously matches with the analytical solution, Eq. (5.13). The second term, called a *parasitic term*, will die out as x_n increases if A is a positive constant, but if A is negative, it will grow exponentially with x_n. Note that we get this peculiar behavior independent of h; smaller step size is of no benefit in eliminating the error.

A numerical example that demonstrates the instability of Milne's method is given in the next section. Such a demonstration by a numerical example is less conclusive than the analytical analysis given above, but is much easier to grasp.

5.7 ADAMS–MOULTON METHOD

A method that does not have the same instability problem as the Milne method, but is about as efficient, is the Adams–Moulton method. It also assumes a set of starting values already calculated by some other technique. Here we take a cubic through four points, from x_{n-3} to x_n, and integrate over one step, from x_n to x_{n+1}. This is the same as the method described in Section 5.5. For $dy/dx = f(x, y)$,

$$\int_{x_n}^{x_{n+1}} \left(\frac{dy}{dx}\right) dx = \int_{x_n}^{x_{n+1}} f(x, y) \, dx = \int_{x_n}^{x_{n+1}} P_3(x) \, dx,$$

$$y_{n+1} - y_n = h(f_n + \tfrac{1}{2}\Delta f_{n-1} + \tfrac{5}{12}\Delta^2 f_{n-2} + \tfrac{3}{8}\Delta^3 f_{n-3}) + \tfrac{251}{720}h^5 y^v(\xi), \qquad (5.20)$$

$$x_{n-3} < \xi < x_{n+1}.$$

The integration formula can be derived by the methods of Chapter 4 or by the method of undetermined coefficients. Alternatively, it can be developed by writing $P_3(x)$ as a Newton–Gegory backwards-interpolating polynomial (fitting at x_n, x_{n-1}, x_{n-2}, and x_{n-3}) and integrating. In the Adams–Moulton method we continue by correcting y_{n+1} before calculating the next step. Using Eq. (5.20) as a predictor formula, we can compute a nearly correct value of f_{n+1}. If we now approximate $f(x, y)$ as a cubic that fits over the range from x_{n-2} to x_{n+1}, and integrate from x_n to x_{n+1}, we will not be extrapolating the polynomial, and will have a more favorable error term. The result is:

$$y_{n+1} - y_n = h(f_{n+1} - \tfrac{1}{2}\Delta f_n - \tfrac{1}{12}\Delta^2 f_{n-1} - \tfrac{1}{24}\Delta^3 f_{n-2}) - \tfrac{19}{720}h^5 y^v(\xi_1), \quad (5.21)$$

$$x_{n-2} < \xi_1 < x_{n+1}.$$

The Adams–Moulton method consists of using Eq. (5.20) as a predictor and Eq. (5.21) as a corrector. We illustrate it with the same example as before, $dy/dx = x + y$, $y(0) = 1$. A difference table (Table 5.11) is computed to assist us (though in a computer program we would use a different scheme: see Eqs. (5.22)).

By the predictor formula,

$$y(0.4) = 1.39971 + 0.1[1.69971 + \tfrac{1}{2}(0.25691) + \tfrac{5}{12}(0.02445) + \tfrac{3}{8}(0.00233)]$$

$$= 1.58363.$$

Then f at $x = 0.4$ is computed and the difference table is calculated. The corrector formula then gives

$$y(0.4) = 1.39971 + 0.1[1.98363 - \tfrac{1}{2}(0.28392) - \tfrac{1}{12}(0.02701) - \tfrac{1}{24}(0.00256)]$$

$$= 1.58364.$$

Table 5.11

	x	y	f	Δf	$\Delta^2 f$	$\Delta^3 f$
	0	1.00000	1.00000			
				0.21034		
	0.1	1.11034	1.21034		0.02212	
Starting				0.23246		0.00233
values	0.2	1.24280	1.44280		0.02445	
				0.25691		(0.00256)
	0.3	1.39971	1.69971		(0.02701)	0.00257
				(0.28392)	0.02702	(0.00283)
Predicted	0.4	(1.58363)	(1.98363)	0.28393	(0.02985)	
Corrected		1.58364	1.98364	(0.31378)		
Predicted	0.5	(1.79742)	(2.29742)			
Corrected		1.79743				

Analytical value at $x = 0.5$ is 1.79744.

The computations are continued in the same manner to get $y(0.5)$. The corrected value almost agrees to five decimals with the predicted value. Comparing error terms of Eqs. (5.20) and (5.21), and assuming that the two fifth-derivative values are equal, we see that the true value should lie between the predicted and corrected values, with the error in the corrected value being about:

$$\frac{19}{251 + 19} \quad \text{or} \quad \frac{1}{14.2}$$

times the difference between the predicted and corrected values. A frequently used criterion for accuracy of the Adams–Moulton method with four starting values is that the corrected value is not in error by more than one in the last place if the difference between predicted and corrected values is less than 14 in the last decimal place.* If this degree of accuracy is inadequate, we know that h is too large.

Equations (5.20) and (5.21) are not well suited to computer utilization because calculating and storing the difference tables is wasteful of both time and memory space. Each of these differences is a linear function of the various f-values, however. Expressing the differences in terms of the f's gives an alternative form for the Adams–Moulton method that is usually employed in a computer program.

$$\textit{Predictor:} \quad y_{n+1} = y_n + \frac{h}{24}(55f_n - 59f_{n-1} + 37f_{n-2} - 9f_{n-3}).$$

$$\tag{5.22}$$

$$\textit{Corrector:} \quad y_{n+1} = y_n + \frac{h}{24}(9f_{n+1} + 19f_n - 5f_{n-1} + f_{n-2}).$$

These same formulas are derived directly by the method of undetermined coefficients as Eqs. (B.28) and (B.29) of Appendix B.

These equations are better suited to machine calculation and to digital computer programming. Without such calculation aids, the large coefficients and alternating signs lead to large round-off errors unless extra guard figures are carried.

Adams–Moulton formulas employing more or less than four starting values can be derived in analogous fashion. The fourth-order formulas we have given are widely used, especially in combination with Runge–Kutta, because both kinds of methods then have local errors of $O(h^5)$.

When the predicted and corrected values agree to as many decimals as the desired accuracy, we can save computational effort by increasing the step size. As mentioned in connection with Milne's method, we can conveniently double the step size, after we have seven equispaced values, by omitting every second one. When the difference between predicted and corrected values reaches or exceeds the accuracy criterion, we should decrease step size. If we interpolate two additional y-values with a fourth-degree polynomial, where the error will be

---

* The convergence criterion of Section 5.8 should also be met.

$O(h^5)$, consistent with the rest of our work, we can readily halve the step size.*
Convenient formulas for this are

$$y_{n-1/2} = \frac{1}{128}[35y_n + 140y_{n-1} - 70y_{n-2} + 28y_{n-3} - 5y_{n-4}],$$

$$y_{n-3/2} = \frac{1}{64}[-y_n + 24y_{n-1} + 54y_{n-2} - 16_{n-3} + 3y_{n-4}]. \tag{5.23}$$

Use of these values with y_n, y_{n-1} gives four values of the function at intervals of $\Delta x = h/2$.

The efficiency of both the Adams–Moulton and Milne methods is about twice that of the Runge–Kutta technique, because only two new function evaluations are needed per step rather than four, and the error terms are similar. In addition, these multistep methods have built in accuracy criteria so that step-size control can readily be secured.

We have stressed that the advantage of the Adams–Moulton method over that of Milne is that it is a stable method rather than an unstable one. An analytical proof of the stability of Adams–Moulton for the equation $y' = Ay$ is similar to the analysis of Section 5.6—an equation results with parasitic terms that do die out as h gets small regardless of the sign of A. Such a treatment is not entirely satisfying because we can't prove stability by examining only one case (proving that an assertion is *not* true is always easier because we need to find only one counterexample).

It is much clearer to compare the stability of Adams–Moulton with Milne through a numerical example. The table in Fig. 5.2 presents results from a computer program that solved $y' = -10y$, $y(0) = 1$ (for which the analytical solution is $y = e^{-10x}$) over the interval $x = 0$ to $x = 2$. The first part of the table is computed with $h = 0.04$. In the second part, for which only partial output is given, $h = 0.004$.

Observe that the results with the Milne method grow to have very large relative errors. In fact, near $x = 2.0$, the relative error is practically 100 percent. The oscillatory behavior of the solution is also characteristic of instability. Even with the smaller h, the Milne method blows up. The errors are even greater near $x = 2.0$ in spite of the smaller step size. (The Milne solution when x is small is considerably more accurate when $h = 0.004$. Results are not shown for this.)

The results by the Adams–Moulton method do not show this anomalous behavior. The relative error, while growing to some degree, still stays manageable (8 percent at $x = 2.0$ with $h = 0.04$, < 0.1 percent at $x = 2.0$ with $h = 0.004$). The expected decrease of error with decrease in step size is realized. The oscillation of values we notice in the results by Milne is absent. In sum, we conclude that the Adams–Moulton method gives good results, particularly at the smaller value of h, while the results of the Milne method are hopeless.

---

* An alternative but computationally more expensive way to get intermediate points would be to compute them with Runge–Kutta formulas.

	SOLUTION USING MILNE METHOD		SOLUTION USING ADAMS–MOULTON METHOD	

VALUES WITH H = 0.04

X	Y	ERROR	Y	ERROR
0.000000E 00	0.100000E 01	0.000000E 00	0.100000E 01	0.000000E 00
0.400000E−01	0.670320E 00	0.000000E 00	0.670320E 00	0.000000E 00
0.800000E−01	0.449329E 00	0.000000E 00	0.449329E 00	0.000000E 00
0.120000E 00	0.301195E 00	0.000000E 00	0.301195E 00	0.000000E 00
0.160000E 00	0.201667E 00	0.229836E−03	0.201552E 00	0.344634E−03
0.200000E 00	0.135271E 00	0.640750E−04	0.134873E 00	0.462234E−03
0.240000E 00	0.904572E−01	0.260949E−03	0.902753E−01	0.442863E−03
0.280000E 00	0.607594E−01	0.507981E−04	0.604129E−01	0.397373E−03
0.320000E 00	0.405497E−01	0.212613E−03	0.404273E−01	0.335068E−03
0.360000E 00	0.273069E−01	0.169165E−04	0.270547E−01	0.269119E−03
0.400000E 00	0.181599E−01	0.155795E−03	0.181052E−01	0.210475E−03
0.440000E 00	0.122874E−01	−0.100434E−04	0.121160E−01	0.161357E−03
0.480000E 00	0.811849E−02	0.111297E−03	0.810813E−02	0.121657E−03
0.520000E 00	0.554193E−02	−0.253431E−04	0.542601E−02	0.905804E−04
0.560000E 00	0.361742E−02	0.804588E−04	0.363111E−02	0.667726E−04
0.599999E 00	0.251041E−02	−0.316396E−04	0.242996E−02	0.488090E−04
0.639999E 00	0.160174E−02	0.598307E−04	0.162614E−02	0.354277E−04
0.679999E 00	0.114626E−02	−0.324817E−04	0.108822E−02	0.255615E−04
0.719999E 00	0.700667E−03	0.459237E−04	0.728243E−03	0.183482E−04
0.759999E 00	0.530960E−03	−0.305050E−04	0.487344E−03	0.131114E−04
0.799999E 00	0.299215E−03	0.362506E−04	0.326133E−03	0.933232E−05
0.839999E 00	0.252197E−03	−0.273281E−04	0.218250E−03	0.661935E−05
0.879999E 00	0.121497E−03	0.292376E−04	0.146054E−03	0.468052E−05
0.919999E 00	0.124879E−03	−0.238383E−04	0.977399E−04	0.330036E−05
0.959999E 00	0.437888E−04	0.239405E−04	0.654081E−04	0.232129E−05
0.999999E 00	0.658748E−04	−0.204745E−04	0.437714E−04	0.162897E−05
0.104000E 01	0.106330E−04	0.197998E−04	0.292920E−04	0.114074E−05
0.108000E 01	0.378257E−04	−0.174260E−04	0.196024E−04	0.797314E−06
0.112000E 01	−0.280455E−05	0.164789E−04	0.131180E−04	0.556327E−06
0.116000E 01	0.239183E−04	−0.147521E−04	0.877864E−05	0.387549E−06
0.120000E 01	−0.762395E−05	0.137682E−04	0.587471E−05	0.269573E−06
0.124000E 01	0.165678E−04	−0.124492E−04	0.393138E−05	0.187252E−06
0.128000E 01	−0.876950E−05	0.115303E−04	0.263090E−05	0.129902E−06
0.132000E 01	0.123371E−04	−0.104865E−04	0.176061E−05	0.900100E−07
0.136000E 01	−0.842895E−05	0.966946E−05	0.117821E−05	0.622986E−07
0.140000E 01	0.965523E−05	−0.882369E−05	0.788466E−06	0.430742E−07
0.144000E 01	−0.755810E−05	0.811550E−05	0.527645E−06	0.297529E−07
0.148000E 01	0.779361E−05	−0.741997E−05	0.353103E−06	0.205325E−07
0.152000E 01	−0.656401E−05	0.681446E−05	0.236298E−06	0.141571E−07

Figure 5.2 Comparison of relative error in results obtained by Milne and Adams–Moulton methods.

It is worth remarking that we usually don't have analytical answers to compare with our numerical results. Observing oscillatory behavior in itself does not mean instability, for the correct solution may *be* oscillatory. For this reason, the usual practice in numerical analysis is to entirely avoid methods that are sometimes unstable, even though they might be more accurate in some instances.

5.8 CONVERGENCE CRITERIA

In Section 5.3, we recorrected in the modified Euler method until no further change in y_{n+1} resulted. Usually this requires one more calculation than would otherwise be needed if we could predict whether the recorrection would make a

| | SOLUTION USING MILNE METHOD | | SOLUTION USING ADAMS–MOULTON METHOD | |

VALUES WITH H = 0.04 CONTINUED

X	Y	ERROR	Y	ERROR
0.156000E 01	0.640522E−05	−0.623733E−05	0.158132E−06	0.975331E−08
0.160000E 01	−0.561102E−05	0.572355E−05	0.105823E−06	0.671417E−08
0.164000E 01	0.531755E−05	−0.524211E−05	0.708171E−07	0.461910E−08
0.168000E 01	−0.475746E−05	0.480803E−05	0.473911E−07	0.317556E−08
0.172000E 01	0.443906E−05	−0.440517E−05	0.317144E−07	0.218119E−08
0.176000E 01	−0.401659E−05.	0.403931E−05	0.212234E−07	0.149759E−08
0.180000E 01	0.371683E−05	−0.370160E−05	0.142028E−07	0.102763E−08
0.184000E 01	−0.338345E−05	0.339366E−05	0.950460E−08	0.704578E−09
0.188000E 01	0.311713E−05	−0.311028E−05	0.636053E−08	0.482931E−09
0.192000E 01	−0.284671E−05	0.285129E−05	0.425650E−08	0.330768E−09
0.196000E 01	0.261645E−05	−0.261337E−05	0.284847E−08	0.226487E−09
0.200000E 01	−0.239358E−05	0.239564E−05	0.190621E−08	0.155008E−09

VALUES NEAR X = 0.2 WITH H = 0.004

X	Y	ERROR	Y	ERROR
0.191192E 01	−0.734619E−05	0.735117E−05	0.496895E−08	0.425615E−11
0.191592E 01	0.744385E−05	−0.743908E−05	0.477411E−08	0.405009E−11
0.191992E 01	−0.752346E−05	0.752805E−05	0.458691E−08	0.392220E−11
0.192392E 01	0.762249E−05	−0.761808E−05	0.440706E−08	0.379785E−11
0.192792E 01	−0.770495E−05	0.770919E−05	0.423425E−08	0.368061E−11
0.193192E 01	0.780545E−05	−0.780138E−05	0.406822E−08	0.350298E−11
0.193592E 01	−0.789078E−05	0.789469E−05	0.39087CE−08	0.339284E−11
0.193992E 01	0.799286E−05	−0.798910E−05	0.375544E−08	0.328626E−11
0.194392E 01	−0.808104E−05	0.808465E−05	0.360819E−08	0.318456E−11
0.194792E 01	0.818480E−05	−0.818133E−05	0.346671E−08	0.303002E−11
0.195192E 01	−0.827585E−05	0.827918E−05	0.333078E−08	0.293365E−11
0.195592E 01	0.838139E−05	−0.837819E−05	0.320017E−08	0.284017E−11
0.195992E 01	−0.847532E−05	0.847839E−05	0.307469E−08	0.274958E−11
0.196392E 01	0.858274E−05	−0.857979E−05	0.295413E−08	0.261657E−11
0.196792E 01	−0.867956E−05	0.868240E−05	0.283830E−08	0.253331E−11
0.197192E 01	0.878896E−05	−0.878623E−05	0.272701E−08	0.245226E−11
0.197592E 01	−0.888869E−05	0.889132E−05	0.262008E−08	0.237388E−11
0.197992E 01	0.900017E−05	−0.899765E−05	0.251735E−08	0.225930E−11
0.198392E 01	−0.910284E−05	0.910526E−05	0.241864E−08	0.218714E−11
0.198792E 01	0.921648E−05	−0.921415E−05	0.232380E−08	0.211720E−11
0.199192E 01	−0.932212E−05	0.932435E−05	0.223268E−08	0.204925E−11
0.199592E 01	0.943801E−05	−0.943586E−05	0.214514E−08	0.195066E−11
0.199992E 01	−0.954665E−05	0.954871E−05	0.206103E−08	0.188805E−11

Figure 5.2 (continued)

significant change. In the methods of Milne and Adams–Moulton, we usually do not recorrect, but use a value of h small enough that this is unnecessary. We now look for a criterion to show how small h should be in the Adams–Moulton method, for $dy/dx = f(x, y)$, so that recorrections are not necessary. Let

y_p = value of y_{n+1} from predictor formula,

y_c = value of y_{n+1} from corrector formula,

y_{cc}, y_{ccc}, etc. = values of y_{n+1} if successive recorrections are made,

y_∞ = value to which successive recorrections converge,

$D = y_c - y_p$.

The change of y_c by recorrecting would be:

$$y_{cc} - y_c = \left(y_n + \frac{h}{24}(9y_c' + 19y_n' - 5y_{n-1}' + y_{n-2}')\right)$$

$$- \left(y_n + \frac{h}{24}(9y_p' + 19y_n' - 5y_{n-1}' + y_{n-2}')\right) \quad (5.24)$$

$$= \frac{9h}{24}(y_c' - y_p').$$

In Eq. (5.24) we have used the subscripts p or c to denote which y-value is used in evaluating the derivative at $x = x_{n+1}$. We now manipulate the difference $(y_c' - y_p')$:

$$y_c' - y_p' = f(x_{n+1}, y_c) - f(x_{n+1}, y_p) = \frac{f(x_{n+1}, y_c) - f(x_{n+1}, y_p)}{(y_c - y_p)}(y_c - y_p)$$

$$= f_y(\xi_1)D, \qquad \xi_1 \text{ between } y_c \text{ and } y_p, \text{ with } D = y_c - y_p.$$

Hence,

$$y_{cc} - y_c = \frac{9hD}{24}f_y(\xi_1)$$

is the difference on recorrecting. If recorrected again, the result is:

$$y_{ccc} - y_{cc} = \frac{9h}{24}(y_{cc}' - y_c')$$

$$= \frac{9h}{24}f_y(\xi_2) \cdot (y_{cc} - y_c)$$

$$= \frac{9h}{24}f_y(\xi_2)\left[\frac{9hD}{24}f_y(\xi_1)\right] \quad (5.25)$$

$$= \left(\frac{9h}{24}\right)^2 [f_y(\xi)]^2 D, \qquad \xi_2 \text{ between } y_c \text{ and } y_{cc}.$$

In Eq. (5.25), we need to impose the restrictions that f be continuous and $f_y(\xi_2)$ have the same sign; ξ lies between the extremes of y_p, y_c, and y_{cc}.

On further recorrections we will have a similar relation. We get y_∞ by adding all the corrections of y_p together:

$$y_\infty = y_p + (y_c - y_p) + (y_{cc} - y_c) + (y_{ccc} - y_{cc}) + \ldots$$

$$= y_p + D + \frac{9hf_y(\xi)}{24}D + \left(\frac{9hf_y(\xi)}{24}\right)^2 D + \left(\frac{9hf_y(\xi)}{24}\right)^3 D + \ldots.$$

The increment to y_p is a geometric series; so, if the ratio is less than unity, which is necessary if a geometric series is to have a sum,

$$y_\infty = y_p + \frac{D}{1 - r}, \qquad r = \frac{9hf_y(\xi)}{24}, \qquad \xi \text{ between } y_p \text{ and } y_\infty.$$

Hence, unless

$$|r| = \frac{h \, |f_y(\xi)|}{24/9} \doteq \frac{h \, |f_y(x_n, y_n)|}{24/9} < 1,$$

the successive recorrections diverge. Our first convergence criterion is

$$h < \frac{24/9}{|f_y(x_n, y_n)|}. \tag{5.26}$$

If we wish to have y_c and y_∞ the same to within one in the Nth decimal place, then:

$$y_\infty - y_c = \left(y_p + \frac{D}{1-r}\right) - (y_p + D) = \frac{rD}{1-r} < 10^{-N}.$$

If $r \ll 1$, the fraction

$$\frac{r}{1-r} \doteq r;$$

and a second convergence criterion, which ensures that the first corrected value is adequate, i.e., it will not be changed in the Nth decimal place by further corrections, is

$$D \cdot 10^N < \left|\frac{1}{r}\right| \doteq \frac{24/9}{h \, |f_y(x_n, y_n)|}. \tag{5.27}$$

For the Adams–Moulton method we have the three criteria given below. If all are met, the corrected value should be good to N decimals:

$$\text{Convergence criteria} \quad \left\{ \begin{array}{l} h < \dfrac{24/9}{|f_y|}, \\[2mm] D \cdot 10^N < \dfrac{24/9}{h \, |f_n|}; \end{array} \right.$$

$$\text{Accuracy criterion} \quad D \cdot 10^N < 14.2.$$

A similar set of criteria for the Milne method are derived in the same way. They are:

$$\text{Convergence criteria} \quad \left\{ \begin{array}{l} h < \dfrac{3}{|f_y|}, \\[2mm] D \cdot 10^N < \dfrac{3}{h \, |f_y|}; \end{array} \right. \tag{5.28}$$

$$\text{Accuracy criterion} \quad D \cdot 10^N < 29.$$

These criteria are for a single first-order equation only. A similar analysis for a system is much more complicated.

We illustrate the use of these criteria with an example. Given that $dy/dx = \sin x - 3y$, $y(0) = 1$. In the neighborhood of the point $(1, 0.3)$, what maximum value of h is permitted if we wish to compute by (a) Milne's method, (b) Adams–Moulton method and get accuracy to five decimals? How close must the predictor and corrector values be?

a) $f_y = -3$,   so $h < \dfrac{3}{|-3|} = 1$

to ensure convergence if we were to recorrect y_c in the Milne method. This requirement is not severe—we certainly would choose Δx smaller than this to give information throughout the range of x values from 0 to 1. Suppose h were taken as 0.2. Then

$$D < \frac{3}{h\,|f_y|\,10^N} = \frac{3}{0.2\,|-3|\,10^5} = 5.0 \times 10^{-5}.$$

The difference between y_p and y_c cannot exceed this value; if it does, recorrections will be needed. This is more severe than $D < 29 \times 10^{-5}$ required for accuracy to one in the fifth decimal for y_c. We should monitor a Milne program in order to be sure that the difference between y_p and y_c does not exceed 5×10^{-5}. If it should exceed this value, we would need to reduce h so that this criterion could be met.

b)   For Adams–Moulton, we calculate:

$$h < \frac{24/9}{|-3|} = 0.89.$$

If $h = 0.2$,

$$D < \frac{24/9}{0.2\,|-3|\,10^5} = 4.4 \times 10^{-5}.$$

Again, this difference in y_p and y_c is more severe than $D < 14.2 \times 10^{-5}$ and should be used to control the value of h.

5.9 ERRORS AND ERROR PROPAGATION

Our previous error analyses have examined the error of a single step only, the so-called local truncation error of the methods. Since all practical applications of numerical methods to differential equations involve many steps, the accumulation of these errors, termed the global truncation error, is important. We remember that there are several sources of error in a numerical calculation in addition to the truncation error.

Original Data Errors

If the initial conditions are not known exactly (or must be expressed inexactly as a terminated decimal number), the solution will be affected to a greater or lesser degree depending on the sensitivity of the equation. Highly sensitive equations are said to be subject to *inherent instability*.

Round-off Errors

Since we can carry only a finite number of decimal places, our computations are subject to inaccuracy from this source, no matter whether we round or whether we chop off. Carrying more decimal places in the intermediate calculations than we require in the final answer is the normal practice to minimize this, but in lengthy calculations this is a source of error that is most difficult to analyze and control. Furthermore, in a computer program, if we use double precision, we require a longer execution time and also more storage to hold the more precise values. If these values are for a large array, the memory space needed may exceed that available to the program. This type of error is especially acute when two nearly equal quantities are subtracted. Both floating-point and fixed-point calculations in computers are subject to round-off errors.

Truncation Errors of the Method

These are the types of error we have been discussing, because we use truncated series for approximation in our work, when an infinite series is needed for exactness. The choice of method is our best control here, with suitable selection of h.

In addition to these three types of error, when we solve differential equations numerically we must worry about the propagation of previous errors through the subsequent steps. Since we use the end values at each step as the starting values for the next one, it is as if incorrect original data were distorting the later values. (Round-off would almost always produce error even if our method were exact.) This effect we now examine, but only for the very simple case of the Euler method. This will show how such error studies are made, as well as suggesting how difficult is the analysis of more practical methods.

We consider the first-order equation $dy/dx = f(x, y)$, $y(x_0) = y_0$. Let

$$Y_n = \text{calculated value at } x_n,$$

$$y_n = \text{true value at } x_n,$$

$$e_n = y_n - Y_n = \text{error in } Y_n; y_n = Y_n + e_n.$$

By the Euler algorithm,
$$Y_{n+1} = Y_n + hf(x_n, Y_n).$$

By Taylor series,
$$y_{n+1} = y_n + hf(x_n, y_n) + \frac{h^2}{2} y''(\xi_n), \qquad x_n < \xi_n < x_n + h,$$

$$e_{n+1} = y_{n+1} - Y_{n+1} = y_n - Y_n + h[f(x_n, y_n) - f(x_n, Y_n)] + \frac{h^2}{2} y''(\xi_n)$$

$$= e_n + h\frac{f(x_n, y_n) - f(x_n, Y_n)}{y_n - Y_n} (y_n - Y_n) + \frac{h^2}{2} y''(\xi_n)$$

$$= e_n + hf_y(x_n, \eta_n)e_n + \frac{h^2}{2} y''(\xi_n), \qquad \eta_n \text{ between } y_n, Y_n.$$

$$(5.29)$$

In Eq. (5.29) we have used the mean-value theorem, imposing continuity and existence conditions on $f(x, y)$ and f_y. We suppose, in addition, that the magnitude of f_y is bounded by the positive constant K in the region of x,y-space in which we are interested.* Hence,

$$e_{n+1} \le (1 + hK)e_n + \tfrac{1}{2}h^2 y''(\xi_n). \tag{5.30}$$

Here $y(x_0) = y_0$ is our initial condition, which we assume free of error. Since $Y_0 = y_0$, $e_0 = 0$:

$$e_1 \le (1 + hK)e_0 + \tfrac{1}{2}h^2 y''(\xi_0) = \tfrac{1}{2}h^2 y''(\xi_0),$$
$$\le (1 + hK)[\tfrac{1}{2}h^2 y''(\xi_0)] + \tfrac{1}{2}h^2 y''(\xi_1) = \tfrac{1}{2}h^2[(1 + hK)y''(\xi_0) + y''(\xi_1)].$$

Similarly,

$$e_3 \le \tfrac{1}{2}h^2[(1 + hK)^2 y''(\xi_0) + (1 + hK)y''(\xi_1) + y''(\xi_2)],$$
$$e_n \le \tfrac{1}{2}h^2[(1 + hK)^{n-1} y''(\xi_0) + (1 + hK)^{n-2} y''(\xi_1) + \cdots + y''(\xi_{n-1})].$$

If $f_y \le K$ is positive, the truncation error at every step is propagated to every later step after being amplified by the factor $(1 + hf_y)$ each time. Note that as $h \to 0$, the error at any point is just the sum of all the previous errors. If the f_y are negative and of magnitude such that $|hf_y| < 2$, the errors are propagated with diminishing effect.

We now show that the accumulated error after n steps is $O(h)$; i.e., the global error of the simple Euler method is $O(h)$. We assume, in addition to the above, that y'' is bounded, $|y''(x)| < M, M > 0$. Equation (5.30) becomes, after taking absolute values,

$$|e_{n+1}| \le (1 + hK)|e_n| + \tfrac{1}{2}h^2 M.$$

Compare to the second-order difference equation:

$$Z_{n+1} = (1 + hK)Z_n + \tfrac{1}{2}h^2 M,$$
$$Z_0 = 0. \tag{5.31}$$

Obviously the values of Z_n are at least equal to the magnitudes of $|e_n|$. The solution to Eq. (5.31) is (check by direct substitution):

$$Z_n = \frac{hM}{2K}(1 + hK)^n - \frac{hM}{2K}.$$

The Maclaurin expansion of e^{hK} is:

$$e^{hK} = 1 + hK + \frac{(hK)^2}{2} + \frac{(hK)^3}{6} + \cdots,$$

---

* This is essentially the same as the Lipschitz condition which will guarantee existence and uniqueness of a solution.

so that

$$1 + hK < e^{hK} \qquad (K > 0),$$

$$Z_n < \frac{hM}{2K}(e^{hK})^n - \frac{hM}{2K} = \frac{hM}{2K}(e^{nhK} - 1)$$

$$= \frac{hM}{2K}(e^{(x_n - x_0)K} - 1) = O(h).$$

It follows that the global error e_n is $O(h)$.

5.10 SYSTEMS OF EQUATIONS AND HIGHER-ORDER EQUATIONS

We have so far treated only the case of a first-order differential equation. Most differential equations which are the mathematical model for a physical problem are of higher order, or even a *set* of simultaneous higher-order differential equations. For example,

$$\frac{w}{g}\frac{d^2x}{dt^2} + b\frac{dx}{dt} + kx = f(x, t)$$

represents a vibrating system in which a linear spring with spring constant k restores a displaced mass of weight w against a resisting force whose resistance is b times the velocity. The function $f(x, t)$ is an external forcing function acting on the mass.

An analogous second-order equation describes the flow of electricity in a circuit containing inductance, capacitance, and resistance. The external forcing function in this case represents the applied electromotive force. Compound spring–mass systems and electrical networks can be simulated by a system of such second-order equations.

We first show how a higher-order differential equation can be reduced to a system of simultaneous first-order equations. We then show that these can be solved by an application of the methods previously studied. We treat here initial-value problems only, for which n values of the functions or derivatives (with n equal to the order of the system) are all specified at the same (initial) value of the independent variable. When some of the conditions are specified at one value of the independent variable and others at a second value, we call it a *boundary-value problem*. Methods of solving these are discussed in the next chapter.

By solving for the second derivative, we can normally express a second-order equation as:

$$\frac{d^2x}{dt^2} = f\left(x, t, \frac{dx}{dt}\right), \qquad x(t_0) = x_0, \qquad x'(t_0) = x_0'. \qquad (5.32)$$

The initial value of the function x and its derivative are generally specified.

We convert this to a pair of first-order equations by the simple expedient of defining the derivative as a second function. Then, since $d^2x/dt^2 = (d/dt)(dx/dt)$,

$$\frac{dx}{dt} = y, \qquad x(t_0) = x_0,$$

$$\frac{dy}{dt} = f(x, t, y), \qquad y(t_0) = x_0'.$$

This pair of first-order equations is equivalent to the original Eq. (5.32). For even higher orders, each of the lower derivatives is defined as a new function, giving a set of n first-order equations that correspond to an nth-order differential equation. For a system of higher-order equations, each is similarly converted, so that a larger set of first-order equations results.

We now illustrate the application of the various methods to the pair of first-order equations

$$\frac{dx}{dt} = xy + t, \qquad x(0) = 1,$$

$$\frac{dy}{dt} = ty + x, \qquad y(0) = -1. \tag{5.33}$$

Taylor-Series Method

We need the various derivatives $x', x'', x''', \ldots, y', y'', y''', \ldots$, all evaluated at $t = 0$:

$x' = xy + t,$ $x'(0) = (1)(-1) + 0 = -1,$

$y' = ty + x,$ $y'(0) = (0)(-1) + 1 = 1,$

$x'' = xy' + x'y + 1,$ $x''(0) = (1)(1) + (-1)(-1) + 1 = 3,$

$y'' = y + ty' + x',$ $y''(0) = -1 + (0)(1) - 1 = -2,$

$x''' = x'y' + xy'' + x''y + x'y',$ $x'''(0) = -7,$

$y''' = y' + y' + ty'' + x'',$ $y'''(0) = 5,$

etc.;   etc.;

$$x(t) = 1 - t + \tfrac{3}{2}t^2 - \tfrac{7}{6}t^3 + \tfrac{27}{24}t^4 - \tfrac{124}{120}t^5 + \cdots,$$

$$y(t) = -1 + t - t^2 + \tfrac{5}{6}t^3 - \tfrac{13}{24}t^4 + \tfrac{47}{120}t^5 + \cdots. \tag{5.34}$$

At $t = 0.1$, $x = 0.9139$ and $y = -0.9092$.

Equations (5.34) are the solution to the set (5.33). Note that we need to alternate between the functions in getting the derivatives; e.g., we cannot get $x''(0)$ until $y'(0)$ is known; we cannot get $y'''(0)$ until $x''(0)$ is known. After we have obtained the coefficients of the Taylor-series expansions in Eq. (5.34), we can evaluate x and y at any value of t, but the error will depend on how many terms we employ.

Euler Predictor–Corrector

We apply the predictor to each equation; then the corrector can be used. Again note that we work alternately with the two functions.

Take $h = 0.1$. Let p and c subscripts indicate predicted and corrected values, respectively:

$$x_p(0.1) = 1 + 0.1[(1)(-1) + 0] = 0.9,$$
$$y_p(0.1) = -1 + 0.1[(0)(-1) + 1] = -0.9,$$
$$x_c(0.1) = 1 + 0.1\left(\frac{-1 + [(0.9)(-0.9) + 0.1]}{2}\right) = 0.9145,$$
$$y_c(0.1) = -1 + 0.1\left(\frac{1 + [(0.1)(-0.9) + 0.9145]}{2}\right) = -0.9088.$$

In computing $x_c(0.1)$, we used the x_p and y_p. In computing $y_c(0.1)$ after $x_c(0.1)$ is known, we have a choice between x_p and x_c. There is an intuitive feel that one should use x_c, with the idea that one should always use the best available values. This does not always hasten convergence, probably due to compensating errors. Here we have used the best values to date. Recorrecting in the obvious manner gives

$$x(0.1) = 0.9135,$$
$$y(0.1) = -0.9089.$$

We can now advance the solution another step if desired, by using the computed values at $t = 0.1$ as the starting values. From this point we can advance one more step, and so on for any value of t. The errors will be the combination of local truncation error at each step plus the propagated error resulting from the use of inexact starting values.

Runge–Kutta

Again there is an alternation between the x and y calculations. In applying this method, one always uses the previous k-value in incrementing the function values and the value of h to increment the independent variable. As in the previous calculations, we oscillate between computations for x and for y; e.g., we do $k_{1,x}$, then $k_{1,y}$ before doing $k_{2,x}$, etc. The k-values for x and y are:

For x:

$$k_{1,x} = h(xy + t)_{x=1, y=-1, t=0}$$
$$= 0.1[(1)(-1) + 0] = -0.1;$$
$$k_{2,x} = h(xy + t)_{x=1-(0.1/2), y=-1+(0.1/2), t=0+(0.1/2)}$$
$$= 0.1[(0.95)(-0.95) + 0.05] = -0.0852;$$
$$k_{3,x} = h(xy + t)_{x=1-(0.0852/2), y=-1+(0.0902/2), t=0+(0.1/2)}$$
$$= 0.1[(0.9574)(-0.9549) + 0.05] = -0.0864;$$
$$k_{4,x} = h(xy + t)_{x=1-0.0864, y=-1+0.0909, t=0+0.1}$$
$$= 0.1[(0.9136)(-0.9091) + 0.1] = -0.0730.$$

For y:

$$k_{1,y} = h(ty + x)_{x=1,y=-1,t=0}$$
$$= 0.1[(0)(-1) +1] = 0.1;$$

$$k_{2,y} = h(ty + x)_{x=1-(0.1/2),y=-1+(0.1/2),t=0+(0.1/2)}$$
$$= 0.1[(0.05)(-0.95) + 0.95] = 0.0902;$$

$$k_{3,y} = h(ty + x)_{x=1-(0.0852/2),y=-1+(0.0902/2),t=0+(0.1/2)}$$
$$= 0.1[(0.05)(-0.9549) + 0.9574] = 0.0909;$$

$$k_{4,y} = h(ty + x)_{x=1-0.0864,y=-1+0.0909,t=0+0.1}$$
$$= 0.1[(0.1)(-0.9091) + 0.9136] = 0.0823.$$

Then

$$x(0.1) = 1 + \tfrac{1}{6}[-0.1 + 2(-0.0852) + 2(-0.0864) - 0.0730] = 0.9139,$$
$$y(0.1) = -1 + \tfrac{1}{6}[0.1 + 2(0.0902) + 2(0.0909) + 0.0823] = -0.9092.$$

The Runge–Kutta values appear to be correct to four decimals, while the modified Euler values are correct to only three, so $h = 0.1$ is too large for that method.

Advancing the solution by the Runge–Kutta method will again involve using the computed values of x and y as the initial values for another step. The errors here will be much less than those for the Euler predictor–corrector method, and we could use a larger step size and still retain four digits of accuracy. There is no obvious way to monitor the accuracy, however, to determine the step size that we should use.

Adams–Moulton

After getting four starting values, we proceed with the algorithm of Eq. (5.22), again alternately computing x and then y (see Table 5.12).

Table 5.12

	t	x	x'	t	y	y'
Starting values	0	1.0	−1.0	0	−1.0	1.0
	0.1	0.9139	−0.7309	0.1	−0.9092	0.8230
	0.2	0.8522	−0.5108	0.2	−0.8341	0.6854
	0.3	0.8107	−0.3246	0.3	−0.7705	0.5796
Predicted	0.4	(0.7867)	(−0.1638)	0.4	(−0.7167)	(0.5000)
Corrected		0.7865			−0.7167	

In the computations we first get predicted values of x and y:

$$x(0.4)_p = 0.8107 + \frac{0.1}{24}[55(-0.3246) - 59(-0.5108) + 37(-0.7309) - 9(-1.0)]$$

$$= 0.7867;$$

$$y(0.4)_p = -0.7705 + \frac{0.1}{24}[55(0.5796) - 59(0.6854) + 37(0.8230) - 9(1.0)]$$

$$= -0.7167$$

After getting x' and y' at $t = 0.4$, using $x(0.4)_p$ and $y(0.4)_p$, we then correct:

$$x(0.4)_c = 0.8107 + \frac{0.1}{24}[9(-0.1638) + 19(-0.3246) - 5(-0.5108) + (-0.7309)]$$

$$= 0.7865;$$

$$y(0.4)_c = -0.7705 + \frac{0.1}{24}[9(0.5000) + 19(0.5796) - 5(0.6854) + (0.8230)]$$

$$= -0.7167.$$

The close agreement of predicted and corrected values indicates four-decimal-place accuracy.

In this method, as we advance the solution to larger values of t, the comparison between predictor and corrector values tells us whether the step size needs to be changed.

Milne

This method can be applied to a system of first-order equations exactly analogously to the application of Adams–Moulton. We do not illustrate it because it is subject to instability.

5.11 COMPARISON OF METHODS FOR DIFFERENTIAL EQUATIONS

It is appropriate that we summarize the various methods that have been discussed in this chapter and compare them. Table 5.13 compares the accuracy, effort required, stability, and other features of the methods.

The data in Table 5.13 lead us to draw the usual conclusion about the best scheme for solving a differential equation of higher order, or a system of N first-order equations: We begin with a fourth-order Runge–Kutta to get a total of four values for each of the functions (this also allows us to compute four values for each of the derivatives), and then advance the solution with Adams–Moulton. At each step after employing Adams–Moulton, we check* the accuracy and adjust the step size when appropriate.

---

* Ordinarily a weighted average of the N errors is monitored. Alternatively, the maximum error is controlled.

Table 5.13 Comparison of methods for differential equations.

Method	Type	Local error	Global error	Function eval./step	Stability	Ease of changing step size	Recommended?
Modified Euler	Single-step	$O(h^3)$	$O(h^2)$	2	Good	Good	No
Fourth-order Runge–Kutta	Single-step	$O(h^5)$	$O(h^4)$	4	Good	Good	Yes
Milne	Multi-step	$O(h^5)$	$O(h^4)$	2	Poor	Poor	No
Adams–Moulton	Multi step	$O(h^5)$	$O(h^4)$	2	Good	Poor	Yes

There is still a problem during the starting phase when Runge–Kutta is being used, and that is how to monitor the accuracy. It is certainly reasonable to assume that if inaccurate starting values are used, the continuing values will be subject to error. A standard way to determine whether the Runge–Kutta values are sufficiently accurate is to recompute the value at the end of each interval with the step size cut in half. If this makes a change of negligible magnitude, the results are accepted; if not, the step is halved again until the results are satisfactory. This is very expensive, however, because eight additional sets of function evaluations are made just to determine the accuracy. (We may not have to make a new computation after the last one to ascertain its accuracy, of course. The knowledge that the method is of $O(h^5)$ accuracy permits us to anticipate the change that an additional computation would make if we have repeated once already.) There have been several schemes proposed to minimize the effort to determine the error in a Runge–Kutta computation. They demand some additional effort, but fewer than eight additional function evaluations.

The Runge–Kutta–Merson method computes five estimates for Δy in the next step; an estimate of the local error is then available from a weighted sum of the individual estimates:*

$$k_1 = h \cdot f(x_n, y_n),$$
$$k_2 = h \cdot f(x_n + h/3, y_n + k_1/3),$$
$$k_3 = h \cdot f(x_n + h/3, y_n + k_1/6 + k_2/6),$$
$$k_4 = h \cdot f(x_n + h/2, y_n + k_1/8 + 3k_3/8),$$
$$k_5 = h \cdot f(x_n + h, y_n + k_1/2 - 3k_3/2 + 2k_4);$$
$$y_{n+1} = y_n + \frac{k_1 + 4k_4 + k_5}{6} + O(h^5);$$
$$E \doteq \tfrac{1}{30}[2k_1 - 9k_3 + 8k_4 - k_5].$$

The correct step size can be estimated from the error computation together with the knowledge that the local errors vary about as h^5.

A similar technique is the Runge–Kutta–Fehlberg method. Six functional evaluations are required, but we then have an estimate of the error as well:

$$k_1 = h \cdot f(x_n, y_n),$$
$$k_2 = h \cdot f(x_n + h/4, y_n + k_1/4),$$
$$k_3 = h \cdot f(x_n + 3h/8, y_n + 3k_1/32 + 9k_2/32),$$
$$k_4 = h \cdot f(x_n + 12h/13, y_n + 1932k_1/2197$$
$$- 7200k_2/2197 + 7296k_3/2197),$$

---

* In these equations, f represents a multivalued function of N y-values and x.

$$k_5 = h \cdot f(x_n + h, y_n + 439k_1/216 - 8k_2$$
$$+ 3680k_3/513 - 845k_4/4104),$$
$$k_6 = h \cdot f(x_n + h/2, y_n - 8k_1/27 + 2k_2$$
$$- 3544k_3/2565 + 1859k_4/4104 - 11k_5/40);$$
$$y_{n+1} = y_n + (25k_1/216 + 1408k_3/2565 + 2197k_4/4104 - k_5/5);$$
$$E \doteq k_1/360 - 128k_3/4275 - 2197k_4/75240 + k_5/50 + 2k_6/55.$$

The basis for these schemes is to compute two Runge–Kutta estimates for the new value of y_{n+1}, but of different orders of error. The difference in the two estimates can then be used to estimate the error in the lower order y_{n+1}.

When this method is used to start the solution for a multistep method that needs equispaced function values, an additional restriction is imposed; the step size must be uniform, which may mean that closer spaced values may need to be computed than are required to meet the accuracy criterion alone.

A method due to Hamming has been widely accepted. It is available through subroutines in some FORTRAN libraries. It begins the solution with a fourth-order Runge–Kutta, and then continues with a predictor–corrector. The subroutines HPCG and HPCL in the IBM-supplied mathematical subroutines* use this method of Hamming's. The equations employed are:

$$y_{i+1,p} = y_{i-3} + \frac{4h}{3}(2f_i - f_{i-1} + 2f_{i-2}),$$
$$y_{i+1,m} = y_{i+1,p} - \tfrac{112}{121}(y_{i,p} - y_{i,c}),$$
$$y_{i+1,c} = \tfrac{1}{8}[9y_i - y_{i-2} + 3h(f_{i+1,m} + 2f_i - f_{i-1})],$$
$$y_{i+1} = y_{i+1,c} + \tfrac{9}{121}(y_{i+1,p} - y_{i+1,c}).$$

(5.35)

The predictor equation is the Milne predictor. Before correcting, the estimate of y_{i+1} is modified using the difference between the predicted and corrected values in the *previous* interval (this is omitted in the first interval, since these are not available). A corrector formula is used that depends on two previous y-values, though heavily weighted in favor of the last one. Finally, an adjustment is made based on the error estimate computed from the difference between predicted and corrected values. Note that, while two additional equations are employed in each step compared to the predictor–corrector methods previously described, only two evaluations of the derivative function are needed, the same as before. For many applications, it is the evaluation of the derivative function that is costly in computer time—the two extra algebraic steps don't count for much.

The special merits of Hamming's method are stability combined with good accuracy. Like Milne's and Adams', the method has a local error of $O(h^5)$ and a global error of $O(h^4)$.

---

* Described in the IBM publication *System/360 Scientific Subroutine Package*, H20–0205.

The IBM programs are quite sophisticated. They handle a system of first-order differential equations (or higher-order equations rewritten as first-order). The starting values for Hamming's method, four being required, are calculated using a fourth-order Runge–Kutta method, with coefficients chosen to minimize the error-term coefficient. Because accuracy is dependent upon good starting values, the Runge–Kutta calculations are repeated with a smaller step size to ascertain whether the values are of adequate precision.

Another feature of the subroutines is automatic step-size control. The error estimate is monitored at each step, comparing it to a tolerance parameter that is input from the calling program. In a system of differential equations, there is no single error estimate, so a weighted average of all the errors is used; the user has control over these weighting factors. If the error estimate is much below the tolerance value, the step size is doubled whenever seven previous values are available (though another parameter controls the increase of step size to a certain maximum value). If the error tolerance is exceeded, the step size is halved, using an $O(h^5)$ interpolation formula.

Gear (1967) has proposed a predictor–corrector method that has an $O(h^6)$ local error but uses only three previous steps rather than the four previous steps employed by Adams–Moulton and Milne. It obtains its high order of error by using recorrected values of the function and derivative values. The formulas are:

$$y_{n+1,p} = -18y_n + 9y_{n-1,c_1} + 10y_{n-2,c_2} + 9hy'_n + 18hy'_{n-1} + 3hy'_{n-2},$$

$$hy'_{n+1,p} = -57y_n + 24y_{n-1,c_1} + 33y_{n-2,c_2} + 24hy'_n + 57hy'_{n-1} + 10hy'_{n-2},$$

$$F = hy'_{n+1,p} - hf(x_{n+1}, y_{n+1,p}),$$

$$y_{n+1} = y_{n+1,p} - \tfrac{95}{288}F,$$

$$y_{n,c_1} = y_n + \tfrac{3}{160}F,$$

$$y_{n-1,c_2} = y_{n-1,c_1} - \tfrac{11}{1440}F,$$

$$hy'_{n+1} = hy'_{n+1,p} - F.$$

The method is stable and is recommended for equations that are termed "stiff."[*] It is applicable to systems of first-order equations.

Special formulas are available for special situations. Kunz (1957) gives the following predictor–corrector formulas for equations that lack a first-derivative term:

$$y_{n+1,p} = y_n + y_{n-2} - y_{n-3} + \frac{h^2}{4}(5y''_n + 2y''_{n-1} + 5y''_{n-2}) + \frac{17}{240}h^6 y^{vi}(\xi),$$

$$y_{n+1,c} = 2y_n - y_{n-1} + \frac{h^2}{12}(y''_{n+1} + 10y''_n + y''_{n-1}) - \frac{1}{240}h^6 y^{vi}(\xi_1).$$

(5.36)

---

[*] A "stiff" differential equation is one whose general solution has an exponential term such as $c_1 e^{bx}$, $b > 0$, but for the given initial conditions has a coefficient $c_1 = 0$. Round-off errors in the numerical solution may produce a solution in which $c_1 \neq 0$, and this creates a special stability problem.

Hamming (1962) gives this and other similar formulas for this special type of second-order equation.

Fox (1962) discusses the methods of Lanczos and of Clenshaw which involve representing the solution as a Chebyshev polynomial. These methods are beyond the present text.

The Numerov method is useful for second-order linear equations which lack the first derivative. Kunz (1957) discusses this. The method has an error term of $O(h^6)$ and requires only two starting values.

5.12 COMPUTER APPLICATIONS

Program 1 (Fig. 5.3) solves a second-order differential equation by the modified Euler method, after reducing it to a pair of first-order equations. The computer

```
C   PROGRAM FOR EULER PREDICTOR-CORRECTOR SOLUTION TO A
C   SECOND OREDER DIFFERENTIAL EQUATION OR TWO SIMULTANEOUS
C   FIRST ORDER EQUATIONS.
C   DEFINE THE DERIVATIVE FUNCTIONS.  FOR A SECOND ORDER
C   EQUATION, ALWAYS USE FCN1 = VARIABLE THAT REPLACES THE FIRST
C   DERIVATIVE.
        FCN1(X, Y, T) = Y
        FCN2(X, Y, T) = (A*SIN(C*T) - B*Y - XK*X)*32.1725/W
C   READ CONSTANTS OF THE PROBLEM
        READ (5,100) W, B, XK, A, C
100     FORMAT (6F10.0)
        READ (5,100) X0, Y0, T0, TEND, DT, TOL
        WRITE (6,200)
200     FORMAT (1H1, 2X, 4HTIME , 7X, 8HDISTANCE , 15X, 8HVELOCITY /)
        WRITE (6,201) T0, X0, Y0
201     FORMAT (1H , F5.2, 6X, F10.7, 12X, F13.7)
C   COMPUTE BY PREDICTOR FORMULAS
5       DX0 = FCN1(X0, Y0, T0)
        XP = X0 + DT*DX0
        DY0 = FCN2(X0, Y0, T0)
        YP = Y0 + DT*DY0
C   NOW CORRECT.  REPEAT CORRECTIONS UP TO 5 TIMES
        DO 10 I = 1,5
        XC = X0 + DT*(DX0 + FCN1(XP, YP, T0+DT))/2.
        YC = Y0 + DT*(DY0 + FCN2(XC, YP, T0+DT))/2.
        Z = (ABS(XC - XP) + ABS(YC - YP))/2.
        IF (Z .LE. TOL)  GO TO 20
        XP = XC
10      YP = YC
        TOUT = T0 + DT
        WRITE (6,202)  TOUT
202     FORMAT (1H , 7HAT T = , F5.2, 42H TOLERANCE MOT MET WITH FIVE RECO
       +RRECTIONS )
20      IF (T0 .GE. TEND)  STOP
C   RESET VARIABLES, PRINT A NEW LINE AND GO ON TO NEXT STEP
        T0 = T0 + DT
        WRITE (6,201) T0, XC, YC
        X0 = XC
        Y0 = YC
        GO TO 5
        END
```

Figure 5.3 Program 1.

output is the solution to the equation that represents a damped spring system,

$$\frac{W}{g}\frac{d^2x}{dt^2} + b\frac{dx}{dt} + kx = a \sin \omega t.$$

In the program, each of the constants in the differential equation has the same name except that k is called XK and ω is called C. In the program, X0 and T0 are the initial values of displacement and time. The variable y is used for dx/dt, and Y0 is its initial value. The solution is computed from T0 to TEND, using a step size equal to DT. TOL is a value used to determine whether recorrections should be continued or not; they terminate when the change on recorrection is

OUTPUT OF PROGRAM 1

TIME	DISTANCE	VELOCITY
0.00	1.0000000	−2.0000000
0.05	0.8557082	−3.7716170
0.10	0.6396286	−4.8715080
0.15	0.3859004	−5.2775640
0.20	0.1278025	−5.0463050
0.25	−0.1057399	−4.2953530
0.30	−0.2926692	−3.1817970
0.35	−0.4191977	−1.8794170
0.40	−0.4801182	−0.5573578
0.45	−0.4781170	0.6375475
0.50	−0.4223407	1.5934810
0.55	−0.3264835	2.2407770
0.60	−0.2066296	2.5533530
0.65	−0.0791642	2.5452400
0.70	0.0410487	2.2632560
0.75	0.1420566	1.7771860
0.80	0.2157059	1.1688440
0.85	0.2579663	0.5214710
0.90	0.2687673	−0.0894828
0.95	0.2514459	−0.6034617
1.00	0.2118824	−0.9791987
1.05	0.1574954	−1.1964020
1.10	0.0962152	−1.2549200
1.15	0.0355488	−1.1718330
1.20	−0.0181727	−0.9770964
1.25	−0.0603096	−0.7084207
1.30	−0.0881733	−0.4060365
1.35	−0.1010206	−0.1080076
1.40	−0.0998793	0.1536952
1.45	−0.0871460	0.3556897
1.50	−0.0661420	0.4845254
1.55	−0.0406128	0.5366996
1.60	−0.0142567	0.5175981
1.65	0.0096729	0.4396236
1.70	0.0286587	0.3198317
1.75	0.0410915	0.1773944
1.80	0.0463066	0.0312471
1.85	0.0445349	−0.1019768
1.90	0.0367503	−0.2094335
1.95	0.0244435	−0.2828653
2.00	0.0094008	−0.3188706
2.05	−0.0065345	−0.3185668

Figure 5.3 (continued)

less than TOL. The parameters and initial conditions are read in. The value of the function is recorrected until the difference in successive values is less than a certain tolerance. If the tolerance is not met in 5 recorrections, the value is printed together with a message to indicate that the tolerance was not met.

Program 2 (Fig. 5.4) uses a subroutine, RK4TH, to solve a single first-order differential equation of the form $dx/dt = f(x, t)$, using the Runge–Kutta method. Parameters are passed to specify the initial value of the function, X0, and the initial value and the increment for the independent variable, T0 and H. The subroutine returns X, the value of the function at the end of the interval. An external function calculates values of the derivative function.

The ecologist's problem concerning the growth in population of field mice, described at the beginning of this chapter, is solved in Program 2 using $\Delta t = 0.2$. In this problem one of the constants in the differential equation, B, is known as a function of time only through a table of values. To find the proper values of this

```
      SUBROUTINE RK4TH (FCN, TO, H, XO, X)
C  THIS SUBROUTINE ADVANCES THE SOLTUION OF A FIRST ORDER
C  DIFFERENTIAL EQUATION OF THE FORM DX/DT = F(X,T), USING
C  THE RUNGE-KUTTA FOURTH ORDER METHOD.
C  PARAMETERS ARE -
C    FCN    FUNCTION SUBPROGRAM TO COMPUTE DX/DT = F(X,T).
C           THIS MUST BE DECLARED EXTERNAL BY THE CALLING PROGRAM.
C    XO,TO  X AND T VALUES AT THE BEGINNING OF THE INTERVAL
C    H      STEP SIZE, THE VALUE OF DELTA T.
C    X      X VALUE AT THE END OF THE INTERVAL AS RETURNED BY SUBROUTINE
      XK1 = H*FCN(XO,TO)
      XK2 = H*FCN(XO+XK1/2., TO+H/2.)
      XK3 = H*FCN(XO+XK2/2., TO+H/2.)
      XK4 = H*FCN(XO+XK3, TO+H)
      X = XO + (XK1 + 2.*XK2 + 2.*XK3 + XK4)/6.
      RETURN
      END


      FUNCTION FCT (XN, T)
C  THIS FUNCTION SUBPROGRAM COMPUTES VALUES FOR THE DIFFERENTIAL EQUATION
      COMMON /PARAM/ A, BOUT
      FCT = A*XN - BOUT*XN**1.7
      RETURN
      END


C  MAIN PROGRAM TO SOLVE THE FIELD MICE PROBLEM.  IT EMPLOYS TWO
C  SUBROUTINES.  ROUTINE RK4TH SOLVES THE DIFFERENTIAL EQUATION AND ROUTINE
C  INTERP INTERPOLATES IN A TABLE OF VALUES FOR ONE OF THE PARAMETERS.
C  THE PARAMETERS OF THE DIFFERENTIAL EQUATION, A AND B, ARE PASSED TO A
C  FUNCTION SUBPROGRAM THROUGH COMMON.  THIS FUNCTION IS
C  AN EXTERNAL SUBPROGRAM BECAUSE IT'S NAME IS PASSED AS A PARAMETER.
      EXTERNAL FCT
      COMMON /PARAM/ A, BOUT
      DIMENSION B(9), D(10)
      A = 0.9
C  READ IN THE VALUES OF THE CONSTANT B, WHICH ARE DEFINED ONLY AS
C  A TABLE OF VALUES FOR VARIOUS T VALUES.
      READ (5,101) (B(I), I = 1,9)
101   FORMAT (8F10.0)
C  READ IN THE STARTING VALUES, DELTA T, AND FINAL T VALUE
      READ (5,100) TO, XO, H, TF
100   FORMAT (4F10.0)
```

Figure 5.4 Program 2.

```
C   WRITE HEADING AND INITIAL VALUES
        WRITE (6,200)
200     FORMAT (1H1, 49HSOLUTION TO A DIFFERENTIAL EQUATION BY R-K METHOD
       +      // 1H , 4X, 1HT, 12X, 1HN /)
        WRITE (6,201) TO, XO
201     FORMAT (1H , F6.1, F14.1)
C   CALL THE SUBROUTINE RK4TH FOR THE NEXT VALUE AND PRINT IT, BUT
C   MUST GET PROPERLY INTERPOLATED B VALUE FROM INTERP.
5       CALL INTERP (0., 8., 1., 9, B, TO, 3, BOUT)
        CALL RK4TH (FCT, TO, H, XO, X)
C   NOW ADVANCE THE VARIABLES
        TO = TO + H
        XO = X
        WRITE (6,201) TO, X
C   ARE WE DONE?  TEST X VALUE TO SEE.
        IF (TO .LT. TF)  GO TO 5
        STOP
        END
```

OUTPUT FROM PROGRAM 2

SOLUTION TO A DIFFERENTIAL EQUATION BY R-K METHOD

T	N
0.0	100.0
0.2	115.4
0.4	133.2
0.6	153.8
0.8	177.8
1.0	205.8
1.2	238.8
1.4	277.6
1.6	323.6
1.8	378.4
2.0	444.0
2.2	522.8
2.4	617.6
2.6	731.9
2.8	870.0
3.0	1036.8
3.2	1237.8
3.4	1478.3
3.6	1763.4
3.8	2097.1
4.0	2480.5
4.2	2910.5
4.4	3382.0
4.6	3879.9
4.8	4379.2
5.0	4846.2
5.2	5241.4
5.4	5518.5
5.6	5651.1
5.8	5633.4
6.0	5480.9
6.2	5224.1
6.4	4903.9
6.6	4551.8
6.8	4192.6
7.0	3843.7

Figure 5.4 (continued)

constant, a subroutine that does interpolation is called. (This subroutine, INTERP, is the same one presented earlier, in Chapter 3.) The values of the two constants needed by the function are passed through COMMON. (They cannot be passed as parameters because the subroutine RK4TH was written with a call to the function that did not include them as parameters.)

The output from Program 2 shows how the mouse population is expected to vary. Of course, only whole numbers of mice are possible; our mathematical model considers population as a *continuous* variable. We are not surprised to see that the population reaches a maximum as the food supply diminishes with time.

The third program is a subroutine, RKSYST (Fig. 5.5) that solves a system of N simultaneous first-order equations by the Runge–Kutta method. Parameters passed to the subroutine include the name of a function that computes each of the N derivative values, the initial value and the step size for the independent variable, an array of function values at the beginning of the interval, and the number of equations. Values of the functions at the end of the interval are returned in an array. A doubly-dimensioned array (declared of size $4 \times N$ in the main program) is employed to hold intermediate values.

```
      SUBROUTINE RKSYST (DERIVS, TO, H, XO, XEND, XWRK, F, N)
C  THIS SUBROUTINE SOLVES A SYSTEM OF N FIRST ORDER DIFFERENTIAL
C  EQUATIONS BY THE RUNGE-KUTTA METHOD.  THE EQUATIONS ARE OF THE FORM
C       DX1/DT = F1(X,T) DX2/DT = F2(X,T), ETC.
C  PARAMETERS ARE -
C    DERIVS  A SUBROUTINE THAT COMPUTES VALUES OF THE N DERIVATIVES.
C            IT MUST BE DECLARED EXTERNAL BY THE CALLER.  IT IS
C            INVOKED BY THE STAEMENT
C               CALL DERIVS (X, T, F, N)
C    TO    THE INITIAL VALUE OF INDEPENDENT VARIABLE
C    H     THE INCREMENT TO T, THE STEP SIZE
C    X     THE ARRAY THAT HOLDS THE INITIAL VALUES OF THE FUNCTIONS.
C    XEND  AN ARRAY THAT RETURNS THE FINAL VALUES OF THE FUNCTIONS.
C    XWRK  AN ARRAY USED TO HOLD INTERMEDIATE VALUES DURING THE
C            COMPUTATION.  IT MUST BE DIMENSIONED OF SIZE 4 X N IN THE
C            MAIN PROGRAM
C    N     THE NUMBER OF EQUATIONS IN THE SYSTEM BEING SOLVED.
C    F     AN ARRAY THAT HOLDS VALUES OF THE DERIVATIVES.
      DIMENSION XO(N), XEND(N), XWRK(4,N), F(N)
C  GET FIRST ESTIMATE OF THE DELTA X'S
      CALL DERIVS (XO, TO, F, N)
      DO 10 I = 1,N
      XWRK(1,I) = H*F(I)
      XEND(I) = XO(I) + XWRK(1,I)/2.
10    CONTINUE
C  GET THE SECOND ESTIMATE.  THE XEND VECTOR HOLDS THE X-VALUES.
      CALL DERIVS (XEND, TO+H/2., F, N)
      DO 20 I = 1,N
      XWRK(2,I) = H*F(I)
      XEND(I) = XO(I) + XWRK(2,I)/2.
20    CONTINUE
```

Figure 5.5 Program 3.

```
C    REPEAT FOR THIRD ESTIMATE.
         CALL DERIVS (XEND, T0+H/2., F, N)
         DO 30 I = 1,N
            XWRK(3,I) = H*F(I)
            XEND(I) = X0(I) + XWRK(3,I)
30       CONTINUE
C    NOW GET LAST ESTIMATE.
         CALL DERIVS (XEND, T0+H, F, N)
         DO 40 I = 1,N
            XWRK(4,I) = H*F(I)
40       CONTINUE
C    WE COMPUTE THE X AT THE END OF THE INTERVAL FROM A
C    WEIGHTED AVERAGE OF THE FOUR ESTIMATES, THEN RETURN.
         DO 50 I = 1,N
            XEND(I) = X0(I) + (XWRK(1,I) + 2.*XWRK(2,I) +
       +         2.*XWRK(3,I) + XWRK(4,I))/6.
50       CONTINUE
         RETURN
         END
```

Figure 5.5 (continued)

This routine can be used to solve an nth-order differential equation, or a system of higher-order equations, by reducing each of them to a set of simultaneous first-order equations.

EXERCISES

Section 5.2

▶1. Solve the differential equation

$$\frac{dy}{dx} = x + y + xy, \qquad y(0) = 1$$

by Taylor-series expansion to get the value of y at $x = 0.1$ and at $x = 0.5$. Use terms through x^5.

2. The general solution to a differential equation normally defines a family of curves. For the differential equation

$$\frac{dy}{dx} = x^2 y^2,$$

find, using the Taylor-series method, the particular curve that passes through $(1, 0)$. Also find the curve through $(0, 1)$. Compare to the analytical solutions.

▶3. Use the Taylor-series method to get y at $x = 0.2(0.2)0.6$,* given that

$$y'' = xy, \qquad y(0) = 1, \qquad y'(0) = 1.$$

4. A spring system has resistance to motion proportional to the square of the velocity, and its motion is described by

$$\frac{d^2 x}{dt^2} + 0.1 \left(\frac{dx}{dt} \right)^2 + 0.6x = 0.$$

If the spring is released from a point which is a unit distance above its equilibrium point, $x(0) = 1$, $x'(0) = 0$, use the Taylor-series method to write a series expression for the displacement as a function of time, including terms up to t^6.

Section 5.3

▶5. Use the simple Euler method to solve for $y(0.1)$ from

$$\frac{dy}{dx} = x + y + xy, \qquad y(0) = 1,$$

with $h = 0.01$. Comparing your result to the value determined by Taylor series in Exercise 1, estimate how small h would need to be to obtain four-decimal accuracy.

6. Solve the differential equation

$$\frac{dy}{dx} = \frac{x}{y}, \qquad y(0) = 1,$$

by the simple Euler method with $h = 0.1$, to get $y(1)$. Then repeat with $h = 0.2$ to get another estimate of $y(1)$. Extrapolate these results, assuming errors are proportional to step size, and compare to the analytical result. (Analytical result is $y^2 = 1 + x^2$.)

---

* This notation means for $x = 0.2$ through $x = 0.6$ with increments of 0.2.

▶7. Repeat Exercise 5, but with the modified Euler method with $h = 0.025$, so that the solution is obtained after four steps. Comparing with the result of Exercise 5, about how much less effort is it to solve this problem to four decimals with the modified Euler method in comparison to the simple Euler method?

8. Find the solution to

$$\frac{dy}{dt} = y^2 + t^2, \qquad y(1) = 0, \qquad \text{at } t = 2,$$

by the modified Euler method, using $h = 0.1$. Repeat with $h = 0.05$. From the two results, estimate the accuracy of the second computation.

▶9. Solve $y' = \sin x + y$, $y(0) = 2$, by the modified Euler method to get y at $x = 0.1(0.1)0.5$.

10. A sky diver jumps from a plane, and during the time before he opens his parachute, the air resistance is proportional to the $\frac{3}{2}$ power of his velocity. If it is known that the maximum rate of fall under these conditions is 80 mph, determine his velocity during the first 2 seconds of fall using the modified Euler method with $\Delta t = 0.2$. Neglect horizontal drift and assume an initial velocity of zero.

Section 5.4

11. Solve Exercise 5 by the Runge–Kutta method but with $h = 0.1$ so that the solution is obtained in only one step. Carry five decimals, and compare the accuracy and amount of work required with this method against the simple and modified Euler techniques in Exercises 5 and 9.

12. Solve Exercise 8 by the Runge–Kutta method, using $h = 0.2$, 0.1, and 0.05.

▶13. Determine y at $x = 0.2(0.2)0.6$ by the Runge–Kutta technique, given that

$$\frac{dy}{dx} = \frac{1}{x + y}, \qquad y(0) = 2.$$

14. Using the conditions of Exercise 10, determine how long it takes for the jumper to reach 90% of his maximum velocity, by integrating the equation using the Runge–Kutta technique with $\Delta t = 0.5$ until the velocity exceeds this value, and then interpolating. Then use numerical integration on the velocity values to determine the distance he falls in attaining $0.9 v_{max}$.

15. It is not easy to know the accuracy with which the function has been determined by either the Euler methods or the Runge–Kutta method. A possible way to measure accuracy is to repeat the problem with a smaller step size, and compare results. If the two computations agree to n decimal places, one then assumes the values are correct to that many places. Repeat Exercise 14 with $\Delta t = 0.3$, which should give a global error about one-eighth as large, and by comparing results, determine the accuracy in Exercise 14. (Why do we expect to reduce the error eightfold by this change in Δt?)

Section 5.5

▶16. The equation

$$\frac{dy}{dx} = f(x, y) = 2x(y - 1), \qquad y(0) = 0,$$

has initial values as follows:

x	y	f
0	0	0
0.1	−0.01005	−0.20201
0.2	−0.04081	−0.41632
0.3	−0.09417	−0.65650
0.4	−0.17351	−0.93881

a) Using the Adams procedure described in Section 5.5 compute $y(0.5)$ by fitting a quadratic through the last three values of $f(x, y)$. Compare to the exact answer: −0.28403.

b) Repeat using the last four points to fit a cubic.

c) Repeat again, but use five points to fit a quartic.

17. For the differential equation

$$\frac{dy}{dt} = y - t^2, \qquad y(0) = 1,$$

starting values are known:

$$y(0.2) = 1.2186, \qquad y(0.4) = 1.4682, \qquad y(0.6) = 1.7379.$$

Use the Adams method, fitting cubics with the last four (y, t) values and advance the solution to $x = 1.2$. Compare to the analytical solution.

18. For the equation

$$\frac{dy}{dt} = t^2 - t, \qquad y(1) = 0,$$

the analytical solution is easy to find:

$$y = \frac{t^3}{3} - \frac{t^2}{2} + \frac{1}{6}.$$

If we use three points in the Adams method, what error would we expect in the numerical solution? Confirm your expectation by performing the computations.

19. Continue the results of Exercise 13 using Adams' method, employing cubics to extrapolate the y'-values.

Section 5.6

20. For the differential equation,

$$\frac{dy}{dx} = y - x^2, \qquad y(0) = 1,$$

starting values are known:

$$y(0.2) = 1.2186, \qquad y(0.4) = 1.4682, \qquad y(0.6) = 1.7379.$$

Use the Milne method to advance the solution to $x = 1.2$. Carry four decimals and compare to the analytical solution.

▶21. For the differential equation $dy/dx = x/y$, the following values are given:

x	y
0	$\sqrt{1}$
1	$\sqrt{2}$
2	$\sqrt{5}$
3	$\sqrt{10}$

To how many decimal places will Milne's method give the value at $x = 4$? How many decimal places must be carried in the starting values of y to ensure this accuracy?

22. For the equation $y' = y \sin \pi x$, $y(0) = 1$, get starting values by the Runge–Kutta method for $x = 0.2(0.2)0.6$, and advance the solution to $x = 1.0$ by Milne's method.

▶23. Continue the results of Exercise 13 to $x = 2.0$ by the method of Milne. If you find that the corrector formula reproduces the predictor values, double the value of h after sufficient values are available.

24. Check that y_n as defined by Eq. (5.15) is a solution of the difference equation, Eq. (5.14).

25. Perform the long division to show that

$$Z_1 = 1 + Ah + O(h^2), \qquad Z_2 = -\left(1 - \frac{Ah}{3}\right) + O(h^2),$$

as given in Eq. (5.18).

Section 5.7

26. Express the differences in Eqs. (5.20) and (5.21) in terms of functional values to show that Eqs. (5.22) are equivalent.

27. Solve Exercise 17 using the Adams–Moulton method.

28. Repeat Exercise 22 using the Adams–Moulton method.

▶29. For the equation

$$\frac{dy}{dx} = x^3 + y^2, \qquad y(0) = 0,$$

using $h = 0.2$, compute three new values by the Runge–Kutta method (four decimals). Then advance to $x = 1.4$ using the Adams–Moulton method. If you find the accuracy criterion is not met, use Eqs. (5.23) to interpolate additional values so that four-place accuracy is maintained.

30. Derive the interpolation formulas given in Eq. (5.23).

Section 5.8

▶31. Given the linear differential equation $dy/dx = y \sin x$. (a) What is the maximum value of h that ensures convergence of the Adams–Moulton method when continuing applications of the corrector formula are made? (b) If an h one-tenth of this maximum value is used, how close must the predictor and corrector values be so that recorrections are not required? (c) In terms of the maximum h in part (a), what size of h is implied in the accuracy criterion, $D \cdot 10^n < 14.2$?

32. Repeat Exercise 31 for the differential equation

$$\frac{dy}{dx} = x^3 + y^2, \qquad y(0) = 0,$$

in the neighborhood of the point $(1.0, 0.15)$.

33. Repeat Exercise 31, using the Milne method. For part (c), the accuracy criterion is $D \cdot 10^n < 29$.

34. Derive Eq. (5.28).

35. Derive convergence criteria similar to Eqs. (5.26) and (5.27) for the Euler predictor–corrector method. Why can one not derive an accuracy criterion similar to those for the methods of Milne and Adams–Moulton?

Section 5.9

36. Estimate the propagated error at each step when the equation

$$\frac{dy}{dx} = x + y, \qquad y(0) = 1,$$

is solved by the simple Euler method with $h = 0.02$, for $x = 0(0.02)0.1$. (The equation is solved by this method in Section 2.) Compare to the actual errors.

▶37. Follow the propagated error between $x = 1$ and $x = 1.6$ when the simple Euler method is used to solve

$$\frac{dy}{dx} = xy^2, \qquad y(1) = 1.$$

Take $h = 0.1$. Compare to the actual errors at each step. The analytical solution is $y = 2/(3 - x^2)$.

Section 5.10

38. The mathematical model of an electrical circuit is given by the equation

$$0.5\frac{d^2Q}{dt^2} + 6\frac{dQ}{dt} + 50Q = 24 \sin 10t,$$

with $Q = 0$ and $I = dQ/dt = 0$ at $t = 0$. Express as a pair of first-order equations.

▶39. In the theory of beams it is shown that the radius of curvature at any point is proportional to the bending moment:

$$EI\frac{y''}{[1 + (y')^2]^{3/2}} = M(x),$$

where y is the deflection of the neutral axis. In the usual approach, $(y')^2$ is neglected in comparison to unity, but if the beam has appreciable curvature, this is invalid. For the cantilever beam for which $y(0) = y'(0) = 0$, express the equation as a pair of simultaneous first-order equations.

40. The motion of the compound spring system as sketched in Fig. 5.6. is given by the solution of the pair of simultaneous equations

$$m_1\frac{d^2y_1}{dt^2} = -k_1y_1 - k_2(y_1 - y_2), \qquad m_2\frac{d^2y_2}{dt^2} = k_2(y_1 - y_2),$$

where y_1 and y_2 are the displacements of the two masses from their equilibrium positions. The initial conditions are

$$y_1(0) = A, \qquad y_1'(0) = B, \qquad y_2(0) = C, \qquad y_2'(0) = D.$$

Express as a set of first-order equations.

Figure 5.6

▶41. Solve the pair of simultaneous equations

$$dx/dt = xy + t, \qquad x(0) = 0, \qquad dy/dt = x - t, \qquad y(0) = 1,$$

by the modified Euler method for $t = 0.2(0.2)0.6$. (Carry three decimals rounded. Recorrect until reproduced to three decimals.

42. Advance the solution of Exercise 41 to $x = 1.0(h = 0.2)$ by the Adams–Moulton method.

43. Repeat Exercise 42 but use Milne's method.

▶44. Find y at $x = 0.6$, given that

$$y'' = yy', \qquad y(0) = 1, \qquad y'(0) = -1.$$

Begin the solution by the Taylor-series method, getting

$$y(0.1), \qquad y(0.2), \qquad y(0.3).$$

Then advance to $x = 0.6$ employing the Adams–Moulton technique with $h = 0.1$ on the equivalent set of first-order equations.

45. Express the third-order equation

$$y''' + ty'' - ty' - 2y = t, \qquad y(0) = y''(0) = 0, \qquad y'(0) = 1,$$

as a set of first-order equations and solve at $t = 0.2, 0.4, 0.6$ by the Runge–Kutta method $(h = 0.2)$.

46. Using the Adams–Moulton method with $h = 0.5$, advance the solution of Exercise 45 to $t = 1.0$. Estimate the accuracy of the value of y at $t = 1.0$.

Section 5.11

▶47. For a resonant spring system with a periodic forcing function, the differential equation is

$$\frac{d^2x}{dt^2} + 64x = 16 \cos 8t, \qquad x(0) = x'(0) = 0.$$

Determine the displacement at $t = 0.1(0.1)0.8$ by the method of Eq. (5.36), getting the starting values by any other method of this chapter. Compare to the analytical solution $t \sin 8t$.

48. Exercise 39. gives the differential equation for a beam subject to loads; $M(x)$ is determined by taking moments of these loads about the point $x = 0$. If y' is neglected in comparison to unity (the usual assumption), the equation is of the form handled by Eq. (5.36). For a cantilever beam 10 feet long, with a single 1000-lb weight hung from the free end, use Eq. (5.36) to determine the elastic curve. (You can neglect the weight of the beam itself.) Assume that $EI = 20 \times 10^6$, in units consistent with x and y measured in inches and moments in in-lbs. Compare the solution with the analytical solution. You will need starting values for Eq. (5.36). Since y changes very slowly near $x = 0$, any convenient method should be satisfactory.

49. Look up the Numerov method and apply it to Exercise 48.

50. Use Hamming's method to solve one or more of Exercises 42, 44, or 46.

APPLIED PROBLEMS AND PROJECTS

51. In finding the deflection of a beam, if the y' term is not neglected (see Exercises 39 and 48), the specialized method of Eq. (5.36) does not apply. (Of course the analytical method is easy, if y' is neglected in the equation). For the values given in Exercise 48, what is the difference in the calculated values of the maximum deflection when the nonlinear relationship (Exercise 39) is used in comparison to the simpler linear equation? For light loads, the error is negligible, of course; only for heavier ones is there a need to use the nonlinear equation. At what value of the load is the error equal to 1% of the true value?

52. Write a program that solves a first-order initial-value problem by the Adams–Moulton method, calling the RK4TH subroutine of the text to obtain starting values. As each new point is computed, monitor the accuracy by comparing the error estimate to a tolerance parameter, printing out a message if the tolerance is not met, but continuing the solution anyway. This might best be done by incorporating the Adams–Moulton method in a subroutine that advances the solution one step.

53. Repeat Problem 52 using Milne's method.

54. Repeat Problem 52 using Hamming's method.

55. The equation $y' = 1 + y^2$, $y(0)$, has the analytical solution $y = \tan x$. The tangent function is infinite at $x = \pi/2$. Use the programs you wrote in Problems 52, 53, and 54 to solve this equation between $x = 0$ and $x = 1.6$ with a step size of 0.1. Compare the results of each program with the analytical solution.

56. What is the behavior of the Runge–Kutta fourth-order method when used over $x = 0$ to $x = 1.6$ on the equation of Problem 55? How does it compare with the multistep methods? How does its behavior change with step size?

57. Enlarge on Problem 52 by modifying your program so that the step size is halved if the error estimate exceeds the tolerance value. Use Eq. (5.23) to interpolate for the new values needed when this is done. If the error estimate is very small, say $\frac{1}{20}$ times the tolerance, your program should double the step size. (Some programs that provide for doubling the step size keep track of how many uniformly spaced values are available and defer the doubling until there are seven. You may wish to do this, but it is a tricky bit of programming. Alternatively, one could always compute two additional values after discovering that the step size can be increased; this guarantees that there are at least seven equispaced values. Another method would be to extrapolate to obtain a value at t_{n-5} when we find, at t_{n+1}, that the error is $\frac{1}{20}$ of the tolerance.)

58. In an electrical circuit (Fig. 5.7) containing resistance, inductance, and capacitance (and every circuit does), the voltage drop across the resistance is iR (i is current in amperes, R is resistance in ohms), across the inductance is $L(di/dt)$—(L is inductance in henries), and across the capacitance is q/C (q is charge in the capacitor in coulombs, C is capacitance in farads). We then can write, for the voltage difference between points A and B,

$$V_{AB} = L\frac{di}{dt} + Ri + \frac{q}{C}.$$

Differentiating with respect to t and remembering that $dq/dt = i$, we have a second-order differential equation:

$$L\frac{d^2 i}{dt^2} + R\frac{di}{dt} + \frac{1}{C}i = \frac{dV}{dt}.$$

If the voltage V_{AB} (which previously has been 0 volts) is suddenly brought to 15 volts (let us say, by connecting a battery across the terminals) and maintained steadily at 15 volts (so $dV/dt = 0$), current will flow through the circuit. Use an appropriate numerical method to determine how the current varies with time between 0 and 0.1 sec if $C = 1000$ microfarads, $L = 50$ millihenries, and $R = 4.7$ ohms; use Δt of 0.002 sec. Also determine how the voltage builds up across the capacitor during this time. You may want to compare the computations with the analytical solution.

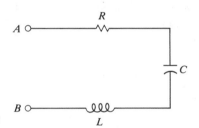

Figure 5.7

59. Repeat Problem 58 but let the voltage source be a 60-Hz sinusoidal input:

$$V_{AB} = 15\sin(120\pi t).$$

How closely does the voltage across the capacitor resemble a sine wave during the last full cycle of voltage variation?

60. After the voltages have stabilized in Problem 58 (15 V across the capacitor), the battery is shorted so that the capacitor discharges through the resistance and inductor. Follow the current and the capacitor voltages for 0.1 sec, again with $\Delta t = 0.002$ sec. The oscillations of decreasing amplitude are called *damped oscillations*. If the calculations are repeated but with the resistance value increased, the oscillations will be damped out more quickly; at $R = 14.14$ ohms the oscillations should disappear; this is called "*critical damping*." Perform numerical computations with values of R increasing from 4.7 to 22 ohms to confirm that critical damping occurs at 14.14 ohms.

61. In Chapter 2, electrical networks were discussed, but these were simple ones with resistance only. A more realistic situation is to have RLC circuits combined into a network. Using either the Kirkhoff voltage law or the Kirkhoff current law, we can set up a set of simultaneous equations, but now these are simultaneous differential equations. For example, for the circuit in Fig. 5.8, which has two voltage sources, we have:

$$(L_1 + L_3)\frac{d^2 i_1}{dt^2} + (R_1 + R_3)\frac{di_1}{dt} + (C_1 + C_3)i_1 - L_3\frac{d^2 i_2}{dt^2} - R_3\frac{di_2}{dt} - C_3 i_2 = e_1(t),$$

$$(L_2 + L_3)\frac{d^2 i_2}{dt^2} + (R_2 + R_3)\frac{di_2}{dt} + (C_2 + C_3)i_2 - L_3\frac{d^2 i_1}{dt^2} - R_3\frac{di_1}{dt} - C_3 i_1 = e_2(t).$$

In the equations, i_1 and i_2 represent the currents in each of the loops. Solve the equations for i_1 and i_2 between $t = 0$ and $t = 0.2$ sec, if

$$e_1(t) = 100 \sin (120\pi t), \qquad e_2(t) = 0,$$

given that:

$$R_1 = 22 \text{ ohms}, \qquad C_1 = C_2 = C_3 = 10 \ \mu\text{f},$$
$$R_2 = 4.7 \text{ ohms}, \qquad L_1 = 2.5 \text{ mH},$$
$$R_3 = 47 \text{ ohms}, \qquad L_2 = L_3 = 0.5 \text{ mH}.$$

Figure 5.8

62. A Foucault pendulum is one free to swing in both the x- and y-directions. It is frequently displayed in science museums to exhibit the rotation of the earth, which causes the pendulum to swing in directions that continuously vary. The equations of motion are

$$\ddot{x} - 2\omega \sin \psi \dot{y} + k^2 x = 0,$$
$$\ddot{y} + 2\omega \sin \psi \dot{x} + k^2 y = 0,$$

when damping is absent (or compensated for.) In these equations the dots over the variable represent differentiation with respect to time. Here ω is the angular velocity of the earth's rotation ($7.29 \times 10^{-5} \ \text{sec}^{-1}$), ψ is the latitude, $k^2 = g/\ell$ where ℓ is the length of the pendulum. How long will it take a 10-meter-long pendulum to rotate its plane of swing by 45° at the latitude where you live? How long if located in Quebec, Canada?

63. The *Handbook of Physics* (Condon and Odishaw, 1967, p. 4–42) discusses Duffing's equation for the flux, ϕ, in a transformer. This nonlinear differential equation is:

$$\ddot{\phi} + \omega_0^2 \phi + b\phi^3 = \frac{\omega}{N} E \cos \omega t.$$

In this equation, $E \sin \omega t$ is the sinusoidal source voltage and N is the number of turns in the primary winding, while ω_0 and b are parameters of the transformer design. Make a plot of ϕ versus t (and compare to the source voltage) if $E = 165$, $\omega = 120\pi$, $N = 600$, $\omega_0^2 = 83$ and $b = 0.14$. For approximate calculations, the nonlinear term,

$b\phi^3$, is sometimes neglected. Evaluate your results to determine whether this makes a significant error in the results.

64. Ethylene oxide is an important raw material for the manufacture of organic chemicals. It is produced by reacting ethylene and oxygen together over a silver catalyst. Laboratory studies have been reported by Shen-Wu Wan in *Ind. Eng. Chem.* **45,** 234 (1953).

It is planned to use this process commercially by passing the gaseous mixture through tubes filled with catalyst. Wan's studies show that the reaction rate varies with pressure, temperature, and concentrations of ethylene and oxygen, according to this equation:

$$r = 1.7 \times 10^6 e^{-9716/T} \left(\frac{P}{14.7} \right) c_E^{0.328} c_O^{0.672},$$

where

$$r = \text{reaction rate (units of ethylene oxide formed per lb of catalyst per hour),}$$

$$T = \text{temperature, } °K \, (°C + 273),$$

$$P = \text{absolute pressure (lb/in}^2\text{),}$$

$$C_E = \text{concentration of ethylene,}$$

$$C_O = \text{concentration of oxygen.}$$

Under the planned conditions, the reaction will occur, as the gas flows through the tube, according to the equation

$$\frac{dx}{dL} = 6.42r,$$

where

$$x = \text{fraction of ethylene converted to ethylene oxide,}$$

$$L = \text{length of reactor tube (ft).}$$

The reaction is strongly exothermic, so that it is necessary to cool the tubular reactor to prevent overheating. (Excessively high temperatures produce undesirable side reactions.) The reactor will be cooled by surrounding the catalyst tubes with boiling coolant under pressure so that the tube walls are kept at 225°C. This will remove heat proportional to the temperature difference between the gas and the boiling water. Of course, heat is generated by the reaction. The net effect can be expressed by this equation for the temperature change per foot of tube, where B is a design parameter:

$$\frac{dT}{dL} = 24320r - B(T - 225).$$

For preliminary computations, it has been agreed that we can neglect the change in pressure as the gases flow through the tubes; we shall use the average pressure of $P = 22 \, \text{lb/in}^2$ absolute. We will also neglect the difference between the catalyst temperature (which should be used to find the reaction rate) and the gas temperature.

You are to compute the length of tubes required for 65% conversion of ethylene if the inlet temperature is 250°C. Oxygen is consumed in proportion to the ethylene converted; material balances show that the concentrations of ethylene and oxygen vary with x, the fraction of ethylene converted, as follows:

$$c_E = \frac{1 - x}{4 - 0.375x},$$

$$c_O = \frac{1 - 1.125x}{4 - 0.375x}.$$

The design parameter B will be determined by the diameter of tubes that contain the catalyst. (The number of tubes in parallel will be chosen to accommodate the quantities of materials flowing through the reactor.) The tube size will be chosen to control the maximum temperature of the reaction, as set by the minimum allowable value of B. If the tubes are too large in diameter (for which the value of B is small), the temperatures will run wild. If the tubes are too *small* (giving a large value to B), so much heat is lost that the reaction tends to be quenched. In your studies, vary B to find the least value that will keep the maximum temperature below 300°C. Permissible values for the parameter B are from 1.0 to 10.0.

In addition to finding how long the tubes must be, we need to know how the temperature varies with x and with the distance along the tubes. In order to have some indication of the controllability of the process, you are also asked to determine how much the outlet temperature will change for a 1°C change in the inlet temperature, using the value of B as determined above.

Chapter Six

Boundary-Value Problems and Characteristic-Value Problems

In all differential equations of order greater than one, two or more values must be known to evaluate the constants in the particular function that satisfies the differential equation. In the problems discussed in Chapter 5, these several values were all specified at the same value of the independent variable, generally at the start or *initial value*. For that reason, such problems are termed initial-value problems.

For an important class of problems, the several values of the function or its derivatives are not all known at the same point, but rather at two different values of the independent variable. Because these values of the independent variable are usually the endpoints (or *boundaries*) of some domain of interest, problems with this type of conditions are classed as *boundary-value problems*. Determining the deflection of a simply supported beam is a typical example, where the conditions specified are the deflections and second derivatives of the elastic curve at the supports. Heat-flow problems fall in this class when the temperatures or temperature gradients are given at two points. A special case of boundary-value problem is involved in vibration problems.

6.1 THE "SHOOTING METHOD"

Suppose we wish to solve the second-order boundary value problem

$$\frac{d^2x}{dt^2} - \left(1 - \frac{t}{5}\right)x = t, \qquad x(1) = 2, \qquad x(3) = -1. \tag{6.1}$$

Note that if $x'(1)$ were given in addition to $x(1)$, this would be an initial-value problem. There is a way to adapt our previous methods to this problem, as illustrated in Fig. 6.1. We know the value of x at $t = 1$, and at $t = 3$, as given by the dots. The curve that represents x between these two points is desired. We

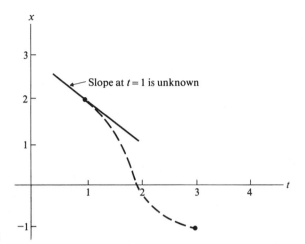

Figure 6.1

anticipate that some such curve, such as the dotted line, exists;* its slope and curvature are interrelated to x and t by the differential Eq. (6.1).

If we assume the slope of the curve at $t = 1$, say $x'(1) = -1.5$, we could solve the equation as an initial-value problem using this assumed value. The test of our assumption is whether we calculate x at $t = 3$ to match the known value, $x(3) = -1$. In Table 6.1 we show the results of this computation employing the

Table 6.1 Solving a second-order equation by the shooting method:

$$x'' = t + \left(1 - \frac{t}{5}\right)x, \qquad x(1) = 2, \qquad x(3) = -1.$$

Time	Distance	Velocity	Distance	Velocity	Distance	Velocity
$\Delta t = 0.2$:	Assume $x'(1) = -1.5$		Assume $x'(1) = -3.0$		Assume $x'(1) = -3.500$	
1.00	2.000	−1.500	2.000	−3.000	2.000	−3.500
1.20	1.751	−0.987	1.449	−2.510	1.348	−3.018
1.40	1.605	−0.478	0.991	−2.068	0.787	−2.599
1.60	1.561	0.043	0.619	−1.655	0.305	−2.221
1.80	1.625	0.594	0.328	−1.252	−0.104	−1.867
2.00	1.803	1.186	0.118	−0.844	−0.443	−1.521
2.20	2.105	1.832	−0.007	−0.417	−0.712	−1.167
2.40	2.542	2.542	−0.045	0.040	−0.908	−0.794
2.60	3.128	3.324	0.013	0.539	−1.026	−0.391
2.80	3.880	4.185	0.175	1.087	−1.060	0.054
3.00	4.811	5.128	0.453	1.693	−1.000	0.547

* The existence of a solution to a boundary-value problem cannot be taken for granted, however. For example $y'' + y = 0$, $y(0) = 1$, $y(\pi) = 0$ has no solution.

computer program of Section 5.12 (Fig. 5.3), modified for this equation, using the modified Euler method with $\Delta t = 0.2$. Since the result of this gives $x(3) = 4.811$, and not the desired $x(3) = -1$, we assume another value for $x'(1)$ and repeat. Since the calculated value of $x(3)$ is too high, we assume a smaller value for the slope, say $x'(1) = -3.0$. The second computation gives $x(3) = 0.453$, which is better but still too high. After these two trials, we linearly interpolate (here extrapolate) for a third trial. At $x'(1) = -3.500$, we get the correct value of $x(3)$. In some problems more attempts are needed to get the correct solution.

The method we have illustrated is called the *shooting method* because it resembles an artillery problem. One sets the elevation of the gun and fires a preliminary round at the target. After successive shots have straddled the target, one zeros in on it by using intermediate values of the gun's elevation. This corresponds to our using assumed values of the initial slope and interpolating based on how close we come to $x(3)$.

You may wonder whether it was only a lucky accident that gave correct results on our third trial, using the initial slope $x'(1) = -3.500$, extrapolated from our earlier guesses. This desirable result will always be true when the boundary-value problem is *linear*, as in this case. (A differential equation is linear when the coefficients of each derivative term and the function are not functions of x. They may be functions of t, as in this example, however.) The reason for this is that, if there are two different functions of x that satisfy the differential equation (but agreeing with different boundary conditions, of course), then a linear combination of them is also a solution. This is quite easy to show.

Suppose that $x_1(t)$ satisfies $x'' + Fx' + Gx = H$, where F, G, and H are functions of t only. Suppose also that $x_2(t)$ is a solution. Then:

$$y = \frac{c_1 x_1 + c_2 x_2}{c_1 + c_2}$$

will be a solution. We show that this will always be a solution by the following:
Since x_1 and x_2 are solutions,

$$x_1'' + Fx_1' + Gx_1 = H,$$

and

$$x_2'' + Fx_2' + Gx_2 = H.$$

Substituting y into the differential equation, with

$$y' = \frac{c_1 x_1' + c_2 x_2'}{c_1 + c_2},$$

$$y'' = \frac{c_1 x_1'' + c_2 x_2''}{c_1 + c_2},$$

we get

$$\frac{c_1 x_1'' + c_2 x_2''}{c_1 + c_2} + F\frac{c_1 x_1' + c_2 x_2'}{c_1 + c_2} + G\frac{c_1 x_1 + c_2 x_2}{c_1 + c_2}$$

$$= \frac{c_1 x_1'' + c_1 F x_1' + c_1 G x_1}{c_1 + c_2} + \frac{c_2 x_2'' + c_2 F x_2' + c_2 G x_2}{c_1 + c_2}$$

$$= \frac{c_1 H}{c_1 + c_2} + \frac{c_2 H}{c_1 + c_2} = H.$$

With two functions, x_1 and x_2, both agreeing with $x(1) = 2.0$ at the left boundary and differing at $t = 3.0$, we can take a linear combination, so the correct value of $x(3.0)$ results.

> Let G1 = first guess at initial slope;
>
> Let G2 = second guess at initial slope;
>
> Let R1 = first result at endpoint (using G1);
>
> Let R2 = second result at endpoint (using G2).

With D = the desired value at the endpoint, we can write:

$$\begin{matrix}\text{Extrapolated estimate} \\ \text{for initial slope}\end{matrix} = \text{G1} + \frac{\text{G2} - \text{G1}}{\text{R2} - \text{R1}}(\text{D} - \text{R1}).$$

In the above example, we get

$$-1.5 + \frac{(-3.0) - (-1.5)}{0.453 - 4.811}(-1.0 - 4.811) = -3.500.$$

The consequence of having $(c_1 x_1 + c_2 x_2)/(c_1 + c_2)$ as a solution of the differential equation goes further than just allowing us to interpolate from the two initial guesses of $x'(1)$ to predict exactly the correct value required for $x'(1)$ that is needed in order to give agreement with the righthand boundary condition. This relationship is true *throughout* the interval as well. This means that each of the values of x could be calculated by taking the proper combination of the values obtained from the earlier calculations, with this combination exactly the same one as that which gives the correct value of x at $t = 3$. (The true curve for x vs. t will be intermediate between two curves that bracket the final condition.)

In our example,

$$c_1 x_1 + c_2 x_2 = \text{true values.}$$

At left end,

$$t = 1: \quad c_1(2.0) + c_2(2.0) = 2.0,$$

At right end,

$$t = 3: \quad c_1(4.811) + c_2(0.453) = -1.0.$$

From the first equation, $c_2 = 1 - c_1$, and from the second $c_1 = -0.3334$, giving $c_2 = 1.3334$. We can calculate the correct values of x using these values.

For example,

At $t = 2.0$: $-0.3334(1.803) + 1.3334(0.118) = -0.443$.

We therefore conclude that a second-order linear boundary-value problem can always be solved by doing only two computations using two different guesses for the initial slope.

With higher-order linear boundary-value problems, the situation is somewhat more difficult although, with a third-order equation, we can choose the starting point at that boundary where two of the three conditions are specified. (If this means working backward from the righthand end, Δt will be negative.) For a fourth-order problem, with two conditions specified at each boundary, we seem to have a double trial-and-error situation. However, a combination of solutions also is a solution in this case, permitting us to get the correct solution by combining four trial computations. In these higher-order problems, some of the conditions we must match are values of the derivatives, requiring us to estimate the derivatives of our computed functions using approximation methods of Chapter 4, and these are more subject to errors.

When the differential equation is nonlinear (the coefficients F, G, and/or H in $x'' + Fx' + Gx = H$ are functions of x or x'), the interpolation between two assumed values for the initial slope does not usually give the correct value. Consider

$$x'' - \left(1 - \frac{t}{5}\right)(x')(x) = t, \qquad x(1) = 2, \qquad x(3) = -1.$$

The presence of the $(x')(x)$ term makes it nonlinear. In this case, the values in Table 6.2 result, using the modified Euler method with $\Delta t = 0.2$.

Table 6.2 Values for a nonlinear boundary-value problem by the shooting method:

$$x'' = t + \left(1 - \frac{t}{5}\right)xx', \qquad x(1) = 2, \qquad x(3) = -1.$$

Assumed value for $x'(1)$	Calculated value for $x(3)$
-1.5	-0.016
-3.0	-2.085
-2.213^*	-1.271
-2.0	-0.972
-1.8	-0.642
-2.017^*	-0.998
-2.01	-0.987
-2.02	-1.002
-2.018^*	-1.000

* Interpolated from two previous values.

Table 6.3

t	x	x'
1.000	2.000	−2.018
1.200	1.557	−2.407
1.400	1.053	−2.632
1.600	0.527	−2.625
1.800	0.026	−2.383
2.000	−0.408	−1.959
2.200	−0.747	−1.431
2.400	−0.977	−0.867
2.600	−1.094	−0.307
2.800	−1.101	0.237
3.000	−1.000	0.775

Only when the assumed values are very near to the correct initial slope does one get the correct value by interpolating linearly. (If the computer program that solves the differential equation is available through a time-sharing system, the above trial-and-error search can be conducted quickly and conveniently. This was the case with the data in Table 6.2.)

Note that we are faced with a root-finding problem, in that the value of $x(3)$ is a function of $x'(1)$. The methods of Chapter 1 then apply, although Newton's method requires the *slope* of this function, which is not readily available. Probably Muller's method, applied after three estimates have been made, would be particularly beneficial.

After we know the correct value for $x'(1)$, we solve the problem as an initial-value problem, getting the results shown in Table 6.3.

A summary of the shooting method

To solve a boundary-value problem, create an initial-value problem by assuming a sufficient number of initial values. (Choose the righthand end of the interval as the "initial point," and use a negative value of h if this requires fewer assumed conditions.) Solve this initial-value problem and compare the computed values with the given conditions at the other boundary.

Repeat the solution with varying values of the assumed conditions until agreement is attained at the other boundary. Methods for solving nonlinear systems can be employed to estimate the new assumed values for the successive iterations in an efficient manner.

The shooting method is often quite laborious. Especially with problems of fourth and higher order, the necessity to assume two or more conditions at the starting point (and match with the same number of conditions at the end) is slow and tedious. Although it tends to use considerable computer time, the NLSYST program of Chapter 2 is a convenient tool for such higher-order problems.

6.2 SOLUTION THROUGH A SET OF EQUATIONS

We have seen in Chapter 4 how the derivatives of a function can be approximated by finite difference quotients. If we replace the derivatives in a differential equation by such expressions, we convert it to a difference equation whose solution is an approximation to the solution of the differential equation. This method is sometimes preferred over the shooting method discussed above. Consider the same linear example as before:

$$\frac{d^2x}{dt^2} - \left(1 - \frac{t}{5}\right)x = t, \qquad x(1) = 2, \qquad x(3) = -1. \tag{6.2}$$

Central-difference approximations to derivatives are more accurate than forward or backward approximations $[O(h^2)$ versus $O(h)]$, so we replace the derivatives with:

$$\left.\frac{dx}{dt}\right|_{t=t_i} = \frac{x_{i+1} - x_{i-1}}{2h} + O(h^2),$$

$$\left.\frac{d^2x}{dt^2}\right|_{t=t_i} = \frac{x_{i+1} - 2x_i + x_{i-1}}{h^2} + O(h^2).$$

The quantity h is the constant difference in t-values. Substituting these equivalences into Eq. (6.2), and rearranging, we get

$$\frac{x_{i+1} - 2x_i + x_{i-1}}{h^2} - \left(1 - \frac{t_i}{5}\right)x_i = t_i,$$

$$x_{i-1} - \left[2 + h^2\left(1 - \frac{t_i}{5}\right)\right]x_i + x_{i+1} = h^2 t_i. \tag{6.3}$$

In Eqs. (6.3), we have replaced x by x_i and t by t_i, since these values correspond to the point at which the difference quotients represent the derivatives.* Our problem now reduces to solving Eqs. (6.3) at points in the interval from $t = 1$ to $t = 3$. Let us subdivide the interval into a number of equal subintervals. For example, if $h = \Delta t = \frac{1}{2}$, the points

$$t_1 = 1, \qquad t_2 = 1.5, \qquad t_3 = 2, \qquad t_4 = 2.5, \qquad \text{and} \qquad t_5 = 3$$

subdivide the interval into four subintervals.

---

* Our procedure is to subdivide the interval from the initial to the final value of t into equal subintervals, replacing the differential equation by a difference equation at each of the discrete points where the function is unknown.

We write the difference equation in (6.3) for each of these values of t at which x is unknown, giving

at $t = t_2 = 1.5$: $x_1 - \left[2 + \left(\frac{1}{2}\right)\left(\frac{1}{2}\right)\left(1 - \frac{1.5}{5}\right)\right]x_2 + x_3 = \left(\frac{1}{2}\right)\left(\frac{1}{2}\right)(1.5)$;

at $t = t_3 = 2.0$: $x_2 - \left[2 + \left(\frac{1}{2}\right)\left(\frac{1}{2}\right)\left(1 - \frac{2.0}{5}\right)\right]x_3 + x_4 = \left(\frac{1}{2}\right)\left(\frac{1}{2}\right)(2.0)$;

at $t = t_4 = 2.5$: $x_3 - \left[2 + \left(\frac{1}{2}\right)\left(\frac{1}{2}\right)\left(1 - \frac{2.5}{5}\right)\right]x_4 + x_5 = \left(\frac{1}{2}\right)\left(\frac{1}{2}\right)(2.5)$.

We know $x_1 = 2$ and $x_5 = -1$; hence the difference equation is not written corresponding to $t = 1$ or $t = 3$. Substituting the values $x_1 = 2$ and $x_5 = -1$ and simplifying, we have, in matrix form:

$$\begin{bmatrix} -2.175 & 1 & 0 \\ 1 & -2.150 & 1 \\ 0 & 1 & -2.125 \end{bmatrix}\begin{bmatrix} x_2 \\ x_3 \\ x_4 \end{bmatrix} = \begin{bmatrix} -1.625 \\ 0.5 \\ 1.625 \end{bmatrix}.$$

Solving, we get

$$x_2 = 0.552, \qquad x_3 = -0.424, \qquad \text{and} \qquad x_4 = -0.964.$$

These are reasonably close to the values obtained by the shooting method. We would expect some significant error because our step size, h, is large and the finite differences will be poor approximations to the derivatives. If we take $h = 0.2$ (ten subdivisions) and write the approximating equations, we get:

$$\begin{bmatrix} -2.0304 & 1 & 0 & 0 & 0 & 0 & 0 & 0 & 0 \\ 1 & -2.0288 & 1 & 0 & 0 & 0 & 0 & 0 & 0 \\ 0 & 1 & -2.0272 & 1 & 0 & 0 & 0 & 0 & 0 \\ 0 & 0 & 1 & -2.0256 & 1 & 0 & 0 & 0 & 0 \\ 0 & 0 & 0 & 1 & -2.0240 & 1 & 0 & 0 & 0 \\ 0 & 0 & 0 & 0 & 1 & -2.0224 & 1 & 0 & 0 \\ 0 & 0 & 0 & 0 & 0 & 1 & -2.0208 & 1 & 0 \\ 0 & 0 & 0 & 0 & 0 & 0 & 1 & -2.0192 & 1 \\ 0 & 0 & 0 & 0 & 0 & 0 & 0 & 1 & -2.0176 \end{bmatrix} x = \begin{bmatrix} -1.952 \\ 0.056 \\ 0.064 \\ 0.072 \\ 0.080 \\ 0.088 \\ 0.096 \\ 0.104 \\ 1.112 \end{bmatrix}. \quad (6.4)$$

We observe that the system is tridiagonal and therefore speedy to solve and also economical of memory space to store the coefficients. This will be true even if the value of N is very large, because we only use x_{i-1}, x_i, x_{i+1} in any equation to replace x' or x''. This is one reason why the finite-difference method is widely used to solve second-order linear boundary-value problems.

If we use the Subroutine TRIDG from Chapter 2 to compute the solution to Eqs. (6.4), we get the values shown in Table 6.4, where we also display the results by the shooting method from Section 6.1 for comparison.

Table 6.4 Results for linear boundary-value problem:

$$x'' = t + \left(1 - \frac{t}{5}\right)x, \quad x(1) = 2, \quad x(3) = -1.$$

t	Values from finite-difference method	Values from shooting method
1.0	2.000	2.000
1.2	1.351	1.348
1.4	0.792	0.787
1.6	0.311	0.305
1.8	−0.097	−0.104
2.0	−0.436	−0.443
2.2	−0.705	−0.712
2.4	−0.903	−0.908
2.6	−1.022	−1.026
2.8	−1.058	−1.060
3.0	−1.000	−1.000

The values obtained with $\Delta t = 0.2$ are more accurate. We might anticipate that the error would decrease proportionately to h^2 since the approximation to the derivatives are of $O(h^2)$ error. If we use the values from the shooting method as the exact values for comparison, the errors are reduced only about threefold (comparing the values at $t = 2$) instead of sixfold. However, the shooting-method calculations are themselves imperfect, and this adds an element of doubt to our observation. To observe how the errors decrease when Δt decreases, we look at another example, one where the exact solution is known.

Example Solve

$$\frac{d^2y}{dx^2} = y, \quad y(1) = 1.1752, \quad y(3) = 10.0179.$$

(The analytical solution is $y = \sinh x$, to which we can compare our estimates of the function.)

While it is not strictly necessary, it is common to normalize the function to the interval $(0, 1)$.* This we can do by the change of variable

$$x = (b - a)t + a,$$

where (a, b) is the original interval to be normalized. Letting $x = 2t + 1$, we write

$$\frac{dy}{dx} = \frac{dy}{dt}\frac{dt}{dx} = \frac{1}{2}\frac{dy}{dt},$$

$$\frac{d^2y}{dx^2} = \frac{d}{dt}\left(\frac{1}{2}\frac{dy}{dt}\right)\frac{dt}{dx} = \frac{1}{2^2}\frac{d^2y}{dt^2},$$

and the problem becomes

$$\frac{d^2y}{dt^2} = 4y, \qquad y(0) = 1.1752, \qquad y(1) = 10.0179.$$

Replacing the second derivative by the central-difference approximation, and thus converting the differential equation to a difference equation, we have:

$$\frac{y_{i+1} - 2y_i + y_{i-1}}{h^2} = 4y_i, \qquad i = 1, 2, 3, \cdots, n.$$

Subdividing the interval $(0, 1)$ into four parts, $h = 0.25$, and writing the difference equation at the three internal points where y is unknown, we have the set of equations:

$$\begin{bmatrix} -2.25 & 1 & 0 \\ 1 & -2.25 & 1 \\ 0 & 1 & -2.25 \end{bmatrix} \begin{bmatrix} y_2 \\ y_3 \\ y_4 \end{bmatrix} = \begin{bmatrix} -1.1752 \\ 0 \\ -10.0179 \end{bmatrix}. \tag{6.5}$$

In Eqs. (6.5) we have used $y_1 = 1.1752$, $y_5 = 10.0179$. Again, a tridiagonal matrix of coefficients occurs. The solution to this set of equations is given in Table 6.5.

If we recalculate this problem using varying values of h, we get the results shown in Table 6.6 for y at $x = 2.0$ (equivalent to $t = 0.5$.).

We observe an approximate fourfold decrease in the error when the step size is halved, meaning an $O(h^2)$ error does exist. When we know how errors vary with step size, we can perform an extrapolation similar to that for Romberg

---

*Perhaps the most important reason is to make a computer program more general. Normalizing is really a scaling of the interval so we have an immediate understanding of whether a given value of h is "small" or "large." Also, when we compute the LU equivalent of the coefficient matrix to solve the system, it is then of more general applicability.

Table 6.5

t	x	Calculated y	Exact value, $\sinh x$	Error
0	1.0	(1.1752)	1.1752	—
0.25	1.5	2.1467	2.1293	−0.0174
0.50	2.0	3.6549	3.6269	−0.0280
0.75	2.5	6.0768	6.0502	−0.0266
1.0	3.0	(10.0179)	10.0179	—

integration in Chapter 3, using results for which the step size is halved:

$$\begin{matrix} \text{Improved} \\ \text{value} \end{matrix} = \begin{matrix} \text{More} \\ \text{accurate} \end{matrix} + \frac{1}{3}\left[\left(\begin{matrix} \text{More} \\ \text{accurate} \end{matrix}\right) - \left(\begin{matrix} \text{Less} \\ \text{accurate} \end{matrix}\right)\right].$$

In this instance we get $3.6340 + \frac{1}{3}(3.6340 - 3.6549) = 3.6270$, when results with $\Delta t = \frac{1}{4}$ and $\Delta t = \frac{1}{8}$ are combined.

When the differential equation underlying a boundary-value problem is nonlinear, this method of finite differences runs into a problem in that the resulting system of equations is nonlinear. Consider the nonlinear problem solved by the shooting method in Section 6.1:

$$\frac{d^2x}{dt^2} - \left(1 - \frac{t}{5}\right)\left(\frac{dx}{dt}\right)(x) = t, \qquad x(1) = 2, \qquad x(3) = -1.$$

When finite differences are substituted for the derivatives, we obtain an equivalent set of difference equations as follows:

$$x_{i-1} - \left[2 + \frac{h}{2}\left(1 - \frac{t_i}{5}\right)(x_{i+1} - x_{i-1})\right]x_i + x_{i+1} = h^2 t_i, \qquad i = 1, 2, 3, \ldots n, \qquad (6.6)$$

which is to be written at each point t_i at which x is unknown. In the middle term, products of the x's occur, exhibiting nonlinearity in the difference equations as a consequence of the nonlinearity in the differential equation. The standard elimination method fails for these equations.

Table 6.6

Number of equations	$h = \Delta t$ for normalized interval	Result	Error
1	0.50	3.7310	−0.1041
3	0.25	3.6549	−0.0280
7	0.125	3.6340	−0.0071
∞	0	3.6269	(exact)

In such cases, iteration techniques can be employed. Suppose we obtain some estimate to the x-vector, the solution to the equations. We could use these values as a means of approximating the coefficient of x_i in Eq. (6.6). (Since the values are multiplied by $h/2$, we might hope that errors of the estimate would be diluted in their effect, especially if h is fairly small, in comparison to the number 2, which should dominate in the coefficient of x_i.)

When this was done, using $h = 0.2$ and the initial x_i values estimated from $x = 3.5 - 1.5t$ (a linear relation between $x(1) = 2$ and $x(3) = -1$), the values in Table 6.7 resulted.

Further iterations after the eighth did not change the values. In Table 6.7, we have listed the results by the shooting method for comparison. In both techniques, an iterative process was used; in the finite-difference method, we iterate on the approximations to x; in the shooting method, we use successive estimates to the initial slope. For this example, about the same number of iterations were required, and roughly the same accuracy was achieved. (The correct solution is intermediate between the results for the two methods shown in Table 6.7.)

The choice between the finite-difference method and the shooting method for nonlinear boundary-value problems is not clearcut. The choice depends on whether a reasonably good initial estimate for the x-vector is available and how strongly nonlinear the equation is. When the nonlinearity in the set of equations is not diluted in its effect and when no good initial estimate is available, the algebraic finite-difference method may not even converge. Although the shooting method often involves a somewhat greater computational effort, it is usually more certain of giving a solution.

Table 6.7 Successive approximations to finite-difference equations for

$$x'' - \left(1 - \frac{t}{5}\right)(x')(x) = t, \qquad x(1) = 2, \qquad x(3) = -1.$$

	Approximations to x-vector					
t	Initial estimate	After 1 interation	After 2 iterations	After 4 iterations	After 8 iterations	Results from shooting method
1.0	2.000	(2.000)	(2.000)	(2.000)	(2.000)	(2.000)
1.2	1.700	1.431	1.601	1.558	1.552	1.557
1.4	1.400	0.845	1.110	1.048	1.039	1.053
1.6	1.100	0.279	0.583	0.513	0.503	0.527
1.8	0.800	−0.235	0.077	0.006	−0.005	0.026
2.0	0.500	−0.668	−0.362	−0.430	−0.440	−0.408
2.2	0.200	−0.997	−0.704	−0.766	−0.775	−0.747
2.4	−0.100	−0.204	−0.937	−0.990	−0.998	−0.977
2.6	−0.400	−1.278	−1.061	−1.101	−1.107	−1.094
2.8	−0.700	−1.211	−0.080	−1.104	−1.107	−1.101
3.0	−1.000	(−1.000)	(−1.000)	(−1.000)	(−1.000)	(−1.000)

6.3 DERIVATIVE BOUNDARY CONDITIONS

The conditions which the solution of a differential equation must satisfy need not necessarily be just the value of the function. In many applied problems, some derivative of the function may be known at the boundaries of an interval. In the more general case a linear combination of the function and its derivatives is specified. The finite-difference procedure needs modification for this type of boundary conditions.* We illustrate by an example:

$$\frac{d^2y}{dx^2} = y,$$

$$y'(1) = 1.1752,$$

$$y'(3) = 10.0179.$$

(This problem has the same differential equation as the example of Section 6.2 but, with the values specified for the derivative at $x = 1$ and $x = 3$, it now has the analytical solution $y = \cosh x$.)

We begin just as before. We change variables to make the interval $(0, 1)$ by letting $x = 2t + 1$:

$$\frac{d^2y}{dt^2} = 4y.$$

We now replace the derivative by a central-difference approximation and write the difference equation at each point where y is unknown. With $h = 0.25$,

$$
\begin{aligned}
t = 0\text{:} &\qquad y_\ell - 2.25y_1 + y_2 = 0, \\
t = 0.25\text{:} &\qquad y_1 - 2.25y_2 + y_3 = 0, \\
t = 0.50\text{:} &\qquad y_2 - 2.25y_3 + y_4 = 0, \qquad\qquad (6.7) \\
t = 0.75\text{:} &\qquad y_3 - 2.25y_4 + y_5 = 0, \\
t = 1.00\text{:} &\qquad y_4 - 2.25y_5 + y_r = 0.
\end{aligned}
$$

Two more equations are required than in Section 6.2 because y is unknown at $t = 0$ and $t = 1$ as well as at the interior points. These two equations involve the values y_ℓ, y_r, points one space to the left and to the right of the interval $(0, 1)$. We assume that the domain of the differential equation can be so extended.

Our problem is now that Eqs. (6.7) contain seven unknowns, and we have only five equations. The boundary conditions, however, have not yet been involved. Let us express these also as difference quotients, preferring central-difference approximations of $O(h^2)$ error as used in replacing derivatives in the

---

* With the shooting method, derivative boundary conditions do not require any change in procedure.

original equation:

$$\frac{dy}{dx}\bigg|_{x=1} = \frac{dy}{dt}\frac{dt}{dx}\bigg|_{t=0} \doteq \left(\frac{1}{2}\right)\frac{y_2 - y_\ell}{2h} = 1.1752, \qquad y_\ell = y_2 - (4)(0.25)(1.1752),$$

$$\frac{dy}{dx}\bigg|_{x=3} = \frac{dy}{dt}\frac{dt}{dx}\bigg|_{t=1} \doteq \left(\frac{1}{2}\right)\frac{y_r - y_4}{2h} = 10.0179, \qquad y_r = y_4 + (4)(0.25)(10.0179).$$

(6.8)

The relations of (6.8), when substituted into (6.7), reduce the number of unknowns to five, and we solve the equations in the usual way. The solution is given in Table 6.8.

Table 6.8

t	x	y	$\cosh x$	Error
0	1.0	1.5522	1.5431	−0.0091
0.25	1.5	2.3338	2.3524	0.0186
0.50	2.0	3.6989	3.7622	0.0633
0.75	2.5	5.9887	6.1323	0.1436
1.00	3.0	9.7757	10.0677	0.2920

We observe here that the errors are much greater than for the previous example, being very large at $x = 3.0$. The explanation is that our approximation for the derivative is poor at large x-values. While of $O(h^2)$, the third derivative appearing in the error term is large in magnitude. In the previous example, knowing the function at the endpoints eliminated the errors there.

Repeating the calculations with the step size cut in half reduces the errors about fourfold. For example, at $t = 0.5$, we get 3.7459 (error 0.0163 vs. 0.0633); and at $t = 1.0$, we get 9.9921 (error 0.0756 vs. 0.2920). Halving h again gives another fourfold reduction in errors ($y = 10.0486$ at $t = 1.0$, error of 0.0191). An extrapolation based on errors of $O(h^2)$ will give nearly perfect results.

It is appropriate to summarize the finite-difference method for solving a boundary-value problem.

The finite-difference method for boundary-value problems

To solve a boundary-value problem, replace the differential equation by a finite-difference equation by replacing the derivatives by central-difference quotients. Subdivide the interval into a suitable number of equal subintervals, and write the difference equation at each point where the value of the function is unknown. When the boundary values involve derivatives, this will require that the domain be extended beyond the interval. Utilize the derivative boundary conditions to write difference quotients that permit the elimination of the fictitious points outside the interval.

Solve the system of equations so created to obtain approximate values for the solution of the differential equation at discrete points on the interval. If the original differential equation is nonlinear, the system of equations will also be nonlinear. In such situations, the shooting method will normally be preferred.

We could, of course, use more accurate finite-difference approximations to the derivatives, not only for the boundary values, but for the equation itself. The disadvantage of doing this is that the system of equations is then not tridiagonal, and there is attendant greater difficulty of solution.

If the more general form of boundary condition applies, $ay + by' = c$, equations similar to (6.8) result, and after the boundary conditions have been approximated and the exterior y-values eliminated from the set of equations, the solution of the problem proceeds as before.

Equations of order higher than the second will involve approximations for the third or higher derivatives. Central-difference formulas will involve points more than h away from the point where the derivative is being approximated, and may be unsymmetrical in the case of the odd derivatives. Probably the method of undetermined coefficients is the easiest way to derive the necessary formulas (see Appendix). Again, nontridiagonal systems result. Fortunately, most of the important physical problems are simulated by equations of order two.

6.4 CHARACTERISTIC-VALUE PROBLEMS

Problems in the fields of elasticity and vibration (including applications of the wave equations of modern physics) fall into a special class of boundary-value problems known as *characteristic-value problems*. (Certain problems in statistics also reduce to such problems.) We discuss here only the most elementary forms of characteristic-value problems.

Consider the homogeneous* second-order equation with homogeneous boundary conditions:

$$\frac{d^2y}{dx^2} + k^2y = 0, \qquad y(0) = 0, \qquad y(1) = 0, \tag{6.9}$$

where k^2 is a parameter. We first solve this equation nonnumerically to show that there is a solution for only certain particular, or "characteristic," values of the parameter. The general solution is

$$y = a \sin kx + b \cos kx,$$

which can easily be verified by substituting into the differential equation; the solution contains the two arbitrary constants a and b, because the differential

---

* Homogeneous here means that all the terms are alike in being functions of y or its derivatives.

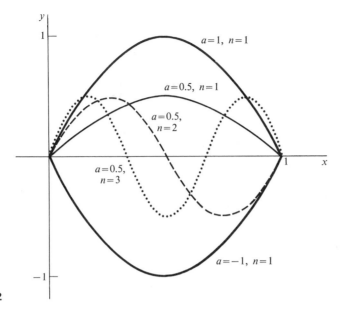

Figure 6.2

equation is second-order. The constants a and b are to be determined to make the general solution agree with the boundary conditions.

At $x = 0$, $y = 0 = a \sin(0) + b \cos(0) = b$. Then b must be zero. At $x = 1$, $y = 0 = a \sin(k)$; we may have either $a = 0$ or $\sin(k) = 0$ to satisfy this condition. The former leads to $y(x) = 0$, which we call the *trivial solution*. This function, y everywhere zero, will always be a solution to any homogeneous differential equation with homogeneous boundary conditions. (The trivial solution is usually of no interest.) To get a nontrivial solution, we must choose the other alternative, $\sin(k) = 0$, which is satisfied only for certain values of k, the characteristic* values of the system. The solution to Eq. (6.9) must then be:

$$k = \pm n\pi, \qquad n = 1, 2, \ldots, \qquad y = a \sin n\pi x. \qquad (6.10)$$

Note that the arbitrary constant a can have any value and still permit the function y to meet the boundary conditions, so that the solution is determined only to within a multiplicative constant. In Fig. 6.2 we sketch several of the solutions as given by Eq. (6.10).

The information of interest in characteristic-value problems are these characteristic values, or *eigenvalues* for the system. If we are dealing with a vibration problem, these give the natural frequencies of the system, which are especially important because, with external loads applied at or near these frequencies, resonance will cause an amplification of motion so that failure is likely. In the field

---

*These are also called "eigenvalues," derived from the German term Eigenwerte.

of elasticity, there is an eigenfunction corresponding to each eigenvalue, and these determine the possible shapes of the elastic curve when the system is in equilibrium. Often the smallest nonzero value of the parameter is especially important; this gives the fundamental frequency of the system.

To illustrate the use of numerical methods, we will solve Eq. (6.9) again. Our attack is the previous one of replacing the differential equation by a difference equation, and writing this at all points into which the x-interval has been subdivided and where y is unknown.[*] Replacing the derivative in (6.9) by a central-difference approximation, we have:

$$\frac{y_{i+1} - 2y_i + y_{i-1}}{h^2} + k^2 y_i = 0. \tag{6.11}$$

Letting $h = 0.2$, and writing out Eq. (6.11) at each of the four interior points we get

$$
\begin{aligned}
y_1 - (2 - 0.04k^2)y_2 + y_3 &= 0, \\
y_2 - (2 - 0.04k^2)y_3 + y_4 &= 0, \\
y_3 - (2 - 0.04k^2)y_4 + y_5 &= 0, \\
y_4 - (2 - 0.04k^2)y_5 + y_6 &= 0.
\end{aligned}
\tag{6.12}
$$

The boundary conditions give $y_1 = y_6 = 0$. Making this substitution and writing in matrix form, after multiplying all equations by -1, we get:

$$
\begin{bmatrix}
2 - 0.04k^2 & -1 & 0 & 0 \\
-1 & 2 - 0.04k^2 & -1 & 0 \\
0 & -1 & 2 - 0.04k^2 & -1 \\
0 & 0 & -1 & 2 - 0.04k^2
\end{bmatrix}
\begin{bmatrix}
y_2 \\
y_3 \\
y_4 \\
y_5
\end{bmatrix} = 0. \tag{6.13}
$$

Note that we can write this as the matrix equation $(A - \lambda I)y = 0$, where

$$
A = \begin{bmatrix}
2 & -1 & 0 & 0 \\
-1 & 2 & -1 & 0 \\
0 & -1 & 2 & -1 \\
0 & 0 & -1 & 2
\end{bmatrix}, \qquad \lambda = 0.04k^2,
$$

and I is the identity matrix of order 4. We shall consider the problem in this form in the next section.

The solution to our characteristic-value problem, Eq. (6.9), reduces to solving the set of equations in (6.12) or (6.13).

---

[*] There is an alternative trial-and-error approach, analogous to the shooting method, but it generally involves more work.

Such a set of homogeneous linear equations has a nontrivial solution only if the determinant of the coefficient matrix is equal to zero. Hence

$$\begin{vmatrix} 2 - 0.04k^2 & -1 & 0 & 0 \\ -1 & 2 - 0.04k^2 & -1 & 0 \\ 0 & -1 & 2 - 0.04k^2 & -1 \\ 0 & 0 & -1 & 2 - 0.04k^2 \end{vmatrix} = 0.$$

Expanding this determinant will give an eighth-degree polynomial in k (this is called the characteristic polynomial of matrix A); the roots of this polynomial will be approximations to the characteristic values of the system. Letting $2 - 0.04k^2 = Z$ makes the expansion simpler, and we get:

$$Z^4 - 3Z^2 + 1 = 0.$$

The roots of this biquadratic are $Z^2 = (3 \pm \sqrt{5})/2$, or

$$Z = 1.618, \quad -1.618, \quad 0.618, \quad -0.618.$$

We will get an estimate of the principal eigenvalues from $2 - 0.04k^2 = 1.618$, giving $k = \pm 3.09$ (compare to $\pm \pi$). The next eigenvalues are obtained from:

$$2 - 0.04k^2 = 0.618, \qquad \text{giving } k = \pm 5.88 \text{ (compare to } \pm 2\pi);$$
$$2 - 0.04k^2 = -0.618, \qquad \text{giving } k = \pm 8.09 \text{ (compare to } \pm 3\pi);$$
$$2 - 0.04k^2 = -1.618, \qquad \text{giving } k = \pm 9.51 \text{ (compare to } \pm 4\pi).$$

Our estimates of the characteristic values get progressively worse. Fortunately the smallest values of k are of principal interest in many applied problems. To improve accuracy, we will need to write our difference equation using smaller values of h. The work soon becomes inordinately long, and we look for a simpler method to find the eigenvalues of a matrix. This is the subject of the next section.

Another reason, beyond the amount of computational effort, that argues against expanding the determinant and then finding the roots of the characteristic equation, is accuracy. Finding the zeros of a polynomial of high degree is a process that is often subject to large errors due to round-off. The process is very ill-conditioned.

6.5 EIGENVALUES OF A MATRIX BY ITERATION

We have seen that particular values of the parameter occurring in a characteristic-value problem are of special interest, and the numerical solution involves finding the eigenvalues of the coefficient matrix of the set of difference equations. To do this, we need to find the values of λ that satisfy the matrix equation

$$Ax = \lambda x$$

where A is a square matrix. In effect, we look for a certain vector, x, which gives a scalar multiple of itself when multiplied by the matrix A. For most vectors chosen at random, this will not be true. If A is $n \times n$ and nonsingular, there are usually exactly n different vectors x which, when multiplied by the matrix A, give a scalar multiple of x. These vectors are called eigenvectors or characteristic vectors. The multipliers λ, which are n in number, are termed the eigenvalues or characteristic values of the matrix A. As we have seen in the previous section, one method is to solve the equivalent matrix equation

$$(A - \lambda I)x = 0$$

by determining the values of λ that make the determinant of $(A - \lambda I)$ equal to zero. This involves expanding the determinant to give a polynomial in λ, the characteristic equation, whose roots must be evaluated. This tells us, of course, that there will be n eigenvalues for the system, but they will not necessarily all be different in value. In fact, it is possible that some eigenvalues will be complex-valued. Solving for the roots of the characteristic equation is laborious.* This section develops an alternative iterative procedure that is less effort.

Consider a simple 3×3 system:

$$\begin{bmatrix} 4 & 1 & 0 \\ 0 & 2 & 1 \\ 0 & 0 & -1 \end{bmatrix}\begin{bmatrix} x_1 \\ x_2 \\ x_3 \end{bmatrix} = \lambda \begin{bmatrix} x_1 \\ x_2 \\ x_3 \end{bmatrix}, \qquad \begin{bmatrix} 4 - \lambda & 1 & 0 \\ 0 & 2 - \lambda & 1 \\ 0 & 0 & -1 - \lambda \end{bmatrix}\begin{bmatrix} x_1 \\ x_2 \\ x_3 \end{bmatrix} = 0. \quad (6.14)$$

We wish to find the values of λ and vectors x that satisfy Eq. (6.14). Let us first see what they are.

The characteristic equation is

$$(4 - \lambda)(2 - \lambda)(-1 - \lambda) - 0 = 0, \qquad -\lambda^3 + 5\lambda^2 - 2\lambda - 8 = 0.$$

The roots are

$$\lambda_1 = 4, \qquad \lambda_2 = 2, \qquad \text{and} \qquad \lambda_3 = -1.$$

We can find the eigenvectors corresponding to these from the three sets of equations:

$$\begin{bmatrix} 4 & 1 & 0 \\ 0 & 2 & 1 \\ 0 & 0 & -1 \end{bmatrix}\begin{bmatrix} x_1 \\ x_2 \\ x_3 \end{bmatrix} = 4\begin{bmatrix} x_1 \\ x_2 \\ x_3 \end{bmatrix}, \qquad \begin{cases} 4x_1 + x_2 = 4x_1 \\ 2x_2 + x_3 = 4x_2, \\ \quad -x_3 = 4x_3 \end{cases} \qquad \begin{cases} x_1 \ \text{arbitrary} \\ x_2 = 0 \\ x_3 = 0 \end{cases}$$

$$\begin{bmatrix} 4 & 1 & 0 \\ 0 & 2 & 1 \\ 0 & 0 & -1 \end{bmatrix}\begin{bmatrix} x_1 \\ x_2 \\ x_3 \end{bmatrix} = 2\begin{bmatrix} x_1 \\ x_2 \\ x_2 \end{bmatrix}, \qquad \begin{cases} 4x_1 + x_2 = 2x_1 \\ 2x_2 + x_3 = 2x_2, \\ \quad -x_3 = 2x_3 \end{cases} \qquad \begin{cases} x_1 = x_2/2 \\ x_2 \ \text{arbitrary} \\ x_3 = 0 \end{cases}$$

$$\begin{bmatrix} 4 & 1 & 0 \\ 0 & 2 & 1 \\ 0 & 0 & -1 \end{bmatrix}\begin{bmatrix} x_1 \\ x_2 \\ x_3 \end{bmatrix} = -1\begin{bmatrix} x_1 \\ x_2 \\ x_3 \end{bmatrix}, \qquad \begin{cases} 4x_1 + x_2 = -x_1 \\ 2x_2 + x_3 = -x_2, \\ \quad -x_3 = -x_3 \end{cases} \qquad \begin{cases} x_1 = x_3/15 \\ x_2 = -x_3/3 \\ x_3 \ \text{arbitrary.} \end{cases}$$

---

* The labor involved is not only in finding the roots of the characteristic polynomial but in evaluating the coefficients in the first place.

We see that any nonzero value can be taken for x_1 in the first vector, for x_2 in the second, and for x_3 in the third. Our eigenvectors are then any multiple of:

$$\begin{bmatrix} 1 \\ 0 \\ 0 \end{bmatrix}, \quad \begin{bmatrix} -\frac{1}{2} \\ 1 \\ 0 \end{bmatrix}, \quad \text{and} \quad \begin{bmatrix} \frac{1}{15} \\ -\frac{1}{3} \\ 1 \end{bmatrix}$$

In the matrix of Eq. (6.14), the eigenvalues 4, 2, −1, are equal to the values of the diagonal elements. This is no coincidence; in fact, for any triangular matrix, the eigenvalues are equal to the values on the diagonal. On reflection, this is obvious because the determinant of a triangular matrix is the product of its diagonal elements. When we write this for the triangular matrix $A - \lambda I$, we have the characteristic equation already in factored form.

Unfortunately, we cannot find the eigenvalues of a general matrix by simply reducing it to triangular form by Gaussian elimination, as one at first might hope. The reason is that row reduction changes the eigenvalues. For example, consider the matrices A and B:

$$A = \begin{bmatrix} 2 & 1 \\ 1 & 2 \end{bmatrix}, \quad B = \begin{bmatrix} 2 & 1 \\ 0 & \frac{3}{2} \end{bmatrix}.$$

Matrix B is derived from A by row reduction: it has eigenvalues 2 and $\frac{3}{2}$, while the eigenvalues of A are 3 and 1.

We now illustrate an iterative procedure for determining the eigenvalue of largest magnitude of Eq. (6.14). This is called the *power method*. We start with an arbitrary vector and multiply it by the matrix repeatedly. Using

$$\begin{bmatrix} 1 \\ 1 \\ 1 \end{bmatrix},$$

we have the results below. At each step, we normalize the vector by making its largest component equal to unity.

$$\begin{bmatrix} 4 & 1 & 0 \\ 0 & 2 & 1 \\ 0 & 0 & -1 \end{bmatrix} \begin{bmatrix} 1 \\ 1 \\ 1 \end{bmatrix} = \begin{bmatrix} 5 \\ 3 \\ -1 \end{bmatrix} = 5 \begin{bmatrix} 1 \\ 0.6 \\ -0.2 \end{bmatrix},$$

$$\begin{bmatrix} 4 & 1 & 0 \\ 0 & 2 & 1 \\ 0 & 0 & -1 \end{bmatrix} \begin{bmatrix} 1 \\ 0.6 \\ -0.2 \end{bmatrix} = \begin{bmatrix} 4.6 \\ 1.0 \\ 0.2 \end{bmatrix} = 4.6 \begin{bmatrix} 1 \\ 0.217 \\ 0.0435 \end{bmatrix},$$

$$\begin{bmatrix} 4 & 1 & 0 \\ 0 & 2 & 1 \\ 0 & 0 & -1 \end{bmatrix} \begin{bmatrix} 1 \\ 0.2174 \\ 0.0435 \end{bmatrix} = \begin{bmatrix} 4.2174 \\ 0.4783 \\ -0.0435 \end{bmatrix} = 4.2174 \begin{bmatrix} 1 \\ 0.1134 \\ -0.0103 \end{bmatrix},$$

$$\begin{bmatrix} 4 & 1 & 0 \\ 0 & 2 & 1 \\ 0 & 0 & -1 \end{bmatrix} \begin{bmatrix} 1 \\ 0.1134 \\ -0.0103 \end{bmatrix} = \begin{bmatrix} 4.1134 \\ 0.2165 \\ 0.0103 \end{bmatrix} = 4.1134 \begin{bmatrix} 1 \\ 0.0526 \\ 0.0025 \end{bmatrix},$$

$$\begin{bmatrix} 4 & 1 & 0 \\ 0 & 2 & 1 \\ 0 & 0 & -1 \end{bmatrix} \begin{bmatrix} 1 \\ 0.0526 \\ 0.0025 \end{bmatrix} = \begin{bmatrix} 4.0526 \\ 0.1077 \\ -0.0025 \end{bmatrix} = 4.0526 \begin{bmatrix} 1 \\ 0.0266 \\ -0.0006 \end{bmatrix},$$

$$\vdots \qquad\qquad \vdots \qquad\qquad\qquad \vdots$$

We see that the successive values of the product vector approach closer and closer to multiples of the eigenvector

$$\begin{bmatrix} 1 \\ 0 \\ 0 \end{bmatrix}.$$

The normalization factors approach the value of the largest eigenvalue.

The method is slow to converge if the magnitudes of the largest eigenvalues are nearly the same.

With more realistically sized matrices, the iteration will take many steps and requires a computer for its practical execution. The method will converge only when the largest eigenvalue in modulus is uniquely determined.

The power method works because the eigenvectors are a set of *basis vectors*. By this we mean that they are a set of linearly independent vectors that *span the space*; i.e., any n-component vector can be written as a unique linear combination of them. Let $v^{(0)}$ be any vector and $x_1, x_2, \ldots x_n$ be eigenvectors. Then

$$v^{(0)} = c_1 x_1 + c_2 x_2 + \ldots + c_n x_n.$$

If we multiply v by the matrix A, we have (since the x_i are eigenvectors with corresponding eigenvalues λ_i),

$$\begin{aligned} v^{(1)} = A v^{(0)} &= c_1 A x_1 + c_2 A x_2 + \ldots + c_n A x_n \\ &= c_1 \lambda_1 x_1 + c_2 \lambda_2 x_2 + \ldots + c_n \lambda_n x_n. \end{aligned} \qquad (6.15)$$

Upon repeated multiplication by A we get, after m times:

$$v^{(m)} = A^m v^{(0)} = c_1 \lambda_1^m x_1 + c_2 \lambda_2^m x_2 + \ldots + c_n \lambda_n^m x_n. \qquad (6.16)$$

If one eigenvalue, say λ_1, is larger in magnitude than all the rest, the values of λ_i^m, $i \neq 1$, will be negligibly small in comparison to λ_1^m when m is large and

$$A^m v \to c_1 \lambda^m x_1, \qquad \text{or} \qquad A^m v \to \text{Multiple of eigenvector } x_1,$$

with the normalization factor of λ_1. This is the principle behind the power method.

In applying Eq. (6.16) to find the dominant eigenvalue (the eigenvalue whose magnitude exceeds all others), we normalize at each step; this in effect is just scaling the c's after each multiplication, and will not change the relation. If some other eigenvalue than λ_1 were the dominant one, we would merely renumber the eigens. It is easy to see why the rate of convergence depends strongly on the ratio of the two eigenvalues of largest magnitude.

Obviously, if we are able to chose the starting vector, $v^{(0)}$, close to the eigenvector x_1, we will converge more rapidly; the coefficients c_i, $i \neq 1$, in Eq. (6.15) will be small in this case. We usually have no knowledge of this eigenvector, so we start with an arbitrary vector with all components equal to unity. This choice could be a bad one. If the value of c_1 in Eq. (6.15) is zero for this arbitrary choice, the process will never converge. Actually, the process often does converge in spite of this; round-off errors incurred as we repeatedly multiply by A produce components in the direction of x_1 so that we do converge, although slowly.

A special advantage of the power method is that the eigenvector that corresponds to the dominant eigenvalue is generated at the same time. For most methods of determining eigenvalues, a separate computation is needed to obtain the eigenvector. A disadvantage, of course, is that it gives only one eigenvalue. Sometimes the largest eigenvalue is the most important, but if it is not, we must modify the method.

In some problems, the most important eigenvalue and eigenvector is that where the eigenvalue is of least magnitude. We can find this one if we apply the power method to the inverse of A. This is because the inverse matrix has a set of eigenvalues which are the reciprocals of the eigenvalues of A. This is readily shown:

Given that $Ax = \lambda x$.

Multiply by A^{-1}:
$$A^{-1}Ax = A^{-1}\lambda x = \lambda A^{-1}x.$$
From this we see that
$$x = \lambda A^{-1}x \quad \text{or} \quad A^{-1}x = \frac{1}{\lambda}x.$$

(It is inefficient to actually invert A before applying the power method to get the eigenvalue. We use the LU equivalent of A, obtainable at much less effort, and solve for $v^{(n+1)}$ from:
$$LU v^{(n+1)} = Av^{(n+1)} = v^{(n)}, \quad \text{equivalent to} \quad A^{-1}v^{(n)} = v^{(n+1)}. \quad (6.17)$$

The solution of Eq. (6.17) when the LU is known requires the same amount of effort as a matrix multiplication.)

The power method is apt to be slowly convergent. We can accelerate it if we know an approximate value for the eigenvalue. This is based on the result of shifting the eigenvalues. Observe that subtracting a constant from the diagonal elements of A gives a system whose eigenvalues are those of A with the same constant subtracted:

Given $Ax = \lambda x$.

Subtract $sIx = sx$ from both sides:

$$Ax - sIx = \lambda x - sx.$$

$$(A - sI)x = (\lambda - s)x.$$

The above relationship can be applied in two ways. Suppose we wish to determine the value of an eigenvalue near to some number s. We shift the eigenvalues by subtracting s from the diagonal elements; there is then an eigenvalue very near to zero in the shifted matrix. We use the power method on the inverse of the shifted matrix. This is often rapidly convergent because the reciprocal of the very small value is very large, and is usually much larger than the next largest one (for the shifted inverse system). After we obtain it, we reverse the transformations to obtain the desired value for the original matrix. This process is called the *inverse power method*. It is illustrated in the example below.

Another application of shifting is to determine the eigenvalue at the other extreme after the dominant one has been computed. Suppose a matrix has eigenvalues whose values are 8, 4, 1, and -5. After applying the power method to get the one at 8, we shift by subtracting 8. The eigenvalues of the shifted system are then 0, -4, -7, and -13. The power method will get the dominant one, at -13. When we add 8 back to reverse the effect of the shifting, we obtain -5, the eigenvalue of the original matrix at the opposite extreme from 8.

We illustrate the use of the power method and its several variations with the 3×3 matrix:

$$A = \begin{bmatrix} 4 & -1 & 1 \\ 1 & 1 & 1 \\ -2 & 0 & -6 \end{bmatrix}.$$

If we begin with the initial vector $(1, 1, 1)^T$, the regular power method gives the results shown in Table 6.9. The iterations were terminated when the normalization factor changed by less than 0.0001 from the previous value.

Table 6.9

Iteration number	Normalization factor	Resultant vector
1	-8	$(-0.5, -0.375, 1)^T$
2	-5	$(0.125, -0.025, 1)^T$
3	-6.25	$(-0.244, -0.176, 1)^T$
.	.	.
.	.	.
.	.	.
22	-5.76849	$(-0.1157, -0.1306, 1)^T$

After the dominant eigenvalue at -5.76849 was determined, the matrix was shifted by subtracting this value from the elements on the diagonal. For this shifted matrix, the power method gave, after 32 iterations, a normalization factor of 9.23759 with a vector of $(1, 0.31963, -0.21121)^T$. We then know that the original matrix had an eigenvalue at 3.4691 (from $9.23759 + (-5.76849)$). (The corresponding eigenvector is the same as that determined by the power method for the shifted matrix.)

Applying the power method to A^{-1} (through application of the LU equivalent, of course) gave, after nine iterations, the value of 0.76989 with a vector of $(0.4121, 1, -0.1129)^T$. Now we know that the original matrix A had its smallest eigenvalue at $1/0.76989 = 1.29923$. For this 3×3 system, we have computed all its eigens.

The dominant eigenvalue came with some slowness; 22 iterations were needed. If we shift by -5.77 and apply the method to the inverse, only four iterations are required to obtain a value of 672.438 and a vector of $(-0.1157, -0.1306, 1)^T$. This is the same vector as that obtained in the first application of the method. We get the dominant eigenvalue for A by adding -5.77 to the reciprocal of 672.438:

$$\frac{1}{672.438} + (-5.77) = -5.76851.$$

Many iterations have been saved.

The key to the inverse power method is finding an approximate value for the desired eigenvalue. Sometimes the physical problem itself can suggest a value. Applying the standard power method for only a few iterations frequently gives an approximate value (the many successive iterations are needed to refine the value.) Gerschgorin's theorem can give an estimate if there is strong diagonal dominance. This theorem states that an eigenvalue of the matrix A lies in a circle in the complex plane whose center is at the point a_{ii} and whose radius is

$$\sum |a_{ij}|, \qquad j = 1, 2, \ldots, n \qquad \text{and} \qquad j \neq i.$$

For the previous example, since all the eigenvalues are real, Gerschgorin's theorem locates the eigenvalues in the intervals $(2, 6)$, $(-1, 3)$, and $(-8, -4)$.

When a matrix has two eigenvalues of equal largest magnitude,* the power method as described will fail. In the case of two largest eigenvalues of opposite sign, the normalization factor does not converge but oscillates. For example, when the power method is applied to:

$$A = \begin{bmatrix} -10 & -7 & 7 \\ 5 & 5 & -4 \\ -7 & -5 & 6 \end{bmatrix},$$

---

*This may occur if the eigenvalues are complex; for a matrix with real coefficients, complex eigenvalues will come in complex conjugate pairs.

Table 6.10

Iteration number	Normalization factor	Resultant vector
1	-10	$(1, -0.6, 0.6)^T$
2	-1.6	$(1, 0.25, 0.25)^T$
3	-10	$(1, -0.525, 0.675)^T$
4	-1.6	$(1, 0.2031, 0.2031)^T$

which has eigenvalues of 4, -4, and 1, the power method gives the results shown in Table 6.10, and the normalizations continue indefinitely to be -10 and -1.6. In such a case, the magnitude of the largest eigenvalue(s) is given by taking the square root of the product of the alternating normalization factors; in this example, -10 and -1.6 have a product, 16, whose square roots are 4 and -4.

In case all the eigenvalues of a large system are required, the QR algorithm is the preferred method. This is based on the similarity transformation (one that preserves the eigenvalues of a matrix):

$$BAB^{-1}x = \lambda x,$$

where B is any nonsingular matrix. (The reader should easily be able to show that BAB^{-1} has the same eigenvalues as A.) The QR method finds transformation matrices that eliminate elements below the diagonal; symmetric matrices whose inverse is equal to itself are employed. When applied to a general matrix, this is an iterative procedure, so a preliminary transformation is first made (the House-holder method is popular) that eliminates all below-diagonal elements except for those adjacent to the diagonal; the QR method is then applied. [For details, see Stewart (1973).] There are other variations on this theme. For symmetric matrices, the Jacobi method has been widely used.

6.6 PROGRAMS

Program 1 (Fig. 6.3) is an example of how a boundary-value problem can be solved by a computer. In it, a second-order boundary-value problem is solved, using the RKSYST subroutine of Chapter 5 to integrate the differential equation. Two preliminary solutions are obtained, employing two guesses for the initial slope of the function. From these, an improved estimate of the initial slope is computed by interpolation; the process of interpolation based on the last two values is iterated until the calculated value for the function at the end of the interval closely approximates the desired value. The program is a straightforward implementation of the technique of Section 6.1.

If one wishes to program the difference-equation method, the differential equation would be approximated by a system of finite-difference equations and the solution to this system would be obtained by the subroutines of Chapter 2. In

```
C     THIS PROGRAM SOLVES A NON-LINEAR SECOND ORDER BOUNDARY VALUE
C     PROBLEM BY THE SHOOTING METHOD.  IT EMPLOYS RKSYST TO SOLVE THE
C     EQUATIONS WITH TWO ASSUMED VALUES OF THE INITIAL SLOPE.
C     THE EQUATION IS D2X/DT2 - (3X)(DX/DT) + X = T, X(1) = 2, X(3) = 7.
C     IT IS REWRITTEN AS TWO FIRST ORDER EQUAIONS:
C          DX1/DT = X2,               X1(1) = 2,
C          DX2/DT = T - X1 - 3(X1)(X2),   X2(1) = ?
C          WHERE X1 = X AND X2 = DX/DT.
C     THE PROGRAM INTERPOLATES FROM THE INITIAL RESULTS TO FIND A
C     BETTER ESTIMATE OF THE UNKNOWN INITIAL VALUE OF
C     DX/DT = X2(1).   ITERATIONS ARE CONTINUED UNTIL SUCCESSIVE VALUES
C     OF X(3) EQUAL 7 WITHIN A SPECIFIED TOLERANCE.
C     IN THE PROGRAM, VARIABLES X1 AND X2 ARE THE FIRST AND SECOND COMPONENTS
C     OF THE VECTOR XO
      DIMENSION XO(2), XEND(2), XWRK(4,2), F(2)
      EXTERNAL DERIVS
C     INITIALIZE THE VARIABLES WITH A DATA STATEMENT
      DATA H, N, G1, G2, TOL /0.1, 2, -1., 1., 0.001 /
      DATA TSTART, XSTART, D /  1.0, 2.0, 7.0 /
C     DO THE INTEGRATION ONCE.
      XO(1) = XSTART
      TO = TSTART
      XO(2) = G1
      DO 10 I = 1,20
        CALL RKSYST (DERIVS, TO, H, XO, XEND, XWRK, F, N)
C     STEP UP THE VARIABLES FOR THE NEXT INTERVAL.
        XO(1) = XEND(1)
        XO(2) = XEND(2)
        TO = TO + H
10    CONTINUE
C     SAVE THE FIRST RESULT FOR EXTRAPOLATING.
      R1 = XO(1)
C     DO THE INTEGRATION AGAIN WITH THE SECOND ESTIMATE OF DX/DT.
      TO = TSTART
      XO(1) = XSTART
      XO(2) = G2
      DO 20 I = 1,20
        CALL RKSYST (DERIVS, TO, H, XO, XEND, XWRK, F, N)
        XO(1) = XEND(1)
        XO (2) = XEND(2)
        TO = TO + H
20    CONTINUE
C     SAVE THE SECOND CALCULATED VALUE.
      R2 = XO(1)
C     NOW WE EXTRAPOLATE.  THEN REPEAT THE ABOVE CALCULATIONS
C     A MAXIMUM OF 20 TIMES, OR UNTIL WE MATCH DESIRED VALUE
C     OF X1(3).
      DO 40 ITER = 1,20
        IF (ABS(R2 - D) .LE. TOL)  GO TO 99
        TO = TSTART
        XO(1) = XSTART
        XO(2) = G1 + (G2 - G1)/(R2 - R1)*(D - R1)
        G1 = G2
        G2 = XO(2)
        R1 = R2
        DO 30 I = 1,20
          CALL RKSYST (DERIVS, TO, H, XO, XEND, XWRK, F, N)
```

Figure 6.3 Program 1

```
                XO(1) = XEND(1)
                XO(2) = XEND(2)
                TO = TO + H
30        CONTINUE
          R2 = XO(1)
40        CONTINUE
C   WHEN WE HAVE A NORMAL TERMINATION OF THE LOOP, WE DIDN'T CONVERGE.
C   PRINT A MESSAGE, THEN THE FINAL VALUE.
          WRITE (6,200)
200       FORMAT (1HO, 91H WE DID NOT MEET TOLERANCE CRITERION IN 20 ITERATI
         +ONS. FINAL VALUE AT END OF INTERVAL WAS )
          GO TO 100
C   WE DID CONVERGE IF WE COME HERE.  PRINT MESSAGE AND VALUE.
99        WRITE (6,201) ITER
201       FORMAT(1HO, 28H TOLERANCE CRITERION MET IN ,I3,
         +    49H ITERATIONS.  FINAL VALUE AT END OF INTERVAL WAS )
100       WRITE (6,202) R2, G2
202       FORMAT (1HO, 20X, F7.4,
         +    30H USING INITIAL SLOPE VALUE OF ,F7.4)
C   NOW RECOMPUTE AND WRITE OUT THE LAST SET OF COMPUTATIONS.
          TO = TSTART
          XO(1) = XSTART
          XO(2) = G2
          WRITE (6,203) TO, XO
203       FORMAT (1HO, 23H LAST COMPUTATIONS WERE /
         +    1HO, 5X, 4HTIME, 5X, 8HX1 VALUE, 5X, 8HX2 VALUE /
         +    1HO, 6X, F3.1, 6X, F7.4, 6X, F7.4)
          DO 50 I = 1,20
             CALL RKSYST (DERIVS, TO, H, XO, XEND, XWRK, F, N)
             XO(1) = XEND(1)
             XO(2) = XEND(2)
          TO = TO + H
          WRITE (6,204) TO, XO
204          FORMAT (1H , 6X, F3.1, 6X, F7.4, 6X, F7.4)
50        CONTINUE
          STOP
          END


          SUBROUTINE DERIVS (X, T, F, N)
C   THIS SUBROUTINE COMPUTES DERIVATIVE VALUES FOR THE RK ROUTINE.
          DIMENSION X(N), F(N)
          F(1) = X(2)
          F(2) = T - X(1) - 3.*X(1)*X(2)
          RETURN
          END
```

Figure 6.3 (continued)

case the boundary-value problem is nonlinear, some iterative method normally would be required.

A subroutine POWER, which finds the largest eigenvalue of an $N \times N$ matrix (together with its corresponding eigenvector) is given in Program 2 (Fig. 6.4). To use it, the matrix and an initial estimate of the eigenvector are passed as parameters. When no estimate of the eigenvector is known, one usually sends a vector which has each element equal to unity, although there are some matrices for which this will not produce convergence. In such cases, one should call the subroutine with a different starting vector, although it must be remembered that the power method fails for some matrices.

In subroutine **POWER**, the starting vector is multiplied successively by the matrix in each iteration, with the product of each multiplication being normalized. When the method converges, the normalized product vectors approach the eigenvector and the normalization factors approach the eigenvalue. The subroutine terminates iterations when successive values of the normalization factor differ by less than a stated tolerance value. Normalization is done by dividing each element of the vector by the largest element, so the normalized vectors always have unity as their largest element.

```
PROGRAM 1 OUTPUT

TOLERANCE CRITERION MET IN  6 ITERATIONS.   FINAL VALUE AT END OF INTERVAL WAS
                    7.0000 USING INITIAL SLOPE VALUE OF 76.9695

LAST COMPUTATIONS WERE
     TIME      X1 VALUE        X2 VALUE

      1.0       2.0000          76.9695
      1.1       6.6453          14.9032
      1.2       7.1277           5.8486
      1.3       7.2761           2.3699
      1.4       7.3204           0.8685
      1.5       7.3237           0.2186
      1.6       7.3101          −0.0579
      1.7       7.2895          −0.1726
      1.8       7.2663          −0.2178
      1.9       7.2422          −0.2338
      2.0       7.2180          −0.2374
      2.1       7.1941          −0.2360
      2.2       7.1706          −0.2324
      2.3       7.1475          −0.2280
      2.4       7.1249          −0.2232
      2.5       7.1028          −0.2183
      2.6       7.0813          −0.2133
      2.7       7.0602          −0.2082
      2.8       7.0396          −0.2032
      2.9       7.0196          −0.1980
      3.0       7.0000          −0.1929
```

Figure 6.3 (concluded)

```
      SUBROUTINE POWER (A, X, C, TOL, NLIMIT, N, NDIM, XWRK)
C THIS SUBROUTINE COMPUTES THE LARGEST EIGENVALUE AND ITS CORRESPONDING
C EIGENVECTOR BY THE POWER METHOD.
C PARAMETERS ARE -
C    A     AN N X N MATRIX WHOSE EIGENS ARE BEING DETERMINED
C    X     ESTIMATE FOR THE EIGENVECTOR USED TO BEGIN ITERATIONS.
C          IF NO APPROXIMATION TO THIS IS KNOWN, THE USUSAL CHOICE
C          IS A VECTOR WIHT ALL COMPONENTS EQUAL TO UNITY.
C          X ALSO RETURNS THE FINAL EIGENVECTOR TO THE CALLER.
C    C     RETURNS THE VALUE OF THE EIGENVALUE.
C    TOL   TOLERANCE VALUE USED TO DETERMINE CONVERGENCE.  ITERATIONS
C          CONTINUE UNTIL SUCCESSIVE ESTIMATES OF THE EIGENVALUE
C          ARE THE SAME WITHIN TOL IN VALUE.
C    NLIMIT LIMIT TO THE NUMBER OF ITERATIONS IF NOT CONVERGENT.
C    N     SIZE OF MATRIX AND THE VECTOR X.
C    NDIM  FIRST DIMENSION OF MATRIX A IN THE CALLING PROGRAM.
C    XWRK  VECTOR USED TO STORE INTERMEDIATE VALUES.  MUST BE
C          DIMENSIONED TO HOLD AT LEAST N ELEMENTS IN THE MAIN
C          PROGRAM.
      DIMENSION A(NDIM, N), X(N), XWRK(N)
C BEGIN THE ITERATIONS.  GET THE PRODUCT OF A AND X, THEN
C NORMALIZE.  THE NORMALIZATION FACTORS SHOULD CONVERGE TO THE
C EIGENVALUE, C, ON REPEATED MULTIPLICATIONS.  WE STORE THE
C CURRENT VALUE OF C FOR COMPARISON WITH THE NEXT VALUE TO TEST CONVERGENCE.
C TO BEGIN, WE MAKE SAVE = 09
      SAVE = 0.
      DO 50 ITER = 1,NLIMIT
        DO 20 IROW = 1,N
          XWRK(IROW) = 0.0
          DO 10 JCOL = 1,N
            XWRK(IROW) = XWRK(IROW) + A(IROW,JCOL)*X(JCOL)
10        CONTINUE
20      CONTINUE
C FIND THE LARGEST ELEMENT OF THE PRODUCT VECTOR FOR NORMALIZING.
        C = 0.0
        DO 30 IROW = 1,N
          IF (ABS(C) .LT. ABS(XWRK(IROW)))  C = XWRK(IROW)
30      CONTINUE
C NOW NORMALIZE THE PRODUCT VECTOR AND PUT INTO X FOR
C THE NEXT ITERATION.
        DO 40 IROW = 1,N
          X(IROW) = XWRK(IROW)/C
40      CONTINUE
C SEE IF TOLERANCE IS MET.  IF SO, WE ARE DONE.
C IF NOT, WE CONTINUE THE ITERATIONS.
        IF (ABS(C - SAVE) .LE. TOL)  RETURN
        SAVE = C
50    CONTINUE
C IF WE DO NOT MEET THE TOLERANCE FOR CONVERGENCE, PRINT
C A MESSAGE AND RETURN LAST VLAUES CALCULATED.
      WRITE (6,200) NLIMIT
200   FORMAT (1H0, 28H CONVERGENCE NOT REACHED IN , I5,
     +  11H ITERATIONS )
      RETURN
      END
```

Figure 6.4 Program 2

EXERCISES

Section 1

▶1. Solve the boundary-value problem

$$y'' + xy' - 3y = 4.2x,$$
$$y(0) = 0, \; y(1) = 1.9$$

by the shooting method, assuming two values of the initial slope, which is near unity. Use $h = 0.25$ and use the modified Euler method, carrying three decimals. Compare to the analytical solution $y = x^3 + 0.9x$.

2. Show that the solution to Exercise 1 found by integrating with an initial slope interpolated from the results of the first two trials can be obtained by interpolating from the trial solutions themselves.

3. In Exercise 1, the exact (analytical) value of the initial slope is 0.9. Using $h = 0.25$, you probably find that $y(1) = 1.9$ is achieved with a different value of $y'(0)$. Explain this discrepancy.

4. The shooting method can work backwards through the interval as well as forward. Solve Exercise 1 by moving backward from $x = 1$ (Δx is then negative), and use an assumed value for the slope at $x = 1$.

▶5. Use the shooting method to solve

$$y'' - yy' = e^x,$$
$$y(0) = 1, \; y(1) = -1.$$

Use the computer programs of Chapter 5, or another one that you have written, to solve the problem with assumed values of the initial slope. Use $h = 0.2$.

6. Exercise 5 solves a nonlinear boundary-value problem. The solution will not be exact because the solution of a differential equation by a numerical method has errors that depend on the step size. If the step size is reduced, the errors will decrease. Vary the step size in Exercise 5 until you believe that the value of y at $x = 0.5$ is accurate to within 5×10^{-5}. Justify your belief that you have attained this accuracy.

7. Use the shooting method to solve

$$\frac{d^3 y}{dt^3} + 10 \frac{dy}{dt} - 5y^3 = t^2 - ty,$$
$$y(0) = 0, \qquad y(2) = -1, \qquad y''(2) = 0.$$

Section 2

8. Given the boundary problem

$$\frac{d^2 y}{d\theta^2} + y = 0, \qquad y(0) = 0, \qquad y\left(\frac{\pi}{2}\right) = 1.$$

a) Normalize to the interval $[0, 1]$ by an appropriate change of variable.

b) Solve the normalized problem by the finite-difference method, replacing the derivative by a difference quotient of error $O(h^2)$, and then solving the set of equations. Use $h = 0.25$.

▶ c) Solve the original equation (without normalizing) by finite difference. Take $\Delta\theta = \pi/8$.

d) Compare your solutions to the analytical solution $y = \sin\theta$.

e) Repeat (b) with $h = 0.5$, and extrapolate the value at the midpoint.

9. Solve the boundary value problem

$$y'' + xy' - xy = 2x, \qquad y(0) = 1, \qquad y(1) = 0.$$

Take $h = 0.2$.

10. Solve Exercise 1 by replacing the derivatives with finite-difference quotients.

11. Solve Exercise 5 by rewriting the differential equation as a set of difference equations.

12. Solve Exercise 7 through a set of difference equations.

Section 3

▶ 13. Repeat Exercise 8, except for the boundary-value problem with derivative conditions:

$$\frac{d^2 y}{d\theta^2} + y = 0, \qquad y'(0) = 0, \qquad y'(\pi/2) = 1.$$

In part (d), compare to $y = -\cos\theta$.

14. Solve

$$\frac{d^2 y}{d\theta^2} + y = 0, \qquad y'(0) + y(0) = 1, \qquad y'\left(\frac{\pi}{2}\right) + y\left(\frac{\pi}{2}\right) = 0.$$

Subdivide the interval into four equal parts.

15. The most general form of conditions that can be specified for a second-order boundary-value problem is a linear combination of the function and its derivative at both ends of the interval. Set up the equations to solve

$$y'' - xy' + x^2 y = x^3, \qquad \begin{cases} y(0) + y'(0) - y(1) - y'(1) = 1, \\ y(0) + y'(0) + y(1) + y'(1) = 2. \end{cases}$$

Use $h = \frac{1}{3}$.

16. Solve the third-order boundary-value problem

$$y''' - y' = e^x, \qquad y(0) = 0, \qquad y(1) = 1, \qquad y'(1) = 0.$$

Use $h = 0.2$. Approximate the third derivative in terms of the average third central difference

$$y_0''' = \frac{f_2 - 2f_1 + 2f_{-1} - f_{-2}}{2h^3} + O(h^2)$$

at $x = 0.4$, 0.6, 0.8. Using the derivative condition will eliminate the assumed function value at $x = 1.2$, but we are short one equation. Obtain this by writing the equation at $x = 0.2$ using the unsymmetrical approximation for y''':

$$y_0''' = \frac{-f_3 + 6f_2 - 12f_1 + 10f_0 - 3f_{-1}}{2h^3} + O(h^2).$$

▶17. Solve the fourth-order problem

$$y^{iv} - y''' + y = x^2, \qquad y(0) = 0, \qquad y'(0) = 0, \qquad y(1) = 2, \qquad y'(1) = 0.$$

Use symmetrical expressions for the derivatives.

18. Solve the differential equation in Exercise 17 subject to the conditions

$$y(0) = y'(0) = y''(0) = 0, \qquad y(1) = 2.$$

Unsymmetrical expressions may not be required!

19. Solve the nonlinear problem

$$y'' = 2 - \frac{4y^2}{\sin^2 x}, \qquad y'(1) = 0.9093, \qquad y(2) = 0.8268,$$

by a set of difference equations. Compare to the analytical solution, $y = \sin^2 x$.

Section 4

20. Consider the characteristic-value problem with k restricted to real values:

$$\frac{d^2 y}{dx^2} - k^2 y = 0, \qquad y(0) = 0, \qquad y(1) = 0.$$

a) Show analytically that there is no solution except the trivial solution $y = 0$.

b) Show, by setting up the set of difference equations corresponding to the differential equation with $h = 0.2$, that there are no real values of k for which a solution to the difference equations exists.

▶21. Consider the characteristic-value problem

$$y'' + 2y' + k^2 y = 0, \qquad y(0) = y(1) = 0.$$

Find the principal eigenvalue and compare to $k = \pm\sqrt{1 + \pi^2} = \pm 3.297$.

a) Use $h = \frac{1}{2}$.

b) Use $h = \frac{1}{3}$.

c) Use $h = \frac{1}{4}$.

d) Assuming errors are proportional to h^2, extrapolate from parts (a) and (c), then from parts (b) and (c), to get improved values for the principal eigenvalue.

22. Find the principal eigenvalue of

$$y'' + k^2 x^2 y = 0, \qquad y(0) = y(1) = 0.$$

23. Using the principal value, $k = 3.297$, in Exercise 21, find y as a function of x over the interval $[0, 1]$. This function is the corresponding eigenfunction.

▶24. The second eigenvalue in Exercise 21 is $k = \sqrt{1 + 4\pi^2} = 6.3623$. Find the corresponding eigenfunction.

Section 5

25. Find the larger eigenvalue and the corresponding eigenvector by the power method:

a) $\begin{bmatrix} 1 & 0 \\ 1 & 1 \end{bmatrix}$
b) $\begin{bmatrix} 1 & 2 \\ 3 & 4 \end{bmatrix}$
c) $\begin{bmatrix} 1 & 1 \\ 1 & -1 \end{bmatrix}$

(In part (c) the two eigenvalues are equal but of opposite sign.)

▶26. Find the dominant eigenvalue and its corresponding eigenvector:

$$\begin{bmatrix} 1 & 1 & 2 \\ 0 & 1 & 3 \\ 1 & 1 & 1 \end{bmatrix}.$$

27. Use iteration to find all the eigenvalues of the matrices for Exercise 21 (a), (b), (c).

28. Invert the matrices in Exercises 25 and 26; then use the power method to get the smallest eigenvalues of the original matrices. Repeat, except avoid the inversion by using the LU decomposition of the matrix. Compare the effort in the two cases.

▶29. Find all the eigenvalues of the matrix in Exercise 26. Then invert the matrix and again find the eigenvalues. Show that the two sets of eigenvalues are reciprocals of each other.

30. After finding the dominant eigenvalue in Exercise 26, subtract that value from each of the diagonal elements and use the power method. Compare the value obtained with the second largest eigenvalue as determined in Exercise 29.

APPLIED PROBLEMS AND PROJECTS

31. If a cantilever beam of length L, which bends due to a uniform load of w lb/ft, is also subject to an axial force P at its free end (see Fig. 6.5), the equation of its elastic curve is

$$EI\frac{dy^2}{dx^2} = Py - \frac{wx^2}{2}.$$

For this equation, the origin O has been taken at the free end. At the point $x = L$, $dy/dx = 0$; at $(0, 0)$, $y = 0$. Solve this boundary-value problem by the shooting

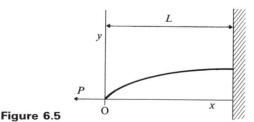

Figure 6.5

method for a $2'' \times 4'' \times 10$-ft wooden beam for which $E = 12 \times 10^5$ lb/in^2. Find y versus x when the beam has the $4''$ dimension vertical with $w = 25$ lb/ft and a tension force of $P = 500$ lb. Also solve for the deflections if the beam is turned so that the $4''$ dimension is horizontal.

32. Solve Problem 31 by replacing the differential equation by difference equations. Compare the solutions by each method.

33. In Problem 31, y'' is used although the radius of curvature should be employed in the equation:

$$EI\frac{y''}{[1 + (y')^2]^{3/2}} = Py - \frac{wx^2}{2}.$$

If the deflection of the beam is small, the difference is negligible, but in the second part of Problem 31 at least, this is not true. Furthermore, if there is considerable bending of the beam, the horizontal distance from the origin to the wall is less than L, the original length of the beam. Solve Problem 31 taking these factors into account, and determine by how much the deflections differ from those previously calculated.

34. A cylindrical pipe has a hot fluid flowing through it. Since the pressure is very high, the walls of the pipe are thick. For such a situation, the differential equation that relates temperatures in the metal wall to radial distance is:

$$r\frac{d^2u}{dr^2} + \frac{du}{dr} = 0,$$

where

r = radial distance from the centerline,

u = temperature.

Solve for the temperatures within a pipe whose inner radius is 1 cm and whose outer radius is 2 cm if the fluid is at 540°C and the temperature of the outer circumference is 20°C.

35. The pipe in Problem 34 is insulated to reduce the heat loss. The insulation used has properties such that the gradient du/dr at the outer circumference is proportional to the difference in temperatures from the outer wall to the surroundings:

$$\frac{du}{dr}\bigg|_{r=2} = 0.083[u(2) - 20].$$

Solve Problem 34 with this boundary condition.

36. When air flows at a velocity of u_∞ past a flat plate placed horizontally to the flow, experiments show that the air velocity at the surface of the plate is zero and that a boundary layer within which the velocities are less than u_∞ builds up as flow progresses along the plate (Fig. 6.6). This is a typical problem of interest to aeronautical engineers. To solve for the velocities within the boundary layer, a stream function f is defined for which

$$\frac{d^3f}{dz^3} + \frac{f}{2}\frac{d^2f}{dz^2} = 0,$$

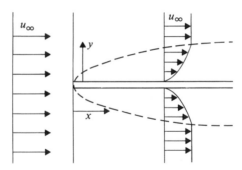

Figure 6.6

where z is a dimensionless distance involving x, y, u_∞, and the viscosity of air. This is a boundary-value problem because $f = 0$ and $f' = 0$ at $z = 0$, while $f' = 1$ at $z = \infty$.

In solving this problem, one does not use $z = \infty$ as a boundary; rather, the integration is taken to a value of z large enough so that there is an insignificant change when it is increased. Solve this problem to find f and df/dz as functions of z.

37. A simple spring–mass system obeys the equation

$$d^2y/dt^2 + \alpha^2 y = 0,$$

where the positive constant α^2 equals k/m, the ratio of the spring constant to the mass. If it is known that $y = 0$ at $t = 0$ and again at $t = 1.26$ sec (so that its period is 2.52 sec), this is a characteristic-value problem that has a solution for only certain values of α. These characteristic values are discussed in Section 6.4 for this equation. Suppose, however, that the spring is not an ideal one with constant k, but that the spring force and elongation vary with y according to the equation $k = b(1 - y^{0.1})$. If it is still true that $y = 0$ at $t = 0$ and also at $t = 1.26$, is this a characteristic-value problem requiring that the ratio k/m be at certain fixed values in order to have a nontrivial solution? If it is, find these values.

38. A shaft rotates in fixed bearings (Fig. 6.7) so that at these points $y = 0$ and $dy/dx = 0$. The governing equation is

$$EI\frac{d^4y}{dx^4} = \frac{w}{g}\omega^2 y,$$

where E is Young's modulus, I is the moment of inertia of the cross-sectional area of the shaft, y is the displacement from the horizontal, x is the distance measured from the point midway between the bearings, w is the weight of the shaft per unit length, g is the acceleration of gravity, $2L$ is the length of the shaft between the bearings, and ω

Figure 6.7

is the rotational speed. At certain critical speeds, the shaft will distort from a straight line. These critical speeds are very important to know in designing the machine of which the shaft is a part. For parameter values given below in consistent units, find the first two values of critical rotational speeds. Also find the curve, $y(x)$, at these speeds.

$$E = 2 \times 10^6 \, \text{lb/in}^2 \qquad g = 386.4 \ \text{in.} \, \text{sec}^2$$

$$I = 0.46 \, \text{in}^4 \qquad\qquad L = 16 \, \text{in.}$$

$$w = 0.025 \, \text{lb/in.}$$

Chapter Seven

Numerical Solution of Elliptic Partial-Differential Equations

7.1 EQUILIBRIUM TEMPERATURES IN A HEATED SLAB

Many important scientific and engineering problems fall into the field of partial differential equations. Here is a situation that is typical, in which the temperatures are a function of the coordinates of position of the point in question.

A piece of metal is 12″ × 3″ × 6′. Three feet of the slab is kept inside a furnace but half of the slab protrudes (see Fig. 7.1). In order to decrease heat losses to the air, the protruding half is covered with one-inch thickness of insulation. If the furnace is maintained at 950°F, will all points of the metal reach a temperature of 800°F or higher, in spite of heat loss through the insulation? Such a question might arise in heat-treating the slab when the only furnace available to heat the metal is too small to contain the whole slab. The sketch in Fig. 7.1 will explain the problem.

You will recognize that this is a boundary-value problem: The temperature is known for the part of the slab inside the furnace, and something is known about the temperature gradients for the insulated part. The problem differs from those discussed in the previous chapter in that more than one independent variable is involved in specifying the coordinates of points within the slab. (Specifically, these variables are the x-, y-, and z-coordinates.)

This chapter presents methods to solve such problems.

Many physical phenomena are a function of more than one independent variable and must be represented by a partial differential equation, by which term we mean an equation involving partial derivatives. Most scientific problems have mathmatical models that are second-order equations, with the highest order of derivative being the second. If u is a function of the two independent variables x and y, there are three second-order partial derivatives:

$$\frac{\partial^2 u}{\partial x^2}, \qquad \frac{\partial^2 u}{\partial x\, \partial y}, \qquad \frac{\partial^2 u}{\partial y^2}.$$

Portion of slab inside furnace

12″

3″

3′

6′

Sheets of insulation around
the metal slab, 1″ thick

Figure 7.1

(We shall treat only of functions for which the order of differentiation is unimportant, so $\partial^2 u/\partial y\,\partial x = \partial^2 u/\partial x\,\partial y$.) Depending on the values of the coefficients of the second-derivative terms, partial differential equations are classified as elliptic, parabolic, or hyperbolic. The most important distinctions to us are the kinds of problems and the nature of boundary conditions which lead to one type of equation or another. Steady-state potential distribution problems fall in the class of elliptic equations, which we discuss first. The equilibrium temperatures attained by the metal piece protruding from the furnace is such a potential distribution problem.

7.2 EQUATION FOR STEADY-STATE HEAT FLOW

We derive the relationship for temperature, u, as a function of the two space variables, x and y, for the equilibrium temperature distribution on a flat plate (Fig. 7.2). Consider the element of the plate whose surface area is $dx\,dy$. We assume that heat flows only in the x- and y-directions and not in the perpendicular direction. If the plate is very thin, or if the upper and lower surfaces are both

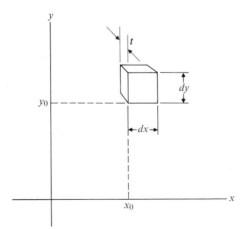

Figure 7.2

well insulated, the physical situation will agree with our assumption. Let t be the thickness of the plate.

Heat flows at a rate proportional to the cross-sectional area, to the temperature gradient ($\partial u/\partial x$ or $\partial u/\partial y$), and to the thermal conductivity k, which we shall assume constant at all points. The flow of heat is from high to low temperature, of course, meaning opposite to the direction of increasing gradient. We use a minus sign in the equation to account for this:

Rate of heat flow into element at $x = x_0$, in the x-direction:

$$-k(t\,dy)\frac{\partial u}{\partial x}.$$

The gradient at $x_0 + dx$ is the gradient at x_0 plus the increment in gradient over the distance dx:

Gradient at $x_0 + dx$:

$$\frac{\partial u}{\partial x} + \frac{\partial}{\partial x}\left(\frac{\partial u}{\partial x}\right)dx,$$

Rate of heat flow out of element at $x = x_0 + dx$:

$$-k(t\,dy)\left[\frac{\partial u}{\partial x} + \frac{\partial}{\partial x}\left(\frac{\partial u}{\partial x}\right)dx\right].$$

Net rate of heat flow into element in x-direction:

$$-k(t\,dy)\left[\frac{\partial u}{\partial x} - \left(\frac{\partial u}{\partial x} + \frac{\partial}{\partial x}\left(\frac{\partial u}{\partial x}\right)dx\right)\right] = kt\,dx\,dy\,\frac{\partial^2 u}{\partial x^2}.$$

Similarly, in the y-direction we have the following.

Net rate of heat flow into element in y-direction:

$$-k(t\,dx)\left[\frac{\partial u}{\partial y} - \left(\frac{\partial u}{\partial y} + \frac{\partial}{\partial y}\left(\frac{\partial u}{\partial y}\right)dy\right)\right] = kt\,dx\,dy\,\frac{\partial^2 u}{\partial y^2}.$$

The total heat flowing into the elemental volume by conduction is the sum of these net flows in the x- and y-directions. If there were heat generated within the element, this would be added to the heat entering by conduction. The sum must be equal to the rate of heat lost by other mechanisms, or else there will be a build-up of heat within the element, causing its temperature to increase with time. In this chapter we will consider only the case where the temperatures do not change with time, the *steady-state* case.

If there is equilibrium as to temperature distribution (i.e., steady state), the total rate of heat flow into the element plus heat generated must be zero. Hence,

$$kt(dx)(dy)\left(\frac{\partial^2 u}{\partial x^2} + \frac{\partial^2 u}{\partial y^2}\right) + Qt(dx)(dy) = 0,$$

where Q is the rate of heat generation per unit volume. (Obviously, if there is heat generation, there must be heat flow *from* the element by conduction that just balances it.) Q will often be a function of x and y.

If $Q = 0$, we have

$$kt \, dx \, dy\left(\frac{\partial^2 u}{\partial x^2} + \frac{\partial^2 u}{\partial y^2}\right) = 0,$$

$$\frac{\partial^2 u}{\partial x^2} + \frac{\partial^2 u}{\partial y^2} = \nabla^2 u = 0.$$

(7.1)

The operator

$$\nabla^2 = \left(\frac{\partial^2}{\partial x^2} + \frac{\partial^2}{\partial y^2}\right)$$

is called the *Laplacian*, and Eq. (7.1) is called *Laplace's equation*. For three-dimensional heat flow problems, we would have, analogously,

$$\nabla^2 u = \left(\frac{\partial^2}{\partial x^2} + \frac{\partial^2}{\partial y^2} + \frac{\partial^2}{\partial z^2}\right)u = 0.$$

(7.2)

Equation (7.2), which has been derived with reference to heat flow, applies as well to steady-state diffusion problems (where u is now concentration of material) and to electrical-potential distribution (where u is electromotive force); in fact, Laplace's equation holds for the steady-state distribution of the potential of any quantity where the rate of flow is proportional to the gradient, and where the proportionality constant does not vary with position or the value of u. In our examples, we shall generally use terminology corresponding to heat flow as being more closely related to the average student's everyday experience.

With Laplace's equation, we assume that no heat is being generated at points in the plate, as by electric heaters imbedded in it, nor does any removal take place, as by cooling coils or other means. In the presence of such "sources" or "sinks," we would not equate the net flow into the element to zero, as in Eq. (7.1), but the net flow into the element of volume would equal the net rate of heat removal from the element, if steady state applies. Assuming this removal rate to be a function of the location of the element in the x,y-plane, $f(x, y)$, we would have, with Q equal to the rate of heat generation per unit area,

$$\nabla^2 u = -\frac{1}{k} Q(x, y) = f(x, y).$$

(7.3)

This equation is called *Poisson's equation*. Our numerical methods of solving elliptic differential equations apply equally well to both Laplace's and Poisson's equations. Analytical methods find the latter type considerably more difficult, or when the boundary conditions are complex for Laplace's equation. These two equations include most of the physical applications of elliptic partial differential equations.

The distinction as to elliptic, parabolic, or hyperbolic for second-order partial-differential equations depends on the coefficients of the second-derivative terms. We can write any such second-order equations (in two independent variables) as:

$$A\frac{\partial^2 u}{\partial x^2} + B\frac{\partial^2 u}{\partial x\,\partial y} + C\frac{\partial^2 u}{\partial y^2} + D\left(x, y, u, \frac{\partial u}{\partial x}, \frac{\partial u}{\partial y}\right) = 0.$$

Depending on the value of $B^2 - 4AC$, we classify the equation as

Elliptic, if $B^2 - 4AC < 0$;

Parabolic, if $B^2 - 4AC = 0$;

Hyperbolic, if $B^2 - 4AC > 0$.

If the coefficients A, B, and C are functions of x, y, and/or u, the equation may change from one classification to another at various points in the domain.

For Laplace's and Poisson's equations, $B = 0$, $A = C = 1$, so these are always elliptic. We study the other types in later chapters.

Equation (7.2) or Eq. (7.3) describes how u varies within the interior of a closed region. The function u is determined on the boundaries of the region by boundary conditions. The boundary conditions may specify the value of u at all points on the boundary, or some combination of the potential and the normal derivative.

7.3 REPRESENTATION AS A DIFFERENCE EQUATION

Our scheme for solving all kinds of partial differential equations will be to replace the derivatives by difference quotients, converting the equation to a difference equation.* We then write the difference equation corresponding to each point at the intersections (nodes) of a gridwork that subdivides the region of interest at which the function values are unknown. Solving these equations simultaneously gives values for the function at each node which approximate the true values. We begin with the two-dimensional case.

Approximating derivatives by difference quotients was the subject matter of Chapter 4. They may also be derived by the method of undertermined coefficients (Appendix B). We rederive the few relations we need independently.

Let $h = \Delta x$ = equal spacing of gridwork in the x-direction (see Fig. 7.3). We assume that the function $f(x)$ has a continuous fourth derivative. By Taylor

---

* This is exactly the same as the method of solving a boundary-value problem through replacing it by difference equations, as discussed in Chapter 6.

series,

$$f(x_n + h) = f(x_n) + f'(x_n)h + \frac{f''(x_n)}{2}h^2 + \frac{f'''(x_n)}{6}h^3 + \frac{f^{iv}(\xi_1)}{24}h^4,$$

$$x_n < \xi_1 < x_n + h,$$

$$f(x_n - h) = f(x_n) - f'(x_n)h + \frac{f''(x_n)}{2}h^2 - \frac{f'''(x_n)}{6}h^3 + \frac{f^{iv}(\xi_2)}{24}h^4,$$

$$x_n - h < \xi_2 < x_n.$$

It follows that

$$\frac{f(x_n + h) - 2f(x_n) + f(x_n - h)}{h^2} = f''(x_n) + \frac{f^{iv}(\xi)}{12}h^2, \qquad x_n - h < \xi < x_n + h.$$

A subscript notation is convenient:

$$\frac{f_{n+1} - 2f_n + f_{n-1}}{h^2} = f''_n + O(h^2). \tag{7.4}$$

In Eq. (7.4) the subscripts on f indicate the x-values at which it is evaluated. The order relation $O(h^2)$ signifies that the error approaches proportionality to h^2 as $h \to 0$.

Similarly, the first derivative is approximated,

$$\frac{f(x_n + h) - f(x_n - h)}{2h} = f'(x_n) + \frac{f'''(\xi)}{6}h^2, \qquad x_n - h < \xi < x_n + h,$$

$$\frac{f_{n+1} - f_{n-1}}{2h} = f'_n + O(h^2). \tag{7.5}$$

(The first derivative could also be approximated by the forward or backward difference, but this would have an error of $O(h)$. We prefer the more accurate central-difference approximation.)

When f is a function of both x and y, we get the second partial derivative with respect to x, $\partial^2 u/\partial x^2$, by holding y constant and evaluating the function at three points where x equals x_n, $x_n + h$, and $x_n - h$. The partial derivative $\partial^2 u/\partial y^2$ is similarly computed, holding x constant. We require that fourth derivatives with respect to both variables exist.

To solve the Laplace equation on a region in the xy-plane, we subdivide the region with equispaced lines parallel to the x- and y-axes. Consider a portion of the region near (x_i, y_j). (see Fig. 7.3). We wish to approximate

$$\nabla^2 u = \frac{\partial^2 u}{\partial x^2} + \frac{\partial^2 u}{\partial y^2} = 0.$$

Replacing the derivatives by difference quotients which approximate the deriva-

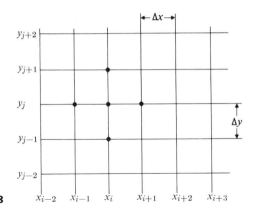

Figure 7.3

tives at the point (x_i, y_j), we get

$$\nabla^2 u(x_i, y_j) = \frac{u(x_{i+1}, y_j) - 2u(x_i, y_j) + u(x_{i-1}, y_i)}{(\Delta x)^2}$$

$$+ \frac{u(x_i, y_{j+1}) - 2u(x_i, y_j) + u(x_i, y_{j-1})}{(\Delta y)^2} = 0.$$

It is convenient to let double subscripts on u indicate the x- and y-values:

$$\nabla^2 u_{i,j} \doteq \frac{u_{i+1,j} - 2u_{i,j} + u_{i-1,j}}{(\Delta x)^2} + \frac{u_{i,j+1} - 2u_{i,j} + u_{i,j-1}}{(\Delta y)^2} = 0.$$

It is common to take $\Delta x = \Delta y = h$, resulting in considerable simplification, whence

$$\nabla^2 u_{i,j} = \frac{1}{h^2} [u_{i+1,j} + u_{i-1,j} + u_{i,j+1} + u_{i,j-1} - 4u_{i,j}] = 0. \qquad (7.6)$$

Note that five points are involved in the relationship of Eq. (7.6), points to the right, left, above, and below the central point (x_i, y_j). It is convenient to represent the relationship pictorially, where the linear combination of u's is represented symbolically. Equation (7.6) becomes:

$$\nabla^2 u_{i,j} = \frac{1}{h^2} \left\{ \begin{matrix} & 1 & \\ 1 & -4 & 1 \\ & 1 & \end{matrix} \right\} u_{i,j} = 0.$$

The representation of the Laplacian operator by the pictorial operator that represents the linear combination of u-values is fundamental to our work in elliptic partial-differential equations. The approximation has $O(h^2)$ error, provided that u is sufficiently smooth.

7.4 LAPLACE'S EQUATION ON A RECTANGULAR REGION

A typical steady-state heat-flow problem is the following: A thin steel plate is a 10 cm $\times$ 20 cm rectangle. If one of the 10-cm edges is held at 100°C and the other three edges are held at 0°C, what are the steady-state temperatures at interior points? For steel, $k = 0.16$ cal/sec $\cdot$ cm^2 $\cdot$ C/cm. We can state the problem mathematically in this way:

Find $u(x, y)$ such that

$$\frac{\partial^2 u}{\partial x^2} + \frac{\partial^2 u}{\partial y^2} = 0, \tag{7.7}$$

with

$$u(x, 0) = 0,$$
$$u(x, 10) = 0,$$
$$u(0, y) = 0,$$
$$u(20, y) = 100.$$

In this statement of the problem, we imagine one corner of the plate at the origin, with boundary conditions as sketched in Fig. 7.4.

A problem with temperatures known on each boundary is said to have *Dirichlet boundary conditions*. This problem can be solved analytically, giving $u(x, y)$ as an infinite Fourier series, but we will use it to illustrate the numerical technique. We replace the differential equation by a difference equation, as described in the previous section:

$$\frac{1}{h^2}[u_{i+1,j} + u_{i-1,j} + u_{i,j+1} + u_{i,j+1} - 4u_{ij}] = 0,$$

which relates the temperature at the point (x_i, y_j) to the temperatures at four neighboring points, each the distance h away from (x_i, y_j). An approximation to the solution of Eq. (7.7) results when we select a set of such points and find the solution to the set of difference equations that result. Suppose we choose $h = 5$ cm. (In normal practice we would choose a smaller mesh size.) Figure 7.5 shows the notation we use.

Figure 7.4

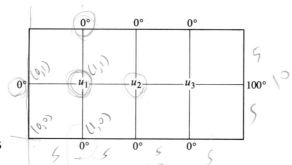

Figure 7.5

The system of equations is:

$$\frac{1}{h^2}\left(u_{(i+1)j} + u_{(i-1)j} + u_{i(j+1)} + u_{i(j+1)} - 4u_{ij}\right) = 0.$$

$$\frac{1}{5^2}(0 + 0 + u_2 + 0 - 4u_1) = 0,$$

$$\frac{1}{5^2}(u_1 + 0 + u_3 + 0 - 4u_2) = 0, \qquad (7.8)$$

$$\frac{1}{5^2}(u_2 + 0 + 100 + 0 - 4u_3) = 0.$$

The pictorial operator of Section 7.3,

$$\nabla^2 u_{ij} = \frac{1}{h^2}\left\{\begin{matrix} & 1 & \\ 1 & -4 & 1 \\ & 1 & \end{matrix}\right\} u_{ij} = 0,$$

makes the equations very easy to write. Observe that the physical properties of the material do not enter the relation,* and that the constant $1/h^2 = 1/5^2$ can be divided out of Eqs. (7.8).

The solution to the set of equations is easy when there are only three of them:

$$u_1 = 1.786, \qquad u_2 = 7.143, \qquad u_3 = 26.786.$$

Of course, with such a coarse grid as that shown in Fig. (7.5), we don't expect high accuracy. Redoing the problem with a smaller value for h should improve the precision. Equations (7.9) and Fig.7.6 result when we make $h = 2.5$ cm. (The equations are expressed in matrix form. We use v instead of u for the temperature variable in order to distinguish it from the previous solution.)

---

* If the conductivity varied at different regions of the plate, this would not be true.

$$
\begin{bmatrix}
-4 & 1 & & 1 & & & & & & & & & & & & & & & & & \\
1 & -4 & 1 & & 1 & & & & & & & & & & & & & & & & \\
 & 1 & -4 & & & 1 & & & & & & & & & & & & & & & \\
1 & & & -4 & 1 & & 1 & & & & & & & & & & & & & & \\
 & 1 & & 1 & -4 & 1 & & 1 & & & & & & & & & & & & & \\
 & & 1 & & 1 & -4 & & & 1 & & & & & & & & & & & & \\
 & & & 1 & & & -4 & 1 & & 1 & & & & & & & & & & & \\
 & & & & 1 & & 1 & -4 & 1 & & 1 & & & & & & & & & & \\
 & & & & & 1 & & 1 & -4 & & & 1 & & & & & & & & & \\
 & & & & & & 1 & & & -4 & 1 & & 1 & & & & & & & & \\
 & & & & & & & 1 & & 1 & -4 & 1 & & 1 & & & & & & & \\
 & & & & & & & & 1 & & 1 & -4 & & & 1 & & & & & & \\
 & & & & & & & & & 1 & & & -4 & 1 & & 1 & & & & & \\
 & & & & & & & & & & 1 & & 1 & -4 & 1 & & 1 & & & & \\
 & & & & & & & & & & & 1 & & 1 & -4 & & & 1 & & & \\
 & & & & & & & & & & & & 1 & & & -4 & 1 & & 1 & & \\
 & & & & & & & & & & & & & 1 & & 1 & -4 & 1 & & 1 & \\
 & & & & & & & & & & & & & & 1 & & 1 & -4 & & & 1 \\
 & & & & & & & & & & & & & & & 1 & & & -4 & 1 & \\
 & & & & & & & & & & & & & & & & 1 & & 1 & -4 & 1 \\
 & & & & & & & & & & & & & & & & & 1 & & 1 & -4
\end{bmatrix}
\begin{bmatrix}
v_1 \\ v_2 \\ v_3 \\ v_4 \\ v_5 \\ v_6 \\ v_7 \\ v_8 \\ v_9 \\ v_{10} \\ v_{11} \\ v_{12} \\ v_{13} \\ v_{14} \\ v_{15} \\ v_{16} \\ v_{17} \\ v_{18} \\ v_{19} \\ v_{20} \\ v_{21}
\end{bmatrix}
=
\begin{bmatrix}
0 \\ 0 \\ 0 \\ 0 \\ 0 \\ 0 \\ -100 \\ 0 \\ 0 \\ 0 \\ 0 \\ 0 \\ -100 \\ 0 \\ 0 \\ 0 \\ 0 \\ 0 \\ 0 \\ 0 \\ -100
\end{bmatrix}
\tag{7.9}
$$

$$21 \times 21 \qquad 21 \times 1 \qquad 21 \times 1$$

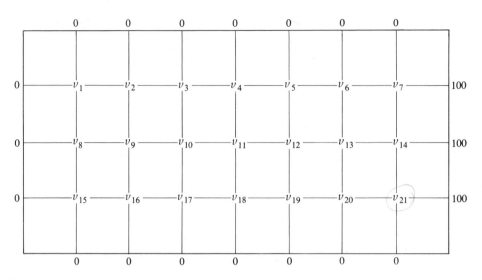

Figure 7.6

We can solve Eqs. (7.9) by elimination. It turns out that the largest element always occurs on the diagonal, so pivoting isn't required. The solution vector is:

$$
\begin{array}{lll}
v_1 = 0.3530, & v_8 = 0.4988, & v_{15} = 0.3530, \\
v_2 = 0.9132, & v_9 = 1.2894, & v_{16} = 0.9132, \\
v_3 = 2.0103, & v_{10} = 2.8323, & v_{17} = 2.0103, \\
v_4 = 4.2957, & v_{11} = 6.0193, & v_{18} = 4.2957, \\
v_5 = 9.1531, & v_{12} = 12.6537, & v_{19} = 9.1531, \\
v_6 = 19.6631, & v_{13} = 26.2893, & v_{20} = 19.6631, \\
v_7 = 43.2101, & v_{14} = 53.1774, & v_{21} = 43.2101.
\end{array}
$$

Symmetry in the system is apparent by inspection, and we could have reduced the number of equations by taking this into account. The regular pattern of coefficients in the matrix of Eqs. (7.9) would have been lost, however.

We now are able to examine the effect of grid size, h. Three points are identical in the two sets of equations. We show these results side by side in Table 7.1 together with the analytical results.

The answers with $h = 2.5$ cm show significant improvement, as expected. In fact, since the error in approximating the derivatives by central-difference quotients is of $O(h^2)$, we might anticipate that the errors in the solution through sets of difference equations would also vary as h^2; this is roughly true. When we have such knowledge of the effect of h, an extrapolation can be made. (See Sections 4.4 and 4.8.) When this is done, temperatures within 3 percent of the analytical values are obtained.

Table 7.1

	With h = 5 cm		With h = 2.5 cm		Analytical value
	Value	Error	Value	Error	
u_1	1.786	−0.692	1.289	−0.195	1.0943
u_2	7.143	−1.655	6.019	−0.531	5.4885
u_3	26.786	−0.692	26.289	−0.195	26.0944

In principle, then, we can use the numerical method to solve the steady-state heat-flow equation to any desired accuracy; all we need do is continue to make h smaller. In practice, however, the above procedure runs into severe difficulties. It is very apparent that the number of equations increases inordinately fast. With $h = 1.25$, we would have 105 discrete interior points; with $h = 0.625$, we have 465, etc. Storing a matrix with 105 rows and 105 columns would require 105^2 or about 11,000 words (44,000 bytes) of computer memory; with 465 equations we would need 216,000 words (865,000 bytes)! Few computer systems allow us such a generous partition, and overlaying memory space from disk storage would be extremely time-consuming. There must be a better way, you are thinking, and there is.

The remedy to the problem of over-large matrices is to take advantage of the sparseness exhibited by Eqs. (7.9). That system of 21 equations has nonzero terms in only 85 out of 441 elements, or 19 percent. The larger matrices that result when h is smaller have an even smaller proportion of nonzero terms (4% and 1% with 105 and 465 equations, respectively). Unfortunately, the systems are not tridiagonal. If they were, we could work with only the nonzero terms, as we found in Chapter 2, With banded matrices such as we have here, none of the zero elements outside the bands changes, but the elements between the side bands do fill up with nonzeros. The preferred method for large numbers of points in our mesh is *iteration*, which we discuss next.

7.5 ITERATIVE METHODS FOR LAPLACE'S EQUATION

When the steady-state heat-flow equation in two dimensions,

$$\nabla^2 u = \frac{\partial^2 u}{\partial x^2} + \frac{\partial^2 u}{\partial y^2} = 0,$$

is approximated by the pictorial relation

$$\nabla^2 u = \frac{1}{h^2} \left\{ \begin{matrix} & 1 & \\ 1 & -4 & 1 \\ & 1 & \end{matrix} \right\} u_{ij} = 0,$$

at a set of grid points (x_i, y_j), a set of simultaneous linear equations results, as Eqs. (7.8) and (7.9) illustrate. Proper ordering of the equations gives a *diagonally dominant* system. (Diagonal dominance means that the magnitude of the coefficient on the diagonal is not less than the sum of the magnitudes of the other coefficients in the same row.) As we saw in Chapter 2, such systems can be solved by Gauss–Seidel iteration.

When applied to Laplace's equation, the iterative technique is called *Liebmann's method*. We illustrate this procedure with the same problem as in the last section. The difference equations, when $h = 5$ cm, are

$$
\begin{aligned}
-4u_1 + u_2 \quad\quad &= 0, \\
u_1 - 4u_2 + u_2 &= 0, \\
u_2 - 4u_3 &= -100.
\end{aligned}
\tag{7.10}
$$

The first step in applying Liebmann's method is to rearrange these equations into a new form by solving each equation for the variable on the diagonal:

$$
\begin{aligned}
u_1 &= \frac{u_2}{4}, \\
u_2 &= \frac{u_1 + u_3}{4}, \\
u_3 &= \frac{u_2 + 100}{4}.
\end{aligned}
\tag{7.11}
$$

Observe that the pictorial operator can give this rearrangement directly:

$$
\left\{
\begin{array}{ccc}
 & 1 & \\
1 & -4 & 1 \\
 & 1 &
\end{array}
\right\} u_{ij} = 0 \rightarrow u_{ij} = \frac{u_{i+1,j} + u_{i-1,j} + u_{i,j+1} + u_{i,j-1}}{4}.
\tag{7.12}
$$

We solve Eqs. (7.12) by beginning with some initial approximation $\bar{u}$ to u, the solution vector, substituting values from this approximation into the righthand sides of Eqs. (7.11), and generating new components of $\bar{u}$ that converge to the solution. (Convergence is usually faster if each new component of $\bar{u}$ is used as soon as it is available.)

The better the beginning approximation, the more rapid the convergence, but, because the system is diagonally dominant, even a poor initial vector will eventually converge. Inspection of the problem underlying the equations lets us guess at reasonably good starting values: we know that u_1 will be small, that u_2 will be about four times as large as u_1, and that u_3 will be a little more than one fourth of 100. Suppose we begin with $u_3 = 30$, $u_2 = 7.5$, $u_1 = 2$. Performing the iterations gives the values shown in Table 7.2.

Table 7.2

u_1	u_2	u_3
	Initial values	
2	7.5	30
1.875	7.969	26.992
1.992	7.246	26.812
1.812	7.156	26.789
1.789	7.144	26.786
1.786	7.143	26.786
1.786	7.143	26.786

The iterations converge as anticipated and to the same values as in the previous section when we solved the equations by elimination. We terminate the iterations when all components of $\bar{u}$ converge to constant values. For only three equations, iteration is surely more computational effort than is elimination, but for large systems there are important advantages.

The most obvious advantage is a greatly reduced demand for computer memory space. In our example problem, if h is 1.25 cm, there are 105 interior points at grid intersections. To store the full 105×105 matrix, we need 44,000 bytes, not even taking into account the righthand sides and the solution vector. With Liebmann's method, we need to store only the 105 values of u_{ij} plus 44 boundary values; this requires about 600 bytes. The difference in storage requirements is even more striking when $h = 0.625$ cm. The impossible 865,000 bytes for the full matrix is reduced to a very moderate 2300 bytes.

With large systems there is also a very significant reduction in execution time as well, although the magnitude of this depends on how good a starting vector we can supply, and very strongly on the precision we require. It is customary to terminate the iterations when the largest change in any component of $\bar{u}$ meets a supplied tolerance value. This tolerance should be chosen with consideration for the importance of precision in the final answers (many engineering applications do not need extreme accuracy) and only after determining whether the parameters of the problem (boundary conditions, dimensions, material properties, etc.) are themselves known with great accuracy.

Since the speed of convergence depends on the initial value of $\bar{u}$, it is frequently worthwhile to spend some effort in obtaining a good initial estimate. Some guidelines can be helpful. It is obvious, since the temperature at each point is the *average* of temperatures at surrounding points, that no interior point can have a temperature greater than the hottest boundary point, or a temperature *lower* than the coldest. Sometimes the simple expedient of setting all interior points equal to the mean of all the boundary temperatures is used to start iterations. Interior points near cold edges have low temperatures and those near hot edges have higher temperatures. This may let us make good guesses from which to begin iterations. A more refined technique solves the problem with a coarse mesh, getting a quick solution; these values then are used as a basis to fill in initial values at intermediate points.

When our sample problem was solved with $h = 2.5$ (21 interior points), employing the average of boundary values as the starting temperatures, these results were obtained after 21 iterations. (The iterations were stopped when the largest change in any $\bar{u}$ component was less than 0.001.)

$$
\begin{array}{lll}
u_1 = 0.354, & u_8 = 0.500, & u_{15} = 0.354, \\
u_2 = 0.914, & u_9 = 1.291, & u_{16} = 0.914, \\
u_3 = 2.012, & u_{10} = 2.834, & u_{17} = 2.011, \\
u_4 = 4.297, & u_{11} = 6.021, & u_{18} = 4.296, \\
u_5 = 9.154, & u_{12} = 12.655, & u_{19} = 9.154, \\
u_6 = 19.664, & u_{13} = 26.290, & u_{20} = 19.663, \\
u_7 = 43.210, & u_{14} = 53.178, & u_{21} = 43.210.
\end{array}
$$

Comparison with the values in Section 7.4 shows that the same results are obtained as when the equations are solved by elimination.

Liebmann's method essentially takes care of the problem of memory requirement for the steady-state heat-flow problem (though three-dimensional problems can still be demanding). The chief drawback in Liebmann's method is slow convergence, which is especially acute when there are a larger number of points, for then each iteration is lengthy and, in addition, more iterations are required to reach a given tolerance.

The relaxation method of Southwell, as discussed in Section 2.9, would be a way of attaining faster convergence in the iterative method. (In fact, Southwell developed his method to solve potential problems.) As pointed out in the discussion of that section, relaxation is not adapted to computer solution of sets of equations. Based on Southwell's technique, the use of an over-relaxation factor can give significantly faster convergence, however. Since we handle each equation in a standard and repetitive order, this method is called *successive over-relaxation*, frequently abbreviated as S.O.R.

To show how successive over-relaxation can be applied to Laplace's equation, we begin with Eq. (7.12), adding superscripts to show that a new value is computed from previous iterates,

$$
u_{ij}^{(k+1)} = \frac{u_{i+1,j}^{(k)} + u_{i-1,j}^{(k)} + u_{i,j+1}^{(k)} + u_{i,j-1}^{(k)}}{4}.
$$

We now both add and subtract $u_{ij}^{(k)}$ on the righthand side, getting:

$$
u_{ij}^{(k+1)} = u_{ij}^{(k)} + \left[\frac{u_{i+1,j}^{(k)} + u_{i-1,j}^{(k)} + u_{i,j+1}^{(k)} + u_{i,j-1}^{(k)} - 4u_{ij}^{(k)}}{4} \right]. \tag{7.13}
$$

Table 7.3 Effect* of over-relaxation factor on speed of convergence.

$h = 2.5$ cm (21 interior points)		$h = 1.25$ cm (105 interior points)	
ω	Number of iterations	ω	Number of iterations
1.00	20	1.00	70
1.10	15	1.10	58
1.20	13	1.20	46
1.30	12	1.30	35
1.40	15	1.40	29
1.50	18	1.50	26
1.60	23	1.60	28
		1.70	36
ω_{opt} at about 1.3		ω_{opt} at about 1.5	

* The example problem using S.O.R. with iterations was continued until maximum change in any component of $\bar{u}$ was less than 0.001.

(The numerator term will be zero when final values, after convergence, are used. The term in brackets is what Southwell called the "residual," which he "relaxed" to zero by changes in the temperature at (x_i, y_j).)

We can consider the bracketed term in Eq. (7.13) to be an adjustment to the old value, $u_{ij}^{(k)}$, to give the new and improved value, $u_{ij}^{(k+1)}$. If, instead of adding just the bracketed term, we add a larger value (thus "over-relaxing"), we get faster convergence. We modify Eq. (7.13) by including an over-relaxation factor ω to get the new iterating relation

$$u_{ij}^{(k+1)} = u_{ij}^{(k)} + \omega\left[\frac{u_{i+1,j}^{(k)} + u_{i-1,j}^{(k)} + u_{i,j+1}^{(k)} + u_{i,j-1}^{(k)} - 4u_{ij}^{(k)}}{4}\right]. \qquad (7.14)$$

Maximum acceleration is obtained for some optimum value of ω. This optimum value will always lie between 1.0 and 2.0.

Table 7.3 shows some results that demonstrate how S.O.R. can speed up the convergence in our example problem for $h = 2.5$ (21 interior points) and for $h = 1.25$ (105 interior points.)

The optimum value for ω is not always predictable in advance. There are methods of using the results of the first few iterations to find a value of ω that is near optimum.* We do not discuss these here. For a rectangular region having constant boundary conditions (Dirichlet conditions), a reasonable estimate of the

* Forsythe and Wasow (1960), Fox (1962), and Varga (1962) present details.

optimum ω can be determined as the smaller root of the quadratic equation

$$\left(\cos\frac{\pi}{p} + \cos\frac{\pi}{q}\right)^2 \omega^2 - 16\omega + 16 = 0, \tag{7.15}$$

where p and q are the number of mesh divisions on each side of the rectangular region. Solving the above equation for ω gives:

$$\omega_{opt} = \frac{4}{2 + \sqrt{4 - c^2}},$$

with

$$c = \left(\cos\frac{\pi}{p} + \cos\frac{\pi}{q}\right).$$

For our example problem, Eq. (7.15) predicts $\omega_{opt} = 1.267$ for $h = 2.5$ and $\omega_{opt} = 1.532$ for $h = 1.25$. We see from Table 7.3 that there is good agreement with actual results.

To show the range of values for ω_{opt} that are customary in steady-state problems, we show in Table 7.4 solutions to Eq. (7.15) for several values of p and q where $p = q$ (a square region):

Table 7.4

Value of $p = q$	ω_{opt}
2	1.000
3	1.072
5	1.260
10	1.528
20	1.729
100	1.939
∞	2.000

7.6 THE POISSON EQUATION

The methods of the previous section are readily applied to Poisson's equation. We illustrate with an analysis of torsion in a rectangular bar subject to twisting. The torsion function ϕ satisfies the Poisson equation:

$$\nabla^2\phi + 2 = 0, \qquad \phi = 0 \text{ on boundary.} \tag{7.16}$$

The tangential stresses are proportional to the partial derivatives of ϕ for a twisted prismatic bar of constant cross section. Let us find ϕ over the cross section of a rectangular bar 6 by 8 inches in size.

Subdivide the cross section into 2-inch squares, so that there are six interior points, as in Fig. 7.7. In terms of difference quotients, Eq. (7.16) becomes:

$$\frac{1}{h^2}\begin{Bmatrix} & 1 & \\ 1 & -4 & 1 \\ & 1 & \end{Bmatrix}\phi_{i,j} + 2 = 0,$$

or   (7.17)

$$\frac{1}{4}\begin{Bmatrix} & 1 & \\ 1 & -4 & 1 \\ & 1 & \end{Bmatrix}\phi_{i,j} + 2 = 0.$$

(The function ϕ is dimensional; with our choice of h, ϕ will have square inches as units.)

The set of equations, when (7.17) is applied at each interior point, is:

$$
\begin{aligned}
0 \;\; + 0 \;\;\; + \phi_{12} + \phi_{21} - 4\phi_{11} + 8 &= 0, \\
\phi_{11} + 0 \;\;\; + 0 \;\;\; + \phi_{22} - 4\phi_{12} + 8 &= 0, \\
0 \;\; + \phi_{11} + \phi_{22} + \phi_{31} - 4\phi_{21} + 8 &= 0, \\
\phi_{21} + \phi_{12} + 0 \;\;\; + \phi_{32} - 4\phi_{22} + 8 &= 0, \\
0 \;\; + \phi_{21} + \phi_{32} + 0 \;\;\; - 4\phi_{31} + 8 &= 0, \\
\phi_{32} + \phi_{22} + 0 \;\;\; + 0 \;\;\; - 4\phi_{32} + 8 &= 0.
\end{aligned}
$$

(7.18)

Figure 7.7 $\nabla^2\phi + 2 = 0$

1.4	2.6	3.0	2.6	1.4
2.5	4.6	5.0	4.6	2.5
3.0	5.2	5.5	5.2	3.0
3.5	5.7	6.0	5.7	3.5
3.0	5.2	5.5	5.2	3.0
2.5	4.6	5.0	4.6	2.5
1.4	2.6	3.0	2.6	1.4

2.042	3.047	3.353	3.047	2.043
3.123	4.794	5.319	4.794	3.123
3.657	5.686	6.335	5.686	3.657
3.818	5.960	6.647	5.960	3.818
3.657	5.686	6.335	5,686	3.657
3.124	4.794	5.319	4.794	3.124
2.043	3.048	3.354	3.048	2.043

(a) Initial values of ϕ for iteration   (b) Final value for ϕ by iterative methods

Figure 7.8   Torsion in a rectangular bar; mesh size of 1 inch. (a) Initial values of ϕ for iteration; (b) final values for ϕ by iteration methods.

Symmetry considerations show that $\phi_{11} = \phi_{12} = \phi_{31} = \phi_{32}$ and $\phi_{21} = \phi_{22}$, so only two unknowns are left in (7.18) after substitutions:

$$\phi_{21} - 3\phi_{11} + 8 = 0,$$
$$2\phi_{11} - 3\phi_{21} + 8 = 0.$$

Obviously these would be solved by elimination; $\phi_{11} = 4.56$, $\phi_{21} = 5.72$.

Now let us solve the problem with a 1-inch-square mesh using iteration. To use Liebmann's method, we need initial estimates of ϕ. As shown in Fig. 7.8(a), we estimate these, using our previous results as a guide. We converge in 25 iterations (tolerance $= 0.001$) to the values shown in Fig. 7.8(b), employing Eq. (7.19):*

$$\phi_{ij} = \tfrac{1}{4}(\phi_{i+1,j} + \phi_{i-1,j} + \phi_{i,j+1} + \phi_{i,j-1} + 2). \qquad (7.19)$$

As we have seen before, applying an over-relaxation factor should speed up the convergence. Accordingly, we rewrite Eq. (7.19) with an over-relaxation factor:

$$\phi_{ij}^{(k+1)} = \phi_{ij}^{(k)} + \frac{\omega}{4}(\phi_{i+1,j}^{(k)} + \phi_{i-1,j}^{(k)} + \phi_{i,j+1}^{(k)} + \phi_{i,j-1}^{(k)} - 4\phi_{ij}^{(k)} + 2). \qquad (7.20)$$

The optimum over-relaxation factor calculated from Eq. (7.15) is 1.383. Using this in Eq. (7.20), we converge in 13 iterations to the same set of values as before and as tabulated in Fig. 7.8(b). Using S.O.R., rather than the standard Liebmann's method, cuts down the number of iterations by nearly 50 percent.

---

* Equation (7.19) differs from Eq. (7.12), the Liebmann method for Laplace's equation, only in that the $f(x, y)$ term of Poisson's equation must be included.

7.7 DERIVATIVE BOUNDARY CONDITIONS

In the previous examples, we solved for the steady-state temperatures at interior points of a rectangular plate, with the boundary temperatures being specified. In many problems, instead of knowing the boundary temperatures, we know the temperature gradient in the direction normal to the boundary, as for example when heat is being lost from the surface by radiation and conduction.

Example Consider a rectangular plate within which heat is also being uniformly generated at each point at a rate of Q cal/cm^3 · sec. The plate is of steel for which $k = 0.16$ cal/sec · cm^2 · °C/cm, and is 4 by 8 cm, 1 cm thick. For this situation, Poisson's equation holds in the form

$$\nabla^2 u = -\frac{Q}{k} = \frac{1}{h^2} \left\{ \begin{matrix} & 1 & \\ 1 & -4 & 1 \\ & 1 & \end{matrix} \right\} u_{i,j}. \tag{7.21}$$

Suppose that $Q = 5$ for our example. The top and bottom faces are perfectly insulated so that no heat is lost, while the upper and lower edges lose heat so that

$$\frac{\partial u}{\partial y} = -15°C/cm$$

in the outward direction, and the right and left edges are held at a constant temperature $u = 20°C$. We shall find the steady-state temperatures at points in the plate.

In Fig. 7.9 we sketch the plate, with a gridwork to give 21 interior points with $h = 1$ cm. In this problem, the upper and lower edge temperatures are also

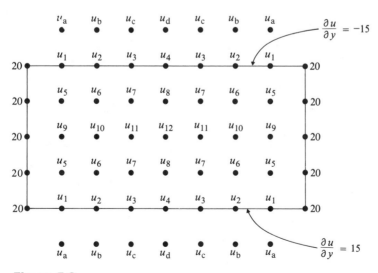

Figure 7.9

unknown, increasing the total number of points at which u is to be determined to 35. Because of symmetry, the equations can be written in terms of 12 quantities; these are indicated on the diagram.

We now write Eq. (7.21) at each unknown point, which is our general procedure in elliptic partial-differential equations, except that at the upper and lower edges we cannot form the five-point combination—there are not enough points to form the star. We get around this by the device of extending our network to a row of exterior points. We utilize these fictitious exterior points to include in the set of equations those whose central point is on the upper or lower edge:

$$
\begin{aligned}
(20 + u_a + u_2 + u_5 - 4u_1) &= -5/0.16, \\
(u_1 + u_b + u_3 + u_6 - 4u_2) &= -5/0.16, \\
(u_2 + u_c + u_4 + u_7 - 4u_3) &= -5/0.16, \\
(u_3 + u_d + u_3 + u_8 - 4u_4) &= 5/0.16, \\
(20 + u_1 + u_6 + u_9 - 4u_5) &= -5/0.16, \\
(u_5 + u_2 + u_7 + u_{10} - 4u_6) &= -5/0.16, \\
(u_6 + u_3 + u_8 + u_{11} - 4u_7) &= -5/0.16, \\
(u_7 + u_4 + u_7 + u_{12} - 4u_8) &= -5/0.16, \\
(20 + u_5 + u_{10} + u_5 - 4u_9) &= -5/0.16, \\
(u_9 + u_6 + u_{11} + u_6 - 4u_{10}) &= -5/0.16, \\
(u_{10} + u_7 + u_{12} + u_7 - 4u_{11}) &= -5/0.16, \\
(u_{11} + u_8 + u_{11} + u_8 - 4u_{12}) &= -5/0.16.
\end{aligned}
\tag{7.22}
$$

It would appear that we have not helped ourselves by the fictitious points outside the plate. We have made it possible to write twelve equations, but four new unknowns have been introduced. However, we have not yet utilized the gradient conditions. Write the derivative conditions as central-difference quotients, as discussed in Chapter 4:

$$
\begin{aligned}
-\left(\frac{\partial u}{\partial y}\right)_1 &= \frac{u_5 - u_a}{2(1)} = 15, & u_a &= u_5 - 30, \\
-\left(\frac{\partial u}{\partial y}\right)_2 &= \frac{u_6 - u_b}{2(1)} = 15, & u_b &= u_6 - 30, \\
-\left(\frac{\partial u}{\partial y}\right)_3 &= \frac{u_7 - u_c}{2(1)} = 15, & u_c &= u_7 - 30, \\
-\left(\frac{\partial u}{\partial y}\right)_4 &= \frac{u_8 - u_d}{2(1)} = 15, & u_d &= u_8 - 30.
\end{aligned}
\tag{7.23}
$$

We write these approximations for the gradient condition choosing the proper order of points so that the outward normal is negative, since heat is flowing outwardly. Using a central-difference approximation of error $O(h^2)$ makes these compatible with our other difference-quotient approximations. Each of the difference quotients allows us to write the fictitious temperature values in terms of points within the rectangle.

We solve the set of equations in (7.22) after eliminating the fictitious points by using Eq. (7.23). Elimination, iteration (Liebmann's method), or relaxation may be used. The solution of the equations is left as an exercise.

We can apply the same method to the more general boundary condition,

$$au - b\frac{\partial u}{\partial x} = c,$$

where a, b, and c are constants, in an obvious application of the relationships of Eq. (7.23).

7.8 IRREGULAR REGIONS

When the boundary of the region is not such that the network can be drawn to have the boundary coincide with the nodes of the mesh, we must proceed differently at points near the boundary. Consider the general case of a group of five points whose spacing is nonuniform, arranged in an unequal-armed star. We represent each distance by $\theta_i h$, where θ_i is the fraction of the standard spacing h that the particular distance represents (Fig. 7.10). Along the line from u_1 to u_0 to u_3, we may approximate the first derivatives:

$$\left(\frac{\partial u}{\partial x}\right)_{1-0} \doteq \frac{u_0 - u_1}{\theta_1 h}; \qquad \left(\frac{\partial u}{\partial x}\right)_{0-3} \doteq \frac{u_3 - u_0}{\theta_3 h}.$$

Since

$$\frac{\partial^2 u}{\partial x^2} = \frac{\partial}{\partial x}\left(\frac{\partial u}{\partial x}\right),$$

we have

$$\frac{\partial^2 u}{\partial x^2} \doteq \frac{(u_3 - u_0)/\theta_3 h - (u_0 - u_1)/\theta_1 h}{\frac{1}{2}(\theta_1 + \theta_3)h} = \frac{2}{h^2}\left[\frac{u_1 - u_0}{\theta_1(\theta_1 + \theta_3)} + \frac{u_3 - u_0}{\theta_3(\theta_1 + \theta_3)}\right]. \quad (7.24)$$

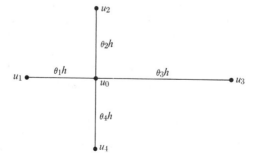

Figure 7.10

Similarly,

$$\frac{\partial^2 u}{\partial y^2} \doteq \frac{2}{h^2}\left[\frac{u_2 - u_0}{\theta_2(\theta_2 + \theta_4)} + \frac{u_4 - u_0}{\theta_4(\theta_2 + \theta_4)}\right]. \tag{7.25}$$

The expressions in Eqs. (7.24) and (7.25) have errors $O(h)$, which introduce larger errors in the computations than for points that are arranged in an equal-armed star. Combining, we get:

$$\nabla^2 u = \frac{\partial^2 u}{\partial x^2} + \frac{\partial^2 u}{\partial y^2}$$

$$= \frac{2}{h^2}\left[\frac{u_1}{\theta_1(\theta_1 + \theta_3)} + \frac{u_2}{\theta_2(\theta_2 + \theta_4)} + \frac{u_3}{\theta_3(\theta_1 + \theta_3)} + \frac{u_4}{\theta_4(\theta_2 + \theta_4)}\right.$$

$$\left. - \left(\frac{1}{\theta_1\theta_3} + \frac{1}{\theta_2\theta_4}\right)u_0\right]. \tag{7.26}$$

We use the operator of Eq. (7.26) for points adjacent to boundary points when the boundary points do not coincide with the mesh, instead of our standard operator. If the boundary conditions involve normal derivatives, great complications arise, especially for curved boundaries. Fox (1962) gives relationships for this situation.

Let us illustrate this procedure by an example. A semicircular plate of radius a has the temperature at the base (the straight side) held at $0°$ while the circumference is held at $c°$. We desire the steady-state temperatures. The analytical solution to this problem is given by the infinite series

$$u(r, \theta) = \frac{4c}{\pi} \sum_{n=1}^{\infty} \frac{1}{2n - 1}\left(\frac{r}{a}\right)^{2n-1} \sin(2n - 1)\theta, \tag{7.27}$$

where (r, θ) are the polar coordinates of a point on the plate. (For points near the circumference, several hundred terms are needed in order to compute temperatures with any degree of accuracy.) There is right-to-left symmetry in the temperatures.

The finite-difference method superimposes a gridwork on the plate. Suppose we take these values for the parameters of the problem: $a = 1$, $c = 100$. With $h = 0.2$, the diagram of Fig. 7.11 results; there are 17 unknowns after we utilize the left-to-right symmetry. For points u_2, u_3, u_{12}, and u_{17}, the grid does not coincide with the boundary. It is easy to find by analytic geometry that the short arms at u_2 and u_{17} have a length of $0.8990h$ and those at u_3 and u_{12} have a length

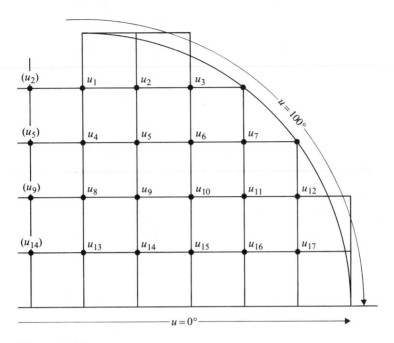

Figure 7.11

of $0.5826h$. Applying Eq. (7.26) we get these pictorial operators:

At point 2: $\quad \dfrac{1}{(0.2)^2}\left\{1 \quad \begin{matrix} 1.1715 \\ -4.2247 \\ 1.0532 \end{matrix} \quad 1\right\} u_2 = 0;$

At point 3: $\quad \dfrac{1}{(0.2)^2}\left\{1 \quad \begin{matrix} 2.1691 \\ -5.4328 \\ 1.2637 \end{matrix} \quad 1\right\} u_3 = 0;$

At point 12: $\quad \dfrac{1}{(0.2)^2}\left\{\begin{matrix} 1 \\ 1.2637 \quad -5.4328 \quad 2.1691 \\ 1 \end{matrix}\right\} u_{12} = 0;$

At point 17: $\quad \dfrac{1}{(0.2)^2}\left\{\begin{matrix} 1 \\ 1.0532 \quad -4.2247 \quad 1.1715 \\ 1 \end{matrix}\right\} u_{17} = 0.$

(When computing these operators, it is useful to observe that the central number is the negative of the sum of the other four values. This helps check for accuracy.)

At all the other points, the standard operator applies. The matrix for the set of equations is given in (7.28).

Row \ Col	1	2	3	4	5	6	7	8	9	10	11	12	13	14	15	16	17	RHS
1	−4	2		1														−100
2	1	−4.2247	−5.4328															−117.15
3		1	0	−4	2													−316.91
4	1		1	1	−4	1												
5		1	1	1	1.0532													
6					2	1.2637		1										−200
7					1	1	−4	1	1									
8				1			0	−4	2				1					
9							1	1	−4	1				1				
10							1		1	−4	1				1			
11								1		−4	1					1		−316.91
12											1.2637	−5.4328					1	
13												0	−4	2				−0
14													1	−4	1			−0
15														1	−4	1		−0
16															1	−4	1	−0
17																1.0532	−4.2247	−117.15

$$(7.28)$$

Table 7.5 Comparison of results for example by finite differences, and analytical solution

x	y	Point number	u, finite differences	u, series solution, Eq. (7.27)
0	0.8	1	86.053	85.906
0.2	0.8	2	87.548	87.417
0.4	0.8	3	92.124	92.094
0	0.6	4	69.116	68.807
0.2	0.6	5	70.773	70.482
0.4	0.6	6	75.994	75.772
0.6	0.6	7	85.471	85.405
0	0.4	8	48.864	48.448
0.2	0.4	9	50.436	50.000
0.4	0.4	10	55.606	55.151
0.6	0.4	11	65.891	65.593
0.8	0.4	12	84.189	84.195
0	0.2	13	25.466	25.133
0.2	0.2	14	26.501	26.109
0.4	0.2	15	30.102	29.527
0.6	0.2	16	38.300	37.436
0.8	0.2	17	57.206	57.006

Observe how sparse matrix (7.28) is. For a larger number of equations, the density of nonzero values will be even less, for no row ever has more than five nonzero values.

When the equations of (7.28) are solved, the values in Table 7.5 result. The analytical results given by Eq. (7.27) are included for comparison, using values of r and θ to coincide with the grid points. In general, the agreement is good. One might expect greater errors at points where the less accurate unequal-arm operators were used. This is not the case for this example, however.

If our mesh of points is chosen very fine (as we would do to get high accuracy), there is an even simpler way to handle irregular boundaries. One uses the closest mesh point as the boundary point, thus in effect perturbing the actual region to one that coincides with the network and then the standard operator of Eq. (7.21) applies everywhere. This introduces some error, of course, but often its effect is no worse than the $O(h)$ operator of (7.26). It is also usually easier to program a computer solution using this perturbed-region technique.

7.9 LAPLACIAN OPERATOR IN NONRECTANGULAR COORDINATES

It is occasionally useful to have a finite-difference approximation to the Laplacian for an equispaced triangular network of points. To derive this, we need the formulas for a rotation of axes from analytical geometry. The point (x, y), written

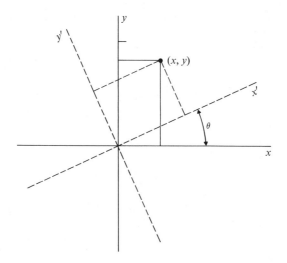

Figure 7.12

in terms of its coordinates with respect to the pair of x',y'-axes which are rotated an angle θ from the x,y-system, is (see Fig. 7.12):

$$x = x' \cos \theta - y' \sin \theta,$$
$$y = x' \sin \theta + y' \cos \theta. \tag{7.29}$$

We first compute $\partial u / \partial x'$:

$$\frac{\partial u}{\partial x'} = \frac{\partial u}{\partial x}\frac{\partial x}{\partial x'} + \frac{\partial u}{\partial y}\frac{\partial y}{\partial x'} = \frac{\partial u}{\partial x} \cos \theta + \frac{\partial u}{\partial y} \sin \theta = u_x \cos \theta + u_y \sin \theta.$$

Then we get

$$\frac{\partial^2 u}{\partial x'^2} = \cos \theta (u_{xx} \cos \theta + u_{xy} \sin \theta) + \sin \theta (u_{yx} \cos \theta + u_{yy} \sin \theta)$$

$$= u_{xx} \cos^2 \theta + 2u_{xy} \sin \theta \cos \theta + u_{yy} \sin^2 \theta.$$

For an equispaced triangular network, the connections between nodes make angles of 0°, 60°, and 120° with the horizontal (Fig. 7.13). Call these the a-, b-, and c-axes.

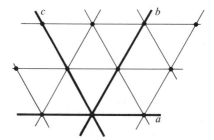

Figure 7.13

For $\theta = 0°$:

$$\frac{\partial^2 u}{\partial a^2} = u_{xx};$$

For $\theta = 60°$:

$$\frac{\partial^2 u}{\partial b^2} = \frac{1}{4} u_{xx} + \frac{\sqrt{3}}{2} u_{xy} + \frac{3}{4} u_{yy};$$

For $\theta = 120°$:

$$\frac{\partial^2 u}{\partial c^2} = \frac{1}{4} u_{xx} - \frac{\sqrt{3}}{2} u_{xy} + \frac{3}{4} u_{yy}.$$

Adding, we obtain:

$$\frac{\partial^2 u}{\partial a^2} + \frac{\partial^2 u}{\partial b^2} + \frac{\partial^2 u}{\partial c^2} = \frac{3}{2} (u_{xx} + 0 + u_{yy}) = \frac{3}{2} \nabla^2 u.$$

Laplace's equation, $\nabla^2 u = 0$, can be represented by

$$\frac{2}{3} \left(\frac{\partial^2 u}{\partial u^2} + \frac{\partial^2 u}{\partial b^2} + \frac{\partial^2 u}{\partial c^2} \right) = 0.$$

Using finite-difference quotients to approximate the partial derivatives gives a pictorial operator for a triangular network:

$$\nabla^2 u = \frac{2}{3h^2} \left\{ \begin{matrix} & 1 & 1 & \\ 1 & -6 & 1 \\ & 1 & 1 & \end{matrix} \right\} u.$$

Note that, for Laplace's equation, we have that the potential at every point is the arithmetic average of the potentials at its six equidistant neighbors. We observe this rule of averages even in the three-dimensional situation below.

For circular regions, one may derive a finite-difference approximation to the Laplacian in polar coordinates. Consider the group of points that are the nodes of a polar-coordinate network (Fig. 7.14):

$$\nabla^2 u = \frac{\partial^2 u}{\partial r^2} + \frac{1}{r} \frac{\partial u}{\partial r} + \frac{1}{r^2} \frac{\partial^2 u}{\partial \theta^2}$$

$$= \frac{u_3 - 2u_0 + u_1}{(\Delta r)^2} + \frac{1}{r_0} \frac{u_3 - u_1}{2\Delta r} + \frac{1}{r_0^2} \frac{u_2 - 2u_0 + u_4}{(\Delta \theta)^2}.$$

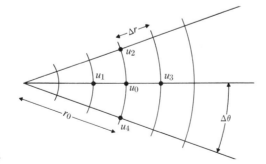

Figure 7.14

No "standard" operator can be written for this, so finding the coefficients and solving the set of equations by iteration when the problem is in polar coordinates is awkward, but it helps to group the various coefficients of each term.

Calling u_0 the central point, u_1 the point nearer the center, u_3 the point farther from the center, and u_2 and u_4 the points on either side at the same radial distance from the center, as shown in Fig. 7.14, we get, for Laplace's equation

$$
\nabla^2 u = \frac{1}{(\Delta r)^2}\left[\left(1 - \frac{\Delta r}{2r_0}\right)u_1 + \left(1 + \frac{\Delta r}{2r_0}\right)u_3 + \left(\frac{\Delta r}{r_0\,\Delta\theta}\right)^2 u_2 \right.
$$
$$
\left. + \left(\frac{\Delta r}{r_0\,\Delta\theta}\right)^2 u_4 - \left(2 + 2\left(\frac{\Delta r}{r_0\,\Delta\theta}\right)^2\right)u_0\right] = 0.
$$

(7.30)

Let us illustrate the Laplacian in polar coordinates by solving again the example presented in Section 7.8. In this we computed the steady-state temperatures on a semicircular plate of radius equal to one, with its base held at 0° and its circumference at 100°. We superimpose a polar-coordinate system; with $\Delta r = 0.2$ and $\Delta\Theta = \pi/8$, Fig. 7.15 results. There are sixteen u-values to be determined, after we allow for left-to-right symmetry.

When we compute the coefficients for Eq. (7.30) applied to each of the sixteen points, the matrix representation of matrix (7.31) results. Table 7.6 compares the results by finite differences with those computed from the infinite-series solution, Eq. (7.27).

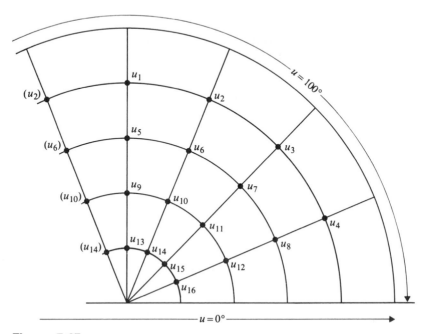

Figure 7.15

Row \ Col	1	2	3	4	5	6	7	8	9	10	11	12	13	14	15	16	RHS
1	-2.8106	0.8106			0.875												-112.5
2	0.4053	-2.8106	0.4053			0.875											-112.5
3		0.4053	-2.8106	0.4053			0.875										-112.5
4			0.4053	-2.8106	0			0.875									-112.5
5	1.1667			0	-3.4410	1.4410			0.8333								0
6		1.1667			0.7205	-3.4410	0.7205			0.8333							0
7			1.1667			0.7205	-3.4410	0.7205			0.8333						0
8				1.667			0.7205	-3.4410	0			0.8333					0
9					1.25			0	-5.2423	3.2423			0.75				0
10						1.25			1.6211	-5.2423	1.6211			0.75			0
11							1.25			1.6211	-5.2423	1.6211			0.75		0
12								1.25			1.6211	-5.2423	0			0.75	0
13									1.5			0	-14.9691	12.9691			0
14										1.5			6.4855	-14.9691	6.4855		0
15											1.5			6.4855	-14.9691	6.4855	0
16												1.5			6.4855	-14.9691	0

$$(7.31)$$

The agreement in Table 7.6 is about as good as that observed in Table 7.5 with a rectangular grid. However, one point, number 4, stands out in Table 7.6 as having an unusually large error. Large errors often occur in finite difference solutions near a discontinuity in the boundary conditions, and this could possibly be the explanation. On the other hand, this phenomenon is absent in Table 7.1, possibly due to a cancellation of errors.

We observe another sparse matrix in (7.31), this time with a band structure. The structure will depend on the ordering of the points, however.

7.10 THE LAPLACIAN OPERATOR IN THREE DIMENSIONS

Writing a difference equation to approximate the three-dimensional Laplacian is straightforward. We use triple subscripts to indicate the spatial position of points, and take the mesh distance the same in each direction:

$$\nabla^2 u = \frac{\partial^2 u}{\partial x^2} + \frac{\partial^2 u}{\partial y^2} + \frac{\partial^2 u}{\partial z^2}$$

$$= \frac{u_{i+1,j,k} - 2u_{i,j,k} + u_{i-1,j,k}}{h^2}$$

$$+ \frac{u_{i,j+1,k} - 2u_{i,j,k} + u_{i,j-1,k}}{h^2} + \frac{u_{i,j,k+1} - 2u_{i,j,k} + u_{i,j,k-1}}{h^2}$$

$$= \frac{1}{h^2} \left\{ \begin{array}{c} \nearrow^1 \\ 1 \underline{\quad} -6 \underline{\quad} 1 \\ \nearrow \\ 1 \end{array} \right\} u_{i,j,k}.$$

We see again, for Laplace's equation $\nabla^2 u = 0$, that the potential u is the arithmetic average of its six nearest neighboring values. The set of equations for this case is more extensive and more tedious to solve, but in principle the methods are unchanged. In hand calculations using iterative methods, keeping track of the successive values of u is awkward. An isometric projection of the points is generally recommended over the use of superimposed sheets of paper. In a computer, triple subscripting solves the problem, but large storage requirements are imposed by problems encountered in practice.

In three-dimensional problems, it is easy to exceed the memory space of even large computer systems. For example, if the volume under consideration has 100 points on a side, a total of $100^3 = 1,000,000$ points are involved. If we were to represent the coefficients of the one million equations in a square matrix, no computer system could hold all the values in real memory. Virtual memory systems could perhaps accommodate that many values, but the speed of computation would be excruciatingly slow.

Table 7.6 Results for example by finite differences in polar coordinates, and analytical solution.

r	θ	Point number	u, finite differences	u, series solution Eq. (7.27)
0.8	$\pi/2$	1	85.661	85.906
0.8	$3\pi/8$	2	84.425	84.792
0.8	$\pi/4$	3	79.459	80.394
0.8	$\pi/8$	4	63.679	66.186
0.6	$\pi/2$	5	68.371	68.808
0.6	$3\pi/8$	6	66.126	66.673
0.6	$\pi/4$	7	58.058	58.862
0.6	$\pi/8$	8	39.165	39.624
0.4	$\pi/2$	9	48.048	48.448
0.4	$3\pi/8$	10	45.543	45.938
0.4	$\pi/4$	11	37.457	37.730
0.4	$\pi/8$	12	22.374	22.252
0.2	$\pi/2$	13	25.003	25.133
0.2	$3\pi/8$	14	23.301	23.393
0.2	$\pi/4$	15	18.245	18.241
0.2	$\pi/8$	16	10.147	10.066

Of course, there are at most seven nonzero coefficients in any of these one million equations, but even 8,000,000 numerical values (the righthand sides are needed too) will exceed the memory capacity. Obviously, the problem must be cut down in size. However, even only 30 points on each side of a cubical volume results in 27,000 equations, with about 216,000 nonzero numerical values. While such a number of coefficients can be stored in large real memories or in virtual memory systems, lots of computer time is involved in iterating through the large set of equations. Unfortunately, many more iterations are required for convergence where the number of equations is great. It is little wonder that three-dimensional problems, while perfectly tractable in principle, are terribly expensive to solve.

7.11 MATRIX PATTERNS, SPARSENESS, AND THE A.D.I. METHOD

All of the examples that have been presented in this chapter illustrate the fact that the coefficient matrix is sparse when an elliptic partial-differential equation is solved by the finite-difference method. Especially in a three-dimensional case, the number of nonzero coefficients is a small fraction of the total.

The relative sparseness increases as the number of equations increases. In the two-dimensional example of Section 7.4, the 21×21 coefficient matrix of Eq. (7.9) has 81% of its positions filled with zeros. If symmetry were taken into account, there would have been a 14×14 coefficient matrix with 71% zeros.

Decreasing the value of h so as to have 105 points within the region gives rise to a coefficient matrix wtih 96% zeros. For a $30 \times 30 \times 30$ three-dimensional case, there would be 90,176 nonzero coefficients, but this is only 0.012% of the 729×10^6 values in the coefficient matrix!

In Chapter 2 it was shown that iterative methods are usually preferred for sparse matrices unless they have a tridiagonal structure. The reason for this is that elimination does not preserve the sparseness (unless the matrix is tridiagonal), so that one cannot work only with nonzero terms. This is illustrated by Fig. 7.16. The original matrix in (a) is shown in partially triangularized form in (b). The shaded portion of the matrix, which originally had 50% zeros, has lost all but one of these.

Frequently the coefficient matrix has a band structure, as illustrated by Eqs. (7.28) and (7.31). Equation (7.31) illustrates a special regularity for the nonzero elements; the ordering of the points must be carefully done to attain this. A band structure is worth working for, however, for elimination does not introduce nonzero terms outside of the limits defined by the original bands. Zeros in the

$$
\begin{bmatrix}
3 & 1 & 0 & 0 & 1 & 0 & 0 & 2 \\
1 & 2 & 1 & 0 & 0 & 1 & 0 & 0 \\
0 & 2 & 3 & 1 & 0 & 0 & 1 & 0 \\
1 & 0 & 0 & 2 & 0 & 0 & 1 & 1 \\
0 & 1 & 1 & 3 & 0 & 0 & 1 & 0 \\
0 & 2 & 0 & 1 & 0 & 2 & 0 & 3 \\
2 & 0 & 0 & 3 & 0 & 1 & 1 & 0 \\
0 & 1 & 0 & 1 & 0 & 0 & 2 & 3
\end{bmatrix}
$$

(a)

$$
\begin{bmatrix}
3 & 1 & 0 & 0 & 1 & 0 & 0 & 2 \\
0 & 1.667 & 1 & 0 & -0.333 & 1 & 0 & -0.666 \\
0 & 0 & 1.8 & 1 & 0.4 & -1.2 & 1 & 0.8 \\
0 & 0 & 0.6 & 2 & -0.4 & 0.2 & 1 & 0.2 \\
0 & 0 & 0.4 & 2.4 & 0.2 & -0.6 & 1 & 0.4 \\
0 & 0 & -1.2 & 1 & 0.4 & 0.8 & 0 & 3.8 \\
0 & 0 & 0.4 & 3 & -0.8 & 1.4 & 1 & -1.6 \\
0 & 0 & -0.6 & 1 & 0.2 & -0.6 & 2 & 3.5
\end{bmatrix}
$$

(b)

Figure 7.16

gaps between the parallel lines are not preserved, however, so that "tightest" possible bandedness is preferred. Sometimes it is possible to so order the points that a pentadiagonal matrix results.

The best of the band structures is tridiagonal, with corresponding economy of storage and speed of solution, as discussed in Chapter 2. A method for the steady-state heat equation called the *alternating direction implicit* (A.D.I.) method results in tridiagonal matrices and is of growing popularity. It was initially developed for unsteady-state problems, which are the subject of Chapter 8.

In the A.D.I. method, we write the finite-difference approximation to Laplace's equation in this fashion:

$$\nabla^2 u = \frac{u_{i,j-1}^{(n)} - 2u_{i,j}^{(n)} + u_{i,j+1}^{(n)}}{(\Delta x)^2} + \frac{u_{i-1,j}^{(n+1)} - 2u_{i,j}^{(n+1)} + u_{i+1,j}^{(n+1)}}{(\Delta y)^2} = 0. \quad (7.32)$$

The superscripts indicate the iteration number. Note that when we are calculating the $(n + 1)$st iterate, we use, for the approximation to $\partial u/\partial x$, a set of values in a horizontal row that are already known, the results of the nth iteration. By writing the equations in the order of the points in each column, we obtain a tridiagonal matrix when the known values of $u_{i,j-1}^{(n)}$, $u_{i,j}^{(n)}$, and $u_{i,j+1}^{(n)}$ are carried over to the righthand sides.

The solution for steady-state values of $u_{i,j}$ is iterative, with the $(k + 1)$st and $(k + 2)$nd values computed from:*

$$u_{i,j}^{(k+1)} = u_{i,j}^{(k)} + \rho(u_{i-1,j}^{(k+1)} - 2u_{i,j}^{(k+1)} + u_{i+1,j}^{(k+1)}) + \rho(u_{i,j-1}^{(k)} - 2u_{i,j}^{(k)} + u_{i,j+1}^{(k)}), \quad (7.33)$$

followed by

$$u_{i,j}^{(k+2)} = u_{i,j}^{(k+1)} + \rho(u_{i-1,j}^{(k+1)} - 2u_{i,j}^{(k+1)} + u_{i+1,j}^{(k+1)}) + \rho(u_{i,j-1}^{(k+2)} - 2u_{i,j}^{(k+2)} + u_{i,j+1}^{(k+2)}). \quad (7.34)$$

The relations are implicit but, by properly ordering the equations, we obtain tridiagonal coefficient matrices. This requires that Eq. (7.33) proceed down each column of points in turn, while Eq. (7.34) processes them row-wise. (This is the reason the method is called the *alternating direction implicit.* The proper choice of the parameter ρ can accelerate the convergence; for fastest convergence of large systems, a sequence of values of ρ is employed.†

The iterations begin with an initial estimate of the $u^{(0)}$ vector. The iterations are terminated when successive calculations agree within a given tolerance. Since

---

* The basis for these iteration relations will become clear in the next chapter when we discuss the A.D.I. method for parabolic partial-differential equations. The method applied to elliptic equations is really a special case of that technique.

† Birkhoff, Varga, and Young give an extensive survey of A.D.I. methods for elliptic equations and present many examples in *Advances in Computers*, Vol. 3, p. 189–273. (Academic Press, New York, 1962.)

there is a bias in the calculated results in the first, third, fifth, ... iterates, one usually ignores these and records the results of only the even-numbered iterations.

Except for the additional work of recalculating righthand sides, this method has a minimum of computational effort, equivalent to S.O.R., but converging faster. The coefficient matrices are always the same, so the reduction step need be performed only once, if one uses the LU equivalent or some similar procedure.

Example A square plate conducts only in the x- and y-directions. Two adjoining edges are kept at $0°$ and the other edges are kept at $100°$. A grid work is drawn with four interior nodes (Fig. 7.17). Use the A.D.I. method to estimate the temperatures at these four nodes.

It will be convenient to use v for the temperatures when ordering the points columnwise and u for the temperatures when ordering the points row-wise. Equations (7.33) and (7.34) become, in matrix form,

$$
\begin{bmatrix}
(1/\rho)+2 & -1 & & \\
-1 & (1/\rho)+2 & 0 & \\
& 0 & (1/\rho)+2 & -1 \\
& & -1 & (1/\rho)+2
\end{bmatrix}
\begin{bmatrix} v_1 \\ v_2 \\ v_3 \\ v_4 \end{bmatrix}
=
\begin{bmatrix}
100+0 & +[(1/\rho)-2]u_1 + & u_2 \\
0+0 & +[(1/\rho)-2]u_3 + & u_4 \\
100+u_1 & +[(1/\rho)-2]u_2 + & 100 \\
0+u_3 & +[(1/\rho)-2]u_4 + & 100
\end{bmatrix},
$$

$$
\begin{bmatrix}
(1/\rho)+2 & -1 & & \\
-1 & (1/\rho)+2 & 0 & \\
& 0 & (1/\rho)+2 & -1 \\
& & -1 & (1/\rho)+2
\end{bmatrix}
\begin{bmatrix} u_1 \\ u_2 \\ u_3 \\ u_4 \end{bmatrix}
=
\begin{bmatrix}
0+100 & +[(1/\rho)-2]v_1 + & v_2 \\
100+100 & +[(1/\rho)-2]v_3 + & v_4 \\
0+v_1 & +[(1/\rho)-2]v_2 + & 0 \\
100+v_3 & +[(1/\rho)-2]v_4 + & 0
\end{bmatrix}.
$$

The results of successive iterations, beginning with $u^{(0)} = (0, 0, 0, 0)$ and $\rho = 0.9$, are shown in Table 7.7.

For this example, the optimum value of ρ is about 1.00. Figure 7.18 shows how the errors after 6 iterations vary with ρ.

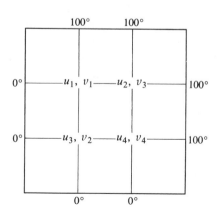

Figure 7.17

Table 7.7

Iteration number	u_1	u_2	u_3	u_4	v_1	v_2	v_3	v_4
1	0.00	0.00	0.00	0.00	35.85	11.52	83.21	58.89
2	49.86	75.47	24.26	49.86				
3					50.27	25.26	74.73	49.71
4	50.00	74.98	25.01	50.00				
5					49.99	24.99	75.01	50.01
6	50.00	75.00	25.00	50.00				
7					50.00	25.00	75.00	50.00

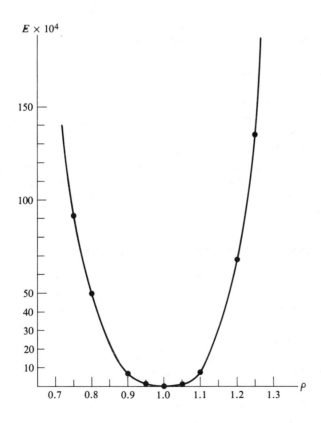

Table 7.8 Reduction in errors with iteration number—comparison of methods in example problem.

Iteration number	Average error		
	A.D.I.	Liebmann's	S.O.R.
1	11.185	17.970	15.055
2	0.3725	5.272	2.930
3	0.2725	1.318	0.2675
4	0.0075	0.330	0.0275
5	0.0065	0.082	0.0232
6	0.0002	0.020	0.0022

With the alternating direction implicit method, the errors show a most interesting pattern as the iterations proceed. Table 7.8 gives the average error at each iteration. Note that the error after an odd number of iterations is only slightly reduced over that of the previous iteration. This is true generally for A.D.I. methods. It is as if there is a bias in the values from the odd iterations which is compensated for in the succeeding even iterate.

A very important phenomenon occurs with the A.D.I. method. The equations break up into independent subsets, here into sets of two. This can often help in the computer solution of very large problems.

Table 7.8 also compares the errors when the example problem is solved by Liebmann's method and by S.O.R., using $\omega = 1.072$ (this value is the calculated ω_{opt} from Eq. (7.15)). In this example, the A.D.I. method is clearly preferable as judged by the rate of convergence, particularly as errors of the A.D.I. would decrease even faster if ρ had been taken at 1.0.

For elliptic problems in three dimensions, the A.D.I. method has great advantages. The directions alternate cyclically through the three coordinates, with new values of u being used in only one direction for each step, keeping the coefficient matrices tridiagonal. The references should be consulted for details.

7.12 COMPUTER PROGRAMS FOR POISSON'S EQUATION

Since Laplace's equation is a special case of Poisson's equation, we present two computer programs for the latter, in two dimensions. They can be applied to Laplace's equation by defining $f(x, y) = 0$.

The first (Fig. 7.19) is a subroutine that forms the coefficient matrix for the Laplacian on a rectangular region. The parameters passed to the subroutine LAPMTX include the matrix A into which the coefficients are placed, NDIM, the first dimension of matrix A in the calling program, and N and M, the number of intervals in the x- and y-directions. Only nonzero values are inserted, so the calling program must first set all the values of A to zero. The subroutine returns arrays similar to the coefficient matrices shown in Eqs. (7.9).

```
      SUBROUTINE LAPMTX (A, NDIM, N, M)
C  THIS SUBROUTINE FORMS THE COEFFICIENT MATRIX FOR THE LAPACIAN
C  ON A RECTANGULAR REGION THAT HAS N POINTS IN THE X DIRECTION
C  AND M POINTS IN THE Y DIRECTION.  IT ASSUMES THAT THE A MATRIX
C  ALREADY HAS ZEROS EVERYWHERE.
C  PARAMETERS ARE -
C     A      MATRIX THAT RETURNS COEFFICIENTS, MUST CONTAIN ZEROS TO START.
C     NDIM   FIRST DIMENSION OF A IN THE MAIN PROGRAM
C     N      NUMBER OF GRID POINTS IN THE X DIRECTION
C     M      NUMBER OF GRID POINTS IN THE Y DIRECTION
      REAL*4  A(NDIM,NDIM)
C  FIRST WE COMPUTE SOME CONSTANTS
      NL1 = N - 1
      NP1 = N + 1
      MSIZE = N*M
      MSLN = MSIZE - N
C  PUT -4 ON THE DIAGONAL
      DO 10 I = 1,MSIZE
      A(I,I) = -4.
10    CONTINUE
C  NOW PUT 1 ABOVE AND BELOW THE DIAGONAL
      DO 15 I = 2,MSIZE
      A(I-1,I) = 1.
      A(I,I-1) = 1.
15    CONTINUE
C  NOW REPLACE SOME ONES WITH ZEROS
      DO 20 I = N,MSLN,N
      A(I,I+1) = 0.
      A(I+1,I) = 0.
20    CONTINUE
C  PUT IN THE SIDE BANDS AND GO BACK
      DO 30 I = NP1,MSIZE
      A(I,I-N) = 1.
      A(I-N,I) = 1.
30    CONTINUE
      RETURN
      END
```

Figure 7.19 Program 1

This routine is used in connection with a linear equation solver such as the subroutine ELIM of Chapter 2; the main program must establish the righthand-side values based on the boundary conditions and $f(x, y)$ values, combining them with the matrix produced by LAPMTX and sending them to the linear equation solver to obtain the solution vector, which will then be the calculated steady-state potential values at each node of the grid. The difficulty with this technique is the large memory requirements that may be needed to store the values of matrix A.

Program 2 (Fig. 7.20) illustrates how successive over-relaxation can be used on problems involving Poisson's equation. The program as shown is applied to Laplace's equation by defining the $f(x, y)$ term in Poisson's equation as zero.

The constants of the problem are set by DATA statements and DO loops; these would be modified for a different problem. As shown, they apply to the example problem of Section 7.5 in which the steady-state temperatures within a rectangular region were determined with 105 interior points. The value of the over-relaxation factor was varied to generate the data of Table 7.3. The program can be readily adapted to other problems with a rectangular region and Dirichlet conditions.

```
C   A PROGRAM TO SOLVE POISSONS EQUATION ON A RECTANGULAR REGION
C   THE ACCELERATED LIEBMANNS (OVER/RELAXATION) METHOD IS USED.
C   PARAMETERS ARE -
C     NWIDE      NUMBER OF NODES IN X-DIRECTION
C     NHIGH      NUMBER OF NODES IN Y-DIRECTION
C     F(X,Y)     R.H.S. FUNVTION FOR POISSON EQUATION
C     TOL        TOLERANCE TO STOP ITERATIONS
C     W          OVER-RELAXATION FACTOR
C     H          MESH SIZE
      DIMENSION U(100,100)
C   DEFINE THE FUNCTION WITH AN ARITHMETIC STATEMENT FUNCTION
      F(X,Y) = 0.0
C   USE A DATA STATEMENT TO INITIALIZE SOME DATA
      DATA NWIDE, NHIGH, TOL, W, H /16,8, 0.0010, 1.0, 2.5/
      NHP1 = NHIGH + 1
      NWP1 = NWIDE + 1
C   USE DO LOOPS TO ESTABLISH BOUNDARY VALUES
      DO 1 I = 1,NHP1
        U(I,1) = 0.0
        U(I,NWP1) = 100.0
1     CONTINUE
      DO 2 I = 2,NWIDE
        U(1,I) = 0.0
        U(NHP1,I) = 0.0
2     CONTINUE
C   COMPUTE MEAN VALUE OF THE BOUNDARY VALUES TO USE FOR INITIAL VALUE
C   OF INTERIOR POINTS
5     SUM = 0.
      DO 10 I = 1,NHP1
        SUM = SUM + U(I,1) + U(I,NWP1)
10    CONTINUE
      DO 20 I = 2,NWIDE
        SUM = SUM + U(1,I) + U(NHP1,I)
20    CONTINUE
      UAVG = SUM/FLOAT(2*NWP1 + 2*(NHIGH - 1))
      X = 0.
      Y = 0.
      DO 30 I = 2,NHIGH
        DO 30 J = 2,NWIDE
      U(I,J) = UAVG + H*H*F(X,Y)
30    CONTINUE
C   PRINT A HEADING
      WRITE (6,199) W
199   FORMAT (1H1, 42HITERATIONS WITH OVER-RELAXATION FACTOR OF , F5.2)
C   BEGIN THE ITERATIONS.  LIMIT TO 100 ITERATIONS OR
C     UNTIL THE MAX CHANGE IN U VALUES .LE. TOL
      DO 50 KNT = 1,100
        CHGMAX = 0.
        DO 40 I = 2,NHIGH
          DO 35 J = 2,NWIDE
            RESID = W/4.*(U(I+1,J)+U(I-1,J)+U(I,J+1)+U(I,J-1)-4.*U(I,J)
     +             + H*H*F(X,Y))
            IF(CHGMAX .LT. ABS(RESID))  CHGMAX = ABS(RESID)
            U(I,J) = U(I,J) + RESID
35        CONTINUE
40      CONTINUE
        IF (CHGMAX .LT. TOL)  GO TO 55
50    CONTINUE
55    WRITE (6,200) KNT, CHGMAX
200   FORMAT (1H0, 13HAFTER ITER NO , IO, 20H  MAX CHANGE IN U = ,
     +  F8.4, 13H U MATRIX IS /)
      DO 45 I = 1,NHP1
45    WRITE (6,201) (U(I,J), J = 1,NWP1)
201   FORMAT (1H , 10F13.3)
      W = W + 0.1
      IF (W .LT. 1.8)  GO TO 5
      STOP
      END
```

Figure 7.20 Program 2

Program 2 begins the iterations by setting all interior points at the average potential of the boundary points. If better starting values are known, they might be read in rather than computing an average potential. If $f(x, y)$ is variable over the region, appropriate values of the x- and y-coordinates of each point will need to be defined within the DO loops that end on statements 35 and 40, so that the arithmetic-statement function will be correctly evaluated.

If one does not wish to investigate the effect of ω on the number of iterations required, the program would be run with only one value for ω, presumably that calculated from Eq. (7.15).

The A.D.I. method is the basis for Program 3 (see Fig. 7.21). It solves for the steady-state potential from Laplace's equation based on a rectangle. The size of the region (as specified by M and N, the number of intervals in the x- and y-directions), the acceleration factor ρ, and constant boundary conditions are defined by DATA statements; DATA statements are also used to insert ones into the off-diagonal elements of the u- and v-coefficient matrices. Program 3 solves a problem with 105 interior points.

The u- and v-coefficient matrices are established, as are the boundary-value terms for computing the righthand sides for equations analogous to those of Section 7.11. The LU decompositions of the coefficient matrices are then found, in preparation for solving the tridiagonal systems.

After this preliminary work, the righthand-side vector for the first traverse is computed using initial estimates of the potential; then the equations are solved for the first iterate of the values of the potential at the interior points in the rectangular region.

These first calculated values are then used to compute the righthand-side vector for the vertical traverse; the alternate set of equations is solved for the second iteration.

Iterations are continued, alternating the direction of traversing the points, until the final solution is obtained. A better way to terminate the iterations would be to compare the change in components of the solution vector against a tolerance criterion, rather than to perform a fixed number of iterations as done here.

```
C    THIS PROGRAM SOLVES A STEADY STATE HEAT FLOW PROBLEM BY THE
C    ALTERNATING DIRECTION IMPLICIT METHOD.
C    PARAMETERS ARE -
C       U       VECTOR OF POTENTIAL VALUES FOR FOR ODD TRAVERSES
C       V       SAME FOR EVEN TRAVERSES
C       UCOEF   MATRIX OF COEFFICIENTS FOR ODD TRAVERSES
C       VCOEF   MATRIX OF COEFFICIENTS FOR EVEN TRAVERSES
C       BCNDU   VECTOR OF BOUNDARY VALUES FOR UCOEFF VALUES
C       BCNDV   SAME FOR VCOEF
C       TOP     VECTOR OF BOUNDARY VALUES ACROSS TOP
C       BOT     SAME FOR BOTTOM OF REGION
C       LFT     SAME FOR LEFT HAND EDGE
C       RT      SAME FOR RIGHT HAND EDGE
C       M       NUMBER OF ROWS OF NODE POINTS
C       N       NUMBER OF COLUMNS OF NODE POINTS
C       RHO     ACCELERATION FACTOR
C    SET UP THE VECTORS AND MATRICES
         REAL UCOEF(500,3), VCOEF(500,3), BCNDU(500), BCNDV(500)
         REAL TOP(100), BOT(100), LFT(100), RT(100), U(500), V(500)
C    INITIALIZE SOME VALUES WITH DATA STATEMENTS
         DATA UCOEF, VCOEF /3000*-1.0/
         DATA M, N, RHO /7, 15, 1.0/
         DATA V /500*0.0/
         DATA RT, LFT, TOP, BOT /100*100.0, 300*0.0/
         DATA BCNDU, BCNDV /1000*0.0/
C    ESTABLISH THE COEFFICIENT MATRICES
C    BY OVER-WRITING ON THE DIAGONAL AND CERTAIN OFF-DIAGONAL TERMS.
         MSIZE = N*M
         DO 10 I = 1,MSIZE
           UCOEF(I,2) = 1./RHO + 2.0
           VCOEF(I,2) = 1./RHO + 2.0
10       CONTINUE
         ML1 = M - 1
         NL1 = N - 1
         MSL1 = MSIZE - 1
         DO 20 I = N,MSL1,N
           UCOEF(I,3) = 0.0
           UCOEF(I+1,1) = 0.0
20       CONTINUE
         DO 30 I = M, MSL1, M
           VCOEF(I,3) = 0.0
           VCOEF(I+1,1) = 0.0
30       CONTINUE
C    NOW GET VALUES INTO THE BCOND VECTORS
         DO 40 I = 1,N
           BCNDU(I) = TOP(I)
           J = MSIZE - N + I
           BCNDU(J) = BOT(I)
40       CONTINUE
         DO 45 I = 1,M
           J = (I-1)*N + 1
           BCNDU(J) = BCNDU(J) + LFT(I)
           J = I*N
           BCNDU(J) = BCNDU(J) + RT(I)
45       CONTINUE
```

Figure 7.21 Program 3

```
          DO 50 I = 1,M
          BCNDV(I) = LFT(I)
            J = MSIZE - M + I
            BCNDV(J) = RT(I)
50        CONTINUE
          DO 55 I = 1,N
            J = (I-1)*M + 1
            BCNDV(J) = BCNDV(J) + TOP(I)
            J = I*M
            BCNDV(J) = BCNDV(J) + BOT(I)
55        CONTINUE
C   WE NOW GET THE LU DECOMPOSITIONS OF THE TWO COEFFICIENT MATRICES
C   THESE ARE VERY EASY TO COMPUTE.  WE STORE THEM BACK
C   IN THE ORIGINAL VECTOR SPACE.
          DO 60 I = 2,MSIZE
            UCOEF(I-1,3) = UCOEF(I-1,3)/UCOEF(I-1,2)
            UCOEF(I,2) = UCOEF(I,2) - UCOEF(I,1)*UCOEF(I-1,3)
            VCOEF(I-1,3) = VCOEF(I-1,3)/VCOEF(I-1,2)
            VCOEF(I,2) = VCOEF(I,2) - VCOEF(I,1)*VCOEF(I-1,3)
60        CONTINUE
C   NOW WE BEGIN THE ITERATIONS, LIMIT THEM TO ITMAX IN NUMBER.
          ITMAX = 30
          DO 190 KNT = 2,ITMAX,2
C   COMPUTE R.H.S FOR THE U EQUATIONS
C   STORE THESE IN THE U VECTOR.
C   FIRST DO THE TOP AND BOTTOM SETS OF TERMS
          DO 65 I = 1,N
            J = (I-1)*M + 1
            U(I) = (1.0/RHO-2.0)*V(J) + V(J+1) + BCNDU(I)
            K = MSIZE - N + I
            J = I*M
            U(K) = V(J-1) + (1.0/RHO - 2.0)*V(J) + BCNDU(K)
65        CONTINUE
C   NOW DO THE INTERMEDIATE ONES
          DO 75 I = 2,ML1
            DO 70 J = 1,N
              K = (I-1)*N + J
              L = I + (J-1)*M
              U(K) = V(L-1) + (1.0/RHO-2.0)*V(L) + V(L+1) + BCNDU(K)
70          CONTINUE
75        CONTINUE
C   NOW GET SOLUTION FOR THE HORIZONTAL TRAVERSE.  FIRST Y = L(-1)*B.
          U(1) = U(1)/UCOEF(1,2)
          DO 80 I = 2,MSIZE
            U(I) = (U(I) - UCOEF(I,1)*U(I-1))/UCOEF(I,2)
80        CONTINUE
C   READY NOW TO GET X = U(-1)*Y.
          DO 90 I = 1,MSL1
            JROW = MSIZE - I
            U(JROW) = U(JROW) - UCOEF(JROW,3)*U(JROW+1)
90        CONTINUE
C   WE DO THE SAME FOR THE VERTICAL TRAVERSE - U AND V EXCHANGE ROLES.
C   COMPUTE R.H.S. FOR THE V EQUATIONS, STORE IN V
C   DO THE TOP AND BOTTOM SETS
          DO 95 I = 1,M
            J = (I-1)*N + 1
            V(I) = (1.0/RHO-2.0)*U(J) + U(J+1) + BCNDV(I)
```

Figure 7.21 (continued)

```
            K = MSIZE - M + I
            J = I*N
            V(K) = U(J-1) + (1.0/RHO-2.0)*U(J) + BCNDV(K)
95    CONTINUE
C     DO THE INTERMEDIATE ROWS
         DO 105 I = 2,NL1
            DO 100 J = 1,M
               K = (I-1)*M + J
               L = I + (J-1)*N
               V(K) = U(L-1) + (1.0/RHO-2.0)*U(L) + U(L+1) + BCNDV(K)
100         CONTINUE
105      CONTINUE
C     GET THE SOLUTION - FIRST Y = L(-1)*B
         V(1) = V(1)/VCOEF(1,2)
         DO 110 I = 2,MSIZE
            V(I) = (V(I) - VCOEF(I,1)*V(I-1))/VCOEF(I,2)
110      CONTINUE
C     THEN X = U(-1)*Y
         DO 120 I = 1,MSL1
            JROW = MSIZE - I
            V(JROW) = V(JROW) - VCOEF(JROW,3)*V(JROW+1)
120      CONTINUE
C     PRINT OUT THE LATEST RESULT
         WRITE (6,202) KNT, (V(I), I = 1,MSIZE)
202      FORMAT(/// 1H , 24H AFTER ITERATION NUMBER , I3/(1H , 15F8.4))
C     END OF THE ITERATION LOOP.
190      CONTINUE
         STOP
         END
```

Figure 7.21 (concluded)

EXERCISES

Section 7.2

1. In Section 7.2, it is shown that flow of heat is a phenomenon for which the rate of flow is equal to

$$-kA\frac{\partial u}{\partial x},$$

where k is a constant, A is the cross-sectional area for heat flow, and $\partial u/\partial x$ is the temperature gradient. Suppose that k, instead of being constant, is a function of the position, (x, y). Derive the equation for net flow of heat into the element $dx\, dy$. Does this reduce to Laplace's equation?

2. Repeat Exercise 1, but for k varying with temperature, such as $k = a + bu + cu^2$. Does the equation for net heat flow now reduce to Laplace's equation? How does the equation differ from that in Exercise 1?

Section 7.3

▶3. The mixed second derivative $\partial^2 u/(\partial x\, \partial y)$ can be considered as

$$\frac{\partial}{\partial x}\left(\frac{\partial u}{\partial y}\right) = \frac{\partial^2 u}{\partial x\, \partial y} = \frac{\partial}{\partial y}\left(\frac{\partial u}{\partial x}\right).$$

Show that, in terms of finite-difference quotients with $\Delta x = \Delta y$, this derivative can be approximated by the pictorial operator

$$\frac{1}{4h^2}\begin{Bmatrix}-1 & & 1\\ 1 & & -1\end{Bmatrix}u_{i,j} + O(h^2).$$

4. If

$$\frac{d^2u}{dx^2} = \frac{-u_{i+2} + 16u_{i+1} - 30u_i + 16u_{i-1} - u_{i-2}}{12h^2} + O(h^4),$$

find the fourth-order operator for the Laplacian. Assume that the function u has a continuous sixth derivative.

5. In some cases it is necessary to approximate a differential equation that contains the third derivative, $\partial^3 u/\partial x^3$, by a difference equation. Set up an expression to approximate this third derivative in terms of finite differences. Can you find an expression that uses function values symmetrical to the point at which the derivative is being evaluated? (In finding the expressions, you may wish to compare the Taylor-series method of Section 7.3 with the methods of Chapter 3 and the method of undetermined coefficients in the Appendix.)

6. Suppose we have a rectangular plate with top and bottom edges that are perfectly insulated ($\partial u/\partial y = 0$) while the righthand edge is held at 100°, and the left edge at 0°. If heat flows in only two directions, it is obvious that the temperatures vary linearly in the x-direction, and are constant along vertical lines.

   a) Show that such a temperature distribution satisfies Eq. (7.6).

   b) Show that this temperature distribution also obeys the relationship derived in Exercise 4. What about points adjacent to the edges of the plate?

Section 7.4

▶7. Set up equations analogous to Eqs. (7.8) for the example problem of Section 7.4, but with a grid spacing of $3\frac{1}{3}$ cm. Solve the set of equations by elimination.

8. Solve for the steady-state temperatures in a rectangular plate, 12 by 15 inches, if one 15-inch edge is held at 100° while the other edges are all held at 20°. The material is aluminum. Take $\Delta x = \Delta y = 3$ inches, and consider heat to flow only in the lateral directions. Sketch in the approximate location of the 50° isothermal curve.

9. Solve for the temperatures in the plate of Exercise 8 when the edge temperatures are held as shown in Fig. 7.22.

Figure 7.22

►10. The region for which we can solve Laplace's equation does not have to be rectangular. We can apply the methods of Section 7.4 for any region so long as the meshes of our network coincide with the boundary. Solve for the steady-state potentials at the eight interior points shown in Fig. 7.23.

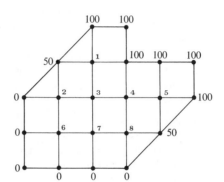

Figure 7.23

11. What fraction of the elements in the coefficient matrix of Exercise 8 are nonzero? What fraction if $h = k = 1$ inch? What fraction if $h = k = 0.1$ inch? How many equations must be solved in the latter case? Will your computer system handle that many coefficients in memory at the same time?

Section 7.5

►12. Repeat Exercise 7, but use Liebmann's method.

13. Solve Exercise 9 by Liebmann's method with all the elements of the initial u vector equal to zero. Repeat with all the elements of the initial vector equal to 300, the upper bound to the steady-state temperatures. Repeat once more with the elements of the initial vector all equal to the arithmetic average of the boundary temperatures. Compare the number of iterations needed to reach a given tolerance for convergence in each case.

14. Solve Exercise 8 by Liebmann's method. Before you begin, make some quick hand computations to allow good estimates of starting values for the initial u vector.

►15. Repeat Exercise 7, employing successive over-relaxation. Vary the over-relaxation factor to determine the optimum value. How does this compare to that predicted by Eq. (7.15)?

16. Repeat Exercise 13, but use S.O.R. with the value of ω_{opt} as given by Eq. (7.15).

17. Repeat Exercise 14, using S.O.R. with the value of ω_{opt} as given by Eq. (7.15).

Section 7.6

▶18. Find the torsion function ϕ for a 2×2 in. square bar.

    a)   Subdivide the cross section into nine equal squares, so that there are four interior points. Because of symmetry, all four values of ϕ are the same.

    b)   Then subdivide the cross section into 36 equal squares so that there are 25 interior points. Use the results of (a) to estimate starting values for iteration.

19. Solve for the torsion in a hollow square bar, 5 inches in outside dimension, and with walls 2 inches thick (so that the inside hole is 1 inch on a side). On the inner surface as well as on the outer, $\phi = 0$.

20. Solve for the torsion in a prismatic bar of a cross section similar to the region of Exercise 10. Assume the eight points are 0.5 m apart.

21. Solve $\nabla^2 u = f(x, y)$ on the square region bounded by

$$x = 0, \qquad x = 1, \qquad y = 0, \qquad y = 1,$$

with $f(x, y) = xy$. Use $h = \frac{1}{3}$. Take $u = 0$ on the boundary.

▶22. Repeat Exercise 18(b), but use S.O.R. Vary the over-relaxation factor to find the optimum. How does this compare to the value from Eq. (7.15)?

23. Repeat Exercise 19, using S.O.R. Vary ω to find the optimum value.

24. Repeat Exercise 21, using $f(x, y) = (1 - x)(1 - y)$. Compare your results to those of Exercise 21.

Section 7.7

25. Solve the set of equations in Eqs. (7.22), using (7.23) to eliminate u_a, u_b, u_c, and u_d.

    a)   Use elimination.

    b)   Use Liebmann's method.

    c)   Use S.O.R.

▶26. Solve the example problem of Section 7.7, except that the 8 cm edges are held at 20° and the 4 cm edges have an outward gradient of $-15°C/cm$.

27. Suppose the outward normal gradient on all the edges of the example problem of Section 7.7 is $-15°C/cm$. Can the problem be solved?

28. Solve for the steady-state temperatures of the region in Exercise 10, except that the plate is insulated at each exterior point marked zero. The edge temperatures are maintained at the temperatures given at the other exterior points.

29. Consider a region which is obtained from Exercise 10 by reflecting the figure across both of the edges marked zero (see Fig. 7.24). Show that the steady potentials at points in the original region are the same as for Exercise 28.

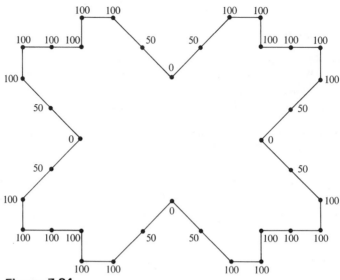

Figure 7.24

Section 7.8

▶ 30. Find the potential distribution in a region whose shape is a 3–4–5 right triangle. The potential is maintained at zero except on the hypotenuse, where it is 50. Use a square mesh with $h = 1$.

31. A coaxial cable has a circular outer conductor 10 cm in diameter, and a square inner conductor 3 cm on a side. (The square conductor is concentric.) The outer conductor is at zero volts while the inner is at 100 volts. Find the potential between the conductors. (Note that only one octant needs to be calculated, because of symmetry.) Use a grid of 1-cm squares.

32. If a hollow shaft has an outer circular cross section of diameter 10 in. and a concentric square inner cross section which is 3 inches on a side, what is the value of the torsion function at the nodes of a 1-inch grid?

Section 7.9

33. Using the Laplacian operator for an equispaced triangular network, determine the potential in the region shown in Fig. 7.25, if $\nabla^2 p = 0$.

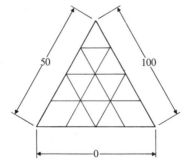

Figure 7.25

34. Find the solution to the Poisson equation over an equilateral triangle 5 inches on a side if

$$\nabla^2 u = k, \qquad u = 0 \text{ on boundary,}$$

and values of k are as shown in Fig. 7.26.

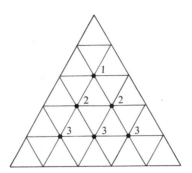

Figure 7.26

▶35. Use the Laplacian in polar coordinates to set up the set of difference equations to solve

$$\nabla^2 u = 0.2,$$

on a semicircular region whose radius is 4. Take $\Delta r = 1$ and $\Delta\theta = \pi/6$. Boundary conditions are $u = 10$ on the straight edge and $u = 0$ on the curved boundary.

Section 7.10

▶36. A cube is 3 cm along its edge. Two opposite faces are held at 100°, the other four faces at 0°. Find the internal temperatures at the nodes of a 1-cm network.

37. Repeat Exercise 36 except that two adjacent faces are held at 100° while the rest are held at 0°.

38. Repeat Exercise 36 except that one of the 0° faces has been insulated.

Section 7.11

▶39. Solve Exercise 9 by the A.D.I. method, using $\rho = 1.0$. Begin with the initial u vector having all elements equal to the arithmetic average of the boundary values. Compare the number of iterations to those required with Liebmann's method (Exercise 13) and S.O.R. using ω_{opt} (Exercise 16).

40. Solve the sets of equations in (7.22) and (7.23) by A.D.I. with $\rho = 1.0$. Compare the number of iterations required with the number needed by S.O.R. as performed in Exercise 25(c).

41. Repeat Exercise 18, but use the A.D.I. method. Vary the value of ρ to find the optimum value.

42. Repeat Exercise 21, using the A.D.I. method. Vary the value of ρ to find the optimum value.

PROJECTS AND APPLIED PROBLEMS

43. Modify and use the program for Poisson's equation in Section 7.12 to solve:

$$\nabla^2 u = xy(x - 2)(y - 2),$$

on the region $0 \leqslant x \leqslant 2$, $0 \leqslant y \leqslant 2$, with $u = 0$ on all boundaries except for $y = 0$, where $u = 1.0$.

44. Modify Program 2 so that it permits boundary conditions of the form

$$au + b\frac{\partial u}{\partial x} = c \quad \text{and} \quad a'u + b'\frac{\partial u}{\partial y} = c'.$$

Test it with the equation in Problem 43, modified in that the boundary where $y = 0$ has $\partial u/\partial y = 1$.

45. Repeat Problem 44, except make the modifications to Program 3. You will need to make changes to permit a Poisson-equation problem as well as derivative boundary conditions.

46. A classic problem in elliptic partial-differential equations is to solve $\nabla^2 u = 0$ on a region defined by $0 \leqslant x \leqslant \pi$, $0 \leqslant y \leqslant \infty$, with boundary condition of $u = 0$ at $x = 0$, at $x = \pi$, and at $y = \infty$. The boundary at $y = 0$ is held at $u = F(x)$. This can be quite readily solved by the method of separation of variables, to give the series solution

$$u = \sum_{n=1}^{\infty} B_n e^{-ny} \sin nx,$$

with

$$B_n = 2\int_0^\pi F(x) \sin nx \, dx.$$

Solve this equation numerically for various definitions of $F(x)$. (You will need to redefine the region so that $0 \leqslant y \leqslant M$, where M is large enough that changes in u with y at $y = M$ are negligible.) Compare your results to the series solution. You might try:

$F(x) = 100;$ $F(x) = 100 \sin x;$ $F(x) = 400x(1 - x);$ $F(x) = 100(1 - |2x - 1|)$.

47. The equation

$$2\frac{\partial^2 u}{\partial x^2} + \frac{\partial^2 u}{\partial y^2} - \frac{\partial u}{\partial x} = 2$$

is an elliptic equation. Solve it on the unit square, subject to $u = 0$ on the boundaries. Approximate the first derivative by a central-difference approximation. Investigate the effect of size of Δx on the results, to determine at what size reducing it does not have further effect.

48. Solve the steady-state heat-flow problem that was discussed at the beginning of the chapter. Assume that the metal is aluminum and that the insulation is mineral wool. Look up the thermal properties in handbooks.

Solve it first as a three-dimensional problem. Then, considering that the plate is not very thick relative to its length and that aluminum is a good conductor so that there is little temperature variation across the thickness, solve it as a two-dimensional problem. Do you reach the same conclusion about the feasibility of the scheme?

Consider carefully how you will handle the problem of dissimilar materials—metal and insulation. Can you approximate the effect of the insulation by some proper value of the gradient at the surface of the aluminum?

Chapter Eight

Parabolic Partial-Differential Equations

The previous chapter discussed the solution of partial-differential equations that were time-independent. Such steady-state problems were described by elliptic equations. Unsteady-state problems in which the function is dependent on time are of great importance. We study these in this chapter.

The problem of unsteady-state flow of heat is a physical situation that can be represented by a parabolic partial-differential equation. The simplest situation is for flow of heat in one direction. Imagine a rod, uniform in cross section and insulated around its perimeter so that heat flows only longitudinally. Consider a differential portion of the rod, dx in length with cross-sectional area A (see Fig. 8.1).

We let u represent the temperature at any point in the rod, whose distance from the left end is x. Heat is flowing from left to right under the influence of the temperature gradient $\partial u/\partial x$. Make a balance of the rate of heat flow into and out of the element. Use k for thermal conductivity, cal/g · cm^2 · °C/cm, which we assume is constant,

Rate of flow of heat in: $-kA\dfrac{\partial u}{\partial x}$;

Rate of flow of heat out: $-kA\left(\dfrac{\partial u}{\partial x} + \dfrac{\partial}{\partial x}\left(\dfrac{\partial u}{\partial x}\right)dx\right).$

The difference between the rate of flow in and the rate of flow out is the rate at which heat is being stored in the element. If c is the heat capacity, cal/g · °C, and ρ is the density, g/cm^3, we have, with t for time,

$$-kA\frac{\partial u}{\partial x} - \left(-kA\left(\frac{\partial u}{\partial x} + \frac{\partial}{\partial x}\left(\frac{\partial u}{\partial x}\right)dx\right)\right) = c\rho(A\,dx)\frac{\partial u}{\partial t}.$$

Figure 8.1

Simplifying, we have

$$k \frac{\partial^2 u}{\partial x^2} = c\rho \frac{\partial u}{\partial t}.$$

This is the basic mathematical model for unsteady-state flow. We have derived it for heat flow, but it applies equally to diffusion of material, flow of fluids (under conditions of laminar flow), flow of electricity in cables (the telegraph equations), etc.

The differential equation is classed as parabolic because comparing to the standard form of an equation in two independent variables, x and t,

$$A \frac{\partial^2 u}{\partial x^2} + B \frac{\partial^2 u}{\partial x \, \partial t} + C \frac{\partial^2 u}{\partial t^2} + D\left(x, t, u, \frac{\partial u}{\partial x}, \frac{\partial u}{\partial t}\right) = 0,$$

we find that $B^2 - 4AC = 0$.

In two- or three-space dimensions, analogous equations apply:

$$k\left(\frac{\partial^2 u}{\partial x^2} + \frac{\partial^2 u}{\partial y^2}\right) = c\rho \frac{\partial u}{\partial t},$$

$$k\left(\frac{\partial^2 u}{\partial x^2} + \frac{\partial^2 u}{\partial y^2} + \frac{\partial^2 u}{\partial z^2}\right) = c\rho \frac{\partial u}{\partial t}.$$

The function that we call the solution to the problem not only must obey the differential equation given above, but must also satisfy an initial condition and a set of boundary conditions. For the one-dimensional heat-flow problem that we first consider, the initial condition will be the initial temperatures at all points along the rod,

$$u(x, t)|_{t=0} = u(x, 0) = f(x).$$

The boundary conditions will describe the temperature at each end of the rod as functions of time. Our first examples will consider the case where these temperatures are held constant:

$$u(0, t) = c_1,$$
$$u(L, t) = c_2.$$

More general (and more practical) boundary conditions will involve not only the temperature but the temperature gradients; and these may vary with time,

$$A_1 u(0, t) + B_1 \frac{\partial u(0, t)}{\partial x} = F_1(t),$$

$$A_2 u(L, t) + B_2 \frac{\partial u(L, t)}{\partial x} = F_2(t).$$

8.1 THE EXPLICIT METHOD

Our approach to solving parabolic partial-differential equations by a numerical method is to replace the partial derivatives by finite-difference approximations. For the one-dimensional heat-flow equation,

$$\frac{\partial^2 u}{\partial x^2} = \frac{c\rho}{k} \frac{\partial u}{\partial t}, \tag{8.1}$$

we can use the relations

$$\left. \frac{\partial^2 u}{\partial x^2} \right|_{\substack{x = x_i \\ t = t_i}} = \frac{u_{i+1}^i - 2u_i^i + u_{i-1}^i}{(\Delta x)^2} + O(\Delta x)^2, \tag{8.2}$$

and

$$\left. \frac{\partial u}{\partial t} \right|_{\substack{x = x_i \\ t = t_i}} = \frac{u_i^{j+1} - u_i^j}{\Delta t} + O(\Delta t). \tag{8.3}$$

We use subscripts to denote position and superscripts for time. Note that the error terms are of different orders since a forward difference is used in Eq. (8.3). This introduces some special limitations, but when this is done, the procedure is simplified.

Substituting Eqs. (8.2) and (8.3) into (8.1) and solving for u_i^{j+1} gives the equation for the forward-difference method:

$$u_i^{j+1} = \frac{k \Delta t}{c\rho(\Delta x)^2} (u_{i+1}^i + u_{i-1}^i) + \left(1 - \frac{2k \Delta t}{c\rho(\Delta x)^2} \right) u_i^i. \tag{8.4}$$

We have solved for u_i^{j+1} in terms of the temperatures at time t_j in Eq. (8.4) in view of the normally known conditions for a parabolic partial-differential equation. We subdivide the length into uniform subintervals and apply our finite-difference approximation to Eq. (8.1) at each point where u is not known. Equation (8.4) then gives the values of u at each interior point at $t = t_1$ since the values at $t = t_0$ are given by the initial conditions. It then can be used to get values at t_2 using the values at t_1 as initial conditions, so we can step the solution forward in time. At the endpoints, the boundary conditions will determine u.

The relative size of the time and distance steps, Δt and Δx, affects Eq. (8.4). If the ratio of $\Delta t/(\Delta x)^2$ is chosen so that $k\,\Delta t/c\rho(\Delta x)^2 = \frac{1}{2}$, the equation is simplified in that the last term vanishes and we have

$$u_i^{j+1} = \tfrac{1}{2}(u_{i+1}^i + u_{i-1}^i). \tag{8.5}$$

If the value of $k\,\Delta t/cp(\Delta x)^2$ is chosen as less than one-half, there will be improved accuracy* (limited, of course, by the errors dependent upon the size of Δx). If the value is chosen greater than one-half, which would reduce the number of calculations required to advance the solution through a given interval of time, the phenomenon of *instability* sets in. (We show later in this chapter why this ratio affects stability and convergence.)

We illustrate the method with a simple example, varying the value of $k\,\Delta t/cp(\Delta x)^2$ to demonstrate its effect.

Example A large flat steel plate is 2 cm thick. If the initial temperatures (°C) within the plate are given, as a function of the distance from one face, by the equations

$$u = 100x \qquad \text{for } 1 \leqslant x \leqslant 1,$$
$$u = 100(2 - x) \qquad \text{for } 1 \leqslant x \leqslant 2,$$

find the temperatures as a function of x and t if both faces are maintained at 0°C.

Since the plate is large, we can neglect lateral flow of heat relative to the flow perpendicular to the faces and hence use Eq. (8.1) for heat flow in one direction. For steel, $k = 0.13$ cal/sec · cm · °C, $c = 0.11$ cal/g · °C and $\rho = 7.8$ g/cm^3. In order to use Eq. (8.5) as an approximation to the physical problem, we subdivide the total thickness into an integral number of spaces. Let us use $\Delta x = 0.25$, giving eight subdivisions. If we wish to use Eq. (8.5), we then fix Δt by the relation

$$\frac{k\,\Delta t}{cp(\Delta x)^2} = \frac{1}{2}, \qquad \Delta t = \frac{(0.11)(7.8)(0.25)^2}{(2)(0.13)} = 0.206 \text{ second.}$$

The boundary conditions are

$$u(0, t) = 0, \qquad u(2, t) = 0.$$

The initial condition is

$$u(x, 0) = 100x \qquad \text{for } 0 \leqslant x \leqslant 1,$$
$$u(x, 0) = 100(2 - x) \qquad \text{for } 1 \leqslant x \leqslant 2.$$

Our calculations are conveniently recorded in a table, as in Table 8.1, where each row of figures is at a particular time. We begin by filling in the initial conditions along the first row, at $t = 0.0$. The simple algorithm of Eq. (8.5) tells

---

* It can be shown that choosing the ratio equal to $\frac{1}{6}$ is especially advantageous in minimizing the truncation error.

Table 8.1 Numerical solution to heat-flow example.

			Calculated temperatures at:			
t	$x = 0$	$x = 0.25$	$x = 0.50$	$x = 0.75$	$x = 1.0$	$x = 1.25$
0.0	0	25.0	50.0	75.0	100.0	75.0
0.206	0	25.0	50.0	75.0	75.0	75.0
0.412	0	25.0	50.0	62.5	75.0	62.5
0.619	0	25.0	43.75	62.5	62.5	62.5
0.825	0	21.88	43.75	53.12	62.5	53.12
1.031	0	21.88	37.5	53.12	53.12	53.12
1.238	0	18.75	37.5	45.31	53.12	45.31
1.444	0	18.75	32.03	45.31	45.31	45.31
1.650	0	16.16	32.03	38.67	45.31	38.67
1.856	0	16.16	27.34	38.67	38.67	38.67
2.062	0	13.67	27.34	33.01	38.67	33.01

us that, at each interior point, the temperature at any point at the end of a time step is just the arithmetic average of the temperatures at the adjacent points at the beginning of that time step. The end temperatures are given by the boundary conditions. Because the temperatures are symmetrical on either side of the center line, we compute only for $x \leq 1.0$. The temperature at $x = 1.25$ is the same as at $x = 0.75$.

In Fig. 8.2, we compare some of the numerical results with the analytical solution, which is:

$$u = 800 \sum_{n=0}^{\infty} \frac{1}{\pi^2(2n + 1)^2} \cos \frac{(2n + 1)\pi(x - 1)}{2} e^{-0.3738(2n+1)^2 t}.$$

Note that the numerical values are close to and oscillate about the curves which are drawn to represent the analytical solution. In general the errors of Table 8.1 are less than 4%. If the size of Δx were less, the errors would be smaller. Unfortunately, this easy method is not always so accurate.

When the value for $k \, \Delta t / c \rho (\Delta x)^2$ in Eq. (8.4) is other than 0.5, the resulting equation is slightly more complicated, but the related computer program is not significantly slower. The major effect is the size of the time steps; reducing the value requires more successive calculations to reach a given time after the start of heat flow. Increase of the ratio to a value greater than 0.5 would decrease the number of successive calculations and hence reduce the computer time needed to solve a given problem. A most important phenomenon occurs, however, when the ratio is >0.5; as illustrated in Fig. 8.3 where temperature profiles are shown at two different times, $t = 0.99 \sec$ and $t = 1.98 \sec$, very inaccurate results may occur. The curves show the solution from the infinite series. The open circles, with $r = 0.6$, show extreme oscillation: the calculated values are very imprecise, and

Figure 8.2

Temperature, u

$x = 0.75$ cm

$x = 0.25$ cm

Time steps

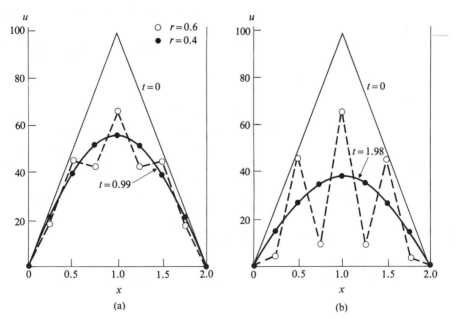

Figure 8.3

○ $r = 0.6$
● $r = 0.4$

$t = 0$

$t = 0.99$

$t = 0$

$t = 1.98$

(a)

(b)

completely impossible behavior is represented. The data for $r = 0.4$ are smooth and almost exactly match the series solution.

The reason for this great difference in behavior is that the choice of the ratio $k \, \Delta t / cp(\Delta x)^2 = 0.6$ introduces instability in which errors grow at an accelerating rate as time increases. (At values of $t > 2$, negative values of u are computed, a patently impossible situation.) As indicated in Fig. 8.3, when this ratio is taken at 0.4 the results are excellent.

In Fig. 8.2 with $k \, \Delta t / cp(\Delta x)^2 = 0.5$, the limiting value to avoid instability was used. Even here, the oscillation of points around the curves indicates its borderline value, and the accuracy is hardly acceptable. The phenomena of stability and convergence, more fully discussed in a later section of this chapter, set a maximum value of 0.5 to the ratio $k \, \Delta t / cp(\Delta x)^2$. When derivative boundary conditions are involved, the ratio must be less than 0.5. Discontinuities in the initial conditions—even a discontinuity in the gradient, $\partial u / \partial x$, at $t = 0$—cause the accuracy to be poor at the maximum value. The next example will demonstrate this.

We will now solve a problem in diffusion, which is governed by the same mathematical equation as is heat conduction in a solid.

Example A hollow tube 20 cm long is initially filled with air containing 2% of ethyl alcohol vapors. At the bottom of the tube is a pool of alcohol which evaporates into the stagnant gas above. (Heat transfers to the alcohol from the surroundings to maintain a constant temperature of 30°C, at which temperature the vapor pressure is 0.1 atm.) At the upper end of the tube, the alcohol vapors dissipate to the outside air, so the concentration is essentially zero. Considering only the effects of molecular diffusion, determine the concentration of alcohol as a function of time and the distance x measured from the top of the tube.

Molecular diffusion follows the law

$$\frac{\partial c}{\partial t} = D \frac{\partial^2 c}{\partial x^2},$$

where D is the diffusion coefficient, with units of cm^2/sec in the cgs system. (This is the same as for the ratio k/cp, which is often termed *thermal diffusivity*.) For ethyl alcohol, $D = 0.119$ cm^2/sec at 30°C, and the vapor pressure is such that 10 volume percent alcohol in air is present at the surface.

Subdivide the length of the tube into five intervals, so $\Delta x = 4$ cm. Using the maximum value permitted for Δt yields

$$D \frac{\Delta t}{(\Delta x)^2} = \frac{1}{2} = 0.119 \frac{\Delta t}{(4)^2}, \qquad \Delta t = 67.2 \text{ sec.}$$

Our initial condition is $c(x, 0) = 2.0$. The boundary conditions are

$$c(0, t) = 0.0, \qquad c(20, t) = 10.0.$$

The concentrations are measured by the percent of alcohol vapor in the air.

Table 8.2 Diffusion of alcohol vapors in a tube—solution of the equation $u_i^{j+i} = \frac{1}{2}(u_{i-1}^j + u_{i+1}^j)$.

Time, seconds	Concentration of alcohol at:					
	$x = 0$	$x = 4$	$x = 8$	$x = 12$	$x = 16$	$x = 20$
0	~~2.0~~ 0.0	2.0	2.0	2.0	2.0	~~2.0~~ 10.00
67.2	0.0	1.00	2.00	2.00	6.00	10.0
134.4	0.0	1.00	1.50	4.00	6.00	10.0
201.6	0.0	0.75	2.50	3.75	7.00	10.0
268.8	0.0	1.25	2.25	4.75	6.875	10.0
336.0	0.0	1.125	3.00	4.562	7.375	10.0
403.2	0.0	1.500	2.844	5.188	7.281	10.0
470.4	0.0	1.422	3.344	5.062	7.594	10.0
537.6	0.0	1.672	3.242	5.469	7.531	10.0
604.8	0.0	1.621	3.570	5.386	7.734	10.0
Steady state	0.0	2.0	4.0	6.0	8.0	10.0

The computations are shown in Table 8.2. Again, using Eq. (8.5), each interior value of c is given by the arithmetic average of concentrations on either side in the line above. A little reflection is required to determine the proper concentrations to be used for $x = 0$ and $x = 20$ at $t = 0$. While these are initially at 2%, we assume they are changed instantaneously to 0% and 10% because the effective concentrations acting during the first time interval are not the initial values but the changed values. We have accordingly rewritten these values as shown in the first row of Table 8.2.

As time passes, the concentration will become a linear function of x, from 10% at $x = 20$ to 0% at $x = 0$. The calculated values approach the steady-state concentrations as time passes. However, as Fig. 8.4 shows, the successive calculated values oscillate about the values calculated from the analytical solution.*

While the algorithm of Eq. (8.5) is simple to use, the results leave something to be desired in the way of accuracy. The oscillatory nature of the results, when a smooth trend of the concentration is expected, is also of concern. We can improve accuracy by smaller steps, but if we decrease Δx, the time steps must also decrease, because the ratio $D \, \Delta t/(\Delta x)^2$ cannot exceed $\frac{1}{2}$. Cutting Δx in half will require making Δt one-fourth of its previous value, giving a total of eight times as many calculations. With the mixed order of error terms, it is not obvious what reduction of error this would give, but they should be reduced about fourfold.

We can reduce Δt without change in Δx, however, for the method is stable for any value of the ratio less than $\frac{1}{2}$. Suppose we make $D \, \Delta t/(\Delta x)^2 = \frac{1}{4}$. The basic

---

* In this instance the analytical solution is given by the infinite series shown in Fig. 8.4.

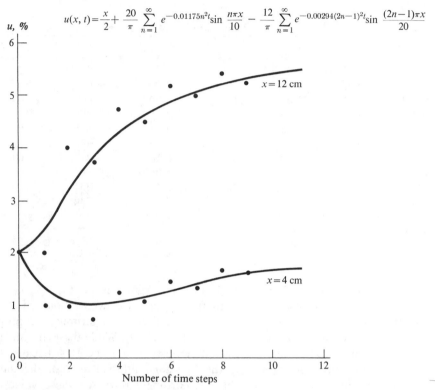

Analytical solution:

$$u(x, t)=\frac{x}{2}+ \frac{20}{\pi} \sum_{n=1}^{\infty} e^{-0.01175n^2 t}\sin \frac{n\pi x}{10} - \frac{12}{\pi} \sum_{n=1}^{\infty} e^{-0.00294(2n-1)^2 t}\sin \frac{(2n-1)\pi x}{20}$$

Figure 8.4

difference equation, (8.4), now becomes

$$u_i^{j+1} = \tfrac{1}{4}(u_{i+1}^{j} + u_{i-1}^{j}) + \tfrac{1}{2}u_i^{j}. \tag{8.6}$$

Table 8.3 summarizes the use of Eq. (8.6) on the same example as before. As shown more clearly by Fig. 8.5, this last calculation causes the calculated concentrations to follow smooth curves, and the error is reduced considerably. The poorest accuracy is near the beginning. The later values are close to the curve.

This initial poor accuracy is a result of the discontinuities in the boundary conditions. The abrupt change of concentrations at the ends of the tube makes the explicit method inaccurate. In a situation with continuity of the initial values of the potential plus continuity of its derivatives and the absence of discontinuities in the boundary conditions, one would get better accuracy throughout.

The method presented in this section is called the *explicit method* because each new value of the potential can be immediately calculated from quantities

Table 8.3 Diffusion of alcohol vapors in a tube: $u_i^{j+1} = \frac{1}{4}(u_{i-1}^j + u_{i+1}^j) + \frac{1}{2}u_i^j$.

Time, seconds	Concentration of alcohol at:					
	$x = 0$	$x = 4$	$x = 8$	$x = 12$	$x = 16$	$x = 20$
0.0	2.0̶ 0.0	2.0	2.0	2.0	2.0	1̶0̶.0̶ 0.0
33.6	0.0	1.50	2.00	2.00	4.00	10.0
67.2	0.0	1.25	1.875	2.50	5.00	10.0
100.8	0.0	1.094	1.875	2.969	5.625	10.0
134.4	0.0	1.015	1.953	3.360	6.055	10.0
168.0	0.0	0.996	2.070	3.682	6.368	10.0
201.6	0.0	1.015	2.204	3.950	6.604	10.0
235.2	0.0	1.058	2.343	4.177	6.790	10.0
268.8	0.0	1.115	2.480	4.372	6.939	10.0
302.4	0.0	1.177	2.612	4.541	7.062	10.0
336.0	0.0	1.241	2.736	4.689	7.166	10.0
369.6	0.0	1.304	2.850	4.820	7.255	10.0
403.2	0.0	1.364	2.956	4.936	7.332	10.0
436.8	0.0	1.421	3.053	5.040	7.400	10.0
470.4	0.0	1.474	3.142	5.133	7.460	10.0
504.0	0.0	1.522	3.223	5.217	7.513	10.0
537.6	0.0	1.567	3.296	5.292	7.561	10.0
Steady state	0.0	2.0	4.0	6.0	8.0	10.0

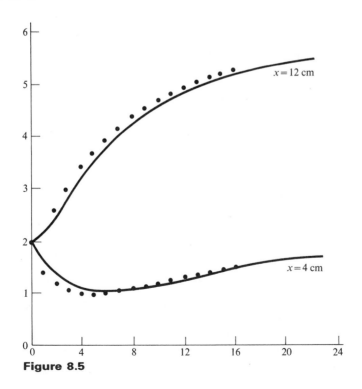

Figure 8.5

that are already known. It is simple and economical of calculation effort but has a severely limited upper value for the ratio $k\,\Delta t/cp(\Delta x)^2$. We can remove this limitation by going to an *implicit method*.

8.2 CRANK–NICOLSON METHOD

When the difference equation, Eq. (8.4), was derived, we noted that a mixed order of error was involved because a forward difference was used to replace the time derivative while a central difference was used for the distance derivative. However, the difference quotient, $(u_i^{j+1} - u_i^j)/\Delta t$, can be considered a central difference if we take it as corresponding to the midpoint of the time interval. Suppose we do consider $(u_i^{j+1} - u_i^j)/\Delta t$ as a central-difference approximation to $\partial u/\partial t$, and equate it to a central-difference quotient for the second derivative with distance, also corresponding to the midpoint in time, by averaging difference quotients at the beginning and end of the time step; then

$$\frac{\partial^2 u}{\partial x^2} = \frac{cp}{k}\frac{\partial u}{\partial t}$$

is approximated by:

$$\frac{1}{2}\left(\frac{u_{i+1}^j - 2u_i^j + u_{i-1}^j}{(\Delta x)^2} + \frac{u_{i+1}^{j+1} - 2u_i^{j+1} + u_{i-1}^{j+1}}{(\Delta x)^2}\right) = \frac{cp}{k}\left(\frac{u_i^{j+1} - u_i^j}{\Delta t}\right).$$

| Central difference at t_j | Central difference at t_{j+1} | Central difference at $t_{j+\frac{1}{2}}$ |

When this is rearranged we get the Crank–Nicolson formula, where

$$r = k\,\Delta t/cp(\Delta x)^2,$$

$$-ru_{i-1}^{j+1} + (2+2r)u_i^{j+1} - ru_{i+1}^{j+1} = ru_{i-1}^j + (2-2r)u_i^j + ru_{i+1}^j. \qquad (8.7)$$

Letting $r = 1$, we get some simplification:

$$-u_{i-1}^{j+1} + 4u_i^{j+1} - u_{i+1}^{j+1} = u_{i-1}^j + u_{i+1}^j.$$

As we will discuss in the next section, one advantage of the Crank–Nicolson method is that it is stable for any value of r, although small values are more accurate.

Equation (8.7) is the usual formula for using the Crank–Nicolson method. Note that the new temperature u_i^{j+1} is not given directly in terms of known temperatures one time step earlier, but is a function of unknown temperatures at adjacent positions as well. It is therefore termed an *implicit method* in distinction from the explicit method of the previous section. With implicit methods, the values of u at $t = t_1$ are not just a function of values at $t = t_0$, but also involve the other u values at the same time step. This requires us to solve a set of simultaneous equations at each time step.

We illustrate the method by solving the problem of diffusion of alcohol vapors, the same as that previously attacked by the explicit method. Again, let us take $\Delta x = 4$ cm and take $r = 1$. Restating the problem, we have

$$\frac{\partial^2 u}{\partial x^2} = \frac{1}{D}\frac{\partial u}{\partial t}, \qquad u(x, 0) = 2.0, \qquad \begin{cases} u(0, t) = 0, \\ u(20, t) = 10.0, \end{cases}$$

$$D = 0.119 \text{ cm}^2/\text{sec}.$$

If $D\,\Delta t/(\Delta x)^2 = r = 1$ and $\Delta x = 4$ cm, then $\Delta t = 134.4$ sec. For $t = 134.4$, at the end of the first time step, we write the equations at each point whose concentration is unknown. The effective concentrations at the two ends, 0% and 10%, respectively, are again used:

$$\begin{aligned}
-0.0 + 4u_1 - u_2 &= 0.0 + 2.0, \\
-u_1 + 4u_2 - u_3 &= 2.0 + 2.0, \\
-u_2 + 4u_3 - u_4 &= 2.0 + 2.0, \\
-u_3 + 4u_4 - 10.0 &= 2.0 + 10.0.
\end{aligned}$$

This method requires more work because, at each time step, we must solve a set of equations similar to the above. Fortunately, the system is tridiagonal. Since we will need to repeatedly solve the system with the same coefficient matrix (we transpose the constant values to the right side in the first and last equations), it is desirable to use an LU method. We then solve by back-substitutions, using the elements of L and U:

$$L = \begin{bmatrix} 1 & 0 & 0 & 0 \\ -1 & 3.75 & 0 & 0 \\ 0 & -1 & 3.733 & 0 \\ 0 & 0 & -1 & 3.732 \end{bmatrix}, \qquad U = \begin{bmatrix} 1 & -0.25 & 0 & 0 \\ 0 & 1 & -0.267 & 0 \\ 0 & 0 & 1 & -0.268 \\ 0 & 0 & 0 & 1 \end{bmatrix}.$$

Remember that, to solve $Ax = b = LUx$, we first solve $Ly = x$ (which requires no reduction of L because it is already triangular) and then we solve $Ux = y$ by back-substitution. Observe that when A is tridiagonal, its zeros are preserved in L and U, and, in fact we need to compute only $2(n - 1)$ new values to get the LU equivalent of the $n \times n$ tridiagonal matrix A. Not only is the computational effort minimized, but we can also minimize computer memory space by storing only the three nonzero values in each row. In fact, as discussed in Chapter 2, we can store the elements of both L and U in place of the original values in A.

Table 8.4   Diffusion of alcohol vapors in a tube—Crank–Nicolson method.

Time, seconds	Calculated concentration values at					
	$x = 0$	$x = 4$	$x = 8$	$x = 12$	$x = 16$	$x = 20$
0.0	0.0	2.0	2.0	2.0	2.0	10.0
134.4	0.0	0.980	2.019	3.072	5.992	10.0
268.8	0.0	1.070	2.363	4.305	6.555	10.0
403.2	0.0	1.276	2.861	4.762	6.962	10.0
537.6	0.0	1.471	3.165	5.115	7.159	10.0

Time, seconds	Analytical values	
	$x = 4$	$x = 12$
134.4	1.078	3.191
268.8	1.108	4.272
403.2	1.340	4.873
537.6	1.543	5.248

In Table 8.4 are listed the results of calculations by the Crank–Nicolson method. For the first line of calculations the b-vector has components 2.0, 4.0, 4.0, and 22.0, and the calculated values are the components of $A^{-1}b$. In computing the second line, the components of b are 2.019, 4.052, 8.011, and 23.072. Succeeding lines use the proper sums of values from the line above to determine the b-vector.

When the calculated values are compared to the analytical results, it is observed that the errors of this method, in this example, are about the same as those made in the explicit method with $D \, \Delta t/(\Delta x)^2 = \frac{1}{4}$. We require only one-fourth as many time steps, however, to reach 537.6 sec. With smaller values of Δx, this freedom of choice in the value of $D \, \Delta t/\Delta x)^2$ is especially valuable.

8.3 DERIVATIVE BOUNDARY CONDITIONS

In heat-conduction problems, the most usual situation at the endpoints is not that they are held at a constant temperature, but that heat is lost by conduction or radiation at a rate proportional to some power of the temperature difference between the surface of the body and its surroundings. This leads to a relationship involving the space derivative of temperature at the surface. In the analytical solution of heat-flow problems through Fourier series, this adds considerable complication in determining the coefficients, but our numerical technique requires only minor modifications.

The rate of heat loss from the surface of a solid is generally expressed as

$$\text{Rate of heat loss} = hA(u - u_0), \tag{8.8}$$

where A is the surface area, u is the surface temperature, u_0 is the temperature of the surrounding medium, and h is a coefficient of heat transfer; h is increased by motion of the surrounding medium. To facilitate heat flow, the surrounding medium is often caused to flow rapidly by mechanical means (as in some salt evaporators) or by proper design (as in vertical-tube heat exchangers in distillation columns). This situation is called *forced convection.*

In other situations, h is also a function of the surface temperature, as in *natural convection,* wherein the motion of surrounding fluid is caused by thermal currents. Another important situation is heat loss by radiation, in which the rate of heat loss is proportional to the fourth power of the temperature difference between the surface and the surrounding surfaces to which heat is being radiated. In both these situations, the rate of heat loss is proportional to some power of the surface temperature. This gives rise to nonlinear equations which are not as readily solved. One usually prefers to force the mathematical model to a linear form even though this is not exactly true. One way to approximate this is to use Eq. (8.8), absorbing the nonlinear aspects into the coefficient h, and to change h appropriately through the progress of the calculations so that it takes on a reasonably correct average value. Our examples will treat only the simpler situations where Eq. (8.8) applies directly.

As we shall see, heat loss from the surface by conduction leads to a derivative boundary condition. Our procedure will be to replace the derivatives in both the differential equation and in the boundary conditions by difference quotients. We illustrate with a simple example.

Example An aluminum cube is 4 by 4 by 4 inches ($k = 0.52$, $c = 0.226$, $\rho = 2.70$ in cgs units). All but one face is perfectly insulated, and the cube is initially at 1000°F. Heat is lost from the uninsulated face to a fluid flowing past it according to the equation

$$\text{Rate of heat loss (in Btu per sec)} = hA(u - u_0),$$

where

$$h = 0.15 \text{ Btu/sec} \cdot \text{ft}^2 \cdot °\text{F},$$
$$A = \text{surface area in ft}^2,$$
$$u = \text{surface temperature, }°\text{F},$$
$$u_0 = \text{temperature of fluid, }°\text{F}.$$

If u_0, the temperature of fluid flowing past the aluminum cube, is constant at 70°F, find the temperatures inside the cube as a function of time. While we could work the problem in cgs units, we elect to use English units, making suitable changes in k, c, and ρ.

Because of the insulation on the lateral faces, again the only direction in which heat will flow is perpendicular to the uninsulated face, and the equation is

$$\frac{\partial u}{\partial t} = \frac{k}{c\rho} \frac{\partial^2 u}{\partial x^2}.$$

The initial condition, with x representing distance from the uninsulated face, is

$$u(x, 0) = 1000.$$

For boundary conditions, at the open surface we have

$$-kA \frac{\partial u}{\partial x}\bigg|_{x=0} = -hA(u - 70),$$

because the rate at which heat leaves the surface must be equal to the rate at which heat flows to the surface. The negative sign on the left is required because heat flows in a direction opposite to a positive gradient; on the right it occurs because heat is being lost in the direction of negative x. At the other side of the cube, the rate of heat flow is zero because of insulation:

$$\frac{\partial u}{\partial x}\bigg|_{x=4} = 0.$$

We plan to use the explicit method with

$$r = k \, \Delta t / c\rho(\Delta x)^2 = \tfrac{1}{4}.^*$$

Suppose we let $\Delta x = 1$ inch. To calculate Δt, we need k, c, and ρ expressed in the units of inches, pounds, seconds, and °F. We first convert units:

$$k = \left(0.52 \frac{\text{cal}}{\text{sec} \cdot \text{cm} \cdot °\text{C}}\right)\left(\frac{1 \, \text{Btu}}{252 \, \text{cal}}\right)\left(\frac{2.54 \, \text{cm}}{1 \, \text{in.}}\right)\left(\frac{1°\text{C}}{1.8°\text{F}}\right)$$

$$= 0.00291 \frac{\text{Btu}}{\text{sec} \cdot \text{in.} \cdot °\text{F}};$$

$$c = \left(0.226 \frac{\text{cal}}{\text{g} \cdot °\text{C}}\right)\left(\frac{1 \, \text{Btu}}{252 \, \text{cal}}\right)\left(\frac{454 \, \text{g}}{1 \, \text{lb}}\right)\left(\frac{1°\text{C}}{1.8°\text{F}}\right)$$

$$= 0.226 \frac{\text{Btu}}{\text{lb} \cdot °\text{F}};$$

$$\rho = \left(2.70 \frac{\text{g}}{\text{cm}^3}\right)\left(\frac{1 \, \text{lb}}{454 \, \text{g}}\right)\left(\frac{2.54 \, \text{cm}}{1 \, \text{in.}}\right)^3$$

$$= 0.0975 \frac{\text{lb}}{\text{in.}^3};$$

$$\Delta t = \frac{c\rho(\Delta x)^2}{4k} = \frac{(0.226)(0.0975)(1)^2}{(4)(0.00291)} = 1.89 \, \text{sec.}$$

---

* The ratio must be smaller than $\tfrac{1}{2}$ to give stability with the derivative end conditions.

For the ratio of values of Δt and Δx that we have chosen, our differential equation becomes

$$u_i^{j+1} = \tfrac{1}{4}(u_{i+1}^j + u_{i-1}^j) + \tfrac{1}{2}u_i^j, \tag{8.9}$$

which is to be applied at every point where u is unknown. In this example, this includes the points at $x = 0$ and $x = 4$ as well as the interior points. To enable us to write Eq. (8.9), we extend the domain of u one step on either side of the boundary. Let x_R be a fictitious point to the right of $x = 4$, and let x_L be a fictitious point to the left of $x = 0$. If x_1 signifies $x = 0$, and if x_5 signifies $x = 4$, we have the relations, from Eq. (8.9):

$$u_1^{j+1} = \tfrac{1}{4}(u_2^j + u_L^j) + \tfrac{1}{2}u_1^j,$$
$$u_5^{j+1} = \tfrac{1}{4}(u_R^j + u_4^j) + \tfrac{1}{2}u_5^j.$$

Now we use the boundary conditions to eliminate the fictitious points, writing them as central-difference quotients:

$$-k\frac{\partial u}{\partial x}\bigg|_{x=0} \doteq -(0.00291)\left(\frac{u_2^j - u_L^j}{2(1)}\right) = -\frac{0.15}{144}(u_1^j - 70);$$

$$-k\frac{\partial u}{\partial x}\bigg|_{x=4} \doteq -(0.00291)\left(\frac{u_R^j - u_4^j}{2(1)}\right) = 0.$$

The 144 factor in the first equation changes h to a basis of in². Solving for u_L and u_R, we have

$$u_L = u_2 - 0.716u_1 + 50.1,$$
$$u_R = u_4,$$

and the set of equations which give the temperatures becomes:

$$
\begin{aligned}
u_1^{j+1} &= 0.32u_1^j + \tfrac{1}{2}u_2^j + 12.5,\\
u_2^{j+1} &= \tfrac{1}{4}(u_1^j + u_3^j) + \tfrac{1}{2}u_2^j,\\
u_3^{j+1} &= \tfrac{1}{4}(u_2^j + u_4^j) + \tfrac{1}{2}u_3^j,\\
u_4^{j+1} &= \tfrac{1}{4}(u_3^j + u_5^j) + \tfrac{1}{2}u_4^j,\\
u_5^{j+1} &= \tfrac{1}{2}u_4^j - \tfrac{1}{2}u_5^j.
\end{aligned}
\tag{8.10}
$$

A similar treatment of the boundary conditions using the Crank–Nicolson method with $r = k\,\Delta t/cp(\Delta x)^2 = 1$ leads to the set of simultaneous equations:

$$
\begin{aligned}
4.716u_1^{j+1} - 2u_2^{j+1} &= -0.716u_1^j + 2u_2^j,\\
-u_1^{j+1} + 4u_2^{j+1} - u_3^{j+1} &= u_1^j + u_3^j,\\
-u_2^{j+1} + 4u_3^{j+1} - u_4^{j+1} &= u_2^j + u_4^j,\\
-u_3^{j+1} + 4u_4^{j+1} - u_5^{j+1} &= u_3^j + u_5^j,\\
-2u_4^{j+1} + 4u_5^{j+1} &= 2u_4^j.
\end{aligned}
\tag{8.11}
$$

For the set of equations in (8.11), Δt will be 7.56 sec. We advance the solution one time step at a time by repeatedly solving either (8.10) or (8.11), using the proper values of u_i^j.

8.4 STABILITY AND CONVERGENCE CRITERIA

We have previously stated that in order to ensure stability and convergence in the explicit method, the ratio $r = k \, \Delta t / c\rho(\Delta x)^2$ must be $\frac{1}{2}$ or less. The implicit Crank–Nicolson method has no such limitation. In this section, these phenomena and criteria will be studied in more detail.

By *convergence*, we mean that the results of the method approach the analytical values as Δt and Δx both approach zero. By *stability*, we mean that errors made at one stage of the calculations do not cause increasingly large errors as the computations are continued, but rather will eventually damp out.

We shall first discuss convergence, limiting ourselves to the simple case of the unsteady-state heat-flow equation in one dimension:*

$$\frac{\partial U}{\partial t} = \frac{k}{c\rho} \frac{\partial^2 U}{\partial x^2}. \tag{8.12}$$

Let us use the symbol U to represent the exact solution to Eq. (8.12), and u to represent the numerical solution. At the moment we assume that u is free of round-off errors, so the only difference between U and u is the error made by replacing Eq. (8.12) by the difference equation. Let $e_i^j = U_i^j - u_i^j$, at the point $x = x_i$, $t = t_j$. By the explicit method, Eq. (8.12) becomes

$$u_i^{j+1} = r(u_{i+1}^j + u_{i-1}^j) + (1 - 2r)u_i^j, \tag{8.13}$$

where $r = k \, \Delta t / c\rho(\Delta x)^2$. Substituting $u = U - e$ into Eq. (8.13), we get:

$$e_i^{j+1} = r(e_{i+1}^j + e_{i-1}^j) + (1 - 2r)e_i^j - r(U_{i+1}^j + U_{i-1}^j) - (1 - 2r)U_i^j + U_i^{j+1}. \tag{8.14}$$

By using Taylor-series expansions, we have:

$$U_{i+1}^j = U_i^j + \left(\frac{\partial U}{\partial x}\right)_{i,j} \Delta x + \frac{(\Delta x)^2}{2} \frac{\partial^2 U(\xi_1, t_j)}{\partial x^2}, \qquad x_i < \xi_1 < x_{i+1},$$

$$U_{i-1}^j = U_i^j - \left(\frac{\partial U}{\partial x}\right)_{i,j} \Delta x + \frac{(\Delta x)^2}{2} \frac{\partial^2 U(\xi_2, t_j)}{\partial x^2}, \qquad x_{i-1} < \xi_2 < x_i,$$

$$U_i^{j+1} = U_i^j + \Delta t \frac{\partial U(x_i, \eta)}{\partial t}, \qquad t_j < \eta < t_{j+1}.$$

---

* We could have treated the simpler equation $\partial U / \partial T = \partial^2 U / \partial X^2$ without loss of generality, for with the change of variables $X = \sqrt{c\rho} \, x$, $T = kt$, the two equations are seen to be identical.

Substituting these into Eq. (8.14) and simplifying, remembering that $r(\Delta x)^2 = k\,\Delta t/c\rho$, we get:

$$e_i^{j+1} = r(e_{i+1}^j + e_{i-1}^j) + (1 - 2r)e_i^j + \Delta t\left[\frac{\partial U(x_i, \eta)}{\partial t} - \frac{k}{c\rho}\frac{\partial^2 U(\xi, t_j)}{\partial x^2}\right], \quad (8.15)$$

$$t_j \leqslant \eta \leqslant t_{j+1}, \qquad x_{i-1} \leqslant \xi \leqslant x_{i+1}.$$

Let E^j be the magnitude of the maximum error in the row of calculations for $t = t_j$, and let $M > 0$ be an upper bound for the magnitude of the expression in brackets in Eq. (8.15). If $r \leqslant \frac{1}{2}$, all the coefficients in Eq. (8.15) are positive (or zero) and we may write the inequality

$$|e_i^{j+1}| \leqslant 2rE^j + (1 - 2r)E^j + M\,\Delta t = E^j + M\,\Delta t.$$

This is true for all the e_i^{j+1} at $t = t_{j+1}$, so

$$E^{j+1} \leqslant E^j + M\,\Delta t.$$

Since this is true at each time step,

$$E^{j+1} \leqslant E^j + M\,\Delta t \leqslant E^{j-1} + 2M\,\Delta t \leqslant \ldots \leqslant E^0 + (j + 1)M\,\Delta t = E^0 + Mt_{j+1}$$
$$= Mt_{j+1},$$

because E^0, the errors at $t = 0$, are zero, since U is given by the initial conditions.

Now, as $\Delta x \to 0$, $\Delta t \to 0$ if $k\,\Delta t/c\rho(\Delta x)^2 \leqslant \frac{1}{2}$, and $M \to 0$, because, as both Δx and Δt get smaller,

$$\left[\frac{\partial U(x_i, \eta)}{\partial t} - \frac{k}{c\rho}\frac{\partial^2 U(\xi, t_j)}{\partial x^2}\right] \to \left(\frac{\partial U}{\partial t} - \frac{k}{c\rho}\frac{\partial^2 U}{\partial x^2}\right)_{i,j} = 0.$$

This last is by virtue of Eq. (8.12), of course. Consequently, we have shown that the explicit method is convergent for $r \leqslant \frac{1}{2}$, because the errors approach zero as Δt and Δx are made smaller.

For the solution to the heat-flow equation by the Crank–Nicolson method, the analysis of convergence may be made by similar methods. The treatment is more complicated, but it can be shown that each E^{j+1} is no greater than a finite multiple of E^j plus a term that vanishes as both Δx and Δt become small, and this is independent of r. Hence, since the initial errors are zero, the finite-difference solution approaches the analytical solution as $\Delta t \to 0$ and $\Delta x \to 0$, requiring only that r stay finite.

Let us begin our discussion of stability with a numerical example. Since the heat-flow equation is linear, if two solutions are known, their sum is also a solution. We are interested in what happens to errors made in one line of the computations as the calculations are continued, and because of the additivity feature, the effect of a succession of errors is just the sum of the effects of the

Table 8.5 Propagation of errors—Explicit method.

t	End-point x_1	x_2	x_3	x_4	End-point x_5
t_0	0	0	0	0	0
t_1	0	e	0	0	0
t_2	0	0	$0.50e$	0	0
t_3	0	$0.25e$	0	$0.25e$	0
t_4	0	0	$0.25e$	0	0
t_5	0	$0.125e$	0	$0.125e$	0
t_6	0	0	$0.125e$	0	0
t_7	0	$0.062e$	0	$0.062e$	0
t_8	0	0	$0.062e$	0	0

individual errors. We follow, then, a single error,* which most likely occurred due to round-off. If this single error does not grow in magnitude, we will call the method *stable*, for then the cumulative effect of all errors affects the later calculations no more than a linear combination of the previous errors would.

Table 8.5 illustrates the principle. We have calculated for the simple case where the boundary conditions are fixed, so that the errors at the endpoints are zero. We assume that a single error of size e occurs at $t = t_1$ and $x = x_2$. The explicit method, $k \, \Delta t / c\rho(\Delta x)^2 = \frac{1}{2}$, was used. The original error quite obviously dies out. As an exercise, it is left to the student to show that with $r > 0.5$, errors have an increasingly large effect on later computations. Table 8.6 shows that errors damp out for the Crank–Nicolson method with $r = 1$ even more rapidly than in the explicit method with $r = 0.5$.

Table 8.6 Propagation of errors—Crank–Nicolson method.

t	x_1	x_2	x_3	x_4	x_5
t_0	0	0	0	0	0
t_1	0	e	0	0	0
t_2	0	$0.071e$	$0.286e$	$0.071e$	0
t_3	0	$0.107e$	$0.143e$	$0.107e$	0
t_4	0	$0.049e$	$0.092e$	$0.049e$	0
t_5	0	$0.036e$	$0.053e$	$0.036e$	0
t_6	0	$0.022e$	$0.033e$	$0.022e$	0
t_7	0	$0.013e$	$0.020e$	$0.013e$	0
t_8	0	$0.008e$	$0.013e$	$0.008e$	0

---

* A computation made assuming that each of the interior points has an error equal to e at $t = t_1$ demonstrates the effect more rapidly.

In order to discuss stability in a more analytical sense, we need some material from linear algebra. In Chapter 6 we discussed eigenvalues and eigenvectors of a matrix. We recall that for the matrix A and vector x, if

$$Ax = \lambda x,$$

then the scalar λ is an eigenvalue of A and x is the corresponding eigenvector. If the N eigenvalues of the N-by-N matrix A are all different, then the corresponding N eigenvectors are linearly independent, and any N-component vector can be written uniquely in terms of them.

Consider an unsteady-state heat-flow problem with fixed boundary conditions. Suppose we subdivide into $N + 1$ subintervals so there are N unknown values of the temperature being calculated at each time step. Think of these N values as the components of a vector. Our algorithm for the explicit method [(Eq. (8.4)] can be written as the matrix equation:*

$$\begin{bmatrix} u^{j+1} \\ u_2^{j+1} \\ \cdot \\ \cdot \\ \cdot \\ u_N^{j+1} \end{bmatrix} = \begin{bmatrix} (1-2r) & r & & & \\ r & (1-2r) & r & & \\ & & \cdot & & \\ & & & \cdot & \\ & & & r & (1-2r) \end{bmatrix} \begin{bmatrix} u_1^{j} \\ u_2^{j} \\ \cdot \\ \cdot \\ \cdot \\ u_N^{j} \end{bmatrix} \qquad (8.16)$$

or

$$u^{j+1} = Au^j,$$

where A represents the coefficient matrix and u^j and u^{j+1} are the vectors whose N components are the successive calculated values of temperature. The components of u^0 are the initial values from which we begin our solution. The successive rows of our calculations are

$$u^1 = Au^0,$$
$$u^2 = Au^1 = A^2u^0,$$
$$\cdot$$
$$\cdot$$
$$\cdot$$
$$u^j = Au^{j-1} = A^2u^{j-2} = \ldots = A^ju^0.$$

(The superscripts on the A are here exponents; on the vectors they indicate time.)

Suppose that errors are introduced into u^0, so that it becomes $\bar{u}^0$. We shall follow the effects of this error through the calculations. The successive lines of calculation are now

$$\bar{u}^j = A\bar{u}^{j-1} = \ldots = A^j\bar{u}^0.$$

---

* A change of variable is required to give boundary conditions of $u = 0$ at each end. This can always be done for fixed end conditions.

Let us define the vector e^j as $u^j - \bar{u}^j$ so that e^j represents the errors in u^j caused by the errors in $\bar{u}^0$. We have

$$e^j = u^j - \bar{u}^j = A^j u^0 - A^j \bar{u}^0 = A^j e^0. \tag{8.17}$$

This shows that errors are propagated by using the same algorithm as that by which the temperatures are calculated, as was implicitly assumed earlier in this section.

Now the N eigenvalues of A are distinct (see below) so that its N eigenvectors $x_1, x_2, \ldots, x_N$ are independent, and

$$Ax_1 = \lambda_1 x_1,$$
$$Ax_2 = \lambda_2 x_2,$$
$$\cdot$$
$$\cdot$$
$$\cdot$$
$$Ax_N = \lambda_N x_N.$$

We now write the error vector e^0 as a linear combination of the x_i:

$$e^0 = c_1 x_1 + c_2 x_2 + \ldots + c_N x_N,$$

where the c's are constants. Then e^1 is, in terms of the x_i,

$$e^1 = Ae^0 = \sum_{i=1}^{N} Ac_i x_i = \sum_{i=1}^{N} c_i Ax_i = \sum_{i=1}^{N} c_i \lambda_i x_i,$$

and for e^2,

$$e^2 = Ae^1 = \sum_{i=1}^{N} Ac_i \lambda_i x_i = \sum_{i=1}^{N} c_i \lambda_i^2 x_i.$$

(Again, the superscripts on vectors indicate time; on λ they are exponents.) After j steps, Eq. (8.17) can be written

$$e^j = \sum_{i=1}^{N} c_i \lambda_i^j x_i.$$

If the magnitudes of all of the eigenvalues are less than or equal to unity, errors will not grow as the computations proceed; i.e., the computational scheme is stable. This then is the analytical condition for stability: that the largest eigenvalue of the coefficient matrix for the algorithm be one or less in magnitude.

The eigenvalues of matrix A [Eq. (8.16)] can be shown to be (note they are all distinct):

$$1 - 4r \sin^2 \frac{n\pi}{2(N+1)}, \qquad n = 1, 2, \ldots, N.$$

We will have stability for the explicit scheme if

$$-1 \leq 1 - 4r \sin^2 \frac{n\pi}{2(N+1)} \leq 1.$$

The limiting value of r is given by

$$-1 \leqslant 1 - 4r \sin^2 \frac{n\pi}{2(N+1)}$$

$$r \leqslant \frac{1/2}{\sin^2\left(n\pi/2(N+1)\right)}.$$

Hence, if $r \leqslant \frac{1}{2}$, the explicit scheme is stable.

The Crank–Nicholson scheme, in matrix form, is

$$
\begin{bmatrix}
(2+2r) & -r & & & \\
-r & (2+2r) & -r & & \\
 & & \cdot & & \\
 & & & \cdot & \\
 & & & -r & (2+2r)
\end{bmatrix}
\begin{bmatrix}
u_1^{j+1} \\
u_2^{j+1} \\
\cdot \\
\cdot \\
u_N^{j+1}
\end{bmatrix}
=
\begin{bmatrix}
(2-2r) & r & & & \\
 & r & (2-2r)r & & \\
 & & \cdot & & \\
 & & & \cdot & \\
 & & & & r(2-2r)
\end{bmatrix}
\begin{bmatrix}
u_1^{j} \\
u_2^{j} \\
\cdot \\
\cdot \\
u_N^{j}
\end{bmatrix},
$$

or

$$A u^{j+1} = B u^{j}.$$

We can write

$$u^{j+1} = (A^{-1}B)u^{j}, \tag{8.18}$$

so that stability is given by the magnitudes of the eigenvalues of $A^{-1}B$. These are

$$\frac{2 - 4r \sin^2\left(n\pi/2(N-1)\right)}{2 + 4r \sin^2\left(n\pi/2(N-1)\right)}, \qquad n = 1, 2, \ldots, N.$$

Clearly all the eigenvalues are no greater than one in magnitude for any positive value of r.

With derivative boundary conditions, a similar analysis shows that the Crank–Nicolson method is stable for any positive value of r. For the explicit scheme, $r = \frac{1}{2}$ leads to instability with a finite surface coefficient. Smith (1965) shows that the limitation on r for stability is

$$r \leqslant \frac{1}{2 + P\,\Delta x},$$

where P is the ratio of surface coefficient to conductivity, h/k.

8.5 PARABOLIC EQUATIONS IN TWO OR MORE DIMENSIONS

In principle, we can readily extend the above methods to higher space dimensions, especially when the region is rectangular. The heat-flow equation in two directions is:

$$\frac{\partial u}{\partial t} = \frac{k}{c\rho}\left(\frac{\partial^2 u}{\partial x^2} + \frac{\partial^2 u}{\partial y^2}\right).$$

Taking $\Delta x = \Delta y$, and letting $r = k\,\Delta t / c\rho(\Delta x)^2$, we find that the explicit scheme becomes

$$u_{i,j}^{k+1} - u_{i,j}^{k} = r(u_{i+1,j}^{k} - 2u_{i,j}^{k} + u_{i-1,j}^{k} + u_{i,j+1}^{k} - 2u_{i,j}^{k} + u_{i,j-1}^{k})$$

or

$$u_{i,j}^{k+1} = r(u_{i+1,j}^{k} + u_{i-1,j}^{k} + u_{i,j+1}^{k} + u_{i,j-1}^{k}) + (1 - 4r)u_{i,j}^{k}.$$

In this scheme, the maximum value permissible for r in the simple case of constant end conditions is $\frac{1}{4}$. (Note that this corresponds again to the numerical value that gives a particularly simple formula.) In the more general case with $\Delta x \neq \Delta y$, the criterion is

$$\frac{k\,\Delta t}{c\rho[(\Delta x)^2 + (\Delta y)^2]} \leq \frac{1}{8}.$$

The analogous equation in three dimensions, with equal grid spacing each way, has the coefficient $(1 - 6r)$, and $r \leq \frac{1}{6}$ is required for convergence and stability.

The difficulty with the use of the explicit scheme is that the restrictions on Δt require inordinately many rows of calculations. One then looks for a method in which Δt can be made larger without loss of stability. In one dimension, the Crank–Nicolson method was such a method. In the two-dimensional case, using averages of central-difference approximations to give $\partial^2 u/\partial x^2$ and $\partial^2 u/\partial y^2$ at the mid-value of time, we get

$$u_{i,j}^{k+1} - u_{i,j}^{k} = \frac{r}{2}[u_{i+1,j}^{k+1} - 2u_{i,j}^{k+1} + u_{i-1,j}^{k+1} + u_{i+1,j}^{k} - 2u_{i,j}^{k} + u_{i-1,j}^{k}$$

$$+ u_{i,j+1}^{k+1} - 2u_{i,j}^{k+1} + u_{i,j-1}^{k+1} + u_{i,j+1}^{k} - 2u_{i,j}^{k} + u_{i,j-1}^{k}].$$

The problem now is that a set of $(M)(N)$ simultaneous equations must be solved at each time step, where M is the number of unknown values in the x-direction and N in the y-direction. Furthermore, the coefficient matrix is no longer tridiagonal, so the solution to each set of equations is slower and memory space to store the elements of the matrix becomes exorbitant.

The advantage of a tridiagonal matrix is retained in the alternating-direction-implicit scheme (A.D.I.) proposed by Peaceman and Rachford (1955). It is widely used in modern computer programs for the solution of parabolic partial differential equations. In this method, we approximate $\nabla^2 u$ by adding a central-difference approximation to $\partial^2 u/\partial x^2$ written at the beginning of the interval to a similar expression for $\partial^2 u/\partial y^2$ written at the end:

$$u_{i,j}^{k+1} - u_{i,j}^{k} = r[\underbrace{u_{i+1,j}^{k} - 2u_{i,j}^{k} + u_{i-1,j}^{k}}_{\text{From } \partial^2 u/\partial x^2 \text{ at start}} + \underbrace{u_{i,j+1}^{k+1} - 2u_{i,j}^{k+1} + u_{i,j-1}^{k+1}}_{\text{From } \partial^2 u/\partial y^2 \text{ at end}}]. \quad (8.19)$$

The obvious bias in this formula is balanced by reversing the order for the second-derivative approximations in the next time span:

$$u_{i,j}^{k+2} - u_{i,j}^{k+1} = r[\underbrace{u_{i+1,j}^{k+2} - 2u_{i,j}^{k+2} + u_{i-1,j}^{k+2}}_{\text{From } \partial^2 u/\partial x^2 \text{ at end}} + \underbrace{u_{i,j+1}^{k+1} - 2u_{i,j}^{k+1} + u_{i,j-1}^{k+1}}_{\text{From } \partial^2 u/\partial y^2 \text{ at start}}]. \quad (8.20)$$

We illustrate the A.D.I. method with a very simple example.

Example A square plate of steel is 15 cm on a side. Initially, all points are at 0°C. Follow the interior temperatures if two adjacent sides are suddenly brought to 100°C and held at that temperature. The plate is insulated on its flat surfaces, so that heat flows only in the x- and y-directions. We are given that $k = 0.13$, $\rho = 7.8$, $c = 0.11$ in cgs units. Take $\Delta x = \Delta y = 5$ cm, so that there are four interior points. Label the points as shown in Fig. 8.6, with u being used for horizontal traverses, v for vertical traverses.

Equations (8.19) and (8.20) become, with $r = k \, \Delta t/cp(\Delta x)^2$,

$$\begin{bmatrix} (1/r) + 2 & -1 & 0 & 0 \\ -1 & (1/r) + 2 & 0 & 0 \\ 0 & 0 & (1/r) + 2 & -1 \\ 0 & 0 & -1 & (1/r) + 2 \end{bmatrix} \begin{bmatrix} u_1 \\ u_2 \\ u_3 \\ u_4 \end{bmatrix} = \begin{bmatrix} 0 + 100 + [(1/r) - 2]v_1 + v_2 \\ 100 + 100 + [(1/r) - 2]v_3 + v_4 \\ 0 + v_1 + [(1/r) - 2]v_2 + 0 \\ 100 + v_3 + [(1/r) - 2]v_4 + 0 \end{bmatrix},$$

$$\begin{bmatrix} (1/r) + 2 & -1 & 0 & 0 \\ -1 & (1/r) + 2 & 0 & 0 \\ 0 & 0 & (1/r) + 2 & -1 \\ 0 & 0 & -1 & (1/r) + 2 \end{bmatrix} \begin{bmatrix} v_1 \\ v_2 \\ v_3 \\ v_4 \end{bmatrix} = \begin{bmatrix} 100 + 0 + [(1/r) - 2]u_1 + u_2 \\ 0 + 0 + [(1/r) - 2]u_3 + u_4 \\ 100 + u_1 + [(1/r) - 2]u_2 + 100 \\ 0 + u_3 + [(1/r) - 2]u_4 + 100 \end{bmatrix}$$

(Note the similarity to the equations for the A.D.I. method for steady-state temperatures in Section 7.12. This is hardly surprising, for the temperatures for the unsteady-state problem will eventually reach the steady state. One can think of this as the rationale behind the A.D.I. method for elliptic equations.)

We solve for the temperature history by using the above equations in succession. One normally discards the alternate computatons, beginning with the first, because they tend to be inaccurate.

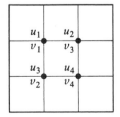

Figure 8.6

Table 8.7 and Fig. 8.7 show the results, with $\Delta t = 16.5$ sec, corresponding to $r = 0.1$.

The compensation of errors produced by this alternation of direction gives a scheme that is convergent and stable for all values of r, although accuracy requires that r not be too large. The three-dimensional analog alternates three ways, returning to each of the three formulas after every third step. (Unfortunately the three-dimensional case is not stable for all fixed values of $r > 0$. A variant due to Douglas (1962) is unconditionally stable, however.) When the formulas are rearranged, in each case tridiagonal coefficient matrices result.

Note that the equations can be broken up into two independent subsets, each containing only two equations. This is always true in the A.D.I. method; each row gives a set independent of the equations from the other rows. For columns, the same thing occurs. For very large problems, this is important, for it permits the ready overlay of main memory in solving the independent sets.

Table 8.7 Temperature changes in a 15 cm × 15 cm steel plate, computed by the A.D.I. method.

t, sec	Temperatures at:			
	u_1 or v_1	u_2 or v_3	u_3 or v_2	u_4 or v_4
0.0	0.000	0.000	0.000	0.000
16.5*	8.392	17.482	0.699	9.790
33.0	16.529	30.515	2.543	16.529
49.5*	22.306	40.298	4.932	22.923
66.0	27.594	47.741	7.466	27.594
82.5*	31.532	53.444	9.891	31.803
99.0	35.000	57.863	12.138	35.001
115.5*	37.668	61.308	14.148	37.788
132.0	39.959	64.017	15.901	39.959
148.5*	41.758	66.159	17.410	41.811
165.0	43.278	67.864	18.693	43.278
181.5*	44.489	69.225	19.776	44.512
198.0	45.500	70.318	20.683	45.500
214.5*	46.314	71.197	21.440	46.324
231.0	46.988	71.907	22.068	46.988
247.5*	46.533	72.482	22.589	47.538
264.0	46.984	72.948	23.019	47.984
.	.	.	.	.
.	.	.	.	.
.	.	.	.	.
∞	50.00	75.00	25.00	50.00

* Calculations are less accurate at these times.

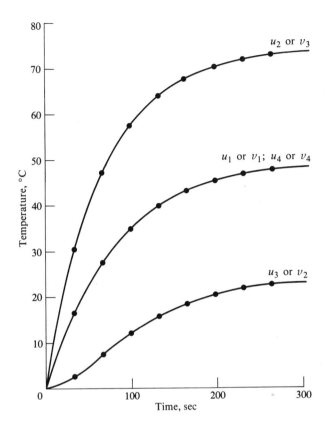

Figure 8.7

When the region in which the heat-flow equation is to be satisfied is not rectangular, one may perturb the boundary to make it agree with a square mesh or interpolate from boundary points to estimate u at adjacent mesh points as discussed in Chapter 7 for elliptic equations.

The frequency with which circular or spherical regions occur makes it worthwhile to mention the heat equation in polar and spherical coordinates. The basic equation

$$\frac{\partial u}{\partial t} = \frac{k}{c\rho} \nabla^2 u$$

becomes, in polar coordinates (r, θ),

$$\frac{\partial u}{\partial t} = \frac{k}{c\rho}\left(\frac{\partial^2 u}{\partial r^2} + \frac{1}{r}\frac{\partial u}{\partial r} + \frac{1}{r^2}\frac{\partial^2 u}{\partial \theta^2}\right),$$

and in spherical coordinates (r, ϕ, θ),

$$\frac{\partial u}{\partial t} = \frac{k}{c\rho}\left(\frac{\partial^2 u}{\partial r^2} + \frac{2}{r}\frac{\partial u}{\partial r} + \frac{1}{r^2}\frac{\partial^2 u}{\partial \theta^2} + \frac{\cot \phi}{r}\frac{\partial u}{\partial \theta} + \frac{1}{r^2 \sin^2 \theta}\frac{\partial^2 u}{\partial \phi^2}\right).$$

Using finite-difference approximations to convert these to difference equations is straightforward except at the origin where $r = 0$. For this point, consider $\nabla^2 u$ in rectangular coordinates, so that, in two dimensions,

$$\partial^2 u = \frac{u_{i+1,j} + u_{i-1,j} + u_{i,j+1} + u_{i,j-1} - 4u_{i,j}}{(\Delta r)^2}, \qquad u_{ij} \text{ at } r = 0.$$

This is exactly the same as the expression for the Laplacian in Chapter 7, Eq. (7.6). This expression for $\nabla^2 u$ is obviously independent of the orientation of the axes. We get the best value by using the average value of all points which are a distance Δr from the origin, so that for $r = 0$,

$$\nabla^2 u = \frac{4(u_{av} - u_0)}{(\Delta r)^2} \qquad \text{at } r = 0.$$

The corresponding relation for spherical coordinates is

$$\nabla^2 u = \frac{6(u_{av} - u_0)}{(\Delta r)^2} \qquad \text{at } r = 0.$$

8.6 PROGRAMS TO SOLVE PARABOLIC EQUATIONS

Three computer programs are presented as examples of how parabolic partial-differential equations in one or two space dimensions can be solved on a computer. Program 1 (Fig. 8.8) uses the explicit method of Section 1; Program 2 (Fig. 8.9) employs the implicit Crank–Nicolson method of Section 2; Program 3 (Fig. 8.10) solves the unsteady-state heat-flow equation in two space dimensions by the A.D.I. method. In the first program, both ends are held at a known temperature, these end temperatures are allowed to vary with time. In the second, however, derivative end conditions are permitted.

For Program 1, the fundamental difference equation which approximates the differential equation is Eq. (8.4):

$$u_i^{j+1} = r(u_{i+1}^j + u_{i-1}^j) + (1 - r)u_i^j,$$

where

$$r = \frac{k \, \Delta t}{c\rho(\Delta x)^2}.$$

This relation is applied at each interior point. The end conditions, $u(0, t)$ and $u(L, t)$, are defined by arithmetic statement functions. At the end of each time step, the interior temperatures are computed and a line of temperatures is printed.

The program is tested by computing the temperatures in a 10-cm-long bar with $k = 0.53$, $c = 0.226$, $\rho = 2.70$, $\Delta x = 1.0$, $r = 0.25$ (all in cgs units). The bar is initially at 20°C at all points, and temperatures are caused to change by

```
C   THIS PROGRAM COMPUTES ONE DIMENSIONAL UNSTEADY STATE POTENTIAL
C   FLOW PROBLEMS BY THE EXPLICIT METHOD.  THE POTENTIAL AT EACH
C   END IS SPECIFIED AS A FUNCTION OF TIME THROUGH ARITHMETIC
C   STATEMENT FUNCTIONS.
C   PARAMETERS ARE -
C     U,V    VALUES OF THE POTENTIAL AT NODES
C     T      TIME
C     TF     FINAL TIME FOR WHICH VALUES ARE COMPUTED
C     DT     DELTA T
C     LEN    LENGTH
C     DX     DELTA X
C     N      NUMBER OF X INTERVALS
C     K      CONDUCTIVITY
C     C      HEAT CAPACITY
C     RHO    DENSITY
C     RATIO  RATIO OF K(DT)/C(RHO)(DX)**2
        DIMENSION U(500), V(500)
        REAL LEN, K
C   WE DEFINE SOME CONSTANTS WITH A DATA STATEMENT
        DATA T, TF, LEN, N, K, C, RHO, RATIO /0.0, 20.0, 10.0, 10, 0.53,
       +  0.226, 2.70, 0.25/
        FLFT(T) = 0.0
        FRT(T) = 100.0
C   READ IN THE INITIAL TEMPERATURES AND WRITE THEM OUT
        NP1 = N + 1
        DX = LEN/N
        DT = RATIO*C*RHO*DX*DX/K
        READ (5,200) (U(I), I = 1,NP1)
200     FORMAT (8F10.0)
        WRITE (6,201) T, LEN, DX, T, (U(I), I = 1,NP1)
201     FORMAT (1H1, 53H POTENTIAL VALUES IN ONE DIMENSION BY EXPLICIT MET
       +HOD / 1H , 11H    FOR X = , F4.1, 8H TO X = , F4.1,
       +    17H WITH DELTA X OF , F6.3// 1H , 8H AT T = , F6.3,
       +    /(1H , 10F13.4))
C   WE GET THE POTENTIAL FROM THE EXPLICIT RELATION, PRINTING
C   EACH SET OF VALUES AS THEY ARE COMPUTED.  TWO ARRAYS ARE
C   USED, HOLDING ALTERNATE SETS OF VALUES.
10      THALF = T + DT/2.0
        U(1) = FLFT(THALF)
        U(N+1) = FRT(THALF)
        DO 20 I = 2,N
          V(I) = RATIO*(U(I+1) + U(I-1)) + (1.0 - 2.0*RATIO)*U(I)
20      CONTINUE
        T = T + DT
        V(1) = FLFT(T)
        V(NP1) = FRT(T)
        WRITE (6,202) T, (V(I), I = 1,NP1)
202     FORMAT (/1H , 15H VALUES AT T = , F8.3/(1H , 10F13.4))
        IF (T .GT. TF)  STOP
C   NOW DO SECOND SET OF VALUES
        THALF = T + DT/2.0
        V(1) = FLFT(THALF)
        V(N+1) = FRT(THALF)
        DO 30 I = 2,N
          U(I) = RATIO*(V(I+1) + V(I-1)) + (1.0 - 2.0*RATIO)*V(I)
30      CONTINUE
        T = T + DT
        U(1) = FLFT(T)
        U(NP1) = FRT(T)
        WRITE (6,202) T, (U(I), I = 1,NP1)
        IF (T .GT. TF)  STOP
        GO TO 10
        END
```

Figure 8.8 Program 1.

OUTPUT FROM PROGRAM 1.

POTENTIAL VALUES IN ONE DIMENSION BY EXPLICIT METHOD
FOR X = 0.0 TO X = 10.0 WITH DELTA X OF 1.000

AT T = 0.000

20.0000	20.0000	20.0000	20.0000	20.0000	20.0000
20.0000					

VALUES AT T = 0.288

0.0000	15.0000	20.0000	20.0000	20.0000	20.0000
100.0000					

VALUES AT T = 0.576

0.0000	12.5000	18.7500	20.0000	20.0000	20.0000
100.0000					

VALUES AT T = 0.863

0.0000	10.9375	17.5000	19.6875	20.0000	20.0000
100.0000					

VALUES AT T = 1.151

0.0000	9.8438	16.4063	19.2188	19.9219	20.0000
100.0000					

VALUES AT T = 1.439

0.0000	9.0234	15.4688	18.6914	19.7656	20.0586
100.0000					

OUTPUT CONTINUED

VALUES AT T = 18.709

0.0000	7.6876	15.5940	23.9194	32.8279	42.4309
100.0000					

VALUES AT T = 18.997

0.0000	7.7423	15.6988	24.0652	33.0015	42.6161
100.0000					

VALUES AT T = 19.285

0.0000	7.7958	15.8012	24.2077	33.1711	42.7968
100.0000					

VALUES AT T = 19.572

0.0000	7.8482	15.9015	24.3469	33.3367	42.9731
100.0000					

VALUES AT T = 19.860

0.0000	7.8995	15.9995	24.4830	33.4983	43.1450
100.0000					

VALUES AT T = 20.148

0.0000	7.9496	16.0954	24.6160	33.6561	43.3128
100.0000					

Figure 8.8 (continued)

20.0000	20.0000	20.0000	20.0000
20.0000	20.0000	20.0000	40.0000
20.0000	20.0000	25.0000	50.0000
20.0000	21.2500	30.0000	56.2500
20.3125	23.1250	34.3750	60.6250
20.9375	25.2344	38.1250	63.9063
52.7748	63.8335	75.5080	87.6344
52.9535	63.9874	75.6209	87.6942
53.1276	64.1373	75.7309	87.7523
53.2973	64.2833	75.8378	87.8089
53.4627	64.4254	75.9419	87.8639
53.6240	64.5639	76.0433	87.9174

Figure 8.8 (concluded)

suddenly cooling one end to 0°C and heating the other end to 100°C. The program computes values until $T > 20$, but only part of the output is shown.

```
C   THIS PROGRAM COMPUTES UNSTEADY STATE POTENTIALS IN   ONE DIMENSION
C   USING CRANK-NICHOLSON IMPLICIT METHOD.   THE POTENTIAL AT
C   EACH END IS DETERMINED BY A RELATION OF THE FORM AU + BU' = C
C   PARAMETERS ARE -
C      U       VALUES OF POTENTIAL AT NODES
C      T       TIME
C      TF      FINAL TIME VALUE FOR WHICH SOLUTION IS DESIRED
C      DT      DELTA T
C      LEN     LENGTH
C      DX      DELTA X
C      N       NUMBER OF X INTERVALS
C      DIF     DIFFUSIVITY, K/C*RHO FOR HEAT FLOW
C      RATIO   RATIO OF DT(DIF)/(DX)**2
C      COEF    COEFFICIENT MATRIX FOR IMPLICIT EQUATIONS
        DIMENSION U(500), COEF(500,3), RHS(500)
        REAL LEN
C   DEFINE SOME CONSTANTS
        DATA T, TF, LEN, N, DIF, RATIO /0.0, 1000.0, 20.0, 20, 0.119, 1.0/
        DATA AL, BL, CL /-0.2, 1.0, -3.0/
        DATA AR, BR, CR /1.0, 0.0, 100.0/
C   READ IN INITIAL VALUES
        NP1 = N + 1
        READ (5,200) (U(I), I = 1,NP1)
200     FORMAT (8F10.0)
        DX = LEN/N
        DT = RATIO*DX*DX/DIF
C   WRITE OUT HEADINGS AND INITIAL VALUES
        WRITE (6,201) T, LEN, DX, T, (U(I), I = 1,NP1)
201     FORMAT (1H1, 53H POTENTIAL VALUES IN ONE DIMENSION BY IMPLICIT MET
       +HOD /1H , 11H      FOR X = , F4.1, 8H TO X = , F4.1,
       +   17H WITH DELTA X OF ,F6.3//1H , 8H AT T = , F7.3, /
       +   (1H , 10F13.4))
C   ESTABLISH COEFFICIENT MATRIX
        IF (BL .EQ. 0.0)  GO TO 10
        COEF(1,2) = 2.0/RATIO + 2.0 - 2.0*AL*DX/BL
        COEF(1,3) = -2.0
        GO TO 20
10      COEF(1,2) = 1.0
        COEF(1,3) = 0.0
20      DO 25 I = 2,N
        COEF(I,1) = -1.0
        COEF(I,2) = 2.0/RATIO + 2.0
        COEF(I,3) = -1.0
25      CONTINUE
        IF (BR .EQ. 0.0)  GO TO 30
        COEF(N+1,1) = -2.0
        COEF(N+1,2) = 2.0/RATIO + 2.0 + 2.0*AR*DX/BR
        GO TO 40
30      COEF(N+1,1) = 0.0
        COEF(N+1,2) = 1.0
```

Figure 8.9 Program 2.

```
C  GET THE LU DECOMPOSITION TO PREPARE FOR SOLVING EQUATIONS
40     DO 50 I = 2,NP1
          COEF(I-1,3) = COEF(I-1,3)/COEF(I-1,2)
          COEF(I,2) = COEF(I,2) - COEF(I,1)*COEF(I-1,3)
50     CONTINUE
C  ESTABLISH THE R.H.S. VECTOR - FIRST THE TOP AND BOTTOM ROWS
55     IF (BL .EQ. 0.0)  GO TO 60
          RHS(1) = (2.0/RATIO - 2.0 + 2.0*AL*DX/BL)*U(1) + 2.0*U(2) -
       +    4.0*CL*DX/BL
          GO TO 70
60     RHS(1) = CL/AL
70     IF (BR .EQ. 0.0)  GO TO 80
          RHS(N+1) = 2.0*U(N) + (2.0/RATIO - 2.0*AR*DX/BR)*U(N+1) +
       +    4.0*CR*DX/BR
          GO TO 90
80     RHS(N+1) = CR/AR
C  NOW FOR THE OTHER ROWS OF THE RHS VECTOR
90     DO 100 I = 2,N
          RHS(I) = U(I-1) + (2.0/RATIO - 2.0)*U(I) + U(I+1)
100    CONTINUE
C  WE GET THE SOLUTION FOR THE CURRENT TIME STEP
          U(1) = RHS(1)/COEF(1,2)
          DO 110 I = 2,NP1
          U(I) = (RHS(I) - COEF(I,1)*U(I-1))/COEF(I,2)
110    CONTINUE
          DO 120 I = 1,N
          JROW = N - I + 1
          U(JROW) = U(JROW) - COEF(JROW,3)*U(JROW+1)
120    CONTINUE
C  WRITE OUT THE SOLUTION JUST FOUND
          T = T + DT
          WRITE (6,202) T, (U(I), I = 1,NP1)
202    FORMAT (/1H , 15H VALUES AT T = , F8.3/(1H , 10F13.4))
          IF (T .LT. TF)  GO TO 55
          STOP
          END
```

Figure 8.9 (continued)

Program 2, using the Crank–Nicolson method, is more elaborate in that derivative end conditions are permitted, in the form

$$au + b\frac{\partial u}{\partial x} = c,$$

where u is the temperature at the end of the bar. This equation can simulate the loss of heat by convection or by imperfect insulation at each end. The program

OUTPUT FROM PROGRAM 2.

POTENTIAL VALUES IN ONE DIMENSION BY IMPLICIT METHOD
FOR X = 0.0 TO X = 20.0 WITH DELTA X OF 1.000

AT T = 0.000

0.0000	10.0000	20.0000	30.0000	40.0000	50.0000
100.0000	90.0000	80.0000	70.0000	60.0000	50.0000
0.0000					

VALUES AT T = 8.403

13.4572	13.6058	20.9659	30.2577	40.0651	50.0026
88.4532	86.9067	79.1736	69.7878	59.9775	50.1222
100.0000					

VALUES AT T = 16.807

16.0850	18.4727	23.3829	31.1954	40.3680	50.0162
84.6067	82.3102	77.0073	69.0244	59.9391	50.8222
100.0000					

VALUES AT T = 25.210

19.2767	21.1530	25.8674	32.6486	.40.9762	50.0444
80.7714	79.2527	74.6255	67.9148	60.0872	52.5873
100.0000					

OUTPUT CONTINUED

VALUES AT T = 974.786

31.6936	35.0348	38.3819	41.7358	45.0972	48.4668
65.4506	68.8752	72.3085	75.7500	79.1992	82.6550
100.0000					

VALUES AT T = 983.189

31,6985	35.0406	38.3886	41.7432	45.1053	48.4754
65.4598	68.8839	72.3167	75.7576	79.2059	82.6608
100.0000					

VALUES AT T = 991.593

31.7033	35.0464	38.3952	41.7506	45.1133	48.4839
65.4686	68.8924	72.3247	75.7649	79.2124	82.6665
100.0000					

VALUES AT T = 999.996

31.7081	35.0521	38.4017	41.7578	45.1211	48.4922
65.4773	68.9008	72.3326	75.7722	79.2189	82.6721
100.0000					

VALUES AT T = 1008.399

31.7128	35.0577	38.4081	41.7649	45.1288	48.5004
65.4859	68.9090	72.3403	75.7793	79.2253	82.6776
100.0000					

Figure 8.9 (continued)

60.0000	70.0000	80.0000	90.0000
40.0000	30.0000	20.0000	10.0000
59.9454	69.7792	79.1713	86.9061
40.5112	31.9226	27.1794	36.7948
59.6862	68.9467	76.9840	82.3038
42.8610	38.5772	43.7572	67.7341
59.1475	67.5826	74.5130	79.2188
47.4619	47.8608	57.3631	75.2801
51.8450	55.2324	58.6291	62.0352
86.1167	89.5832	93.0534	96.5260
51.8540	55.2416	58.6384	62.0444
86.1215	89.5869	93.0558	96.5273
51.8629	55.2507	58.6476	62.0535
86.1262	89.5905	93.0583	96.5285
51.8716	55.2597	58.6566	62.0625
86.1308	89.5940	93.0607	96.5297
51.8802	55.2685	58.6655	62.0713
86.1353	89.5974	93.0630	96.5309

Figure 8.9 (concluded)

```
C   THIS PROGRAM SOLVES THE UNSTEADY STATE HEAT FLOW EQUATION
C   IN TWO SPACE DIMENSIONS USING THE A.D.I METHOD
C   PARAMETERS ARE -
C     U       VECTOR OF TEMPERATURES AFTER ODD TRAVERSES
C     V       TEMPERATURES AFTER EVEN TRAVERSES
C     UCOEF   MATRIX OF COEFFICIENTS FOR THE U'S
C     VCOEF   MATRIX FOR THE V'S
C     BCNDU   VECTOR OF BOUNDARY VALUES FOR THE U EQUATIONS
C     BCNDV   SAME FOR THE V'S
C     TOP     VECTOR TO HOLD BOUNDARY VALUES ACROSS TOP OF REGION
C     BOT     SAME FOR THE VALUES ACROSS THE BOTTOM
C     RT      HOLD THE RIGHT HAND SIDE VALUES
C     LFT     HOLD VALUES FOR THE LEFT HAND EDGE
C     M       NUMBER OF ROWS OF NODES IN THE GRID
C     N       NUMBER OF COLUMNS OF NODES
C     DIFF    THERMAL DIFFUSIVITY = K/C*DENSITY
C     H       THE VALUE OF DELTA X = DELTA Y
C     TIME    THE TIME VARIABLE
C     TMAX    MAXIMUM VALUE OF TIME FOR WHICH COMPUTATIONS ARE DESIRED
C     DT      TIME STEP SIZE, RELATED TO DELTA X THROUGH R
C   ESTABLISH THE MATRICES AND VECTORS, AND PUT IN SOME VALUES WITH DATA
      REAL U(500), V(500), UCOEF(500,3), VCOEF(500,3)
      REAL BCNDU(500), BCNDV(500)
      REAL TOP(100), BOT(100), LFT(100), RT(100)
      DATA UCOEF, VCOEF /3000*-1.0/
      DATA BCNDU, BCNDV /1000*0.0/
      DATA M, N, R /7, 15, 0.5 /
      DATA H, DIFF /0.125, 0.152/
      DATA TMAX /20.0/
      DATA RT, LFT, TOP, BOT /100*100.0, 300*0.0/
C   FOR THE TEST CASE, WE BEGIN WITH TEMPERATURES EVERYWHERE EQUAL TO ZERO
C   ESTABLISH THESE BY A DATA STATEMENT
      DATA V /500*0.0/
C   SET UP THE COEFFICIENT MATRICES BY OVER-WRITING
C   ON THE DIAGONAL AND CERTAIN OFF-DIAGONAL TERMS
      MSIZE = M*N
      DO 10 I = 1,MSIZE
         UCOEF(I,2) = 1./R + 2.0
         VCOEF(I,2) = 1./R + 2.0
10    CONTINUE
      ML1 = M - 1
      NL1 = N - 1
      MSL1 = MSIZE - 1
      DO 20 I = N,MSL1,N
         UCOEF(I,3) = 0.0
         UCOEF(I+1,1) = 0.0
20    CONTINUE
      DO 30 I = M,MSL1,M
         VCOEF(I,3) = 0.0
         VCOEF(I+1,1) = 0.0
30    CONTINUE
C   NOW GET VALUES INTO THE BCND VECTORS
      DO 40 I = 1,N
         BCNDU(I) = TOP(I)
         J = MSIZE - N + I
         BCNDU(J) = BOT(I)
40    CONTINUE
      DO 45 I = 1,M
         J = (I-1)*N + 1
         BCNDU(J) = BCNDU(J) + LFT(I)
         J = I*N
         BCNDU(J) = BCNDU(J) + RT(I)
45    CONTINUE
      DO 50 I = 1,M
         BCNDV(I) = LFT(I)
         J = MSIZE - M + I
         BCNDV(J) = RT(I)
50    CONTINUE
      DO 55 I = 1,N
         J = (I-1)*M + 1
         BCNDV(J) = BCNDV(J) + TOP(I)
         J = I*M
         BCNDV(J) = BCNDV(J) + BOT(I)
55    CONTINUE
```

Figure 8.10 Program 3.

```
C   WE NOW GET THE LU DECOMPOSITIONS OF UCOEF AND VCOEF
        DO 60 I = 2,MSIZE
          UCOEF(I-1,3) = UCOEF(I-1,3)/UCOEF(I-1,2)
          UCOEF(I,2) = UCOEF(I,2) - UCOEF(I,1)*UCOEF(I-1,3)
          VCOEF(I-1,3) = VCOEF(I-1,3)/VCOEF(I-1,2)
          VCOEF(I,2) = VCOEF(I,2) - VCOEF(I,1)*VCOEF(I-1,3)
60      CONTINUE
C   NOW WE DO THE ITERATIONS UNTIL TIME EQUAL TMAX
        TIME = 0.0
        DT = R/DIFF*H*H
62      IF (TIME .GT. TMAX)  STOP
C   COMPUTE THE R.H.S. FOR THE U EQUATIONS
C   STORE IN THE U VECTOR.
C   FIRST DO THE TOP AND BOTTOM ROWS
        DO 65 I = 1,N
          J = (I-1)*M + 1
          U(I) = (1.0/R - 2.0)*V(J) + V(J+1) + BCNDU(I)
          K = MSIZE - N + I
          J = I*M
          U(K) = V(J-1) + (1.0/R - 2.0)*V(J) + BCNDU(K)
65      CONTINUE
C   NOW FOR THE OTHER ONES
        DO 75 I = 2,ML1
          DO 70 J = 1,N
            K = (I-1)*N + J
            L = I + (J-1)*M
            U(K) = V(L-1) + (1.0/R - 2.0)*V(L) + V(L+1) + BCNDU(K)
70        CONTINUE
75      CONTINUE
C   NOW GET THE SOLUTION FOR THE ODD TRAVERSE
        U(1) = U(1)/UCOEF(1,2)
        DO 80 I = 2,MSIZE
          U(I) = (U(I) - UCOEF(I,1)*U(I-1))/UCOEF(I,2)
80      CONTINUE
        DO 90 I = 1,MSL1
          JROW = MSIZE - I
          U(JROW) = U(JROW) - UCOEF(JROW,3)*U(JROW+1)
90      CONTINUE
C   COMPUTE THE R.H.S FOR THE ODD TRAVERSE, STORE IN V.
C   DO THE TOP AND BOTTOM ONES
        DO 95 I = 1,M
          J = (I-1)*N + 1
          V(I) = (1.0/R - 2.0)*U(J) + U(J+1) + BCNDV(I)
          K = MSIZE - M + I
          J = I*N
          V(K) = U(J-1) + (1.0/R - 2.0)*U(J) + BCNDV(K)
95      CONTINUE
C   NOW THE REST OF THE ROWS
        DO 105 I = 2,NL1
          DO 100 J = 1,M
            K = (I-1)*M + J
            L = I + (J-1)*N
            V(K) = U(L-1) + (1.0/R - 2.0)*U(L) + U(L+1) + BCNDV(K)
100       CONTINUE
105     CONTINUE
C   GET THE SOLUTION FOR THE ODD TRAVERSE
        V(1) = V(1)/VCOEF(1,2)
        DO 110 I = 2,MSIZE
          V(I) = (V(I) - VCOEF(I,1)*V(I-1))/VCOEF(I,2)
110     CONTINUE
        DO 120 I = 1,MSL1
          JROW = MSIZE - I
          V(JROW) = V(JROW) - VCOEF(JROW,3)*V(JROW+1)
120     CONTINUE
        TIME = TIME + 2.0*DT
C   PRINT OUT THE LAST RESULT
        WRITE (6,202) TIME, (V(I), I = 1,MSIZE)
202     FORMAT (/// 1H , 10H WHEN T = , F6.3,
     +     14H V VALUES ARE /(1H , 15F8.4))
        GO TO 62
        END
```

Figure 8.10 (continued)

REPRESENTATIVE OUTPUT FROM PROGRAM 3

WHEN T =   0.103 V VALUES ARE

0.0000	0.0000	0.0000	0.0000	0.0000	0.0000	0.0000	0.0000	0.0000	0.0000	0.0000	0.0000
0.0000	0.0000	0.0000	0.0000	0.0000	0.0000	0.0000	0.0000	0.0000	0.0000	0.0000	0.0000
0.0001	0.0001	0.0001	0.0001	0.0001	0.0003	0.0004	0.0004	0.0004	0.0004	0.0004	0.0003
0.0014	0.0014	0.0013	0.0010	0.0039	0.0049	0.0052	0.0053	0.0052	0.0049	0.0039	0.0145
0.0194	0.0184	0.0145	0.0542	0.0687	0.0725	0.0733	0.0725	0.0687	0.0542	0.2022	0.2563
0.2563	0.2022	0.7546	0.9566	1.0097	1.0203	1.0097	0.9566	0.7546	2.8163	3.5699	3.7682
2.8163	10.5105	13.3231	14.0633	14.2113	14.0633	13.3231	10.5105	39.2255	49.7225	52.4849	53.0374

WHEN T =   0.206 V VALUES ARE

0.0000	0.0000	0.0000	0.0000	0.0000	0.0000	0.0000	0.0000	0.0000	0.0000	0.0000	0.0000
0.0001	0.0001	0.0001	0.0001	0.0001	0.0000	0.0002	0.0003	0.0003	0.0004	0.0003	0.0003
0.0012	0.0012	0.0012	0.0010	0.0006	0.0020	0.0034	0.0040	0.0042	0.0040	0.0034	0.0020
0.0140	0.0135	0.0115	0.0067	0.0224	0.0382	0.0448	0.0465	0.0448	0.0382	0.0224	0.0738
0.1465	0.1251	0.0738	0.2393	0.4019	0.4695	0.4866	0.4695	0.4019	0.2393	0.7584	1.2578
1.2578	0.7584	2.3284	3.7901	4.3852	4.5350	4.3852	3.7901	2.3284	6.8156	10.7705	12.3462
6.8156	18.4418	27.6030	31.0749	31.9244	31.0749	27.6031	18.4418	42.7227	56.0189	59.9858	60.8475

WHEN T =   1.850 V VALUES ARE

0.0501	0.0926	0.1210	0.1310	0.1210	0.0926	0.0501	0.1119	0.2067	0.2700	0.2922	0.2700
0.3673	0.4798	0.5193	0.4798	0.3673	0.1988	0.3287	0.6072	0.7931	0.8584	0.7931	0.6072
1.2695	1.3739	1.2695	0.9719	0.5262	0.8268	1.5268	1.9937	2.1574	1.9937	1.5268	0.8268
3.3414	3.0885	2.3665	1.2822	1.9685	3.6294	4.7320	5.1175	4.7320	3.6294	1.9685	2.9997
7.1825	5.5193	2.9997	4.5511	8.3401	10.8140	11.6696	10.8140	8.3401	4.5511	6.9059	12.5524
12.5524	6.9059	10.5604	18.8743	23.9882	25.6955	23.9882	18.8743	10.5604	16.4980	28.4636	35.3076
16.4980	26.9979	43.2305	51.3472	53.7850	51.3472	43.2305	26.9980	48.2849	66.1531	73.1177	74.9977

WHEN T =   1.953 V VALUES ARE

0.0539	0.0996	0.1301	0.1408	0.1301	0.0996	0.0539	0.1196	0.2209	0.2886	0.3124	0.2886
0.3892	0.5085	0.5504	0.5085	0.3892	0.2107	0.3450	0.6374	0.8327	0.9012	0.8327	0.6374
1.3204	1.4290	1.3204	1.0109	0.5473	0.8526	1.5745	2.0560	2.2249	2.0560	1.5745	0.8526
3.4205	3.1616	2.4224	1.3125	2.0025	3.6922	4.8141	5.2063	4.8141	3.6922	2.0025	3.0362
7.2707	5.5868	3.0362	4.5885	8.4092	10.9043	11.7674	10.9043	8.4092	4.5885	6.9421	12.6193
12.6193	6.9422	10.5932	18.9348	24.0673	25.7811	24.0673	18.9348	10.5932	16.5250	28.5136	35.3728
16.5250	27.0172	43.2661	51.3937	53.8354	51.3937	43.2661	27.0172	48.2950	66.1716	73.1420	75.0240

Figure 8.10 (continued)

```
 0.0000   0.0000   0.0000
 0.0000   0.0001   0.0001
 0.0010   0.0013   0.0014
 0.0184   0.0194   0.0196
 0.2705   0.2734   0.2705
 3.8079   3.7682   3.5699
52.4849  49.7225  39.2256
```

```
 0.0000   0.0000   0.0000
 0.0002   0.0006   0.0010
 0.0067   0.0115   0.0135
 0.1251   0.1465   0.1520
 1.4636   1.5157   1.4636
12.7396  12.3462  10.7705
59.9858  56.0189  42.7227
```

```
 0.2067   0.1119   0.1988
 0.3287   0.5262   0.9719
 1.2822   2.3665   3.0885
 5.5193   7.1825   7.7617
16.1636  17.3968  16.1636
37.5042  35.3076  28.4636
73.1177  66.1531  48.2849
```

```
 0.2209   0.1196   0.2107
 0.3450   0.5473   1.0109
 1.3125   2.4224   3.1616
 5.5868   7.2707   7.8572
16.2510  17.4915  16.2510
37.5748  35.3728  28.5136
73.1420  66.1716  48.2950
```

Figure 8.10 (concluded)

generates the coefficients of the set of simultaneous equations that are to be solved, which are, when $r = 1$,

At left end: $\quad (4 - 2\alpha\,\Delta x)u_1^{j+1} - 2u_2^{j+1} = 2\alpha\,\Delta x u_1^j + 2u_2^j - 4\nu\,\Delta x;$

Interior points: $\quad -u_{i-1}^{j+1} + 4u_i^{j+1} - u_{i+1}^{j+1} = u_{i-1}^j + u_{i+1}^j;$

At right end: $\quad -2u_n^{j+1} + (4 + 2\alpha\,\Delta x)u_{n+1}^{j+1} = 2u_n^j - 2\alpha\,\Delta x u_{n+1}^j + 4\nu\,\Delta x.$

In these relations, $\alpha = -a/b$, $\nu = c/b$. The coefficients are compressed into an n-by-4 array and the tridiagonal system is solved by first finding the LU decomposition. Temperatures are printed after each time step. The program calculates temperatures until $T = 1000$ for a bar of length 20 cm whose initial temperatures are given by $u = 100 - 10\,|x - 10|$, with the left end condition of $u_x = 0.2(u - 15)$ and right end condition of $u = 100$. Partial output is shown, at the beginning and then at the end of the time span.

Program 3 illustrates how the A.D.I. method can be implemented with a computer program. Two vectors are used, U and V, to hold the function values as calculated by the alternating horizontal and vertical traverses. The program sets up the coefficient matrices for each traverse, computes the righthand sides from the last computed values of U and V, and solves the system using LU decomposition. Function values are output only after each second set of computations because the odd traverses are less accurate. Representative output is shown for a rectangular region with 105 interior nodes when the boundary is held at 0 on three sides and at 100 on the fourth side. Initially, the potential function is 0 at all points.

EXERCISES

Section 8.1

1. The parameters of the basic equation for unsteady-state heat transfer are dimensional. If it is desired to measure u in °F and x in inches, how must the units of k, c, and ρ be chosen in

$$\frac{\partial^2 u}{\partial x^2} = \frac{c\rho}{k}\frac{\partial u}{\partial t}?$$

▶2. Solve for the temperatures at $t = 2.062$ seconds in the 2-cm-thick steel slab of Section 8.1 if the initial temperatures are given by the relation

$$u(x, 0) = 100 \sin\frac{\pi x}{2}.$$

Use the explicit method with $\Delta x = 0.25$ cm. Compare to the analytical solution: $100e^{-0.3738t}\sin(\pi x/2)$.

3. Solve for the temperatures in a copper rod 10 inches long, with the outer curved surface insulated so that heat flows in only one direction. The initial temperature is linear from 0°C at one end to 100°C at the other, when suddenly the hot end is brought to 0°C, and the cold end is held at 0°C. Use $\Delta x = 1$ inch and an appropriate value of Δt so that $k\,\Delta t/c\rho(\Delta x)^2 = \frac{1}{2}$. Look up values of k, c, and ρ in a handbook. Carry out the solution for 10 time steps.

4. Repeat Exercise 3 with $\Delta x = 0.5$ inch, and compare the temperatures at points 1 inch, 3 inches, and 6 inches from the cold end in the two calculations.

5. Repeat Exercise 3 with $\Delta x = 1$ inch, and Δt such that $k\,\Delta t/c\rho(\Delta x)^2 = \frac{1}{4}$. Compare results with Exercises 3 and 4.

▶ 6. Repeat computations with Δx as given below for the diffusion example of Section 8.1, and compare to the analytical solution at $x = 4$ and $x = 12$ cm. Carry them out until $t = 268.8$ sec.

a) With $\Delta x = 4$ cm, $D\,\Delta t/(\Delta x)^2 = 0.125$;

b) With $\Delta x = 2$ cm, $D\,\Delta t/(\Delta x)^2 = \frac{1}{2}$;

c) With $\Delta x = 1$ cm, $D\,\Delta t/(\Delta x)^2 = \frac{1}{2}$.

If you use the results shown in Section 8.1 of the text, you now have data that illustrate the effect of size of Δx and of r on accuracy. Considering the amount of calculations required, which is the more effective way to improve accuracy?

Section 8.2

7. Solve Exercise 2 by the Crank–Nicolson method, $r = 1$.

8. Solve Exercise 3 by the Crank–Nicolson method, $r = 1$. Compare results with Exercises 3, 4, and 5.

▶ 9. The methods of Sections 8.1 and 8.2 can be applied readily to more complicated situations. For example, if heat is being generated at various points along a bar at a rate which is a function of x, the unsteady-state heat equation becomes

$$k\frac{\partial^2 u}{\partial x^2} - c\rho\frac{\partial u}{\partial t} = f(x).$$

Solve this equation where $f(x) = x$ cal/cm³ · sec., subject to conditions

$$u(0, t) = 0, \qquad u(1, t) = 0, \qquad u(x, 0) = 0.$$

Take $\Delta x = 0.2$, $k = 0.37$, $c\rho = 0.433$ in cgs units. (These are properties of magnesium.) Use the Crank–Nicolson method, $r = 1$, and solve for five time steps.

Section 8.3

10. Use the set of equations in Eq. (8.10) to find the solution to the example in Section 8.3 through eight time steps.

11. Solve the set of equations in Eq. (8.11) to find the solution to the example in Section 8.3 by the Crank–Nicolson method. Compare results with those from Exercise 10.

▶12. Solve the example of Section 8.3 but with two opposite faces of the cube losing heat at a rate equal to $0.15A(u - 70)$, where u is the surface temperature in °F, and A is the area. Use $\Delta x = 1$ inch, and employ the explicit method with $k\,\Delta t/c\rho(\Delta x)^2 = \frac{1}{4}$.

13. Solve Exercise 12 using the Crank–Nicolson method with $r = 1$. Compare results by the two methods of solution.

▶14. Heat is added to one end of a 2-foot-long bar of copper at a rate given by

$$-hA(u - u_0),$$

and lost at the other end through a similar rate equation. Temperatures are $u_0 = $

500°F at the hot end, and $u_0 = 60$°F at the cold end, with resistance to heat flow being such that $h = 0.3$ Btu/sec · ft^2 · °F at both ends. The circumference of the bar is carefully insulated so that heat flows in only one dimension. Find the time required for the midpoint of the bar to reach 200°F. The bar is initially at 60°F at all points.

Section 8.4

▶15. Demonstrate, by performing calculations similar to those given in Table 8.5, that the explicit method is unstable with $k \, \Delta t / c\rho(\Delta x)^2 = 0.6$.

16. Demonstrate, by performing calculations similar to those given in Table 8.5, that the explicit method with $k \, \Delta t / c\rho(\Delta x)^2 = \frac{1}{4}$ has errors that damp out less rapidly than those in Table 8.5, but the method is still stable.

17. Demonstrate, by performing calculations similar to those in Table 8.5, except that at both $x = x_1$ and $x = x_5$ the gradient is zero ($\partial u/\partial x = 0$ instead of $u =$ constant), that the explicit method is still stable with $k \, \Delta t / c\rho(\Delta x)^2 = \frac{1}{2}$. Note, however, how much more slowly an error damps out, and that the error at a later step becomes a linear combination of earlier errors.

18. Demonstrate, by performing calculations similar to those in Table 8.6, that the Crank–Nicolson method is still stable even though the value of $k \, \Delta t / c\rho(\Delta x)^2$ is taken as 10. You will probably wish to use the LU method to solve the system.

▶19. Compute the largest eigenvalue of the coefficient matrix of Eq. (8.16) for $r = 0.5$, then for $r = 0.6$, when $N = 10$. (You may wish to use the power-method program of Chapter 6.) Do you find that the statements of the text about the value of the eigenvalues is confirmed?

20. Repeat Exercise 19 but for Eq. (8.18) using matrix $A^{-1}B$. Use $r = 1.0$ and $r = 2.0$.

Section 8.5

21. A rectangular plate 2 by 3 inches is initially at 50°. At $t = 0$, one 2-inch edge is suddenly raised to 100°, and one 3-inch edge is suddenly cooled to 0°. The temperature on these two edges is then held constant at these temperatures. The other two faces are perfectly insulated. Use a 1-inch grid to subdivide the plate and write the A.D.I. equations for each of the six points where unknown temperatures are involved. Use $r = 2$, and solve the equations for four time intervals.

22. Suppose the cube used for the example problem in Section 8.3 (4 by 4 by 4 inches) had heat flowing in all three directions, such as by having three adjacent faces lose heat by conduction to the flowing liquid. Set up the equations to solve for the temperature at any point. Do this in terms of the surrounding temperatures one Δt previously, using the explicit method with $\Delta x = \Delta y = \Delta z = 1$. How many time steps are needed to reach $t = 15.12$ sec? How many equations are involved in each stage of the calculation?

23. Repeat Exercise 22 for the Crank–Nicolson method, $r = 1$.

24. Repeat Exercise 22 using the A.D.I. method:

APPLIED PROBLEMS AND PROJECTS

25. Modify Program 1 so that end conditions of the form

$$au + b\frac{\partial u}{\partial x} = c$$

are permitted, similar to Program 2. Then use your program to solve Exercise 14. Vary the size of Δx and/or $r = k\,\Delta t/c\rho(\Delta x)^2$ until you are sure that the time for the midpoint temperature to become 200°F is known to within 0.5% relative error.

26. Solve Exercise 14 by Program 2. Critically compare the efficiency of Program 2 in obtaining the solution with the efficiency of the program you wrote in Problem 25.

27. Write a program similar to Program 3 but employing the A.D.I. method in three space dimensions with a parallelopiped-shaped region. Provide for derivative boundary conditions as an alternative to fixed boundary conditions by allowing for conditions of the form

$$au + b\frac{\partial u}{\partial x} = c.$$

28. When steel is forged, billets are heated in a furnace until the metal is of the proper temperature, between 2000°F and 2300°F. It can then be formed by the forging press into rough shapes that are later given their final finishing operations. In order to produce a certain machine part, a billet of size $4'' \times 4'' \times 20''$ is heated in a furnace whose temperature is maintained at 2350°F. You have been requested to estimate how long it will take all parts of the billet to reach a temperature above 2000°F. Heat transfers to the surface of the billet at a very high rate, principally through radiation. It has been suggested that you can solve the problem by assuming that the surface temperature becomes 2250°F instantaneously and remains at that temperature. Using this assumption, find the required heating time.

Since the steel piece is relatively long compared to its width and thickness, it may not introduce significant error to calculate as if it were infinitely long. This will simplify the problem, permitting a two-dimensional treatment rather than three-dimensional. Such a calculation should also give a more conservative estimate of heating time. Compare the estimates from two- and three-dimensional approaches.

29. After you have calculated the answers to Problem 28, your results have been challenged on the basis of assuming constant surface temperature of the steel. Radiation of heat flows according to the equation

$$q = E\sigma(u_F^4 - u_S^4) \text{ Btu/hr} \cdot \text{ft}^2,$$

where E = emissivity (use 0.80), σ is the Stefan–Boltzmann constant (0.171×10^{-8} Btu/hr $\cdot$ ft^2 $\cdot$ °F^4), u_F and u_S are the furnace and surface absolute temperatures, respectively (°F + 460°).

The heat radiating to the surface must also flow into the interior of the billet by conduction, so

$$q = -k\frac{\partial u}{\partial x},$$

where k is the thermal conductivity of steel (use 26.2 Btu/hr $\cdot$ ft^2 $\cdot$ °F/ft) and $(\partial u/\partial x)$ is the temperature gradient at the surface in a direction normal to the surface. Solve the problem with this boundary condition and compare your solution to that of Problem 28. (Observe that this is now a nonlinear problem. Think carefully how your solution can cope with it.)

30. Shipment of liquefied natural gas by refrigerated tankers to industrial nations is becoming an important means of supplying the world's energy needs. It must be stored at the receiving port, however. [A. R. Duffy and his co-workers discuss the storage of liquefied natural gas in underground tanks in *Oil and Gas Journal*, Aug. 1 and Aug. 8, 1966 and in *Chem. Eng. Prog.* **63**, 6, 55 (1967).] A commercial design, based on experimental verification of its feasibility, contemplated a prestressed concrete tank 270 ft in diameter and 61 ft deep, holding some 600,000 bbl of liquefied gas at −258°F. Convection currents in the liquid were shown to keep the temperature uniform at this value, the boiling point of the liquid.

Important considerations of the design are the rate of heat gained from the surroundings (causing evaporation of the liquid gas) and variation of temperatures in the earth below the tank (relating to the safety of the tank, which could be affected by possible settling or frost-heaving.)

The tank itself is to be made of concrete 6″ thick, covered with 8″ of insulation (on the liquid side.) (A sealing barrier keeps the insulation free of liquid; otherwise its insulating capacity would be impaired.) The experimental tests showed that there is a very small temperature drop through the concrete: 12°F. This observed 12°F temperature difference seems reasonable in light of the relatively high thermal conductivity of concrete. We expect then that most of the temperature drop occurs in the insulation or in the earth below the tank.

Since the commercial-design tank is very large, if we are interested in ground temperatures near the center of the tank (where penetration of cold will be a maximum) it should be satisfactory to consider heat flowing in only one dimension, in a direction directly downward from the base of the tank. Making this simplifying assumption, compute how long it will take for the temperature to decrease to 32°F (freezing point of water) at a point 8 ft away from the tank wall. Necessary thermal data is:

	Insulation	Concrete	Earth
Thermal conductivity (Btu/hr $\cdot$ ft $\cdot$ °F)	0.013	0.90	2.6
Density (lb/ft^3)	2.0	150	132
Specific heat (Btu/lb $\cdot$ °F)	0.195	0.200	0.200

Assume that the initial conditions are: Temperature of liquid: −258°F; temperature of insulation: −258°F to 72°F (inner surface to outer); temperature of concrete: 72°F to 60°F; temperature of earth: 60°F.

Chapter Nine

Hyperbolic Partial-Differential Equations

The third classification of partial differential equations, hyperbolic differential equations, includes the "wave equation" which is fundamental to the study of vibrating systems. They are also involved in transport problems (diffusion of matter, neutron diffusion, and radiation transfer), wave mechanics, gas dynamics, supersonic flow, and other important areas. The technology in most of these applications is so complex that their study is beyond our scope. We will settle for simple situations within the present experience of our readers. One such case is the vibrations of a string, held taut by fixed ends, as in a violin string. We outline the derivation of the simple wave equation in one dimension. Imagine an elastic string stretched between two fixed endpoints and set to vibrating by plucking it with a finger.

We make a number of simplifying assumptions: the string is perfectly elastic, and we neglect gravitational forces, so that the only force is the tension force in the direction of the string; the string is uniform in density and thickness and has a weight per unit length of w lb/ft; the lateral displacement of the string is so small that the tension can be considered to be a constant value of T lb; and the slope of the string is hence small enough that $\sin \alpha \doteq \tan \alpha$, where α is the angle of inclination. Figure 9.1 illustrates the problem with the lateral displacement greatly exaggerated.

Figure 9.1

We take $x = 0$ at the left end. L is the total length of the string. Focus attention on the element of length dx between points A and B. The uniform tension T acts at each end. We are interested in how y, the lateral displacement, varies with time t and with distance x along the string.

The forces acting in the y-direction are the vertical components of the two tensions (see Fig. 9.2). We take the upward direction as positive:

Upward force at left end: $-T \sin \alpha_1 \doteq -T \tan \alpha_1 = -T\left(\dfrac{\partial y}{\partial x}\right)_A$;

Upward force at right end: $+T \sin \alpha_2 \doteq +T \tan \alpha_2 = T\left(\dfrac{\partial y}{\partial x}\right)_B$

$$= T\left(\left(\frac{\partial y}{\partial x}\right)_A + \frac{\partial}{\partial x}\left(\frac{\partial y}{\partial x}\right)dx\right);$$

Net force: $T\dfrac{\partial^2 y}{\partial x^2}\,dx.$

Partials are used to express the slope because y is a function of both t (time) and x (horizontal distance.) We use Newton's law, and equate the force to mass times acceleration in the vertical direction. Our simplifying assumptions permit us to use $w\,dx$ as the weight:

$$\left(\frac{w\,dx}{g}\right)\frac{\partial^2 y}{\partial t^2} = T\frac{\partial^2 y}{\partial x^2}\,dx, \qquad \frac{\partial^2 y}{\partial t^2} = \frac{Tg}{w}\frac{\partial^2 y}{\partial x^2}. \tag{9.1}$$

This is the wave equation in one dimension. The conditions imposed on the solution are the end conditions $[ay + b(\partial y/\partial x) = f(t)]$ at $x = 0$ and $x = L$, and the initial conditions at $t = 0$. Initial conditions specifying both $y = f(x)$ and the velocity $\partial y/\partial t = g(x)$ are usual for this problem.

Comparing Eq. (9.1) to the general form of a second-order partial-differential equation,

$$A\frac{\partial^2 y}{\partial t^2} + B\frac{\partial^2 y}{\partial t\,\partial x} + C\frac{\partial^2 y}{\partial x^2} + D\left(t, x, y, \frac{\partial y}{\partial t}, \frac{\partial y}{\partial x}\right) = 0,$$

shows $B^2 - 4AC > 0$, since $B = 0$, $A = 1$, and $C = -Tg/w$. We see that Eq. (9.1) falls in the class of hyperbolic equations.

Element from x to $x+dx$

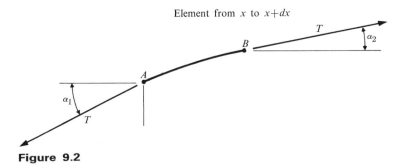

Figure 9.2

9.1 SOLVING THE WAVE EQUATION BY FINITE DIFFERENCES

We attack the problem of solving the one-dimensional wave equation in the usual way by replacing derivatives by difference quotients. We use superscripts to denote time, and subscripts for position:

$$\frac{\partial^2 y}{\partial t^2} = \frac{Tg}{w} \frac{\partial^2 y}{\partial x^2},$$

$$\frac{y_i^{j+1} - 2y_i^j + y_i^{j-1}}{(\Delta t)^2} = \frac{Tg}{w}\left(\frac{y_{i+1}^j - 2y_i^j + y_{i-1}^j}{(\Delta x)^2}\right).$$

Solving for the displacement at the end of the current interval, at $t = t_{j+1}$, we get

$$y_i^{j+1} = \frac{Tg(\Delta t)^2}{w(\Delta x)^2}(y_{i+1}^j + y_{i-1}^j) - y_i^{j-1} + 2\left(1 - \frac{Tg(\Delta t)^2}{w(\Delta x)^2}\right)y_i^j.$$

Note that selecting the ratio $Tg(\Delta t)^2/w(\Delta x)^2 = 1$ gives some simplification, though y_i^{j+1} is still a function of conditions at both t_j and at t_{j-1}.

$$y_i^{j+1} = y_{i+1}^j + y_{i-1}^j - y_i^{j-1}, \qquad \Delta t = \frac{\Delta x}{\sqrt{Tg/w}}. \tag{9.2}$$

Equation (9.2) is the usual way that the one-dimensional wave equation is solved numerically. We are, of course, interested in the validity of choosing the ratio at unity. Our finite-difference replacements to the derivatives do not here have mixed error terms, as they did in the explicit method for the heat equation; but, as we shall later show, after discussing the method of characteristics, if the ratio is greater than one, we cannot be sure of convergence. Stability also sets a limit of unity to the ratio. It is surprising to find that if one uses a value of $w(\Delta t)^2/Tg(\Delta x)^2$ of less than one, the results are less accurate, while with the ratio equal to one we can get exact analytical answers!*

There is still a problem in applying Eq. (9.2). We know y at $t = t_0 = 0$ from the initial condition, but to compute y at $t = \Delta t = t_1$, we need values at $t = t_{-1}$. We may at first be bothered by a need to know displacements *before* the start of the problem; but if we imagine the function $y = y(x, t)$ to be extended backward in time, the term t_{-1} makes good sense. Since we will ordinarily get periodic functions for y versus t at a given point, we can consider zero time as an arbitrary point at which we know the displacement and velocity, within the duration of an ongoing process.

One commonly used way to get values for the fictitious point at $t = t_{-1}$ is by using the specification for the initial velocity $\partial y/\partial t = g(x)$ at $t = 0$. Using a

---

* What this means is that, with $w(\Delta t)^2/Tg(\Delta x)^2$ less than one, errors will decrease as $\Delta x \to 0$; that is, the method is stable. However, with finite values for Δx, the computed values are not exact unless the ratio is equal to one. Our discussion of characteristics later in this chapter will clarify this unusual situation.

central-difference approximation, we have

$$\frac{\partial y}{\partial t}(x_i, 0) \doteq \frac{y_i^1 - y_i^{-1}}{2\Delta t} = g(x_i),$$

$$y_i^{-1} = y_i^1 - 2g(x_i)\Delta t \qquad \text{at } t = 0 \text{ only.} \tag{9.3}$$

Equation (9.3) is valid only at $t = 0$; substituting into Eq. (9.2) gives us, for $t = t_1$,

$$y_i^1 = \tfrac{1}{2}(y_{i+1}^0 + y_{i-1}^0) + g(x_i)\Delta t. \tag{9.4}$$

After computing the first line with Eq. (9.4), we use Eq. (9.2) thereafter.* If the boundary conditions involve derivatives, we rewrite them in terms of difference quotients to incorporate them in our difference equation.

Example   A banjo string is 80 cm long and weighs 1.0 gram. It is stretched with a tension of 40,000 grams. At a point 20 cm from one end it is pulled 0.6 cm from the equilibrium position and then released. Find the displacement of points along the string as a function of time. How long does it take for one complete cycle of motion? From this, compute the frequency with which it vibrates.

Let $\Delta x = 10$ cm. In Table 9.1 we show the results of using Eqs. (9.4) and (9.2). Because the string is just released from its initially displaced position, the initial velocity at all points is zero, and Eq. (9.4) becomes simply

$$y_i^1 = \tfrac{1}{2}(y_{i+1}^0 + y_{i-1}^0).$$

The initial conditions imply that y is linear in x from $y = 0$ at $x = 0$ to $y = 0.6$ at $x = 20$ and also linear to $x = 80$. The size of time steps is given by·

$$\frac{Tg(\Delta t)^2}{w(\Delta x)^2} = 1 = \frac{(40{,}000)(980)(\Delta t)^2}{(1.0/80)(10)^2}, \qquad \Delta t = 0.000179 \text{ sec.}$$

After we have completed calculations for eight time steps, we observe that the y-values are reproducing the original steps but with negative signs and the end of the string reversed; i.e., half a cycle has been completed in eight steps. For a complete cycle, 16 Δt's are needed. The frequency of vibration is then:

$$f = \frac{1}{(16)(0.000179)} = 350 \text{ cycles/sec.}$$

(This value is exactly the same as that given by the standard formula,

$$f = \frac{1}{2L}\sqrt{Tg/w.}\Bigg)$$

The solution to our example as given in Table 9.1 is exactly the analytical solution, as we shall now demonstrate. We shall compare our finite-difference equation (Eq. (9.2)), with the d'Alembert solution to the wave equation.

---

* We shall later discuss a more accurate way to begin the solution. Equation (9.4) is satisfactory when the initial velocity is zero.

Table 9.1 Solution to banjo string example.

t	0	10	20	30	40	50	60	70	80
0	0.0	0.3	0.6	0.5	0.4	0.3	0.2	0.1	0.0
Δt	0.0	0.3	0.4	0.5	0.4	0.3	0.2	0.1	0.0
$2\Delta t$	0.0	0.1	0.2	0.3	0.4	0.3	0.2	0.1	0.0
$3\Delta t$	0.0	−0.1	0.0	0.1	0.2	0.3	0.2	0.1	0.0
$4\Delta t$	0.0	−0.1	−0.2	−0.1	0.0	0.1	0.2	0.1	0.0
$5\Delta t$	0.0	−0.1	−0.2	−0.3	−0.2	−0.1	0.0	0.1	0.0
$6\Delta t$	0.0	−0.1	−0.2	−0.3	−0.4	−0.3	−0.2	−0.1	0.0
$7\Delta t$	0.0	−0.1	−0.2	−0.3	−0.4	−0.5	−0.4	−0.3	0.0
$8\Delta t$	0.0	−0.1	−0.2	−0.3	−0.4	−0.5	−0.6	−0.3	0.0
$9\Delta t$	0.0	−0.1	−0.2	−0.3	−0.4	−0.5	−0.4	−0.3	0.0
$10\Delta t$	0.0	−0.1	−0.2	−0.3	−0.4	−0.3	−0.2	−0.1	0.0
$11\Delta t$	0.0	−0.1	−0.2	−0.3	−0.2	−0.1	0.0	0.1	0.0
$12\Delta t$	0.0	−0.1	−0.2	−0.1	0.0	0.1	0.2	0.1	0.0
$13\Delta t$	0.0	−0.1	0.0	0.1	0.2	0.3	0.2	0.1	0.0
$14\Delta t$	0.0	0.1	0.2	0.3	0.4	0.3	0.2	0.1	0.0
$15\Delta t$	0.0	0.3	0.4	0.5	0.4	0.3	0.2	0.1	0.0
$16\Delta t$	0.0	0.3	0.6	0.5	0.4	0.3	0.2	0.1	0.0

9.2 COMPARISON TO THE D'ALEMBERT SOLUTION

The method of d'Alembert lets us find the analytical solution to the wave equation in one dimension. The one-dimensional wave equation, with $\sqrt{Tg/w} = c$, is

$$\frac{\partial^2 y}{\partial t^2} = c^2 \frac{\partial^2 y}{\partial x^2}. \tag{9.5}$$

By direct substitution, it is readily seen that, for any arbitrary functions F and G, Eq. (9.5) is solved by:

$$y(x, t) = F(x + ct) + G(x - ct). \tag{9.6}$$

The demonstration is easy, since

$$\frac{\partial y}{\partial t} = F' \frac{\partial(x + ct)}{\partial t} + G' \frac{\partial(x - ct)}{\partial t} = cF' - cG',$$

$$\frac{\partial^2 y}{\partial t^2} = c^2 F'' - c(-c)G'' = c^2 F'' + c^2 G''; \tag{9.7}$$

$$\frac{\partial y}{\partial x} = F' \frac{\partial(x + ct)}{\partial x} + G' \frac{\partial(x - ct)}{\partial x} = F' + G',$$

$$\frac{\partial^2 y}{\partial x^2} = F'' + G''. \tag{9.8}$$

Equation (9.5) results immediately from Eqs. (9.7) and (9.8).

The solution to the problem is found then, if we can find a pair of functions of the form of (9.6) whose sum matches the initial conditions and the boundary conditions for the problem. If the initial conditions are:

$$y(x, 0) = f(x),$$

$$\frac{\partial y}{\partial t}(x, 0) = g(x),$$

it is again readily seen that $y(x, t)$ is given by:

$$y(x, t) = \frac{1}{2}[f(x + ct) + f(x - ct)] + \frac{1}{2c}\int_{x-ct}^{x+ct} g(v)\, dv. \qquad (9.9)$$

Note that we can rewrite Eq. (9.9) in the form of Eq. (9.6). In Eq. (9.9) we have changed to the dummy variable v under the integral sign. The demonstration parallels that above when we recall how to differentiate an integral:

$$\frac{\partial}{\partial t}\left[\frac{1}{2c}\int_{x-ct}^{x+ct} g(v)\, dv\right] = \frac{1}{2c}[cg(x + ct) - (-c)g(x - ct)],$$

$$\frac{\partial^2}{\partial t^2}\left[\frac{1}{2c}\int_{x-ct}^{x+ct} g(v)\, dv\right] = \frac{\partial}{\partial t}\left[\frac{1}{2}g(x + ct) + \frac{1}{2}g(x - ct)\right] = \frac{1}{2}cg' - \frac{1}{2}cg' = 0,$$

$$\frac{\partial}{\partial x}\left[\frac{1}{2c}\int_{x-ct}^{x+ct} g(v)\, dv\right] = \frac{1}{2c}[g(x + ct) - g(x - ct)],$$

$$\frac{\partial^2}{\partial x^2}\left[\frac{1}{2c}\int_{x-ct}^{x+ct} g(v)\, dv\right] = \frac{\partial}{\partial x}\left[\frac{1}{2c}g(x + ct) - \frac{1}{2c}g(x - ct)\right] = \frac{1}{2c}g' - \frac{1}{2c}g' = 0.$$

Equation (9.9) gives the value of y at any time t provided that y is known at points ct to the right and left of the point at $t = 0$, and in terms of the integral of the initial velocity between the lateral points. Hence it is useful to find the displacement of points in the interior portions of a vibrating string.

We are now in a position to verify that our numerical procedure, given by Eq. (9.2), gives the exact solution (except for round-off), provided that two lines of correct y-values are known. The algorithm we use for the numerical procedure is:

$$y_i^{j+1} = y_{i-1}^j + y_{i+1}^j - y_i^{j-1}. \qquad (9.10)$$

We now show that the solution to this difference equation is a solution to the differential equation. Consider the function

$$y_i^j = F(x_i + ct_j) + G(x_i - ct_j). \qquad (9.11)$$

Because we use

$$\frac{Tg(\Delta t)^2}{w(\Delta x)^2} = 1 = \frac{c^2(\Delta t)^2}{(\Delta x)^2},$$

then

$$\Delta x = c\Delta t.$$

Write x_i and t_j in terms of starting values, x_0, and $t_0 = 0$:

$$x_i = x_0 + i\,\Delta x,$$

$$ct_j = c(t_0 + j\,\Delta t) = cj\,\Delta t = j\,\Delta x.$$

Substituting into Eq. (9.11), we get,

$$y_i^j = F(x_0 + i\,\Delta x + j\,\Delta x) + G(x_0 + i\,\Delta x - j\,\Delta x)$$
$$= F[x_0 + (i + j)\Delta x] + G[x_0 + (i - j)\Delta x].$$

Use this relation to rewrite the righthand side of the difference equation, Eq. (9.10):

$$y_{i-1}^j + y_{i+1}^j - y_i^{j-1} = F[x_0 + (i - 1 + j)\Delta x] + G[x_0 + (i - 1 - j)\Delta x]$$
$$+ F[x_0 + (i + 1 + j)\Delta x] + G[x_0 + (i + 1 - j)\Delta x]$$
$$- F[x_0 + (i + j - 1)\Delta x] - G[x_0 + (i - j + 1)\Delta x]$$
$$= F[x_0 + (i + j + 1)\Delta x] + G[x_0 + (i - j - 1)\Delta x] = y_i^{j+1}.$$

$$(9.12)$$

Equation (9.12) shows that if the previous two lines of the numerical solution are of the form of Eq. (9.11) (and hence must be exact solutions of the wave equation in view of Eq. (9.6)), it then follows that the values on the next line are exact solutions also, because they also are of the form of Eq. (9.11).

In order that our simple algorithm give the analytical solution, it is then necessary only that two lines of the computation be correct. The first line is correct because it is given by the initial conditions. While Eq. (9.4) is frequently recommended for giving the second line in terms of the initial velocity, it is sometimes inaccurate because it assumes that the initial velocity and the average velocity from t_{-1} to t_1 are the same, and this may not be true.[*]

Equation (9.9) offers a more exact way to get the second line of y-values. If we employ that relationship at $t = \Delta t$, we have

$$y(x, \Delta t) = y_i^1 = \frac{1}{2}[f(x_i + \Delta x) + f(x_i - \Delta x)] + \frac{1}{2c}\int_{x_i - \Delta x}^{x_i + \Delta x} g(v)\,dv$$

$$= \frac{1}{2}[y_{i+1}^0 + y_{i-1}^0] + \frac{1}{2c}\int_{x_i - 1}^{x_i + 1} g(v)\,dv.$$

$$(9.13)$$

---

[*] The relationship of Eq. (9.4) is correct whenever $g(x) = $ constant. It is therefore correct for the case of $g(x) = 0$.

Equation (9.13) differs from Eq. (9.4) only in the last term. We now see why the banjo-string example of Section 9.1 gave correct values in spite of using Eq. (9.4): In the example $g(x)$ was everywhere zero, so the form of the last term was unimportant. In the next example we illustrate the difference between the two different ways to begin the solution.

Example A string whose length is 9 units is initially in its equilibrium position. It is set into motion by striking it so that its initial velocity is given by $\partial y/\partial t = 3 \sin(\pi x/L)$. Take $\Delta x = 1$ unit and assume that $c = \sqrt{Tg/w} = 2$. When the ratio of $c^2(\Delta t)^2/(\Delta x)^2$ is unity, $\Delta t = 0.5$ time units.

If the ends are fixed, find the displacements at $t = 0.5$ time units later. The length is subdivided into nine intervals because $\Delta x = 1$.

The displacements we require are after one time step. Table 9.2 summarizes the computations. We first compute them using Eq. (9.4). The values disagree by several percent from the analytical values, which are given by

$$y(x, t) = \frac{1}{2c}\int_{x-ct}^{x+ct} 3\sin\frac{\pi v}{L}\, dv = \frac{3L}{2c\pi}\left\{\cos\left(\frac{\pi x}{L} - \frac{\pi ct}{L}\right) - \cos\left(\frac{\pi x}{L} + \frac{\pi ct}{L}\right)\right\}.$$

When Δx is cut in half, Eq. (9.4) gives improved values; it now requires two time steps to reach $t = 0.5$, of course.

Using Eq. (9.13), and evaluating the integral with Simpson's rule, we obtain results with $\Delta x = 1$ that are accurate to within one in the fourth decimal place.

Table 9.2 Comparison of ways to begin the solution of the wave equation in one dimension:

$$\frac{\partial^2 y}{\partial t^2} = c^2 \frac{\partial^2 y}{\partial x^2} \quad \text{over } x = 0 \text{ to } x = 9, \text{ with } c = 2;$$

$$y(x, 0) = 0, \quad \frac{\partial y}{\partial t}(x, 0) = 3\sin\left(\frac{\pi x}{9}\right).$$

	Value of displacements at $t = 0.5$			
Using Eq. (9.4): $\Delta x = 1$				
x	1	2	3	4
y	0.5130	0.9642	1.2990	1.4772
Using Eq. (9.4): $\Delta x = 0.5$				
x	1	2	3	4
y	0.5052	0.9495	1.2793	1.4548
Using Eq. (9.13): $\Delta x = 1$ Simpson's-rule integration				
x	1	2	3	4
y	0.5027	0.9448	1.2729	1.4475
Analytical solution:				
x	1	2	3	4
y	0.50267	0.94472	1.27282	1.44740

9.3 STABILITY OF THE NUMERICAL METHOD

Since we will ordinarily solve the one-dimensional wave equation numerically only with

$$\frac{Tg(\Delta t)^2}{w(\Delta x)^2} = 1,$$

it is sufficient to demonstrate stability for that scheme. We assume that a set of errorless computations have been made when a single error of size 1 occurs; we trace the effects of the single error. If the error does not have increasingly great effect on subsequent calculations, we call the method *stable*.

This simple procedure is adequate because, since the problem is linear, the principle of superposition lets us add the effects of all errors together and lets us add these errors to the true solution to obtain the actual results. Table 9.3 demonstrates the principle, assuming that the displacement of the endpoints is specified so that these are always free of error.

Table 9.3 Propagation of single error in numerical solution to wave equation

Initially error-free values	0	0	0	0	0	0	0
	0	0	0	0	0	0	0
Error made here	0	0	1	0	0	0	0
	0	1	0	1	0	0	0
	0	0	0	0	1	0	0
	0	-1	0	0	0	1	0
	0	0	-1	0	0	0	0
	0	0	0	-1	0	-1	0
	0	0	0	0	-2	0	0
	0	0	0	-1	0	-1	0
	0	0	-1	0	0	0	0
	0	-1	0	0	0	1	0

As the arrows indicate, the wave equation propagates disturbances in opposite directions, with reflections occurring at fixed ends, with a reversal of sign on reflection. Stability is demonstrated because the original error does not grow in size.

9.4 METHOD OF CHARACTERISTICS

The properties of the solution to the wave equation are further elucidated by considering the "characteristic curves" of the equation. This also will permit us to extend our numerical method to more general hyperbolic equations.

Consider the second-order partial-differential equation in two variables x and t:

$$au_{xx} + bu_{xt} + cu_{tt} + e = 0. \tag{9.14}$$

Here we have used the subscript notation to represent partial derivatives. The coefficients a, b, c, and e may be functions of x, t, u_x, u_t, and u, so the equation is very general.* We take $u_{xt} = u_{tx}$. To facilitate manipulations, let

$$p = \frac{\partial u}{\partial x} = u_x, \qquad q = \frac{\partial u}{\partial t} = u_t.$$

Write out the differentials of p and q:

$$dp = \frac{\partial p}{\partial x} dx + \frac{\partial p}{\partial t} dt = u_{xx} \, dx + u_{xt} \, dt,$$

$$dq = \frac{\partial q}{\partial x} dx + \frac{\partial q}{\partial t} dt = u_{xt} \, dx + u_{tt} \, dt.$$

Solving these last equations for u_{xx} and u_{tt}, respectively, we have

$$u_{xx} = \frac{dp - u_{xt} \, dt}{dx} = \frac{dp}{dx} - u_{xt} \frac{dt}{dx},$$

$$u_{tt} = \frac{dq - u_{xt} \, dx}{dt} = \frac{dq}{dt} - u_{xt} \frac{dx}{dt}.$$

Substituting in Eq. (9.14) and rearranging, we obtain

$$-au_{xt} \frac{dt}{dx} + bu_{xt} - cu_{xt} + a\frac{dp}{dx} + c\frac{dq}{dt} + e = 0.$$

Now multiplying by $-dt/dx$, we get

$$u_{xt}\left[a\left(\frac{dt}{dx}\right)^2 - b\left(\frac{dt}{dx}\right) + c\right] - \left[a\frac{dp}{dx}\frac{dt}{dx} + c\frac{dq}{dx} + e\frac{dt}{dx}\right] = 0.$$

Suppose, in the x,t-plane, we define curves such that the first bracketed expression is zero. On such curves, the original differential equation is equivalent to setting the second bracketed expression equal to zero. That is, if

$$am^2 - bm + c = 0, \tag{9.15}$$

where $m = dt/dx$, then the solution to the original equation (Eq. (9.14)) can be found by solving

$$am \, dp + c \, dq + e \, dt = 0. \tag{9.16}$$

---

* When the coefficients are independent of the function u or its derivatives, it is linear. If they are functions of u, u_x, or u_t (but not u_{xx} or u_{tt}), it is called quasilinear.

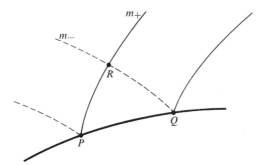

Figure 9.3

We have elected to write Eq. (9.16) in the form of differentials. It will be seen that we reduce the original problem, which is a second-order partial-differential equation, to solving a pair of first-order equations of the form of Eq. (9.16).

The curves whose slope m is given by Eq. (9.15) are called the *characteristics* of the differential equation. Since the equation is a quadratic, it may have one, two, or no real solutions, depending on the value of $b^2 - 4ac$. The value of this discriminant is the usual basis for classifying partial-differential equations. If

$$b^2 - 4ac < 0,$$

the equation is called *elliptic*, and there are no (real) characteristics. If $b^2 - 4ac = 0$, there is a single characteristic at any point, and the equation is termed *parabolic*. When $b^2 - 4ac > 0$, at every point there will be a pair of characteristic curves whose slopes are given by the two distinct, real roots of Eq. (9.15). Such equations are called *hyperbolic*, and our present discussion considers only this type.

Along the characteristics, the solution has special and desirable properties. For example, discontinuities in the initial conditions are propagated along them. On the characteristic curves, the numerical solution can be developed in the general case also.

We shall now outline a method of solving equations of the form of Eq. (9.14) by numerical integration along the characteristics.* We visualize the initial conditions as specifying the function u on some curve in the t,x-plane,† as well as its normal derivative. Consider two points, P and Q on this initial curve (Fig. 9.3). When Eq. (9.14) is hyperbolic, there are two characteristic curves through each point. The rightmost curve through P intersects the leftmost curve through Q, and these curves are such that their slopes are given by the appropriate roots of Eq. (9.15). Call m_+ the values of the slope on curve PR, and m_- the values on curve QR.

---

* We discuss only the solution of hyperbolic differential equations by the method of characteristics, but the technique can also be applied to parabolic equations.
† This curve must not itself be one of the characteristics, or advancing the solution is impossible.

Since these curves are characteristics, the solution to the problem can be found by solving Eq. (9.16) along them.

Our procedure will be to first find point R (perhaps only as a first approximation if a, b, or c involve the unknown function u). This is done by solving the equation

$$\Delta t = \left(\frac{dt}{dx}\right)_{av} \Delta x = m_{av} \, \Delta x, \tag{9.17}$$

applied over the arcs PR and QR simultaneously. When dt/dx is a function of x and/or t, it may be possible to integrate Eq. (9.17) analytically. When dt/dx varies with u, we will use a procedure resembling the Euler predictor–corrector method, by predicting with m_{av} taken as equal to m_+ at P or m_- at Q to start the solution. We correct by using the arithmetic average of m at the endpoints of each arc as soon as the value of m at R can be evaluated.

We then integrate Eq. (9.16) in the form

$$a_{av} m_{av} \, \Delta p + c_{av} \, \Delta q + e_{av} \, \Delta t = 0,$$

starting first from P and then from Q, using the appropriate values of m for each. This will estimate p and q at point R. Finally we evaluate the function u at R from

$$du = \frac{\partial u}{\partial x} \, dx + \frac{\partial u}{\partial t} \, dt,$$

used in the form

$$\Delta u = p_{av} \, \Delta x + q_{av} \, \Delta t.$$

The equation for Δu can be applied to either the change along

$$P \rightarrow R \quad \text{or} \quad Q \rightarrow R;$$

when these do not give the same result, we will *average* the two values. It may be necessary to iterate this procedure to get improved values at R. Average values are used for all varying quantities. The calculations are repeated for a second point S which is the intersection of characteristics through Q and another initial point W, and then continued in a like manner to evaluate u throughout the region in the x,t-plane as desired. We illustrate with three examples, first with dt/dx equal to a constant, then with dt/dx varying with x, and finally a more complex example with dt/dx varying with u.

For the simple wave equation,

$$u_{tt} = c^2 u_{xx},$$

the slopes of the characteristics are

$$m = \pm \frac{1}{c},$$

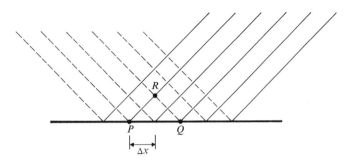

Figure 9.4

and the characteristics are the lines

$$t = \pm \frac{1}{c}(x - x_i).$$

Consider the curves from points P and Q, taken on the line for $t = 0$ (Fig. 9.4). The network of points used in the explicit finite-difference method of Section 9.1 are seen to be the intersections of characteristics through pairs of points spaced $2\,\Delta x$ apart. The finite-difference method, with

$$Tg(\Delta x)^2/w(\Delta t)^2 = 1,$$

will be found to be the equivalent of integration along the characteristics, lending further support to the likelihood that it will give exact answers.

Example 1 Solve

$$\frac{\partial^2 u}{\partial t^2} = 2\frac{\partial^2 u}{\partial x^2} - 4,$$

with initial conditions

$$u = 12x \quad \text{for} \quad 0 \leqslant x \leqslant 0.25,$$
$$u = 4 - 4x \quad \text{for} \quad 0.25 \leqslant x \leqslant 1.0,$$
$$\frac{\partial u}{\partial t} = 0 \quad \text{for} \quad 0 \leqslant x \leqslant 1.0;$$

boundary conditions are $u = 0$ at $x = 0$ and at $x = 1.0$.

Putting the equation into the standard form,

$$a\frac{\partial^2 u}{\partial x^2} + b\frac{\partial^2 u}{\partial x\,\partial t} + c\frac{\partial^2 u}{\partial t^2} + e = 0,$$

gives $a = -2$, $b = 0$, $c = 1$, and $e = 4$. (The equation is linear since a, b, c, and e are independent of u, u_x, and u_t.)

The slopes of the characteristics are the roots of

$$-2m^2 + 1 = 0,$$

$$m = \pm \frac{\sqrt{2}}{2},$$

so the characteristic curves are straight lines in the x,t-plane, as shown in Fig. 9.5. Consider points P, Q, and R—$(0.25, 0)$, $(0.75, 0)$, and $(0.5, 0.1768)$—and solve Eq. (9.16), which is:

$$am_{av}\, \Delta p + c\, \Delta q + e\, \Delta t = 0.$$

Along $P \to R$:

$$-2\left(\frac{\sqrt{2}}{2}\right)\Delta p + \Delta q + 4\Delta t = 0,$$

$$-\sqrt{2}(p_R - p_P) + (q_R - q_P) + 4(0.1768) = 0.$$

Along $Q \to R$:

$$-2\left(\frac{-\sqrt{2}}{2}\right)\Delta p + \Delta q + 4\Delta t = 0,$$

$$\sqrt{2}(p_R - p_Q) + (q_R - q_Q) + 4(0.1768) = 0.$$

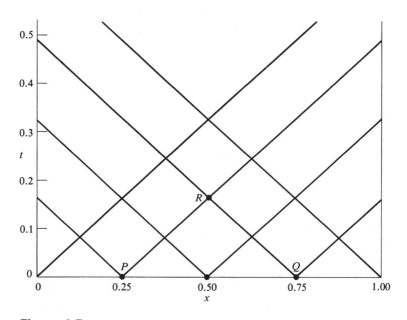

Figure 9.5

Table 9.4

x	0	0.25	0.5	0.75	1.0
$u\,(t = 0)$	0.0	3.0	2.0	1.0	0.0
$u\,(t = 0.1768)$	0.0	0.9375	1.9375	0.9375	0.0
$u\,(t = 0.3535)$	0.0	−1.1875	−0.2500	0.8125	0.0
$u\,(t = 0.5303)$	0.0	−1.3125	−2.4375	−1.3125	0.0

Using

$$p_P = \left(\frac{\partial u}{\partial x}\right)_P = -4,^* \qquad q_P = \left(\frac{\partial u}{\partial t}\right)_P = 0, \qquad p_Q = \left(\frac{\partial u}{\partial x}\right)_Q = -4,$$

$$q_Q = \left(\frac{\partial u}{\partial t}\right)_Q = 0,$$

we find $p_R = -4$, $q_R = -\sqrt{2}/2$ by solving the equations $P \to R$ and $Q \to R$ simultaneously.

Now we evaluate u at point R through its change along $P \to R$:

$$\Delta u = p_{av}\,\Delta x + q_{av}\,\Delta t = -4(0.25) + \left(\frac{0 - \sqrt{2}/2}{2}\right)(0.1768)$$

$$= -1.0625;$$

$$u_R = 3 + (-1.0625) = 1.9375.$$

(If we compute through evaluating Δu along $Q \to R$, we get the same result.)

For this simple problem, the finite-difference method is much simpler, and we expect it to give the same results. Following the procedure of Section 9.1† we compute these results with $\Delta x = 0.25$, $\Delta t = \Delta x/\sqrt{c} = 0.1768$, and obtain Table 9.4. The circled value agrees exactly with that calculated by the method of characteristics.

Example 2 Solve

$$\frac{\partial^2 u}{\partial t^2} = (1 + 2x)\frac{\partial^2 u}{\partial x^2}$$

over $(0, 1)$ with fixed boundaries and the initial conditions:

$$u(x, 0) = 0, \qquad \frac{\partial u}{\partial t}(x, 0) = x(1 - x).$$

---

* The gradient has a discontinuity at $x = 0.25$. The value of $\partial u/\partial x$ for points to the right of P applies for the region PRQ.

† The algorithm is $u_i^{j+1} + (u_{i+1}^j + u_{i-1}^j) - u_i^{j-1} - 4(\Delta t)^2$ with $\Delta t = \Delta x/\sqrt{2}$. For the first time step, $u_i^1 = \frac{1}{2}(u_{i+1}^0 + u_{i-1}^0) - \frac{1}{2}(4)(\Delta t)^2$.

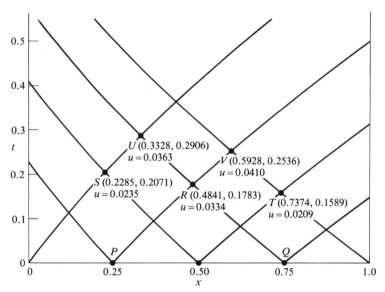

Figure 9.6

For this problem, $a = -(1 + 2x)$, $b = 0$, $c = 1$, $e = 0$, Then $am^2 + bm + c = 0$ gives

$$m = \pm\sqrt{\frac{1}{(1 + 2x)}}.$$

The characteristic curves are found by solving the differential equations $dt/dx = \sqrt{1/(1 + 2x)}$ and $dt/dx = -\sqrt{1/(1 + 2x)}$. Integrating* from the initial point x_0 and t_0, we have:

$$t = t_0 + \sqrt{1 + 2x} - \sqrt{1 + 2x_0} \quad \text{from } m_+,$$

$$t = t_0 - \sqrt{1 + 2x} + \sqrt{1 + 2x_0} \quad \text{from } m_-.$$

Figure 9.6 shows several of the characteristic curves. We select two points on the initial curve for $t = 0$, at $P = (0.25, 0)$ and $Q = (0.75, 0)$, whose characteristics intersect at point R. Solving for the intersection, we find $R = (0.4841, 0.1782)$. We now solve Eq. (9.16) to obtain $p = \partial u/\partial x$ and $q = \partial u/\partial t$ at R:

At point P: $x = 0.25$, $t = 0$, $u = 0$; $p = \left(\dfrac{\partial u}{\partial x}\right)_P = 0$,

$$q = \left(\frac{\partial u}{\partial t}\right)_P = x - x^2 = 0.1875;$$

$$m = \sqrt{1/(1 + 2x)} = 0.9165, \quad a = -(1 + 2x) = -1.5,$$

$$b = 0, \quad c = 1, \quad e = 0.$$

---

* In this example, the integration methods of calculus are easy to use. We could use a numerical method if they were not.

At point Q: $x = 0.75$, $t = 0$, $u = 0$; $p = 0$, $q = x - x^2 = 0.1875$;

$m = -\sqrt{1/(1 + 2x)} = -0.6325$, $a = -2.5$, $b = 0$, $c = 1$,

$e = 0$.

At point R: $x = 0.4841$, $t = 0.1783$, $m_+ = \sqrt{1/(1 + 2x)} = 0.7128$,

$m_- = -\sqrt{1/(1 + 2x)} = -0.7128$, $a = -1.9682$, $b = 0$,

$c = 1$, $e = 0$.

Equation (9.16) becomes, when we use average values for a and m,

$P \rightarrow R$: $-1.7341(0.8146)(p_R - 0) + (1)(q_R - 0.1875) = 0$;

$Q \rightarrow R$: $-2.2341(-0.6726)(p_R - 0) + (1)(q_R - 0.1875) = 0$.

Solving simultaneously, we get $p_R = 0$, $q_r = 0.1875$.

We calculate the change in u along the characteristics:

$P \rightarrow R$: $\Delta u = 0(0.2341) + 0.1875(0.1783) = 0.0334$,

$Q \rightarrow R$: $\Delta u = 0(-0.2659) + 0.1875(0.1783) = 0.0334$,

$u_R = 0 + 0.0334 = 0.0334$.

Figure 9.6 gives the results at several other intersections of characteristics. The student should verify these results to be sure he understands the method of characteristics.

Example 3   Solve the quasilinear equation, with conditions as shown, by numerical integration along the characteristics. (This might be a vibrating string with tension related to the displacement u and subject to an external lateral force.)

$$\frac{\partial^2 u}{\partial x^2} - u\frac{\partial^2 u}{\partial t^2} + (1 - x^2) = 0, \qquad u(x, 0) = x(1 - x), \qquad u_t(x, 0) = 0,$$
$$(9.18)$$
$$u(0, t) = 0, \qquad u(1, t) = 0.$$

We shall advance the solution beyond the start from P, at $x = 0.2$, $t = 0$, and Q, at $x = 0.4$, $t = 0$, to one new point R. Comparing Eq. (9.18) to the standard form,

$$au_{xx} + bu_{xt} + cu_{tt} + e = 0,$$

we have $a = 1$, $b = 0$, $c = -u$, $e = 1 - x^2$. We first compute u, p, and q at points P and Q:

$$u = x(1 - x)$$

(from the initial conditions), so

$$u_P = 0.2(1 - 0.2) = 0.16,$$
$$u_Q = 0.4(1 - 0.4) = 0.24;$$

also,

$$p = \frac{\partial u}{\partial x} = 1 - 2x$$

(by differentiating the initial conditions), so

$$p_P = 1 - 2(0.2) = 0.6,$$
$$p_O = 1 - 2(0.4) = 0.2;$$

and

$$q = \frac{\partial u}{\partial t} = 0$$

(from the initial conditions), so

$$q_P = 0,$$
$$q_O = 0.$$

To locate point R, we need the slope m of the characteristic. Using $am^2 - bm + c = 0$, we get:

$$m = \frac{b \pm \sqrt{b^2 - 4ac}}{2a},$$

$$m = \frac{\pm\sqrt{4u}}{2} = \pm\sqrt{u}.$$

Since m depends on the solution u, we shall need to find point R through the predictor–corrector approach. In the first trial, use the initial values over the whole arc; i.e., take $m_+ = +m_P$ and $m_- = -m_O$:

$$m_+ = \sqrt{u_P} = \sqrt{0.16} = 0.4,$$
$$m_- = \sqrt{u_O} = -\sqrt{0.2} \quad -0.490.$$

We now estimate the coordinates of R by solving simultaneously

$$t_R = m_+(x_R - x_P) = 0.4(x_R - 0.2),$$
$$t_R = m_-(x_R - x_O) = -0.490(x_R - 0.4).$$

These give

$$x_R = 0.310, \qquad t_R = 0.044.$$

We write Eq. (9.16) along each characteristic, again using the initial values of m, since m at R is still unknown:

$$am\,\Delta p + c\,\Delta q + e\,\Delta t = 0,$$

$$(1)(0.4)(p_R - 0.6) + (-0.16)(q_R - 0) + \left(1 - \frac{0.04 + 0.096}{2}\right)(0.044) = 0,$$

$$(1)(-0.490)(p_R - 0.2) + (-0.24)(q_R - 0) + \left(1 - \frac{0.16 + 0.096}{2}\right)(0.044) = 0.$$

In these equations we used the arithmetic average of x^2 in the last terms. Solving simultaneously we get

$$p_R = 0.399, \qquad q_R = -0.246.$$

As a first approximation for u at R, then,

$$\Delta u = p\,\Delta x + q\,\Delta t,$$

$$u_R - 0.16 = \frac{0.6 + 0.399}{2}(0.310 - 0.2) + \frac{0 - 0.246}{2}(0.044 - 0),$$

$$u_R = 0.2095.$$

The last computation was along PR, using average values of p and q. We could have alternatively proceeded along QR. If this is done,

$$u_R - 0.24 = \frac{0.2 + 0.399}{2}(0.310 - 0.4) + \frac{0 - 0.246}{0}(0.044 - 0),$$

$$u_R = 0.2076.$$

The two values should be close to each other. Let us use the average value, 0.2086, as our initial estimate of u_R. We now repeat the work. In getting the coordinates of R, we now use average values of the slopes,

$$t_R = \frac{0.4 + \sqrt{0.2086}}{2}(x_R - 0.2),$$

$$t_R = \frac{-0.490 - \sqrt{0.2086}}{2}(x_R - 0.4),$$

$$x_R = 0.305, \qquad t_R = 0.045;$$

$$(1)\left(\frac{0.4 + \sqrt{0.2086}}{2}\right)(p_R - 0.6) - \left(\frac{0.16 + 0.2086}{2}\right)(q_R - 0)$$

$$+ \left(1 - \frac{0.04 + 0.0930}{2}\right)(0.045) = 0,$$

$$(1)\left(\frac{-0.490 - \sqrt{0.2086}}{2}\right)(p_R - 0.2) - \left(\frac{0.24 + 0.2086}{2}\right)(q_R - 0)$$

$$+ \left(1 - \frac{0.16 + 0.0930}{2}\right)(0.045) = 0,$$

$$p_R = 0.398, \qquad q_R = -0.242;$$

$$u_R = 0.16 + \frac{0.6 + 0.398}{2}(0.305 - 0.2) + \frac{0 - 0.242}{2}(0.045 - 0),$$

$$u_R = 0.2071 \qquad \text{(along } PR\text{)};$$

$$u_R = 0.24 + \frac{0.2 + 0.398}{2}(0.305 - 0.4) + \frac{0 - 0.242}{2}(0.045 - 0),$$

$$u_R = 0.2063 \quad \text{(along } QR\text{)}.$$

The average value is 0.2067.

Another round of calculations gives $u_R = 0.2066$, which checks the previous value sufficiently. This method is, of course, very tedious by hand.

9.5 THE WAVE EQUATION IN TWO SPACE DIMENSIONS

The finite-difference method can be applied to hyperbolic partial-differential equations in two or more space dimensions. A typical problem is the vibrating membrane. Consider a thin flexible membrane stretched over a rectangular frame and set to vibrating. A development analogous to that for the vibrating string gives:

$$\frac{\partial^2 u}{\partial t^2} = \frac{Tg}{w}\left(\frac{\partial^2 u}{\partial x^2} + \frac{\partial^2 u}{\partial y^2}\right),$$

in which u is the displacement, t is the time, x and y are the space coordinates, T is the uniform tension per unit length, g is the acceleration of gravity, and w is the weight per unit area. For simplification, let $Tg/w = \alpha^2$. Replacing each derivative by its central-difference approximation, and using $h = \Delta x = \Delta y$, gives (we recognize the Laplacian on the righthand side):

$$\frac{u_{i,j}^{k+1} - 2u_{i,j}^k + u_{i,j}^{k-1}}{(\Delta t)^2} = \alpha^2 \frac{u_{i+1,j}^k + u_{i-1,j}^k + u_{i,j+1}^k + u_{i,j-1}^k - 4u_{i,j}^k}{h^2}. \quad (9.19)$$

Solving for the displacement at time t_{k+1}, we obtain:

$$u_{i,j}^{k+1} = \frac{\alpha^2(\Delta t)^2}{h^2}\left\{\begin{matrix}& 1 & \\ 1 & 0 & 1 \\ & 1 & \end{matrix}\right\}u_{i,j}^k - u_{i,j}^{k-1} + \left(2 - 4\frac{\alpha^2(\Delta t)^2}{h^2}\right)u_{i,j}^k. \quad (9.20)$$

In Eqs. (9.19) and (9.20), we use superscripts to denote the time. When $\alpha^2(\Delta t)^2/h^2 = \frac{1}{2}$, the last term vanishes and we get:

$$u_{i,j}^{k+1} = \frac{1}{2}\left\{\begin{matrix}& 1 & \\ 1 & 0 & 1 \\ & 1 & \end{matrix}\right\}u_{i,j}^k - u_{i,j}^{k-1}. \quad (9.21)$$

For the first time step, we get displacements from Eq. (9.22), which is obtained by approximating $\partial u/\partial t$ at $t = 0$ by a central-difference approximation involving $u_{i,j}^1$ and $u_{i,j}^{-1}$:

$$u_{i,j}^1 = \frac{1}{4}\left\{\begin{matrix}& 1 & \\ 1 & 0 & 1 \\ & 1 & \end{matrix}\right\}u_{i,j}^0 + (\Delta t)g(x, y). \quad (9.22)$$

In Eq. (9.22), $g(x, y)$ is the initial velocity.

It should not surprise us to learn that this ratio $\alpha^2(\Delta t)^2/h^2 = \frac{1}{2}$ is the maximum value for stability, in view of our previous experience with explicit methods. However, in contrast with the wave equation in one space dimension, we do not get exact answers from the numerical procedure of Eq. (9.21), and we further observe that we must use smaller time steps in relation to the size of the space interval. Therefore we advance in time more slowly. However, the numerical method is straightforward, as the following example will show.

Example   A membrane for which $\alpha^2 = Tg/w = 3$ is stretched over a square frame that occupies the region $0 \leqslant x \leqslant 2$, $0 \leqslant y \leqslant 2$, in the x,y-plane. It is given an initial displacement described by

$$u = x(2 - x)y(2 - y),$$

and has an initial velocity of zero. Find how the displacement varies with time.

We divide the region with $h = \Delta x = \Delta y = \frac{1}{2}$, obtaining nine intermediate points. Initial displacements are calculated from the initial conditions: $u^0(x, y) = x(2 - x)y(2 - y)$; Δt is taken at its maximum value for stability, $h/(\sqrt{2}\alpha) = 0.0204$. The values at the end of one time step are given by:

$$u_{i,j}^1 = \tfrac{1}{4} \left\{ \begin{matrix} & 1 & \\ 1 & 0 & 1 \\ & 1 & \end{matrix} \right\} u_{i,j}^0$$

because $g(x, y)$ in Eq. (9.22) is everywhere zero. For succeeding time steps, Eq. (9.21) is used. Table 9.5 gives the results of our calculations. Also shown in Table 9.5 (in parentheses) are analytical values, computed from the double infinite series:

$$u(x, y, t) = \sum_{m=1}^{\infty} \sum_{n=1}^{\infty} B_{mn} \sin \frac{m\pi x}{a} \sin \frac{n\pi y}{b} \cos \left(\alpha \pi t \sqrt{\frac{m^2}{a^2} - \frac{n^2}{b^2}} \right),$$

$$B_{mn} = \frac{16a^2b^2A}{\pi^6 m^3 n^3} (1 - \cos m\pi)(1 - \cos n\pi),$$

which gives the displacement of a membrane fastened to a rectangular framework, $0 \leqslant x \leqslant a$, $0 \leqslant y \leqslant b$, with initial displacements of $Ax(a - x)y(b - y)$.

We observe that the finite-difference results do not agree exactly with the analytical calculations. The finite-difference values are symmetrical with respect to position and repeat themselves with a regular frequency. The very regularity of the values itself indicates that the finite-difference computations are in error, for they predict that the membrane could emit a musical note. We know from experience that a drum does not give a musical tone when struck; therefore the vibrations do not have a cyclic pattern of constant frequency, as exhibited by our numerical results.

Table 9.5 Displacements of a vibrating membrane—finite-difference method: $\Delta t = h/\sqrt{2}\alpha$.

t	(0.5, 0.5)	(1.0, 0.5)	(1.5, 0.5)	(0.5, 1.0)	(1.0, 1.0)	(1.5, 1.0)	(0.5, 1.5)	(1.0, 1.5)	(1.5, 1.5)
					Grid location				
0	0.5625 (0.5625)	0.750 (0.750)	0.5625	0.750	1.000 (1.000)	0.750	0.5625	0.750	0.5625
0.204	0.375 (0.380)	0.531 (0.536)	0.375	0.531	0.750 (0.755)	0.531	0.375	0.531	0.375
0.408	−0.031 (−0.044)	0.000 (−0.009)	−0.031	0.000	0.062 (0.083)	0.000	−0.031	0.000	−0.031
0.612	−0.375 (−0.352)	−0.531 (−0.539)	−0.375	−0.531	−0.750 (−0.813)	−0.531	−0.375	−0.531	−0.375
0.816	−0.500 (−0.502)	−0.750 (−0.746)	−0.500	−0.750	−1.125 (−1.114)	−0.750	−0.500	−0.750	−0.500
1.021	−0.375 (−0.407)	−0.531 (−0.535)	−0.375	−0.531	−0.750 (−0.691)	−0.531	−0.375	−0.531	−0.375
1.225	−0.031 (−0.015)	0.000 (0.008)	−0.031	0.000	0.062 (0.030)	0.000	−0.031	0.000	−0.031
1.429	0.375 (0.410)	0.531 (0.534)	0.375	0.531	0.750 (0.688)	0.531	0.375	0.531	0.375

Decreasing the ratio of $\alpha^2(\Delta t)^2/h^2$ and using Eq. (9.20) gives little or no improvement in the average accuracy; to approach closely to the analytical results, $h = \Delta x = \Delta y$ must be made smaller. When this is done, Δt will need to decrease in proportion, requiring many time steps and leading to many repetitions of the algorithm and extravagant use of computer time. One remedy is the use of implicit methods, which allow the use of larger ratios of $\alpha^2(\Delta t)^2/h^2$. However, with many nodes in the x,y-grid, this requires large, sparse matrices similar to the Crank–Nicholson method for parabolic equations in two space dimensions. A.D.I. methods have been used for hyperbolic equations—tridiagonal systems result. We do not discuss these methods.

9.6 A PROGRAM FOR THE SIMPLE WAVE EQUATION

Program 1 (Fig. 9.7) uses the easy algorithm of Section 9.1, but starts the solution by use of Eq. (9.13). The integral of the initial velocity is computed by Simpson's $\frac{1}{3}$ rule with the interval from x_{i-1} to x_{i+1} subdivided into 20 panels. The initial displacement $f(x)$ and the initial velocity $g(x)$ are each defined by function subprograms. The program is tested by calculating the example problem of Section 9.2, in which a string initially at rest is set vibrating by giving it an initial velocity of $g(x) = \sin(\pi x/L)$. Input values are $L = 20$, $\Delta x = 1$, $\alpha = \sqrt{Tg/w} = 1$, so that five time steps are here required to reach the time $t = L/4\alpha$.

The first part of the output from the program is shown. Additional values for the displacement of the string will be printed until the time is equal to TLAST.

```
C   THIS PROGRAM COMPUTES SOLUTIONS FOR THE ONE-DIMENSIONAL WAVE
C   EQUATION.  THE INITIAL DISPLACEMENTS OF THE VIBRATING STRING, FROM X = 0
C   TO X = XLEN ARE GIVEN BY F(X), THE INITIAL VELOCITIES ARE GIVEN
C   BY G(X).  F(X) AND G(X) ARE DEFINED BY FUNCTION SUBPROGRAMS.
C   THE END POINTS ARE ASSUMED FIXED AT ZERO DISPLACEMENT .
C   THE RELATION USED IS U(I,J-1) = U(I-1,J) + U(I+1,J) - U(I,J-1) EXCEPT
C   FOR THE FIRST TIME STEP, WHEN THE VALUE IS GIVEN BY
C     U(I,1) = 0.5*(U(I-1,J) + U(I+1,J)) + 0.5/C*INTEGRAL OF INIT VEL
C   PARAMTERS ARE -
C     X      DISTANCE ALONG THE STRING
C     DX     INCREMENTS OF DISTANCE
C     XLEN   TOTAL LENGTH OF STRING
C     N      NUMBER OF SUBDIVISIONS
C     T      TIME
C     TLAST  FINAL VALUE OF TIME FOR WHICH SOLUTION IS DESIRED
C     F(X)   INITIAL DISPLACEMENTS
C     G(X)   INITIAL VELOCITIES
C     TDM    VALUE OF TENSION / MASS * C SQUARED
C     U      DISPLACEMENTS AT EVEN TIME INTERVALS
C     V      DISPLACEMENTS AT ODD TIME INTERVALS
        DIMENSION U(100), V(100)
C   DEFINE SOME DATA STATEMENTS FOR INITIAL VALUES
        DATA X, N, T / 0.0, 20, 0.0 /
        DATA TDM, XLEN, TLAST /3.0, 10.0, 10.0/
C   GET INITIAL DISPLACEMENTS
        NP1 = N + 1
        DX = XLEN/FLOAT(N)
        U(1) = 0.0
        U(NP1) = 0.0
        V(1) = 0.0
        V(NP1) = 0.0
        DO 10 I = 2,N
          X = X + DX
          U(I) = F(X)
10      CONTINUE
C   WRITE OUT THE INITIAL DISPLACEMENTS
        WRITE (6,200) (U(I), I = 1,NP1)
200     FORMAT (1H1, 38H SOLUTION TO VIBRATING STRING PROBLEM
       +     1H0, 27H INITIAL DISPLACEMENTS ARE //(1H , 10F13.4))
```

Figure 9.7 Program 1

```
C   NOW GET DISPLACEMENTS AFTER FIRST TIME STEP.  USE SIMPSON'S RULE
C   TO PERFORM INTEGRATIONS, USING 20 INTERVALS BETWEEN X(N-1) AND X(N+1)
C   WE DO THIS FOR EACH INTERMEDIATE X VALUE
        SUBDX = DX/10.0
        XSUB = DX
        DO 30 I = 2,N
          SUM = 0.0
          XSUB = XSUB - DX
          DO 20 J = 1,19,2
            SUM = SUM + G(XSUB) + 4.*G(XSUB+SUBDX) + G(XSUB+2.*SUBDX)
            XSUB = XSUB + 2.*SUBDX
20        CONTINUE
          V(I) = 0.5*(U(I-1) + U(I+1)) + 0.5/SQRT(TDM)*SUBDX/3.*SUM
30      CONTINUE
C   WRITE OUT DISPLACEMENTS AFTER FIRST TIME STEP
        T = DX/TDM
        WRITE (6,201) T, (V(I), I = 1,NP1)
201     FORMAT (1H0, 8H AT T = , F8.3/(1H , 10F13.4))
C   NOW COMPUTE UNTIL TLAST IS REACHED
35      IF (T .GE. TLAST)  STOP
        DO 40 I = 2,N
          U(I) = V(I-1) + V(I+1) - U(I)
40      CONTINUE
        T = T + DX/TDM
        WRITE (6,201) T, (U(I), I = 1,NP1)
        DO 50 I = 2,N
          V(I) = U(I-1) + U(I+1) - V(I)
50      CONTINUE
        T = T + DX/TDM
        WRITE (6,201) T, (V(I), I = 1,NP1)
        GO TO 35
        END
        FUNCTION F(X)
        F = 0.1*X*(X-6.)*(X-10.)
        RETURN
        END
        FUNCTION G(X)
        G = X/3.
        IF (X .GT. 6.)  G = (10 - X)/2.
        RETURN
        END
```

Figure 9.7 (continued)

OUTPUT FROM PROGRAM 1

SOLUTION TO VIBRATING STRING PROBLEM
INITIAL DISPLACEMENTS ARE

0.0000	2.6125	4.5000	5.7375	6.4000	6.5625	6.3000
2.5000	1.2375	0.0000	−1.1375	−2.1000	−2.8125	−3.2000
0.0000						

AT T =   0.167

0.0000	2.2981	4.2712	5.5943	6.3424	6.5906	6.4137
2.9561	1.7792	0.5973	−0.5448	−1.5420	−2.2891	−2.7113
0.0000						

AT T =   0.333

0.0000	1.6587	3.3924	4.8762	5.7849	6.1936	6.1773
3.3622	2.3159	1.2344	0.1928	−0.7340	−1.4408	−1.8226
0.0000						

AT T =   0.500

0.0000	1.0943	2.2637	3.5830	4.7273	5.3717	5.5910
3.6883	2.8174	1.9114	1.0453	0.2940	−0.2674	−0.5339
0.0000						

AT T =   0.667

0.0000	0.6049	1.2849	2.1148	3.1698	4.1247	4.6547
3.9042	3.2838	2.6283	2.0126	1.5118	1.2008	1.1547
0.0000						

AT T =   0.833

0.0000	0.1906	0.4561	0.8717	1.5122	2.4528	3.3684
4.0100	3.7151	3.3850	3.0948	2.9195´	2.9340	2.8633
0.0000						

AT T =   1.000

0.0000	−0.1488	−0.2227	−0.1465	0.1547	0.7559	1.7320
4.0056	4.1112	4.1816	4.2919	4.5170	4.5820	4.2118
0.0000						

AT T =   1.167

0.0000	−0.4132	−0.7514	−0.9396	−0.9029	−0.5661	0.1156
3.8911	4.4721	5.0180	5.6038	5.9544	5.7948	5.2001
0.0000						

AT T =   1.333

0.0000	−0.6026	−1.1302	−1.5078	−1.6604	−1.5431	−1.1109
3.6665	4.7980	5.8943	6.6805	6.8816	6.5725	5.8283
0.0000						

AT T =   1.500

0.0000	−0.7170	−1.3590	−1.8510	−2.1480	−2.2052	−1.9476
3.3317	5.0887	6.4605	7.1721	7.2987	6.9151	6.0963
0.0000						

AT T =   1.667

0.0000	−0.7564	−1.4378	−1.9992	−2.3958	−2.5525	−2.3944
2.8868	4.9942	6.3665	7.0786	7.2056	6.8225	6.0042
0.0000						

AT T =   1.833

0.0000	−0.7208	−1.3966	−1.9826	−2.4037	−2.5849	−2.4513

Figure 9.7 (continued)

5.6875	4.8000	3.7125
−3.1875	−2.7000	−1.6625

5.8868	5.0849	4.0830
−2.7335	−2.2806	−1.2778

5.8111	5.1698	4.3285
−1.8045	−1.3113	−0.6181

5.4604	5.0547	4.4490
−0.4004	−0.1420	−0.0335

4.8346	4.7396	4.4145
1.1286	0.8774	0.4762

3.9339	4.1944	4.1948
2.4326	1.7467	0.9109

2.7281	3.3891	3.7899
3.4815	2.4660	1.2705

1.1872	2.3237	3.2000
4.2453	3.0053	1.5552

−0.2888	0.9981	2.4248
4.7239	3.3344	1.7348

−1.3000	−0.1877	1.4646
4.9174	3.4534	1.7792

.8464	−0.8335	0.7192
4.8258	3.3622	1.7186

−1.9279	−0.9395	0.5887

Figure 9.7 (concluded)

EXERCISES

Section 9.1

▶1. If the banjo string in the example of Section 9.1 is tightened, its frequency of vibration is increased. Likewise, if the length of the vibrating portion of the string is shortened, as by holding it against one of the frets with the finger, the pitch is raised. What would the frequency be if the tension is made 42,000 grams, and the effective length is 70 cm? Determine the answer by finding the number of Δt steps for the original displacement to be duplicated and compare to

$$f = \frac{1}{2L} \sqrt{\frac{Tg}{w}}.$$

2. A vibrating-string system has $Tg/w = 4$ cm^2/sec^2. Divide the length L into intervals so that $\Delta x = L/8$ cm. Find the displacement for $t = 0$ up to $t = L$ if both ends are fixed, and the initial conditions are:

   a) $y = \dfrac{x(L - x)}{L^2}, \qquad \dfrac{\partial y}{\partial t} = 0.$

   b) The string is displaced $+1$ units at $L/4$ and -1 units at $5L/8$, $\partial y/\partial t = 0$.

   c) $y = 0, \qquad \dfrac{\partial y}{\partial t} = \dfrac{-x(L - x)}{L^2}.$ (Use Eq. (9.3) to begin the solution.)

   d) The string is displaced 1 unit at $L/2$, $\partial y/\partial t = -y$.

   In part (a), compare to the analytical solution:

$$y = \frac{8}{\pi^3} \sum_{n=1}^{\infty} \frac{1}{(2n - 1)^3} \sin\left[(2n - 1)\pi x\right] \cos\left[(4n - 2)\pi t\right].$$

▶3. A function u satisfies the equation

$$\frac{\partial^2 u}{\partial x^2} = \frac{\partial^2 u}{\partial t^2},$$

with boundary conditions $u = 0$ at $x = 0$, $u = 0$ at $x = 1$, for all values of t, and with initial conditions

$$u = \sin \pi x, \qquad \partial u/\partial t = 0 \quad \text{at} \quad t = 0 \quad \text{for} \quad 0 \leqslant x \leqslant 1.$$

Solve by the finite-difference method and show that the results are the same as the analytical solution:

$$U(x, t) = \sin \pi x \cos \pi t.$$

4. The ends of the vibrating string need not be fixed. Solve $\partial^2 y/\partial t^2 = \partial^2 y/\partial x^2$ with initial conditions $y = 0$, $\partial y/\partial t = 0$, for $0 \leqslant x \leqslant 1$, and end conditions

$$y = 0 \quad \text{at} \quad x = 0, \qquad y = \sin \pi t/4, \qquad \partial y/\partial x = 0 \quad \text{at} \quad x = 1.$$

Section 9.2

5. Verify that the analytical solution given in Exercise 3 satisfies (a) the differential equation, (b) the initial conditions, and the boundary conditions, and (c) Eq. (9.9). What are the functions F and G of Eq. (9.6) for this solution?

▶6. Since Eq. (9.4) is sometimes inaccurate when the initial velocity is not zero, the solutions to parts (c) and (d) of Exercise 2 are not exact. Repeat these problems, getting more accurate y-values by using Eq. (9.13) and evaluating the integral by Simpson's $\frac{1}{3}$ rule.

7. A string that weighs w lb/ft is tightly stretched from $x = 0$ to $x = L$ and is initially at rest. Each point of the string is given initially a velocity of:

$$\left.\frac{\partial y}{\partial t}\right|_{t=0} = v_0 \sin^3 \frac{\pi x}{L}.$$

The analytical solution to this problem is

$$y(x, t) = \frac{v_0 L}{12 a \pi}\left(9 \sin \frac{\pi x}{L} \sin \frac{a \pi t}{L} - \sin \frac{3 \pi x}{L} \sin \frac{3 a \pi t}{L}\right),$$

where $a = \sqrt{Tg/w}$, T being the tension and g the acceleration of gravity. When $L = 3$ ft, $w = 0.02$ lb/ft, and $T = 5$ lb, with $v_0 = 1$ ft/sec, the analytical equation predicts $y = 0.081$ in. at the midpoint when $t = 0.01$ sec. Solve the problem numerically to confirm this prediction. Does your solution conform to the analytical solution at other values of x and t?

Section 9.3

8. Demonstrate that if $Tg(\Delta t)^2/w(\Delta x)^2 > 1$, the finite-difference method for the wave equation will propagate an error with increasing effect by tracing the growth of the error similar to that in Table 9.3. Take the simple case with ends fixed. Repeat for the ratio <1.

▶9. Trace the magnitude of an isolated error in solving the wave equation with $Tg(\Delta t)^2/w(\Delta x)^2 = 1$, similar to Table 9.3, but for the case when the lefthand end is fixed and the righthand end moves according to:

$$y(1, t) = \sin \frac{\pi t}{4}, \quad \frac{\partial y}{\partial x} = 0 \quad \text{at} \quad x = 1.$$

Section 9.4

10. For the partial-difference equation

$$au_{xx} + bu_{xt} + cu_{tt} + e = 0,$$

sketch the characteristic curves through the point $x = 0.5$, $t = 0$ when:

a) $a = 1$, $b = 2$, $c = 1$ b) $a = 1$, $b = 2$, $c = -3$

c) $a = 1$, $b = 2$, $c = 3$ d) $a = 1$, $b = 2$, $c = -1$

e) $a = t^2$, $b = xt$, $x = -2x^2$

11. Verify the values at points S, T, U, and V in Fig. 9.6 for Example 2 of Section 9.4 by the method of characteristics.

▶12. The equation

$$\frac{\partial^2 u}{\partial t^2} = (1 + 2x)\frac{\partial^2 u}{\partial x^2}$$

can be solved by the difference-equation method discussed in Sections 9.1 and 9.2. Solve it by that method with initial conditions of

$$u(x, 0) = 0, \qquad \frac{\partial u}{\partial t}(x, 0) = x(1 - x).$$

Compare these results with those from the method of characteristics in Fig. 9.6.

13. Get part of the solution by integrating along characteristics for the equation

$$\frac{\partial^2 u}{\partial x^2} + \frac{\partial^2 u}{\partial x\,\partial t} - \frac{\partial^2 u}{\partial t^2} = 1,$$

subject to initial conditions

$$u = 0, \qquad \frac{\partial u}{\partial t} = x(1 - x).$$

Find the solution at three points in the x,t-plane, where the characteristics through $(0.4, 0)$, $(0.5, 0)$ and $(0.6, 0)$ intersect.

▶14. Solve the equation

$$u_{xx} + xtu_{xt} - u_{tt} = 0,$$

with initial conditions

$$u = 2x, \qquad \frac{\partial u}{\partial t} = 0.$$

Boundary conditions are $u(0, t) = 0$, $u(1, t) = 2$.
Find the solution at several points in the region of the x,t-plane bounded by the lines $x = 0.4$, $x = 0.6$.

15. Continue the solution of Example 3 of Section 9.4,

$$u_{xx} - uu_{tt} + 1 - x^2 = 0, \qquad u(x, 0) = x(1 - x), \qquad u_t(x, 0) = 0,$$
$$u(0, t) = 0, \qquad u(1, t) = 0,$$

by finding the solution at the intersection of characteristics through points at $(0.4, 0)$ and $(0.6, 0)$. Then find the solution at the intersection of characteristics through $(0.2, 0)$ and $(0.6, 0)$ as an example of how the solution can be advanced in time.

Section 9.5

16. Solve the vibrating-membrane example of Section 9.5 but with initial conditions of

$$u(x, y)|_{t=0} = 0, \qquad \frac{\partial u}{\partial t}(x, y)\bigg|_{t=0} = x(2 - x)y(2 - y).$$

▶17. A membrane is stretched over a frame that occupies the region in the x,y-plane bounded by

$$x = 0, \qquad x = 2, \qquad y = 0, \qquad y = 4.$$

Initially the point at $(1, 3)$ is lifted 1 in. above the x,y-plane, then released. If $T = 7$ lb/in. and $w = 0.07$ lb/in.^{2}, find the displacement at the point $(1, 2)$ as a function of time.

18. How do the vibrations in Exercise 17 change if $w = 0.7$, the other parameters being unchanged?

19. The frame holding the membrane in Exercise 17 is distorted by lifting the corner at $(2, 4)$ so it is 1 in. above the x,y-plane. (The frame members stretch at the same time so the corner point moves only vertically.) The membrane is set to vibrating in the same way as in Exercise 17. Follow its vibrations through time. (Assume that the initial positions of points on the membrane lie on intersecting planes.)

APPLIED PROBLEMS AND PROJECTS

20. A vibrating string which has a damping force opposing its motion that is proportional to the velocity, follows the equation

$$\frac{\partial^2 y}{\partial t^2} = \frac{Tg}{w}\frac{\partial^2 y}{\partial x^2} - B\frac{\partial y}{\partial t},$$

where B is the magnitude of the damping force. Solve the problem if the length of the string is 5 ft with $T = 24$ lbs, $w = 0.1$ lb/ft, and $B = 2.0$. Initial conditions are:

$$y(x)|_{t=0} = \frac{x}{3}, \qquad 0 \leqslant x < 3,$$

$$y(x)|_{t=0} = \frac{5}{2} - \frac{x}{2}, \qquad 3 \leqslant x \leqslant 5,$$

$$\frac{\partial y}{\partial t}\bigg|_{t=0} = x(x - 5).$$

Compute a few points of the solution by both difference equations and the method of characteristics. Compare the effort involved in the two methods.

21. A horizontal elastic rod is initially undeformed and is at rest. One end, at $x = 0$, is fixed and the other end, at $x = L$ (when $t = 0$), is pulled with a steady force of F lb/ft^2. It can be shown that the displacements, $y(x, t)$, of points originally at the

point x are given by:

$$\frac{\partial^2 y}{\partial t^2} = a^2 \frac{\partial^2 y}{\partial x^2}, \qquad y(0, t) = 0, \qquad \frac{\partial y}{\partial t}\bigg|_{x=L} = \frac{F}{E},$$

$$y(x, 0) = 0, \qquad \frac{\partial y}{\partial t}\bigg|_{t=0} = 0.$$

where $a^2 = Eg/\rho$; $E =$ Young's modulus (lb/ft^2); $g =$ acceleration of gravity; $\rho =$ density (lb/ft^3). Find y versus t for the midpoint of a 2-ft-long piece of rubber for which $E = 1.8 \times 10^6$ and $\rho = 70$ if $F/E = 0.7$.

22. A circular membrane, when set to vibrating, obeys the equation (in polar coordinates):

$$\frac{1}{r}\frac{\partial}{\partial r}\left(r\frac{\partial u}{\partial r}\right) + \frac{1}{r^2}\frac{\partial^2 u}{\partial \theta^2} = \frac{w}{Tg}\frac{\partial^2 y}{\partial t^2}.$$

A 3-ft-diameter kettledrum is started to vibrating by depressing the center $\frac{1}{2}$ in. If $w = 0.072$ lb/ft^2 and $T = 80$ lb/ft, find how the displacements at 6 in. and 12 in. from the center vary with time. The problem can be solved in polar coordinates, or it can be solved in rectangular coordinates using the method of Section 7.8 to approximate $\nabla^2 u$ near the boundaries.

23. A flexible chain hangs freely, as shown in Fig. 9.8. For small disturbances from its equilibrium position (hanging vertically), the equation of motion is:

$$x\frac{\partial^2 y}{\partial x^2} + \frac{\partial y}{\partial x} = \frac{1}{g}\frac{\partial^2 y}{\partial t^2}.$$

In this equation, x is the distance from the end of the chain, y is the displacement from the equilibrium position, t is the time, and g is the acceleration of gravity. A 10-ft-long chain is originally hanging freely. It is set into motion by striking it sharply at its midpoint, imparting a velocity there of 1 ft/sec. Find how the chain moves as a result of the blow.

Figure 9.8

24. Write a computer program that solves hyperbolic partial-differential equations by the method of characteristics, given the values of u and $\partial u/\partial t$ along a curve that is not one of the characteristics. Your program should be a subroutine that accepts the coordinates of two points on the initial-condition curve and computes the value of u and $\partial u/\partial t$ at the intersection of the characteristic curves through the two points.

Chapter Ten

Curve-Fitting, Splines, and Approximation of Functions

In an early chapter, when we studied how to fit a polynomial to a set of data, we made the implicit assumption that the data were free of error (except round-off) so that it was appropriate to match the interpolating polynomial exactly at the data points. In the case of experimental data, this assumption as to accuracy is often not true. Each data point is subject to experimental errors which, in the case of complicated measurements, may be relatively large in magnitude. Again, in the previous chapter, the true function which relates the data was generally unknown, while, in the case of experimental results, the form of the function is frequently known from the physical laws which apply.

We wish to consider the problem of finding the "best" curve which represents data that are subject to error. "Best" is in quotation marks because the criterion of goodness of fit is to some degree arbitrary, although the least-squares criterion is commonly applied.

A second topic we will consider is spline interpolation. We shall also consider in this chapter the most efficient way in which functions can be represented by a polynomial or a ratio of polynomials over a given range of values of the argument. This is of special importance in connection with the library function subroutines for digital computers.

10.1 LEAST-SQUARES APPROXIMATIONS

Suppose we wish to fit a curve to an approximate set of data, such as from the determination of the effects of temperature on a resistance by a student in his physics laboratory. He has recorded the temperature and resistance measurements as shown in Fig. 10.1, where the graph suggests a linear relationship. We want to suitably determine the constants a and b in the equation relating resistance R and temperature T,

$$R = aT + b, \tag{10.1}$$

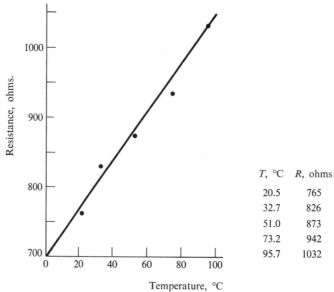

T, °C	R, ohms
20.5	765
32.7	826
51.0	873
73.2	942
95.7	1032

Figure 10.1

so that in subsequent use the resistance can be predicted at any temperatures. The line as sketched by eye represents the data fairly well, but if we replotted the data and asked someone else to draw a line, he would rarely get exactly the same line. One of our requirements for fitting a curve to data is that the process be *unambiguous.* We would also like, in some sense, to minimize the deviations of the points from the line. The deviations are measured by the distances from the points to the line—how these distances are measured depends on whether or not both variables are subject to error. We shall assume that the error of reading the temperatures in Fig. 10.1 is negligible so that all the errors are in the resistance measurements, and use vertical distances. (If both were subject to error, we might use perpendicular distances, and would modify the following. Thereby the problem also becomes considerably more complicated. We shall treat only the simpler case.)

We might first suppose we could minimize the deviations by making their sum a minimum, but this is not an adequate criterion. Consider the case of only two points (Fig. 10.2). Obviously, the best line passes through each, but any line that passes through the midpoint of the segment connecting them has a sum of errors equal to zero.

Then what about making the sum of the magnitudes of the errors a minimum? This also is inadequate, as the case of three points shows (Fig. 10.3). Assume that two of the points are at the same *x*-value (which is not an abnormal situation since frequently experiments are duplicated). The best line obviously will pass through the average of the duplicated tests. However, any line that falls

Figure 10.2

between the dotted lines shown will have the same sum of the magnitudes of the vertical distances. Since we wish an unambiguous result, we cannot use this as a basis for our work.

We might accept the criterion that we make the magnitude of the maximum error a minimum (the so-called *minimax* criterion), but for the problem at hand this is rarely done.* This criterion is awkward because the absolute-value function has no derivative at the origin, and it also is felt to give undue importance to a single large error. The usual criterion is to minimize the sum of the *squares* of the errors, the "least squares" principle.†

In addition to giving a unique result for a given set of data, the least-squares method is also in accord with the *maximum likelihood* principle of statistics. If the measurement errors have a so-called normal distribution and if the standard deviation is constant for all the data, the line determined by minimizing the sum of squares can be shown to have values of slope and intercept which have maximum likelihood of occurrence.

Let Y_i represent an experimental value, and let y_i be a value from the equation

$$y_i = ax_i + b,$$

where x_i is a particular value of the variable assumed free of error. We wish to determine the best values for a and b so that the y's predict the function values

Figure 10.3

* We will use this criterion later in this chapter, however.
† The various criteria for a "best fit" can be described by minimizing a norm of the error vector. Relate each criterion to its corresponding vector norm to review the definition of such norms.

that correspond to x-values. Let $e_i = Y_i - y_i$. The least-squares criterion requires that

$$S = e_1^2 + e_2^2 + \ldots + e_N^2$$

$$= \sum_{i=1}^{N} e_i^2$$

$$= \sum_{i=1}^{N} (Y_i - ax_i - b)^2$$

be a minimum. N is the number of x, Y-pairs. We reach the minimum by proper choice of the parameters a and b, so they are the "variables" of the problem. At a minimum for S, the two partial derivatives $\partial S/\partial a$ and $\partial S/\partial b$ will both be zero. Hence, remembering that the x_i and Y_i are data points unaffected by our choice of values for a and b, we have

$$\frac{\partial S}{\partial a} = 0 = \sum_{i=1}^{N} 2(Y_i - ax_i - b_i)(-x_i),$$

$$\frac{\partial S}{\partial b} = 0 = \sum_{i=1}^{N} 2(Y_i - ax_i - b_i)(-1).$$

Dividing each of these equations by -2 and expanding the summation, we get the so-called normal equations

$$a\sum x_i^2 + b\sum x_i = \sum x_i Y_i,$$

$$a\sum x_i + bN = \sum Y_i. \tag{10.2}$$

All the summations in Eq. (10.2) are from $i = 1$ to $i = N$. Solving these equations simultaneously gives the values for slope and intercept a and b.

For the data in Fig. 10.1 we find that

$$N = 5, \quad \sum T_i = 273.1, \quad \sum T_i^2 = 18{,}607.27, \quad \sum R_i = 4438,$$

$$\sum T_i R_i = 254{,}932.5.$$

Our normal equations are then

$$18{,}607.27a + 273.1b = 254{,}932.5,$$

$$273.1a + 5b = 4438.$$

From these we find $a = 3.395$, $b = 702.2$, and hence write Eq. (10.1) as

$$R = 702 + 3.39T.$$

10.2 FITTING NONLINEAR CURVES BY LEAST SQUARES

In many cases, of course, data from experimental tests are not linear, so we need to fit some other function than a first-degree polynomial to them. Popular forms that are tried are the exponential forms

$$y = ax^b$$

or

$$y = ae^{bx}.$$

We can develop normal equations for these analogously to the above development for a least-squares line by setting the partial derivatives equal to zero. Such nonlinear simultaneous equations are much more difficult to solve* than linear equations. Because of this, the exponential forms are usually linearized by taking logarithms before determining the parameters:

$$\ell n \; y = \ell n \; a + b \, \ell n \; x$$

or

$$\ell n \; y = \ell n \; a + bx.$$

We now fit the new variable $z = \ell n \; y$ as a linear function of $\ell n \; x$ or x as described above. Here we do not minimize the sum of squares of the deviations of Y from the curve, but rather the deviations of $\ell n \; Y$. In effect, this amounts to minimizing the squares of the percentage errors, which itself may be a desirable feature. An added advantage of the linearized forms is that plots of the data on either log-log or semi-log graph paper show at a glance whether these forms are suitable by whether a straight line represents the data when so plotted.

In cases when such linearization of the function is not desirable, or when no method of linearization can be discovered, graphical methods are frequently used; one merely plots the experimental values and sketches in a curve that seems to fit well. Special forms of graph paper, in addition to log-log and semi-log, may be useful (probability, log-probability, etc.). Transformation of the variables to give near linearity, such as by plotting against $1/x$, $1/(ax + b)$, $1/x^2$, and other nonpolynomial forms of the argument may give curves with gentle enough changes in slope to allow a smooth curve to be drawn. S-shaped curves are not easy to linearize; the Gompertz relation

$$y = ab^{c^x}$$

is sometimes employed. The constants a, b, and c are determined by special procedures. Another relation that fits data to an S-shaped curve is

$$\frac{1}{y} = a + be^{-x}.$$

In awkward cases, subdividing the region of interest into subregions with a piecewise fit in the subregions can be used.

---

* They are treated briefly in Chapter 2.

The objection to the last-mentioned methods is the *lack of uniqueness*. Two individuals will usually not draw the same curve through the points. One's judgment is frequently distorted by one or two points that deviate widely from the remaining data. Often one tends to pay too much attention to the extremities in comparison to the points in the central parts of the region of interest.

Further problems are caused if we wish to integrate or differentiate the function. Our discussion of least-squares polynomials is one solution to these difficulties.

Because polynomials can be readily manipulated, fitting such functions to data that do not plot linearly is common. We now consider this case. It will turn out that the normal equations are linear for this situation, which is an added advantage. In the development, we use n as the degree of the polynomial and N as the number of data pairs. Obviously if $N = n + 1$, the polynomial passes exactly through each point and the methods of Chapter 3 apply, so we will always have $N > n + 1$ in the following.

We assume the functional relationship

$$y = a_0 + a_1 x + a_2 x^2 + \ldots + a_n x^n, \tag{10.3}$$

with errors defined by

$$e_i = Y_i - y_i = Y_i - a_0 - a_1 x_i - a_2 x_i^2 - \ldots - a_n x_i^n.$$

We again use Y_i to represent the observed or experimental value corresponding to x_i, with x_i free of error. We minimize the sum of squares,

$$S = \sum_{i=1}^{N} e_i^2 = \sum_{i=1}^{N} (Y_i - a_0 - a_1 x_i - a_2 x_i^2 - \ldots - a_n x_i^n)^2.$$

At the minimum, all the partial derivatives $\partial S/\partial a_0$, $\partial S/\partial a_1, \ldots, \partial S/\partial a_n$ vanish. Writing the equations for these gives $n + 1$ equations:

$$\frac{\partial S}{\partial a_0} = 0 = \sum_{i=1}^{N} 2(Y_i - a_0 - a_1 x_i - \ldots - a_n x_i^n)(-1),$$

$$\frac{\partial S}{\partial a_1} = 0 = \sum_{i=1}^{N} 2(Y_i - a_0 - a_1 x_i - \ldots - a_n x_i^n)(-x_i),$$

$$\vdots$$

$$\frac{\partial S}{\partial a_n} = 0 = \sum_{i=1}^{N} 2(Y_i - a_0 - a_1 x_i - \ldots - a_n x_i^n)(-x_i^n).$$

Dividing each by -2 and rearranging gives the $n+1$ normal equations to be solved simultaneously:

$$a_0 N + a_1 \sum x_i + a_2 \sum x_i^2 + \ldots + a_n \sum x_i^n = \sum Y_i,$$

$$a_0 \sum x_i + a_1 \sum x_i^2 + a_2 \sum x_i^3 + \ldots + a_n \sum x_i^{n+1} = \sum x_i Y_i,$$

$$a_0 \sum x_i^2 + a_1 \sum x_i^3 + a_2 \sum x_i^4 + \ldots + a_n \sum x_i^{n+2} = \sum x_i^2 Y_i, \quad (10.4)$$

$$\vdots$$

$$a_0 \sum x_i^n + a_1 \sum x_i^{n+1} + a_2 \sum x_i^{n+2} + \ldots + a_n \sum x_i^{2n} = \sum x_i^n Y_i.$$

Putting these equations in matrix form shows an interesting pattern in the coefficient matrix.

$$\begin{bmatrix} n & \sum x_i & \sum x_i^2 & \sum x_i^3 & \ldots & \sum x_i^n \\ \sum x_i & \sum x_i^2 & \sum x_i^3 & \sum x_i^4 & \ldots & \sum x_i^{n+1} \\ \sum x_i^2 & \sum x_i^3 & \sum x_i^4 & \sum x_i^5 & \ldots & \sum x_i^{n+2} \\ \vdots & & & & & \\ \sum x_i^n & \sum x_i^{n+1} & \sum x_i^{n+2} & \sum x_i^{n+3} & \ldots & \sum x_i^{2n} \end{bmatrix} a = \begin{bmatrix} \sum Y_i \\ \sum x_i Y_i \\ \sum x_i^2 Y_i \\ \vdots \\ \sum x_i^n Y_i \end{bmatrix}. \quad (10.5)$$

All the summations in Eqs. (10.4) and (10.5) run from 1 to N.

Solving large sets of linear equations is not a simple task. Methods for this are the subject of Chapter 2. These particular equations have an added difficulty in that they have the undesirable property known as *ill-conditioning*. The result of this is that round-off errors in solving then cause unusually large errors in the solutions, which of course are the desired values of the coefficients a_i in Eq. (10.3). Up to $n = 4$ or 5 the problem is not too great (i.e., double-precision arithmetic in computer solutions is only desirable and not essential), but beyond this point special methods are needed. Such special methods use orthogonal polynomials in an equivalent form of Eq. (10.3). We shall not pursue this matter further[*] although we shall treat one form of orthogonal polynomials later in this chapter in connection with representation of functions. From the point of view of the experimentalist, functions more complex than fourth-degree polynomials are rarely needed, and when they are, the problem can often be handled by fitting a series of polynomials to subsets of the data.

We illustrate the use of Eqs. (10.4) to fit a quadratic to the data of Table 10.1. Figure 10.4 shows a plot of the data. (The data are actually a perturbation

---

[*] Ralston (1965) is a good source of further information. The ill-conditioning problem, though very real, is often academic, for it is seldom that a degree above 4 or 5 is needed to give a curve that fits the data with adequate precision.

Table 10.1   Data to illustrate curve-fitting.

x_i	0.05	0.11	0.15	0.31	0.46	0.52	0.70	0.74	0.82	0.98	1.17
Y_i	0.956	0.890	0.832	0.717	0.571	0.539	0.378	0.370	0.306	0.242	0.104

$$\sum x_i = 6.01 \qquad N = 11$$
$$\sum x_i^2 = 4.6545 \qquad \sum Y_i = 5.905$$
$$\sum x_i^3 = 4.1150 \qquad \sum x_i Y_i = 2.1839$$
$$\sum x_i^4 = 3.9161 \qquad \sum x_i^2 Y_i = 1.3357$$

of the relation $y = 1 - x + 0.2x^2$. It will be of interest to see how well we approximate this function.) To set up the normal equations, we need the sums tabulated in Table 10.1. An electronic desk calculator can give these as accumulated totals directly. A computer program at the end of this chapter is designed to set up the matrix and the righthand-side vector. We need to solve the set of

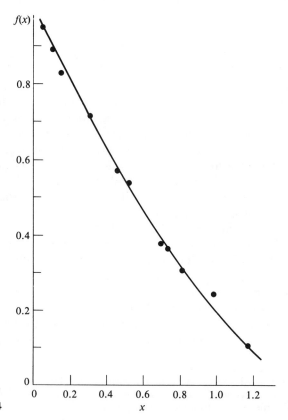

Figure 10.4

equations

$$11a_0 + 6.01a_1 + 4.6545a_2 = 5.905,$$
$$6.01a_0 + 4.6545a_1 + 4.1150a_2 = 2.1839,$$
$$4.6545a_0 + 4.1150a_1 + 3.9161a_2 = 1.3357.$$

The result is $a_0 = 0.998$, $a_1 = -1.018$, $a_2 = 0.225$, so the least-squares method gives

$$y = 0.998 - 1.018x + 0.225x^2.$$

Compare this to $y = 1 - x + 0.2x^2$. We do not expect to reproduce the coefficients exactly because of the errors in the data.

In the general case, we may wonder what degree of polynomial should be used. As we use higher-degree polynomials, we of course will reduce the deviations of the points from the curve until, when the degree of the polynomial, n, equals $N - 1$, there is an exact match (assuming no duplicate data at the same x-value) and we have the interpolating polynomials of Chapter 3. The answer to this problem is found in statistics. One increases the degree of approximating polynomial so long as there is a statistically significant decrease in the variance, σ^2, which is computed by

$$\sigma^2 = \frac{\sum e_i^2}{N - n - 1}. \tag{10.6}$$

For the above example, when the degree of the polynomial made to fit the points is varied from 1 to 7, we obtain the results shown in Table 10.2.

Table 10.2

Degree	Equation	σ^2 (Eq. (10.6))	$\sum e^2$
1	$y = 0.952 - 0.760x$	0.3956	3.560
2	$y = 0.998 - 1.018x + 0.225x^2$	0.2836	2.269
3	$y = 1.004 - 1.079x + 0.351x^2 - 0.069x^3$	0.3023	2.116
4	$y = 0.988 - 0.838x - 0.522x^2 + 1.040x^3$ $- 0.454x^4$	0.3854	2.312
5	$y = 1.031 - 1.704x + 4.278x^2 - 9.477x^3$ $+ 9.394x^4 - 3.290x^5$	0.5529	2.764
6	$y = 1.038 - 1.910x + 5.952x^2 - 15.078x^3$ $+ 18.227x^4 - 9.835x^5 + 1.836x^6$	0.8815	3.526
7	$y = 1.032 - 1.742x + 4.694x^2 - 11.898x^3$ $+ 16.645x^4 - 14.346x^5 + 8.141x^6 - 2.293x^7$	1.6206	4.862

The criterion of Eq. (10.6) chooses the optimum degree as two. This is no surprise, in view of how the data were constructed. It is important to realize that the numerator of Eq. (10.6), the *sum of the deviations squared* of the points from the curve, should continually decrease as the degree of the polynomial is raised. It is the denominator of Eq. (10.6) that makes σ^2 increase as we go above the optimum degree. In this example, this behavior is observed for $n = 3$. Above $n = 3$, a second effect sets in. Due to ill-conditioning, the coefficients of the least-squares polynomials are determined with poor precision. This exaggerates the increases of the values of σ^2.

10.3 FITTING DATA WITH A CUBIC SPLINE

The fitting of a polynomial curve to a set of data may be considered from another point of view, that of the draftsman. We suppose that the data are not necessarily subject to error, so that the least-squares approach is not appropriate, but we wish to fit a "smooth curve" to the points. One may use a French curve on the drafting table, but this is a very subjective operation. Fitting a polynomial of high degree to a set of six or eight points, say, does not appeal, since we do not expect that the functional relationship is that complicated.

Not only do high-degree polynomials seem undesirable on the basis of our intuition; in some cases they show unexpectedly large deviations from a smooth curve through the data. As an extreme case, in which this effect is exaggerated, consider this function:

$$
\begin{aligned}
f(x) &= 0, & -1 \leqslant x \leqslant -0.2, \\
f(x) &= 5|x - 0.2|, & -0.2 < x < 0.2; \\
f(x) &= 0, & 0.2 \leqslant x \leqslant 1.0.
\end{aligned}
$$

Suppose we fit exact polynomials of degree 2, 4, 6, and 8, all agreeing with $f(x)$ over the interval $-1 \leqslant x \leqslant 1$ (employing the methods of Chapter 3). Figure 10.5 shows the results. The wide swings when the degree is high, which are caused by a local "bump" in the function at $x = 0$, but which occur at points far removed from $x = 0$, are strong arguments against using high-degree polynomials in cases such as this, when a function is generally smooth but has some local roughness. (The roughness might be only apparent, being caused by an abnormally large error in one of the points.)

One solution to this problem is fitting subregions of the region $-1 \leqslant x \leqslant 1$ with low-degree polynomials. Figure (10.6) illustrates how quadratics would accomplish this. The difficulty with this approach is that the slope is discontinuous at the points where the quadratics join. If we have a generally smooth function, this is undesirable. We seek a method that retains smoothness where the function is smooth, but still can fit local irregularities without the violent misbehavior exhibited in our example above.

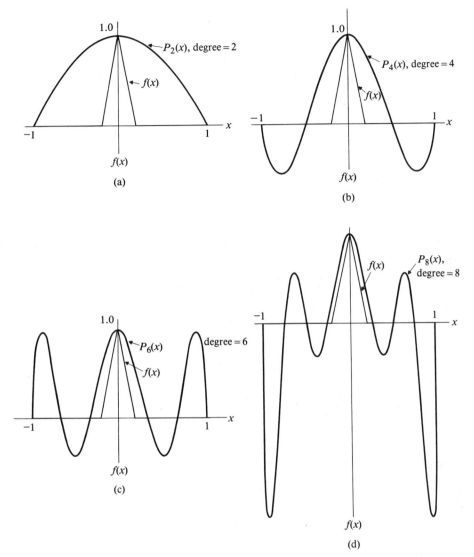

Figure 10.5

One technique that is becoming increasingly important is the so-called *spline fitting* of a curve. The name derives from another draftsman's device. A spline is a flexible strip which can be held by weights so that it passes through each of the given points, but goes smoothly from each interval to the next according to the laws of beam flexure. The present mathematical procedure is an adaptation of this idea. It is particularly advantageous when we want to find derivatives of the data.

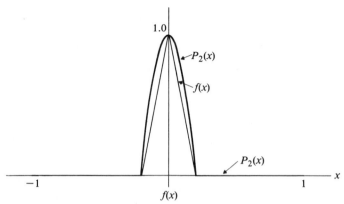

Figure 10.6

The conditions for a cubic spline fit are that we pass a set of cubics through the points, using a new cubic in each interval. To correspond to the idea of the draftsman's spline, we require that both the slope and the curvature be the same for the pair of cubics that join at each point. We now develop the equations subject to these conditions.

Write the cubic for the ith interval, which lies between the points (x_i, y_i) and (x_{i+1}, y_{i+1}) in the form

$$y = a_i(x - x_i)^3 + b_i(x - x_i)^2 + c_i(x - x_i) + d_i. \qquad (10.7)$$

Since it fits at the two endpoints of the interval,

$$y_i = a_i(x_i - x_i)^3 + b_i(x_i - x_i)^2 + c_i(x_i - x_i) + d_i = d_i; \qquad (10.8)$$

$$y_{i+1} = a_i(x_{i+1} - x_i)^3 + b_i(x_{i+1} - x_i)^2 + c_i(x_{i+1} - x_i) + d_i$$

$$= a_i h_i^3 + b_i h_i^2 + c_i h_i + d_i. \qquad (10.9)$$

In the last equation, we use h_i for Δx in the ith interval. We need the first and second derivatives to relate the slopes and curvatures of the joining polynomials, so we differentiate Eq. (10.7):

$$y' = 3a_i(x - x_i)^2 + 2b_i(x - x_i) + c_i, \qquad (10.10)$$

$$y'' = 6a_i(x - x_i) + 2b_i. \qquad (10.11)$$

The mathematical procedure is simplified if we write the equations in terms of the second derivatives of the interpolating cubics. Let S_i represent the second derivative at the point (x_i, y_i) and S_{i+1} at the point (x_{i+1}, y_{i+1}).

From Eq. (10.11) we have

$$S_i = 6a_i(x_i - x_i) + 2b_i$$

$$= 2b_i;$$

$$S_{i+1} = 6a_i(x_{i+1} - x_i) + 2b_i$$

$$= 6a_i h_i + 2b_i.$$

Hence we can write

$$b_i = S_i/2, \tag{10.12}$$

$$a_i = (S_{i+1} - S_i)/6h_i. \tag{10.13}$$

We substitute the relations for a_i, b_i, d_i given by Eqs. (10.8), (10.12), and (10.13) into Eq. (10.9) and then solve for c_i:

$$y_{i+1} = \left(\frac{S_{i+1} - S_i}{6h_i}\right)h_i^3 + \frac{S_i}{2}h_i^2 + c_i h_i + y_i;$$

$$c_i = \frac{y_{i+1} - y_i}{h_i} - \frac{2h_i S_i + h_i S_{i+1}}{6}$$

We now invoke the condition that the slopes of the two cubics that join at (x_i, y_i) are the same. For the equation in the ith interval, Eq. (10.10) becomes, with $x = x_i$,

$$y_i' = 3a_i(x_i - x_i)^2 + 2b_i(x_i - x_i) + c_i = c_i.$$

In the previous interval, from x_{i-1} to x_i, the slope at its right end will be

$$y_i' = 3a_{i-1}(x_i - x_{i-1})^2 + 2b_{i-1}(x_i - x_{i-1}) + c_{i-1}$$
$$= 3a_{i-1}h_{i-1}^2 + 2b_{i-1}h_{i-1} + c_{i-1}.$$

Equating these, and substituting for a, b, c, d their relationships in terms of S and y, we get

$$y_i' = \frac{y_{i+1} - y_i}{h_i} - \frac{2h_i S_i + h_i S_{i+1}}{6}$$
$$= 3\left(\frac{S_i - S_{i-1}}{6h_{i-1}}\right)h_{i-1}^2 + 2\left(\frac{S_{i-1}}{2}\right)h_{i-1} + \frac{y_i - y_{i-1}}{h_{i-1}} - \frac{2h_{i-1}S_{i-1} + h_{i-1}S_i}{6}.$$

On simplifying this equation we get

$$h_{i-1}S_{i-1} + (2h_{i-1} + 2h_i)S_i + h_i S_{i+1} = 6\left(\frac{y_{i+1} - y_i}{h_i} - \frac{y_i - y_{i-1}}{h_{i-1}}\right). \tag{10.14}$$

Equation (10.14) applies at each internal point, from $i = 2$ to $i = n - 1$; n being the total number of points. This gives $n - 2$ equations relating the n values of S_i. We get two additional equations involving S_1 and S_n when we specify conditions pertaining to the end intervals of the whole curve. To some extent these end conditions are arbitrary. Three alternative choices are often used:

1. Take $S_1 = 0$, $S_n = 0$. This is equivalent to assuming that the end cubics approach linearity at their extremities.

2. Take $S_1 = S_2$, $S_n = S_{n-1}$. This is equivalent to assuming that the end cubics approach parabolas at their extremities.

3.   Take S_1 as a linear extrapolation from S_2 and S_3, and S_n as a linear extrapolation from S_{n-1} and S_{n-2}. With this assumption, for a set of data that are fit by a single cubic equation, their cubic splines will all be this same cubic. The other conditions do not have this property.

The relations for end condition 3 are:

At the left end:

$$\frac{S_2 - S_1}{h_1} = \frac{S_3 - S_2}{h_2}, \qquad S_1 = \frac{(h_1 + h_2)S_2 - h_1 S_3}{h_2},$$

At the right end: (10.15)

$$\frac{S_n - S_{n-1}}{h_{n-1}} = \frac{S_{n-1} - S_{n-2}}{h_{n-2}}, \qquad S_n = \frac{(h_{n-2} + h_{n-1})S_{n-1} - h_{n-1}S_{n-2}}{h_{n-2}}.$$

Relation 1, where $S_1 = 0$ and $S_n = 0$, is called a *natural spline*. It is often felt that this flattens the curve too much at the ends; in spite of this, it is frequently used. Relation 3 often suffers from the other extreme, giving too much curvature in the end intervals. Probably the best spline curve is given when one knows the slope of the function at the endpoints, or can make reasonable estimates of these slopes.

If we write the equations for $S_2, S_3, \ldots, S_{n-1}$ (Eq. (10.14)) in matrix form, we get:

$$
\begin{bmatrix}
h_1 & 2(h_1 + h_2) & h_2 & & & \\
 & h_2 & 2(h_2 + h_3) & h_3 & & \\
 & & h_3 & 2(h_3 + h_4) & h_4 & \cdots \\
 & & & & & \\
 & & & h_{n-2} & 2(h_{n-2} + h_{n-1}) & h_{n-1}
\end{bmatrix}
\begin{bmatrix}
S_1 \\ S_2 \\ S_3 \\ S_4 \\ \vdots \\ S_{n-1} \\ S_n
\end{bmatrix}
= 6
\begin{bmatrix}
\dfrac{Y_3 - Y_2}{h_2} - \dfrac{Y_2 - Y_1}{h_1} \\[2mm]
\dfrac{Y_4 - Y_3}{h_3} - \dfrac{Y_3 - Y_2}{h_2} \\[2mm]
\dfrac{Y_5 - Y_4}{h_4} - \dfrac{Y_4 - Y_3}{h_3} \\[2mm]
\vdots \\[2mm]
\dfrac{Y_n - Y_{n-1}}{h_{n-1}} - \dfrac{Y_{n-1} - Y_{n-2}}{h_{n-2}}
\end{bmatrix}
. \qquad (10.16)
$$

In the matrix array above, there are only $n - 2$ equations, but n unknowns. We can eliminate two unknowns (S_1 and S_n) using the above relations that correspond to the end-condition assumptions. In each case we can reduce the S vector to $n - 2$ elements, and the coefficient matrix becomes square, of size $(n - 2) \times (n - 2)$. Furthermore, the matrix becomes tridiagonal, and hence is solved speedily and can be stored economically. A program is given at the end of this chapter that creates this tridiagonal matrix and augments it with the proper righthand-side vector, and then solves the equations for the values of S (the second-derivative values) in each interval.

For each end condition, the coefficient matrices become:

Condition 1 $S_1 = 0$, $S_n = 0$:

$$
\begin{bmatrix}
2(h_1 + h_2) & h_2 & & & \\
h_2 & 2(h_2 + h_3) & h_3 & & \\
 & h_3 & 2(h_3 + h_4) & h_4 & \cdots \\
 & & & & \ddots \\
 & & & & h_{n-2} \quad 2(h_{n-2} + h_{n-1})
\end{bmatrix}
$$

Condition 2 $S_1 = S_2$, $S_n = S_{n-1}$:

$$
\begin{bmatrix}
(3h_1 + 2h_2) & h_2 & & & \\
h_2 & 2(h_2 + h_3) & h_3 & & \\
 & h_3 & 2(h_3 + h_4) & h_4 & \cdots \\
 & & & & \ddots \\
 & & & & h_{n-2} \quad (2h_{n-2} + 3h_{n-1})
\end{bmatrix}
$$

Condition 3 S_1 and S_n are linear extrapolations:

$$
\begin{bmatrix}
\dfrac{(h_1 + h_2)(h_1 + 2h_2)}{h_2} & \dfrac{h_2^2 - h_1^2}{h_2} & & \\
h_2 & 2(h_2 + h_3) & h_3 & \\
 & h_3 & 2(h_3 + h_4) & h_4 \quad \cdots \\
 & & & \ddots \\
 & & \dfrac{h_{n-2}^2 - h_{n-1}^2}{h_{n-2}} & \dfrac{(h_{n-1} + h_{n-2})(h_{n-1} + 2h_{n-2})}{h_{n-2}}
\end{bmatrix}
$$

With condition 3, after solving the set of equations, we compute S_1 and S_n using Eq. (10.15). For each of the cases, the righthand-side vector is the same; it is that given in Eq. (10.16). If the data are equally spaced, the matrices reduce to a simple form.

After the S_1 values are calculated, we get the values of a_i, b_i, c_i, d_i for each interval if we want to interpolate for new values of the function:

$$
a_i = \frac{S_{i+1} - S_i}{6h_i};
$$

$$
b_i = \frac{S_i}{2};
$$

$$
c_i = \frac{Y_{i+1} - Y_i}{h_i} - \frac{2h_i S_i + h_i S_{i+1}}{6};
$$

$$
d_i = Y_i.
$$

Example 1   Fit the data of Table 10.3 by a cubic spline curve. (The true relation is just $y = x^3 - 8$.) Use all three end conditions.

Table 10.3

x	y
0	-8
1	-7
2	0
3	19
4	56

For end condition 1, we solve:

$$\begin{bmatrix} 4 & 1 & \\ 1 & 4 & 1 \\ & 1 & 4 \end{bmatrix} \begin{bmatrix} S_2 \\ S_3 \\ S_4 \end{bmatrix} = \begin{bmatrix} 36 \\ 72 \\ 108 \end{bmatrix},$$

so that

$S_1 = 0,$ $S_2 = 6.4285,$ $S_3 = 10.2857,$ $S_4 = 24.4285,$ $S_5 = 0.$

For end condition 2, we solve:

$$\begin{bmatrix} 5 & 1 & \\ 1 & 4 & 1 \\ & 1 & 5 \end{bmatrix} \begin{bmatrix} S_2 \\ S_3 \\ S_4 \end{bmatrix} = \begin{bmatrix} 36 \\ 72 \\ 108 \end{bmatrix},$$

so that

$S_1 = 4.80,$ $S_2 = 4.80,$ $S_3 = 12.00,$ $S_4 = 19.20,$ $S_5 = 19.20.$

For end condition 3, we solve:

$$\begin{bmatrix} 6 & 0 & \\ 1 & 4 & 1 \\ & 0 & 6 \end{bmatrix} \begin{bmatrix} S_2 \\ S_3 \\ S_4 \end{bmatrix} = \begin{bmatrix} 36 \\ 72 \\ 108 \end{bmatrix},$$

so that

$S_1 = 0.00,$ $S_2 = 6.00,$ $S_3 = 12.00,$ $S_4 = 18.00,$ $S_5 = 24.00.$

The equation of the data, $y = x^3 - 8$, gives second-derivative values that match those using condition three.

The next example is a more realistic situation in which a cubic spline might be used.

Example 2 The data in the following table are from astronomical observations of a type of variable star called a *Cepheid variable*, and represent variations in its apparent magnitude with time:

Time	0.0	0.2	0.3	0.4	0.5	0.6	0.7	0.8	1.0
Apparent magnitude	0.302	0.185	0.106	0.093	0.240	0.579	0.561	0.468	0.302

Use each of the three end conditions to compute cubic splines, and compare the values interpolated from each spline function at intervals of time of 0.05.

The augmented matrices whose solutions give values for $S_2, S_3, \ldots, S_8$ are shown in Table 10.4(a). The computer program at the end of this chapter was used, giving the results shown in Table 10.4(b).

Observe that values between $t = 0.25$ and $t = 0.75$ are almost identical for all three conditions. In the end intervals, however, there are significant differences, especially for condition 3. This condition gives entirely too much curvature for these intervals, particularly for $0 < t < 0.2$. This is well illustrated by Fig. 10.7.

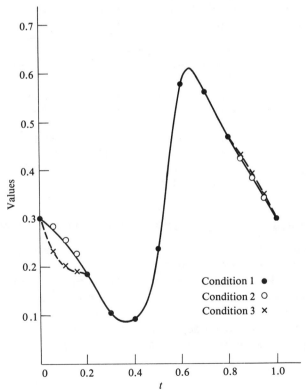

Condition 1 ●
Condition 2 ○
Condition 3 ×

Figure 10.7

Table 10.4(a)

Condition 1			
Matrix coefficients are:			
0.20	0.60	0.10	−1.23
0.10	0.40	0.10	3.96
0.10	0.40	0.10	9.60
0.10	0.40	0.10	11.52
0.10	0.40	0.10	−21.42
0.10	0.40	0.10	−4.50
0.10	0.60	0.20	0.60

Condition 2			
Matrix coefficients are:			
0.20	0.80	0.10	−1.23
0.10	0.40	0.10	3.96
0.10	0.40	0.10	9.60
0.10	0.40	0.10	11.52
0.10	0.40	0.10	−21.42
0.10	0.40	0.10	−4.50
0.10	0.80	0.20	0.60

Condition 3			
Matrix coefficients are:			
0.20	1.20	−0.30	−1.23
0.10	0.40	0.10	3.96
0.10	0.40	0.10	9.60
0.10	0.40	0.10	11.52
0.10	0.40	0.10	−21.42
0.10	0.40	0.10	−4.50
−0.30	1.20	0.20	0.60

10.4 APPLICATIONS OF CUBIC SPLINE FUNCTIONS

In addition to their obvious use for interpolation, splines can be used for finding derivatives and integrals of functions, even when the function is known only as a table of values. These applications are similar to the application of interpolating polynomials as presented in Chapter 4. The smoothness of splines, because of the requirement that each portion have the same first and second derivatives as its neighbor, can give improved accuracy in some cases.

Since the cubic spline approximates $f(x)$ as a cubic, we can write, for the interval $x_i \leqslant x \leqslant x_{i+1}$:

$$f(x) = a_i(x - x_i)^3 + b_i(x - x_i)^2 + c_i(x - x_i) + d_i,$$

Table 10.4(b)

t	Values, Cond. 1	Values, Cond. 2	Values, Cond. 3
0.000	0.302	0.302	0.302
0.050	0.278	0.282	0.238
0.100	0.252	0.256	0.203
0.150	0.222	0.224	0.188
0.200	0.185	0.185	0.185
0.250	0.143	0.142	0.141
0.300	0.106	0.106	0.106
0.350	0.087	0.088	0.088
0.400	0.093	0.093	0.093
0.450	0.133	0.133	0.133
0.500	0.240	0.240	0.240
0.550	0.424	0.424	0.424
0.600	0.579	0.579	0.579
0.650	0.608	0.608	0.608
0.700	0.561	0.561	0.561
0.750	0.511	0.511	0.511
0.800	0.468	0.468	0.468
0.850	0.426	0.426	0.430
0.900	0.385	0.384	0.392
0.950	0.343	0.343	0.350
1.000	0.302	0.302	0.302

where the coefficients are determined as in Section 10.3. The method outlined in the previous section computes S_i and S_{i+1}, the values of the second derivative at each end of the subinterval. From these S-values and the values of $f(x)$, we compute the coefficients of the cubic:

$$a_i = \frac{S_{i+1} - S_i}{6(x_{i+1} - x_i)},$$

$$b_i = \frac{S_i}{2},$$

$$c_i = \frac{f(x_{i+1}) - f(x_i)}{x_{i+1} - x_i} - \frac{2(x_{i+1} - x_i)S_i + (x_{i+1} - x_i)S_{i+1}}{6},$$

$$d_i = f(x_i).$$

Approximating the first and second derivatives is straightforward; we estimate these as the values of the derivatives of the cubic:

$$f'(x) \doteq 3a_i(x - x_i)^2 + 2b_i(x - x_i) + c_i,$$

$$f''(x) \doteq 6a_i(x - x_i) + 2b_i.$$

At the $n + 1$ points x_i where the function is known and the spline matches $f(x)$, these formulas are particularly simple:

$$f'(x_i) \doteq c_i,$$
$$f''(x_i) \doteq 2b_i.$$

(We note that a cubic spline is not useful for approximating derivatives of order higher than the second. A higher degree of spline function would be required for these.)

Approximating the integral of $f(x)$ over the n intervals where $f(x)$ is approximated by the spline is similarly straightforward:

$$\int_{x_1}^{x_{n+1}} f(x)\, dx = \int_{x_1}^{x_{n+1}} P_3(x)\, dx = \sum_{i=1}^{n} \frac{a_i}{4}(x - x_i)^4 + \frac{b_i}{3}(x - x_i)^3 + \frac{c_i}{2}(x - x_i)^2$$

$$+ d_i(x - x_i) \Big]_{x_i}^{x_{i+1}}$$

$$= \sum_{i=1}^{n} \left[\frac{a_i}{4}(x_{i+1} - x_i)^4 + \frac{b_i}{3}(x_{i+1} - x_i)^3 + \frac{c_i}{2}(x_{i+1} - x_i)^2 + d_i(x_{i+1} - x_i) \right].$$

If the intervals are all the same size, $(h = x_{i+1} - x_i)$, this equation becomes:

$$\int_{x_1}^{x_{n+1}} f(x)\, dx = \frac{h^4}{4} \sum_{i=1}^{n} a_i + \frac{h^3}{3} \sum_{i=1}^{n} b_i + \frac{h^2}{2} \sum_{i=1}^{n} c_i + h \sum_{i=1}^{n} d_i.$$

We illustrate the use of splines to compute derivatives and integrals by a simple example.

Example Compute the integral and derivatives of $f(x) = \sin \pi x$ over the interval $0 \leqslant x \leqslant 1$ from the spline that fits at $x = 0, 0.25, 0.5, 0.75,$ and 1.0. (See Table 10.5.) We use end condition 1: $S_1 = 0, S_5 = 0$. Solving for the coefficients of the cubic spline, we get the results shown in Table 10.6.

Table 10.5

i, point number	x	$f(x)$
1	0	0
2	0.25	0.7071
3	0.50	1.0000
4	0.75	0.7071
5	1.0	0

Table 10.6

i	x	S_i	a_i	b_i	c_i	d_i
1	0	0	−4.8960	0	3.1340	0
2	0.25	−7.3440	−2.0288	−3.6720	2.2164	0.7071
3	0.50	−10.3872	2.0288	−5.1936	0	1.0
4	0.75	−7.3440	4.8960	−3.6720	−2.2164	0.7071

The derivatives can now be computed:

At $x = 0.25$: $f'(x) = 2.2164$ $f''(x) = -7.344.$

                     (exact = 2.2214);   (exact = −6.979);

At $x = 0.50$: $f'(x) = 0$ $f''(x) = -10.387$

                     (exact = 0);   (exact = −9.870);

At $x = 0.75$: $f'(x) = -2.2164$ $f''(x) = -7.344$

                     (exact = −2.2214);   (exact = −6.979).

The errors of the first derivatives computed from the cubic spline are smaller than those from a fourth-degree interpolating polynomial; the errors of the second derivatives from the spline are somewhat larger than from the polynomial. We readily compute the integral:

$$\int_0^1 f(x)\, dx = \frac{(0.25)^4}{4}(0) + \frac{(0.25)^3}{3}(-12.5376) + \frac{(0.25)^2}{2}(3.1340) + 0.25(2.4142)$$

$$= 0.6362 \quad \text{(exact} = 0.6366; \quad \text{error} = +0.0004\text{)}.$$

For this example, the accuracy is significantly better than that obtained with Simpson's $\frac{1}{3}$ rule, which gives 0.6381, error = −0.0015.

Cubic splines can also be used to solve ordinary differential equations. To solve a first-order equation, $dy/dt = f(y, t)$ with the initial condition $y(t_0) = y_0$, we write a cubic that matches a Taylor-series expansion of y about the point t_0:

$$S(t) = S(t_0 + h) = y_0 + y_0' h + \frac{y_0''}{2} h^2 + \frac{A}{6} h^3. \tag{10.17}$$

In Eq. (10.17), the coefficient A is to be determined. We evaluate A so that the derivative of $S(t)$ equals the value of the derivative of y at $t_0 + h$ as given by the differential equation:

$$S'(t) = S'(t_0 + h) = y_0' + y_0'' h + \frac{A}{2} h^2 = f(y_{n+1}, t_{n+1})$$

$$= f(S(t_0 + h), t_0 + h) \tag{10.18}$$

$$= f\left(\left(y_0 + y_0' h + \frac{y_0''}{2} h^2 + \frac{A}{6} h^3\right), t_0 + h\right).$$

In Eq. (10.18), we have substituted $S(t_0 + h)$ for y_{n+1}. We must now solve Eq. (10.18) for A. If the differential equation is linear, this equation is linear, and solving for A is easy. If the differential equation is nonlinear, Eq. (10.18) will also be nonlinear in A, but it is usually quite easy to solve for A by iteration. We illustrate the solving of first-order differential equations with two examples.

Example 1 Solve $dy/dt = y + t$, $y(0) = 1$.

We need $y''(t)$, obtained by differentiating dy/dt:

$$y'' = y' + 1.$$

From Eq. (10.17):

$$S(t_0 + h) = y_0 + y_0'h + \frac{y_0''}{2}h^2 + \frac{A}{6}h^3$$

$$= 1 + h + h^2 + \frac{A}{6}h^3,$$

and from Eq. (10.18):

$$S'(t_0 + h) = y_0' + y_0''h + \frac{A}{2}h^2 = f(S(t_0 + h), t_0 + h)$$

$$= 1 + 2h + \frac{A}{2}h^2 = \left(1 + h + h^2 + \frac{A}{6}h^3\right) + (h),$$

which gives, for $h = 0.2$, $A = 2.14286$. Therefore we have

$$y(0.2) = 1 + 0.2 + (0.2)^2 + \frac{2.14286}{6}(0.2)^3$$

$$= 1.24286,$$

which is correct to five digits (error $= -5.16 \times 10^{-5}$).

For $h = 0.4$, $A = 2.30769$ and $y(0.4) = 1.58462$, which is correct to three digits (error $= -9.66 \times 10^{-4}$); and for $h = 0.6$, $A = 2.500$, giving $y(0.6) = 2.0500$ (error $= -5.76 \times 10^{-3}$). These results demonstrate an $O(h^4)$ error.

Example 2 Solve the nonlinear problem $dy/dt = y^2 + t^2$, $y(1) = 0$.

Equation (10.17) gives

$$S(t_0 + h) = 0 + h + h^2 + \frac{A}{6}h^3.$$

Equation (10.18) gives

$$S'(t_0 + h) = 1 + 2h + \frac{A}{2}h^2.$$

Equating $S'(t_0 + h)$ to $f(y_{n+1}, t_{n+1})$, we get this nonlinear equation in A:

$$1 + 2h + \frac{A}{2}h^2 = \left(h + h^2 + \frac{A}{2}h^3\right)^2 + (1 + h)^2$$

or

$$A = \frac{\left(h + h^2 + A\dfrac{h^3}{6}\right)^2 + h^2}{h^2/2}.$$

This readily gives, by iteration, $A = 5.0432$ when $h = 0.2$ and $A = 7.0422$ when $h = 0.4$. The values of y are then:

$$y(0.2) = 0.24672, \qquad \text{error} = -0.00041;$$
$$y(0.4) = 0.63512, \qquad \text{error} = 0.01243.$$

These errors vary about as h^5.

Cubic splines can also be applied to the solution of second-order differential equations, but the results are not as good. As a first step, we rearrange the equation to the form:

$$\frac{d^2y}{dt^2} = f(y, y', t), \qquad y(t_0) = y_0, \qquad y'(t_0) = y_0'.$$

To solve this equation, we find a cubic that approximates the function y in the interval from t_0 to $t_0 + h$:

$$y = a(t - t_0)^3 + b(t - t_0)^2 + c(t - t_0) + d.$$

From the initial conditions, we know that:

$$y(t_0) = y_0 = a(0)^3 + b(0)^2 + c(0) + d = d;$$
$$y'(t_0) = y_0' = 3a(0)^2 + 2b(0) + c = c;$$
$$y''(t_0) = y_0'' = 6a(0) + 2b = 2b.$$

In the last relation, we compute y_0'' from $f(t_0, y_0, y_0')$. In addition to the values of b, c, and d as calculated above, we need the value of a. We do this by requiring that $y''(t_0 + h)$ satisfy the differential equation:

$$y''(t_0 + h) = 6ah + y_0'' = f(y, y', t)_{t=t_0+h} \qquad (10.19)$$

In Eq. (10.19), we evaluate f using the cubic relation, giving an equation that we solve for a.

Now we can determine y, y', and y'' at $t_0 + h$ from the cubic polynomial. Using these values, we step ahead to the next point. The particular merit of this procedure is that only one evaluation of the derivative function is required for each step; but a result with errors that vary about as h^3 is produced.

Example 3 Given $d^2y/dt^2 = y + ty'$, $y(0) = 1$, $y'(0) = -1$. Find $y(0.2)$ and $y(0.4)$ by the method of cubic splines.

$$y(0) = 1, \quad d = 1;$$
$$y'(0) = -1, \quad c = -1;$$
$$y''(0) = 1 + (0)(-1) = 1, \quad b = 0.5;$$
$$6a(0.2) + 1 = [a(0.2)^3 + 0.5(0.2)^2 - 0.2 + 1] + 0.2[3a(0.2)^2 + 2(0.5)(0.2) - 1],$$
$$a = -0.2911;$$
$$y(0.2) = -0.2911(0.2)^3 + 0.5(0.2)^2 + (-1)(0.2) + 1$$
$$= \underline{0.8177}.$$

To get $y(0.4)$, we can advance from $t = 0$ with $h = 0.4$, or we can advance from $t = 0.2$ with $h = 0.2$. Doing the former:

From the above, $\quad d = 1$, $c = -1$, $b = 0.5$;
$$6a(0.4) + 1 = [a(0.4)^3 + 0.5(0.4)^2 - 0.4 + 1] + 0.4[3a(0.4)^2 + 2(0.5)(0.4) - 1],$$
$$a = -0.2836;$$
$$y(0.4) = -0.2836(0.4)^3 + 0.5(0.4)^2 + (-1)(0.5)$$
$$= \underline{0.6618}.$$

We get exactly the same value if we begin at $y = 0.8177$, $t = 0.2$ and use $h = 0.2$, calculating values for d, c, b, and a as we did in the first interval. (The exact values are $y(0.2) = 0.817527$, and $y(0.4) = 0.6617$.)

The method can also be used for second-order boundary-value problems. In this application, the region between the boundaries is subdivided into intervals. (These are usually made of uniform size, but this is not required.) Cubics are then computed which have second derivatives agreeing with the differential equation at each point and which also agree with the boundary conditions at each end. If the equation is of higher order than two, a spline of higher degree can be used.

10.5 CHEBYSHEV POLYNOMIALS

We turn now to the problem of representing a function with minimum error. This is a central problem in the software development of digital computers because it is more economical to compute the values of the common functions using an efficient approximation than to store a table of values and employ interpolation techniques. Since digital computers are essentially only arithmetic devices, the most elaborate function they can compute is a rational function, a ratio of polynomials. We will hence restrict our discussion to representation of functions by polynomials or rational functions.

One way to approximate a function by a polynomial is to use a truncated Taylor series. This is not the best way, in most cases. In order to study better ways, we need to introduce the Chebyshev polynomials.

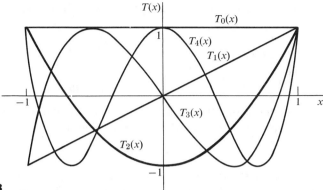

Figure 10.8

The familiar Taylor-series expansion represents the function with very small error near the point of the expansion, but the error increases rapidly (proportional to a power) as we employ it at points farther away. In a digital computer, we have no control over where in an interval the approximation will be used, so the Taylor series is not usually appropriate. We would prefer to trade some of its excessive precision at the center of the interval to reduce the errors at the ends.

We can do this while still expressing functions as polynomials by the use of Chebyshev polynomials. The first few of these are:*

$$T_0(x) = 1,$$
$$T_1(x) = x,$$
$$T_2(x) = 2x^2 - 1,$$
$$T_3(x) = 4x^3 - 3x,$$
$$T_4(x) = 8x^4 - 8x^2 + 1,$$
$$T_5(x) = 16x^5 - 20x^3 + 5x, \tag{10.20}$$
$$T_6(x) = 32x^6 - 48x^4 + 18x^2 - 1,$$
$$T_7(x) = 64x^7 - 112x^5 + 56x^3 - 7x,$$
$$T_8(x) = 128x^8 - 256x^6 + 160x^4 - 32x^2 + 1,$$
$$T_9(x) = 256x^9 - 576x^7 + 432x^5 - 120x^3 + 9x,$$
$$T_{10}(x) = 512x^{10} - 1280x^8 + 1120x^6 - 400x^4 + 50x^2 - 1.$$

The members of this series of polynomials can be generated from the two-term recursion formula

$$T_{n+1}(x) = 2xT_n(x) - T_{n-1}(x), \qquad T_0(x) = 1, \qquad T_1(x) = x. \tag{10.21}$$

Note that the coefficient of x^n in $T_n(x)$ is always 2^{n-1}. In Fig. 10.8 we plot the first four polynomials of Eq. (10.20).

---

* The commonly accepted symbol $T(x)$ comes from the older spelling, Tschebycheff.

These polynomials have some unusual properties. They form an orthogonal set, in that

$$\int_{-1}^{1} \frac{T_n(x)T_m(x)}{\sqrt{1-x^2}}\,dx = \begin{cases} 0, & n \neq m, \\ \pi, & n = m = 0, \\ \pi/2, & n = m \neq 0. \end{cases} \qquad (10.22)$$

The orthogonality of these functions will not be of immediate concern to us.

The Chebyshev polynomials are also terms of Fourier series, since

$$T_n(x) = \cos n\theta, \qquad (10.23)$$

where $\theta = \arccos x$. Observe that $\cos 0 = 1$, $\cos \theta = \cos (\arccos x) = x$.

In order to demonstrate the equivalence of Eq. (10.23) to Eqs. (10.20) and (10.21), we recall some trigonometric identities, such as:

$$\cos 2\theta = 2 \cos^2 \theta - 1,$$
$$T_2(x) = 2x^2 - 1;$$
$$\cos 3\theta = 4 \cos^3 \theta - 3 \cos \theta,$$
$$T_3(x) = 4x^3 - 3x;$$
$$\cos (n + 1)\theta + \cos (n - 1)\theta = 2 \cos \theta \cos n\theta,$$
$$T_{n+1}(x) + T_{n-1}(x) = 2xT_n(x).$$

Because of the relation $T_n(x) = \cos n\theta$, it is apparent that the Chebyshev polynomials have a succession of maximums and minimums of alternating signs, each of magnitude one. Further, since $|\cos n\theta| = 1$ for $n\theta = 0$, π, $2\pi, \ldots$, and since θ varies from 0 to π as x varies from 1 to -1, $T_n(x)$ assumes its maximum magnitude of unity $n + 1$ times on the interval $[-1, 1]$.

Most important for our present application of these polynomials is the fact that, of all polynomials of degree n where the coefficient of x^n is unity, the polynomial

$$\frac{1}{2^{n-1}} T_n(x)$$

has a smaller upper bound to its magnitude in the interval $[-1, 1]$ than any other. Because the maximum magnitude of $T_n(x)$ is one, the upper bound referred to is $1/2^{n-1}$. This is of importance because we shall be able to write power-series representations of functions whose maximum errors are given in terms of this upper bound.

We first prove this assertion about bounds on the magnitude of polynomials. The proof is by contradiction. Let $P_n(x)$ be a polynomial whose leading term* is

---

* We restrict the polynomials to those whose leading term is x^n is so that all are scaled alike.

x^n and suppose that its maximum magnitude on $[-1, 1]$ is less than that of $T_n(x)/2^{n-1}$. Write

$$T^n(x)/2^{n-1} - P_n(x) = P_{n-1}(x),$$

where $P_{n-1}(x)$ is a polynomial of degree $n - 1$ or less, since the x^n terms cancel. The polynomial $T_n(x)$ has $n + 1$ extremes (counting endpoints), each of magnitude one, so $T_n(x)/2^{n-1}$ has $n + 1$ extremes each of magnitude $1/2^{n-1}$, and these successive extremes alternate in sign. By our supposition about $P_n(x)$, at each of these maximums or minimums, the magnitude of $P_n(x)$ is less than $1/2^{n-1}$; hence $P_{n-1}(x)$ must change its sign at least for every extreme of $T_n(x)$, which is then at least $n + 1$ times. $P_{n-1}(x)$ hence crosses the axis at least n times and would have n zeros. But this is impossible if $P_{n-1}(x)$ is only of degree $n - 1$, unless it is identically zero. The premise must then be false and $P_n(x)$ has a larger magnitude than the polynomial we are testing or, alternatively, $P_n(x)$ is exactly the same polynomial.

10.6 APPROXIMATION OF FUNCTIONS WITH ECONOMIZED POWER SERIES

We are now ready to use Chebyshev polynomials to "economize" a power series. Consider the Maclaurin series for e^x:

$$e^x = 1 + x + \frac{x^2}{2} + \frac{x^3}{6} + \frac{x^4}{24} + \frac{x^5}{120} + \frac{x^6}{720} + \cdots .$$

If we would like to use a truncated series to approximate e^x on the interval $[0, 1]$ with a precision of 0.001, we will have to retain terms through that in x^6, since the error after the term in x^5 will be more than $1/720$. Suppose we subtract

$$\left(\frac{1}{720}\right)\left(\frac{T_6}{32}\right)$$

from the truncated series. We note from Eq. (10.20) that this will exactly cancel the x^6 term and at the same time make adjustments in other coefficients of the Maclaurin series. Since the maximum value of T_6 on the interval $[0, 1]$ is unity, this will change the sum of the truncated series by only

$$\frac{1}{720} \cdot \frac{1}{32} < 0.00005,$$

which is small with respect to our required precision of 0.001. Performing the calculations, we have

$$e^x \doteq 1 + x + \frac{x^2}{2} + \frac{x^3}{6} + \frac{x^4}{24} + \frac{x^5}{120} + \frac{x^6}{720} - \frac{1}{720}\left(\frac{1}{32}\right)(32x^6 - 48x^4 + 18x^2 - 1)$$

$$= 1.000043 + x + 0.499219x^2 + \frac{x^3}{6} + 0.043750x^4 + \frac{x^5}{120}. \tag{10.24}$$

Table 10.7 Comparison of errors of economized power series and a Maclaurin series for e^x.

x	e^x	Maclaurin, sixth-degree	Economized, fifth-degree	Economized, fourth-degree	Maclaurin, fourth-degree
0	1.00000	1.00000	1.00004	1.00004	1.00000
0.2	1.22140	1.22140	1.22142	1.22098	1.22140
0.4	1.49182	1.49182	1.49179	1.49133	1.49173
0.6	1.82212	1.82211	1.82208	1.82212	1.82140
0.8	2.22554	2.22549	2.22553	2.22605	2.22240
1.0	2.71828	2.71806	2.71801	2.71749	2.70833
Maximum error		0.00022	0.00027	0.00079	0.00995

This gives a fifth-degree polynomial that approximates e^x on [0, 1] almost as well as the sixth-degree one derived from the Maclaurin series. (The actual maximum error of the fifth-degree expression is 0.000270; for the sixth-degree expression it is 0.000226.) We hence have "economized" the power series in that we get nearly the same precision with fewer terms.

By subtracting $\frac{1}{120}(T_5/16)$ we can economize further, getting a fourth-degree polynomial that is almost as good as the economized fifth-degree one. It is left as an exercise to do this and to show that the maximum error is now 0.000781, so that we have found a fourth-degree power series that meets an error criterion that requires us to use two additional terms of the original Maclaurin series. Because of the relative ease with which they can be developed, such economized power series are frequently used for approximations to functions and are much more efficient than power series of the same degree obtained by merely truncating a Taylor or Maclaurin series. Table 10.7 compares the errors of these power series.

The maximum error in the economized fifth-degree polynomial is only slightly greater than the sixth-degree Maclaurin series. The economized fourth-degree polynomial incurs a maximum error about three and one-half times as much, but still within the 0.001 limit that was initially imposed, and will require significantly reduced computational effort. In addition, there is a proportionately reduced memory-space requirement to store the constants of the polynomial. In contrast, a fourth-degree Maclaurin series has an error nearly ten times greater than the 0.001 tolerance, and its error is over twelve times that of the fourth-degree economized form.

By rearranging the Chebyshev polynomials, we can express powers of x in terms of them:

$$1 = T_0.$$
$$x = T_1,$$
$$x^2 = \tfrac{1}{2}(T_0 + T_2),$$
$$x^3 = \tfrac{1}{4}(3T_1 + T_3),$$

$$x^4 = \tfrac{1}{8}(3T_0 + 4T_2 + T_4),$$

$$x^5 = \tfrac{1}{16}(10T_1 + 5T_3 + T_5),$$

$$x^6 = \tfrac{1}{32}(10T_0 + 15T_2 + 6T_4 + T_6),$$

$$x^7 = \tfrac{1}{64}(35T_1 + 21T_3 + 7T_5 + T_7),$$

$$x^8 = \tfrac{1}{128}(35T_0 + 56T_2 + 28T_4 + 8T_6 + T_8),$$

$$x^9 = \tfrac{1}{256}(126T_1 + 84T_3 + 36T_5 + 9T_7 + T_9).$$

(10.25)

By substituting these identities into an infinite Taylor series and collecting terms in $T_i(x)$, we create a Chebyshev series. For example, we can get the first four terms of a Chebyshev series by starting with the Maclaurin expansion for e^x. Such a series converges more rapidly than does a Taylor series on $[-1, 1]$:

$$e^x = 1 + x + \frac{x^2}{2} + \frac{x^3}{6} + \frac{x^4}{24} + \ldots.$$

Replacing terms by Eq. (10.25), but omitting polynomials beyond $T_3(x)$, since we want only four terms,* we have:

$$e^x = T_0 + T_1 + \tfrac{1}{4}(T_0 + T_2) + \tfrac{1}{24}(3T_1 + T_3) + \tfrac{1}{192}(3T_0 + 4T_2 + \ldots)$$

$$+ \tfrac{1}{1920}(10T_1 + 5T_3 + \ldots) + \tfrac{1}{23,040}(10T_0 + 15T_2 + \ldots) + \ldots$$

$$= 1.2661T_0 + 1.1303T_1 + 0.2715T_2 + 0.0444T_3 + \ldots.$$

In order to compare the Chebyshev expansion with the Maclaurin series, we convert back to powers of x, using Eq. (10.20):

$$e^x = 1.2661 + 1.1303(x) + 0.2715(2x^2 - 1) + 0.0444(4x^3 - 3x) + \ldots$$

$$= 0.9946 + 0.9971x + 0.5430x^2 + 0.1776x^3 + \ldots.$$

(10.26)

Table 10.8 and Fig. 10.9 compare the error of the Chebyshev expansion, Eq. (10.26), with the Maclaurin series, using terms through x^3 in each case. The figure shows how the Chebyshev expansion attains a smaller maximum error by permitting the error at the origin to increase. The errors can be considered to be distributed more or less uniformly throughout the interval. In contrast to this, the Maclaurin expansion, which gives very small errors near the origin, allows the error to bunch up at the ends of the interval.

If the function is to be expressed directly as an expansion in Chebyshev polynomials, the coefficients can be obtained by integration. Based on the orthogonality property, the coefficients are computed from

$$a_i = \frac{2}{\pi} \int_{-1}^{1} \frac{f(x) T_i(x)}{\sqrt{1 - x^2}} \, dx,$$

---

* The number of terms that are employed determines the accuracy of the computed values, of course.

Table 10.8 Comparison of Chebyshev series for e^x with Maclaurin series:
$$e^x = 0.9946 + 0.9971x + 0.5430x^2 + 0.1776x^3;$$
$$e^x = 1 + x + 0.5x^2 + 0.1667x^3.$$

x	e^x	Chebyshev	Error	Maclaurin	Error
-1.0	0.3679	0.3629	0.0050	0.3333	0.0346
-0.8	0.4493	0.4535	-0.0042	0.4347	0.0146
-0.6	0.5488	0.5535	-0.0047	0.5440	0.0048
-0.4	0.6703	0.6713	-0.0010	0.6693	0.0010
-0.2	0.8187	0.8155	0.0032	0.8187	0.0000
0	1.0000	0.9946	0.0054	1.0000	0.0000
0.2	1.2214	1.2172	0.0042	1.2213	0.0001
0.4	1.4918	1.4917	0.0001	1.4907	0.0011
0.6	1.8221	1.8267	-0.0046	1.8160	0.0061
0.8	2.2255	2.2307	-0.0052	2.2053	0.0202
1.0	2.7183	2.7123	0.0060	2.6667	0.0516

and the series is expressed as

$$f(x) = \frac{a_0}{2} + \sum_{i=1}^{\infty} a_i T_i(x).$$

A change of variable will be required if the desired interval is other than $(-1, 1)$. In some cases, the definite integral which defines the coefficients can be profitably evaluated by numerical methods as described in Chapter 4.

Since the coefficients of the terms of a Chebyshev expansion usually decrease even more rapidly than the terms of a Maclaurin expansion, one can get an estimate of the magnitude of the error from the next nonzero term after those

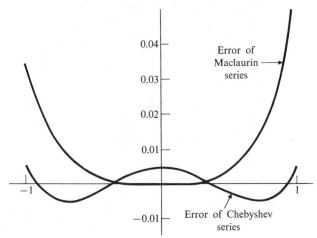

Figure 10.9

which were retained. For the truncated Chebyshev series given by Eq. (10.26), the $T_4(x)$ term would be

$$\tfrac{1}{192}(T_4) + \tfrac{1}{23,040}(6T_4) + \ldots = 0.00525T_4.$$

Since the maximum value of $T_4(x)$ on $(-1, 1)$ is 1.0, we estimate the maximum errors of Eq. (10.26) to be 0.00525. The maximum error in Table 10.8 is 0.0060. This good agreement is caused by the very rapid decrease in coefficients in this example.

The computational economy to be gained by economizing a Maclaurin series, or by using a Chebyshev series, is even more dramatic when the Maclaurin series is slowly convergent. The previous example for $f(x) = e^x$ is a case in which the Maclaurin series converges rapidly. The power of the methods of this section is better demonstrated in the following example.

Example A Maclaurin series for $(1 + x)^{-1}$ is:

$$(1 + x)^{-1} = 1 - x + x^2 - x^3 + x^4 - \ldots \qquad (-1 < x < 1).$$

Table 10.9 compares the accuracy of truncated Maclaurin series with the economized series derived from them.

In Table 10.9, we see that the error of the Maclaurin series is small for $x = 0.2$, and this also would be true for other values near $x = 0$, while the economized polynomial has less accuracy. At $x = 0.8$, the situation is reversed, however. Economized polynomials of degrees 8 and 6, derived from truncated Maclaurin series of degrees 10 and 8, actually have smaller errors than their precursors. Further economization, giving polynomials of degrees 6 and 4, have lesser or only slightly greater errors than their precursors, at significant savings of computational effort and with smaller storage requirements in a computer's memory for the coefficients of the polynomials.

10.7 APPROXIMATION WITH RATIONAL FUNCTIONS

We have seen that expansion of a function in terms of Chebyshev polynomials gives a power-series expansion that is much more efficient on the interval $(-1, 1)$ than the Maclaurin expansion, in that it has a smaller maximum error with a given number of terms. These are not the best approximations for use in most digital computers, however. In this application, we measure efficiency by the computer time required to evaluate the function, plus some consideration of storage requirements for the constants. Since the arithmetic operations of a computer can directly evaluate only polynomials, we limit our discussion of more efficient approximations to rational functions, which are the ratios of two polynomials.

Our discussion of methods of finding efficient rational approximations will be elementary and introductory only. Obtaining truly best approximations is a difficult subject. In its present stage of development it is as much art as science,

Table 10.9 Comparison of Maclaurin and economized series for $(1 + x)^{-1}$.

Maclaurin			Economized					
Degree	Value	Error	Degree*	Value	Error	Degree*	Value	Error*
			x = 0.2					
2	0.840000	0.006667	2	0.758600	−0.0747333			
4	0.833600	0.000267	4	0.764594	−0.068739			
6	0.833344	$11 * 10^{-6}$	6	0.803646	−0.029687			
8	0.833334	$1 * 10^{-6}$	8	0.822786	−0.010547			
10	0.833333	0						
			x = 0.8					
2	0.840000	0.284445	2	0.812600	0.257045			
4	0.737600	0.182045	4	0.678314	0.122759	4	0.658246	0.102691
6	0.672064	0.116509	6	0.628558	0.073003	6	0.598199	0.042644
8	0.630121	0.074566	8	0.602106	0.046551			
10	0.603277	0.047722						

* Economized series were derived from Maclaurin series of corresponding degree.

and requires successive approximations from a "suitably close" initial approximation. Our study will serve to introduce the student to some of the ideas and procedures used. The topic is of great importance, however, since the saving of just 1 millisecond of time in the generation of a frequently used elementary function may save hundreds of dollars' worth of machine time each year.

We start with a discussion of Padé approximations. Suppose we wish to represent a function as the quotient of two polynomials:

$$f(x) \doteq R_N(x) = \frac{a_0 + a_1 x + a_2 x^2 + \ldots + a_n x^n}{1 + b_1 x + b_2 x^2 + \ldots + b_m x^m}, \qquad N = n + m.$$

The constant term in the denominator can be taken as unity without loss of generality, since we can always convert to this form by dividing numerator and denominator by b_0. The constant b_0 will generally not be zero, for, in that case, the fraction would be undefined at $x = 0$. The most useful of the Padé approximations are those with the degree of the numerator equal to, or one greater than, the degree of the denominator. Note that the number of constants in $R_N(x)$ is $N + 1 = n + m + 1$.

The Padé approximations are related to Maclaurin expansions in that the coefficients are determined in a similar fashion to make $f(x)$ and $R_N(x)$ agree at $x = 0$ and also to make the first N derivatives agree at $x = 0$.*

We begin with the Maclaurin series for $f(x)$ (we use only terms through x^N) and write

$$f(x) - R_N(x)$$
$$= (c_0 + c_1 x + c_2 x^2 + \ldots + c_N x^N) - \frac{a_0 + a_1 x + \ldots + a_n x^n}{1 + b_1 x + \ldots + b_m x^m}$$
$$= \frac{(c_0 + c_1 x + \ldots + c_N x^N)(1 + b_1 x + \ldots + b_m x^m) - (a_0 + a_1 x + \ldots + a_n x^n)}{1 + b_1 x + \ldots + b_m x^m}.$$
$$(10.27)$$

The coefficients c_i are $f^{(i)}(0)/(i!)$ of the Maclaurin expansion. Now if $f(x) = R_N(x)$ at $x = 0$, the numerator of Eq. (10.27) must have no constant term. Hence

$$c_0 - a_0 = 0. \qquad (10.28)$$

In order for the first N derivatives of $f(x)$ and $R_N(x)$ to be equal at $x = 0$, the coefficients of the powers of x up to and including x^N in the numerator must all be zero also. This gives N additional equations for the a's and b's. The first n of

---

* A similar development can be derived for the expansion about a nonzero value of x, but the manipulations are not as easy. By a change of variable we can always make the region of interest contain the origin.

these involve a's, the rest only b's and c's:

$$b_1 c_0 + c_1 - a_1 = 0,$$
$$b_2 c_0 + b_1 c_1 + c_2 - a_2 = 0,$$
$$b_3 c_0 + b_2 c_1 + b_1 c_2 + c_3 - a_3 = 0,$$

.
.
.

$$b_m c_{n-m} + b_{m-1} c_{n-m+1} + \ldots + c_n - a_n = 0, \tag{10.29}$$

$$b_m c_{n-m+1} + b_{m-1} c_{n-m+2} + \ldots + c_{n+1} = 0,$$
$$b_m c_{n-m+2} + b_{m-1} c_{n-m+3} + \ldots + c_{n+2} = 0,$$

.
.
.

$$b_m c_{N-m} + b_{m-1} c_{N-m+1} + \ldots + c_N = 0.$$

Note that, in each equation, the sum of the subscripts on the factors of each product is the same, and is equal to the exponent of the x-term in the numerator. The $N + 1$ equations of Eqs. (10.28) and (10.29) give the required coefficients of the Padé approximation. We illustrate by an example.

Example Find arctan $x \doteq R_9(x)$. Use in the numerator a polynomial of degree five.

The Maclaurin series through x^9 is

$$\text{arctan } x \doteq x - \tfrac{1}{3}x^3 + \tfrac{1}{5}x^5 - \tfrac{1}{7}x^7 + \tfrac{1}{9}x^9. \tag{10.30}$$

We form, analogously to Eq. (10.27),

$$f(x) - R_9(x)$$
$$= \frac{(x - \tfrac{1}{3}x^3 + \tfrac{1}{5}x^5 - \tfrac{1}{7}x^7 + \tfrac{1}{9}x^9)(1 + b_1 x + b_2 x^2 + b_3 x^3 + b_4 x^4) - (a_0 + a_1 x + \ldots + a_5 x^5)}{(1 + b_1 x + b_2 x^2 + b_3 x^3 + b_4 x^4)}.$$
$$\tag{10.31}$$

Making coefficients through that of x^9 in the numerator equal to zero, we get

$$a_0 = 0,$$
$$a_1 = 1,$$
$$a_2 = b_1,$$
$$a_3 = -\tfrac{1}{3} + b_2,$$
$$a_4 = -\tfrac{1}{3}b_1 + b_3,$$
$$a_5 = \tfrac{1}{5} - \tfrac{1}{3}b_2 + b_4,$$

$$\tfrac{1}{5}b_1 - \tfrac{1}{3}b_3 = 0,$$
$$-\tfrac{1}{7} + \tfrac{1}{5}b_2 - \tfrac{1}{3}b_4 = 0,$$
$$-\tfrac{1}{7}b_1 + \tfrac{1}{5}b_3 = 0,$$
$$\tfrac{1}{9} - \tfrac{1}{7}b_2 + \tfrac{1}{5}b_4 = 0.$$

Solving first the last four equations for the b's, and then getting the a's, we have

$$a_0 = 0, \quad a_1 = 1, \quad a_2 = 0, \quad a_3 = \tfrac{7}{9}, \quad a_4 = 0, \quad a_5 = \tfrac{64}{945},$$
$$b_1 = 0, \quad b_2 = \tfrac{10}{9}, \quad b_3 = 0, \quad b_4 = \tfrac{5}{21}.$$

A rational function which approximates arctan x is then

$$\arctan x \doteq \frac{x + \tfrac{7}{9}x^3 + \tfrac{64}{945}x^5}{1 + \tfrac{10}{9}x^2 + \tfrac{5}{21}x^4}. \tag{10.32}$$

In Table 10.10 we compare the errors for Padé approximation (Eq. 10.32) to the Maclaurin series expansion (Eq. 10.30). Enough terms are available in the Maclaurin series to give five-decimal precision at $x = 0.2$ and 0.4, but at $x = 1$ (the limit for convergence of the series) the error is sizeable. Even though we used no more information in establishing it, the Padé formula is surprisingly accurate, having an error only 1/275 as large at $x = 1$. It is then particularly astonishing to realize that the Padé approximation is still not the best one of its form, for it violates the minimax principle. If the extreme precision near $x = 0$ is relaxed, we can make the maximum error smaller in the interval.

Before we discuss such better approximations in the form of rational functions, remarks on the amount of effort required for the computation using Eq. (10.32) are in order. If we implement the equation in a computer as it stands, we would, of course, use the constants in decimal form, and we would evaluate the polynomials in nested form:

$$\text{Numerator} = [(0.0677x^2 + 0.7778)x^2 + 1]x,$$
$$\text{Denominator} = (0.2381x^2 + 1.1111)x^2 + 1.$$

Since additions and subtractions are generally much faster than multiplications or divisions, we generally neglect them in a count of operations. We have then three multiplications for the numerator, two for the denominator, plus one to get x^2, and one division, for a total of seven operations. The Maclaurin series is

Table 10.10 Comparison of Padé approximation to Maclaurin series for arctan x.

x	True Value	Padé Eq. (10.32)	Error	Maclaurin Eq. (10.30)	Error
0.2	0.19740	0.19740	0.00000	0.19740	0.00000
0.4	0.38051	0.38051	0.00000	0.38051	0.00000
0.6	0.54042	0.54042	0.00000	0.54067	−0.00025
0.8	0.67474	0.67477	−0.00003	0.67982	−0.00508
1.0	0.78540	0.78558	−0.00018	0.83492	−0.04952

evaluated with six multiplications, using the nested form. If division and multiplication consume about the same time, there is about a standoff in effort, but greater precision for Eq. (10.32).*

Since small differences in effort accumulate for a frequently used function, it is of interest to see if we can further decrease the number of operations to evaluate Eq. (10.32). By means of a succession of divisions we can re-express it in continued-fraction form:

$$\frac{0.0677x^5 + 0.7778x^3 + x}{0.2381x^4 + 1.1111x^2 + 1}$$

$$= \frac{0.2844x^5 + 3.2667x^3 + 4.2x}{x^4 + 4.6667x^2 + 4.2}$$

$$= \frac{0.2844x(x^4 + 11.4846x^2 + 14.7659)}{x^4 + 4.6667x^2 + 4.2} = \frac{0.2844x}{\dfrac{x^4 + 4.6667x^2 + 4.2}{x^4 + 11.4846x^2 + 14.7659}}$$

$$= \frac{0.2844x}{1 - \dfrac{6.879x^2 + 10.5659}{x^4 + 11.4846x^2 + 14.7659}} = \frac{0.2844x}{1 - \dfrac{6.879(x^2 + 1.5497)}{x^4 + 11.4846x^2 + 14.7659}}$$

$$= \frac{0.2844x}{1 - \dfrac{6.8179}{\dfrac{x^4 + 11.846x^2 + 14.7659}{x^2 + 1.5497}}} = \frac{0.2844x}{1 - \dfrac{6.8179}{x^2 + 9.9348 - \dfrac{0.6304}{x^2 + 1.5497}}}.$$

In this last form, we see that three divisions and two multiplications are needed (one multiplication by x and one to get x^2), for a total of five operations. We have saved one step. In most cases there is an even greater advantage to the continued fraction form; in this example the missing powers of x favored the evaluation as polynomials.

The error of a Padé approximation can often be roughly estimated by computing the next nonzero term in the numerator of Eq. (10.32). For the above example, the coefficient of x^{10} is zero, and the next term is

$$(-\tfrac{1}{7}b_4 + \tfrac{1}{9}b_2 - \tfrac{1}{11})x^{11} = [-\tfrac{1}{7}(\tfrac{5}{21}) + \tfrac{1}{9}(\tfrac{10}{9}) - \tfrac{1}{11}]x^{11}$$
$$= -0.0014x^{11}.$$

Dividing by the denominator, we have

$$\text{Error} \doteq \frac{-0.0014x^{11}}{1 + 1.1111x^2 + 0.2381x^4}.$$

---

* On many computers, division is slower than multiplication. This will modify the conclusion reached here.

At $x = 1$ this estimate gives -0.00060, which is about three times too large, but still of the correct order of magnitude. It is not unusual that such estimates be rough; analogous estimates of error by using the next term in a Maclaurin series behave similarly. The validity of the rule of thumb that "next term approximates the error" is poor when the coefficients do not decrease rapidly.

The preference for Padé approximations with the degree of the numerator the same as or one more than the degree of the denominator rests on the empirical fact that the errors are usually less for these. Ralston (1965) gives examples demonstrating this.

One can get somewhat improved rational-function approximations by starting with the Chebyshev expansion and operating analogously to the method for Padé approximations. We illustrate with an approximation for e^x. The Chebyshev series was derived in Section 10.6, Eq. (10.26):

$$e^x \doteq 1.2661\,T_0 + 1.1303\,T_1 + 0.2715\,T_2 + 0.0444\,T_3.$$

Using this approximation, we form the difference

$$f(x) - \frac{P_n(x)}{Q_m(x)}$$

$$= \frac{(1.2661 + 1.303\,T_1 + 0.2715\,T_2 + 0.0444\,T_3)(1 + b_1 T_1) - (a_0 + a_1 T_1 + a_2 T_2)}{1 + b_1 T_1}.$$

Here we have chosen the numerator as a second-degree Chebyshev polynomial and the denominator as first degree. We again make the first $N = n + m$ powers of x in the numerator vanish. Expanding the numerator, we get:

$$\text{Numerator} = 1.2661 + 1.1303\,T_1 + 0.2715\,T_2 + 0.0444\,T_3 + 1.2661 b_1 T_1$$

$$+ 1.1303 b_1 T_1^2 + 0.2715 b_1 T_1 T_2 + 0.0444 b_1 T_1 T_3 - a_0$$

$$- a_1 T_1 - a_2 T_2.$$

Before we can equate coefficients to zero, we need to resolve the products of Chebyshev polynomials that occur. Recalling that $T_n(x) = \cos n\theta$, we can use the trignometric identity

$$\cos n\theta \cos m\theta = \tfrac{1}{2}[\cos(n + m)\theta + \cos(n - m)\theta],$$

$$T_n(x)T_m(x) = \tfrac{1}{2}[T_{n+m}(x) + T_{|n-m|}(x)].$$

The absolute value of the difference $n - m$ occurs because $\cos(z) = \cos(-z)$.

Using this relation we can write the equations:

$$a_0 = 1.2661 + \frac{1.1303}{2} b_1,$$

$$a_1 = 1.1303 + \left(\frac{0.2715}{2} + 1.2661\right)b_1,$$

$$a_2 = 0.2715 + \left(\frac{1.1303}{2} + \frac{0.0444}{2}\right)b_1,$$

$$0 = 0.0444 + \frac{0.2715}{2} b_1.$$

Solving, we get $b_1 = -0.3266$, $a_0 = 1.0815$, $a_1 = 0.6724$, $a_2 = 0.07966$, and

$$e^x \doteq \frac{1.0815 + 0.6724 T_1 + 0.07966 T_2}{1 - 0.3266 T_1}$$

$$= \frac{1.0018 + 0.6724x + 0.1593x^2}{1 - 0.3266x}. \tag{10.33}$$

The last expression results when the Chebyshev polynomials are written in terms of powers of x. In Table 10.11 the error of this rational approximation is compared to the Chebyshev expansion. We see that the maximum error is

Table 10.11 Comparison of rational approximations (Eq. (10.33) with Chebyshev series for e^x.

x	e^x	Chebyshev	Error	Rational function	Error
−1.0	0.3679	0.3629	0.0050	0.3684	−0.0005
−0.8	0.4493	0.4535	−0.0042	0.4486	0.0007
−0.6	0.5488	0.5535	−0.0047	0.5482	0.0006
−0.4	0.6703	0.6713	−0.0010	0.6707	−0.0004
−0.2	0.8187	0.8155	0.0032	0.8201	−0.0014
0	1.0000	0.9946	0.0054	1.0018	−0.0018
0.2	1.2214	1.2172	0.0042	1.2225	−0.0011
0.4	1.4918	1.4917	0.0001	1.4911	0.0007
0.6	1.8221	1.8267	−0.0046	1.8191	0.0030
0.8	2.2255	2.2307	−0.0052	2.2225	0.0031
1.0	2.7183	2.7123	0.0060	2.7230	−0.0047

reduced by 22%. Note that we do not, nevertheless, yet have a "best approximation." The error should reach equal maximums at five points in the interval—instead the error is large near $x = 1$ and too small elsewhere.

The basis for this last statement is the minimax theorem. Based on a theorem due to Chebyshev, we may state a principle whereby we may determine whether the approximation represented by a given polynomial or rational function is optimum, in the sense that it gives the least maximum error of any rational function of the same degree of numerator and denominator on a given interval. An expression is "minimax" if and only if there are at least $n + 2$ maxima in the deviations, and these are all equal in magnitude and of alternating sign, on the interval of approximation. (Here n is the sum of the degrees of numerator and denominator of the rational function.) In the discussion that follows, we shall be referring to the magnitude of the errors and will not be concerned about their sign.

A second important consequence of this principle is that we can put bounds on the error of the minimax expression from the range of the errors of a function that is not minimax. Suppose we have an expression like Eq. (10.33). It has five maxima of alternating sign on the interval $[-1, 1]$, as shown by Table 10.11. This is the correct number, but we know the function isn't minimax because the maxima aren't equal in magnitude. While some of the maxima are not given precisely by the table, we can see that the smallest is 0.0005 (at $x = -1$) and the largest appears to be 0.0047 (at $x = 1$). From this range of maximum error values, we can bound the maximum error of the minimax rational function [of degree $(2, 1)$]: the minimax expression will have a maximum error on $[-1, 1]$ no less than 0.0005 and no greater than 0.0047.

Similarly, the truncated Chebyshev series of degree 3 whose errors are tabulated in Table 10.8 is not truly minimax; it has five alternating maxima to its errors but they are not quite equal in magnitude. From an examination of Table 10.8 we can say, however, that the minimax polynomial of degree 3 will have a maximum error bounded by 0.0047 and 0.0060. (A tighter bound might result from a more careful computation of the errors of Eq. (10.26).) Such a prediction about the amount of improvement that will be provided by a minimax expression can help one decide whether the additional effort to find it is worthwhile.

To obtain the optimum rational function that approximates the function with equal-magnitude errors distributed through the interval is beyond the scope of this text. The approach that is used is to improve an initial estimate of a function, such as Eq. (10.33), by successive trials, often modifying the constants on the basis of experience until eventually one has a satisfactory formula. Hastings (1955) gives examples of the process. Systematic methods of determining the constants in such minimax rational approximations have also been determined. They are iteration methods beginning from an initial "sufficiently good" approximation. They are expensive to compute because the iterations involve solving a set of nonlinear equations. Ralston (1965) describes one such method.

10.8 PROGRAMS

Two computer programs are presented. Program 1 (Fig. 10.10) finds least-squares polynomials that fit a set of x,y-pairs. The program reads in the coordinates of the points and computes the coefficients of the normal equations for polynomials whose degree ranges from MS to MF. Values of these parameters are also read in. To get these normal equations, the terms of the largest augmented matrix [Eq. (10.5)] are computed. (This matrix corresponds to degree MF.) In the process, the powers of x are stored in a vector; this expedites forming the sums. The program recognizes the symmetry of the coefficient matrix which lessens the need to recompute terms.

After the matrix that represents the normal equations has been formed, a subroutine is called to compute the LU equivalent. This subroutine is very similar to one in Chapter 2, except that pivoting is not done. The solution for the coefficients of each of the polynomials is obtained by calling a second subroutine. Using the LU decompostion process avoids triangularizing the normal equations repeatedly for the various degrees of polynomial. The values of Eq. (10.6), here called BETA, are computed and printed to assist the user in selecting the optimum degree of polynomial. The program was run with the input data shown in Fig. 10.4.

The second program (Fig. 10.11) is a subroutine that determines the values of S (second derivatives) for cubic splines that fit a set of x,y-pairs passed to the subroutine in two vectors. Since the solution of the tridiagonal system is so quickly obtained, in-line code within the subroutine was written to accomplish it.

```
C    THIS PROGRAM IS USED IN FITTING A POLYNOMIAL TO A SET OF DATA.
C    THE PROGRAM READS IN N PAIRS OF X AND Y VALUES AND COMPUTES
C    THE COEFFICIENTS OF THE NORMAL EQUATIONS FOR THE LEAST
C    SQUARES METHOD.
C      PARAMETERS ARE -
C        X, Y     ARRAY OF X AND Y VALUES
C        N        NUMBER OF DATA PAIRS
C        MS, MF   THE RANGE OF DEGRE OF POLYNOMIALS TO BE COMPUTED.
C                 THE MAXIMUM DEGREE IS 9.
C        A        AUGMENTED ARRAY OF THE COEFFICIENTS OF THE NORMAL EQUATIONS
C        C        ARRAY OF COEFFICIENTS OF THE LEAST SQUARES POLYNOMIALS
       DIMENSION X(100), Y(100), C(10), A(10,11), XN(100)
C    READ IN N, THEN THE X AND Y VALUES.
C    N IS ON THE FIRST CARD, FOLLOWED BY CARDS WITH ONE X, Y PAIR ON EACH
       READ (5,100) N, (X(I), Y(I), I = 1,N)
100    FORMAT (I5/(2F10.0))
C    READ IN MS, MF.  THE PROGRAM WILL FIND COEFFICIENTS FOR EACH
C    DEGREE OF POLYNOMIAL FROM DEGREE MS TO DEGREE MF.
       READ (5,101) MS, MF
101    FORMAT (2I5)
C    COMPUTE MATRIX OF COEFFIECIENTS AND R.H.S FOR MF DEGREE
C    HOWEVER, FIRST CHECK TO SEE IF MAX DEGREE REQUESTED IS TOO LARGE.
C    IT CANNOT EXCEED N - 1.  IF IT DOES, REDUCE TO EQUAL
C    N - 1 AND PRINT MESSAGE.
       IF (MF .LE. (N - 1))  GO TO 5
       MF = N - 1
       WRITE (6,200) MF
200    FORMAT (1H0, 42HDEGREE OF POLYNOMIAL CANNOT EXCEED N - 1.  /
      +   1H , 47HREQUESTED MAXIMUM DEGREE TOO LARGE - REDUCED TO , I3)
5      MFP1 = MF + 1
       MFP2 = MF + 2
C    PUT ONES INTO A NEW ARRAY.  THIS WILL HOLD POWERS OF THE X VALUES
C    AS WE PROCEED.
       DO 10 I = 1,N
10     XN(I) = 1.
C    COMPUTE FIRST COLUMN AND N+1 ST COLUMN OF A.  I MOVES DOWN THE
C    ROWS, J SUMS OVER THE N VALUES.
       DO 30 I = 1,MFP1
       A(I,1) = 0.
       A(I,MFP2) = 0.
       DO 20 J = 1,N
       A(I,1) = A(I,1) + XN(J)
       A(I,MFP2) = A(I,MFP2) + Y(J)*XN(J)
20     XN(J) = XN(J)*X(J)
30     CONTINUE
C    COMPUTE THE LAST ROW OF A.  I MOVES ACROSS THE COLUMNS,
C    J SUMS OVER N VALUES.
       DO 50 I = 2,MFP1
       A(MFP1,I) = 0.
       DO 40 J = 1,N
       A(MFP1,I) = A(MFP1,I) + XN(J)
40     XN(J) = XN(J)*X(J)
50     CONTINUE
C    NOW FILL IN THE REST OF THE A MATRIX.  I MOVES DOWN THE
C    ROWS, J MOVES ACROSS THE COLUMNS.
       DO 70 J = 2,MFP1
       DO 60 I = 1,MF
60     A(I,J) = A(I+1,J-1)
70     CONTINUE
C    WRITE OUT THE MATRIX OF NORMAL EQUATIONS.
       WRITE (6,201) ((A(I,J), J = 1,MFP2), I = 1,MFP1)
201    FORMAT (1H0, 9F13.5)
C    NOW CALL A SUBROUTINE TO SOLVE THE SYSTEM.  DO THIS FOR EACH DEGREE
C    FROM MS TO MF.
C    GET THE LU DECOMPOISITION OF A
       CALL LUDCMQ (A, MFP1, 10)
C    RESET THE R.H.S. INTO C.  WE NEED TO DO THIS FOR EACH DEGREE.
       MSP1 = MS + 1
       DO 95 I = MSP1,MFP1
       DO 90 J = 1,I
90     C(J) = A(J,MFP2)
       CALL SOLNQ (A, C, I, 10)
       IM1 = I - 1
```

Figure 10.10 Program 1.

```
C   NOW WRITE OUT THE COEFFICIENTS OF THE LEAST SQ POLYNOMIAL
        WRITE (6,202) IM1, (C(J), J = 1,I)
202     FORMAT (1H0, 14HFOR DEGREE OF , I2, 17H COEFFICIENTS ARE,/
       + 1H , 10X, 11F11.3)
C   COMPUTE AND PRINT THE VALUE OF BETA = SUM OF DEV SQUARED/(N - M - 1).
        BETA = 0.0
        DO 94 IPT = 1,N
          SUM = 0.0
          DO 93 ICOEF = 2,I
            JCOEF = I - ICOEF + 2
            SUM = (SUM + C(JCOEF))*X(IPT)
93        CONTINUE
          SUM = SUM + C(1)
          BETA = (Y(IPT) - SUM)**2
94      CONTINUE
        BETA = BETA/(N - I)
        WRITE (6,203) BETA
203     FORMAT (1H , 10X, 9H BETA IS , F10.5)
95      CONTINUE
        STOP
        END


        SUBROUTINE LUDCMP (A, N, NDIM)
        DIMENSION A(NDIM,NDIM)
C   THIS SUBROUTINE FORMS THE LU EQUIVALENT OF THE SQUARE
C   COEFFICIENT MATRIX A.  THE LU, IN COMPACT FORM, IS RETURNED
C   IN THE A MATRIX SPACE.  THE UPPER TRIANGULAR MATRIX U HAS ONES
C   ON ITS DIAGONAL - THESE VALUES ARE NOT INCLUDED IN THE RESULT.
        DO 30 I = 1,N
        DO 30 J = 2,N
        SUM = 0.
        IF (J .GT. I)  GO TO 15
        JM1 = J - 1
        DO 10 K = 1,JM1
10      SUM = SUM + A(I,K)*A(K,J)
        A(I,J) = A(I,J) - SUM
        GO TO 30
15      IM1 = I - 1
        IF (IM1 .EQ. 0.)  GO TO 25
        DO 20 K = 1,IM1
20      SUM = SUM + A(I,K)*A(K,J)
C   TEST FOR SMALL VALUE ON DIAGONAL
25      IF (ABS(A(I,I)) .LT. 1.E-10)  GO TO 99
        A(I,J) =(A(I,J) - SUM)/A(I,I)
30      CONTINUE
        RETURN
99      WRITE (6,100) I
100     FORMAT (1H0, 32H REDUCTION NOT COMPLETED BECAUSE ,
       + 38H SMALL VALUE FOUND FOR DIVISOR IN ROW , I3)
        RETURN
        END


        SUBROUTINE SOLNQ (A, B, N, NDIM)
        DIMENSION A(NDIM,NDIM), B(NDIM)
C   THIS SUBROUTINE FINDS THE SOLUTION TO A SET OF N LINEAR
C   EQUATIONS THAT CORRESPONDS TO THE RIGHT HAND SIDE VECTOR B.
C   THE A MATRIX IS THE LU DECOMPOSITION EQUIVALENT TO
C   THE COEFFICIENT MATRIX OF THE ORIGINAL EQUATIONS, AS
C   PRODUCED BY LUDCMP.  THE SOLUTION VECTOR IS
C   RETURNED IN THE B VECTOR.
C   DO THE REDUCTION STEP
        B(1) = B(1)/A(1,1)
        DO 20 I = 2,N
        IM1 = I - 1
        SUM = 0.
        DO 10 K = 1,IM1
10      SUM = SUM + A(I,K)*B(K)
20      B(I) = (B(I) - SUM)/A(I,I)
```

Figure 10.10 (continued)

```
C  NOW WE ARE READY FOR BACK SUBSTITUTION.  REMEMBER THAT THE
C  ELEMENTS OF U ON THE DIAGONAL ARE ALL ONES.
       DO 40 J = 2,N
       NMJP2 = N - J + 2
       NMJP1 = N - J + 1
       SUM = 0.
       DO 30 K = NMJP2,N
30     SUM = SUM + A(NMJP1,K)*B(K)
40     B(NMJP1) = B(NMJP1) - SUM
       RETURN
       END
```

OUTPUT FROM PROGRAM 1

11.00000	6.01000	4.65450	4.11496	3.91613	3.91865	4.06701	4.33743	5.90500
6.01000	4.65450	4.11496	3.91613	3.91865	4.06701	4.33743	4.72163	2.18387
4.65450	4.11496	3.91613	3.91865	4.06701	4.33743	4.72163	5.22039	1.33572
4.11496	3.91613	3.91865	4.06701	4.33743	4.72163	5.22039	5.84071	0.99948
3.91613	3.91865	4.06701	4.33743	4.72163	5.22039	5.84071	6.59458	0.83031
3.91865	4.06701	4.33743	4.72163	5.22039	5.84071	6.59458	7.49848	0.74023
4.06701	4.33743	4.72163	5.22039	5.84071	6.59458	7.49848	8.57337	0.69612
4.33743	4.72163	5.22039	5.84071	6.59458	7.49848	8.57337	9.84490	0.68281

FOR DEGREE OF 1 COEFFICIENTS ARE
            0.952   -0.760
        BETA IS   0.39565

FOR DEGREE OF 2 COEFICIENTS ARE
            0.998   -1.018   0.225
        BETA IS   0.28363

FOR DEGREE OF 3 COEFFICIENTS ARE
            1.004   -1.079   0.351   -0.069
        BETA IS   0.30229

FOR DEGREE OF 4 COEFFICIENTS ARE
            0.988   -0.838   -0.522   1.040   -0.454
        BETA IS   0.38536

FOR DEGREE OF 5 COEFFICIENTS ARE
            1.031   -1.704   4.278   -9.477   9.394   -3.290
        BETA IS   0.55288

FOR DEGREE OF 6 COEFFICIENTS ARE
            1.038   -1.910   5.952   -15.078   18.227   -9.835   1.836
        BETA IS   0.88149

FOR DEGREE OF 7 COEFFICIENTS ARE
            1.032   -1.742   4.694   -11.898   16.645   -14.346   8.141   2.293
        BETA IS   1.62064

Figure 10.10 (concluded)

```
      SUBROUTINE CUBSPL (X, Y, S, N, IEND, A)
      DIMENSION X(N), Y(N), S(N), A(N,4)
C  THIS ROUTINE COMPUTES THE MATRIX FOR FINDING THE COEFFICIENTS
C  OF A CUBIC SPLINE THROUGH A SET OF DATA.
C  THE SYSTEM THEN IS SOLVED TO OBTAIN THE SECOND DERIVATIVE VALUES.
C    PARAMETERS ARE
C      X, Y     ARRAYS OF X AND Y VALUES TO BE FITTED
C      S        ARRAY OF SECOND DERIVATIVE VALUES AT THE POINTS
C      N        NUMBER OF POINTS
C      IEND     TYPE OF END CONDITION TO BE USED
C                 IEND = 1, LINEAR ENDS, S(1) = S(N) = 0.
C                 IEND = 2, PARABOLIC ENDS, S(1) = S(2), S(N) = S(N-1)
C                 IEND = 3, CUBIC ENDS, S(1), S(N) ARE EXTRAPOLATED
C      A        AUGMENTED MATRIX OF COEFFICIENTS AND R.H.S FOR FINDING S
C
C  COMPUTE FOR THE N-2 ROWS
      NM2 = N - 2
      NM1 = N - 1
      DX1 = X(2) - X(1)
      DY1 = (Y(2) - Y(1))/DX1*6.
      DO 10 I = 1,NM2
        DX2 = X(I+2) - X(I+1)
        DY2 = (Y(I+2) - Y(I+1))/DX2*6.
        A(I,1) = DX1
        A(I,2) = 2.*(DX1 + DX2)
        A(I,3) = DX2
        A(I,4) = DY2 - DY1
        DX1 = DX2
        DY1 = DY2
10    CONTINUE
C  ADJUST FIRST AND LAST ROWS APPROPRIATE TO END CONDITION.
      GO TO (20, 50, 80), IEND
C  FOR IEND = 1, NO CHANGE IS NEEDED.
20    GO TO 100
C  FOR IEND = 2, S(1) = S(2), S(N) = S(N-1), PARABOLIC ENDS.
50    A(1,2) = A(1,2) + X(2) - X(1)
      A(NM2,2) = A(NM2,2) + X(N) - X(NM1)
      GO TO 100
C  FOR IEND = 3, CUBIC ENDS, S(1), S(N) ARE EXTRAPOLATED.
80    DX1 = X(2) - X(1)
      DX2 = X(3) - X(2)
      A(1,2) = (DX1 + DX2)*(DX1 + 2.*DX2)/DX2
      A(1,3) = (DX2*DX2 - DX1*DX1)/DX2
      DXN2 = X(NM1) - X(NM2)
      DXN1 = X(N) - X(NM1)
      A(NM2,1) = (DXN2*DXN2 - DXN1*DXN1)/DXN2
      A(NM2,2) = (DXN1 + DXN2)*(DXN1 + 2.*DXN2)/DXN2
      A(NM2,2) = (DXN1 + DXN2)*(DXN1 + 2.*DXN2)/DXN2
      GO TO 100
```

Figure 10.11 Program 2

```
C   NOW WE SOLVE THE TRIDIAGONAL SYSTEM.   FIRST REDUCE.
100     DO 110 I = 2,NM2
            A(I,2) = A(I,2) - A(I,1)/A(I-1,2)*A(I-1,3)
            A(I,4) = A(I,4) - A(I,1)/A(I-1,2)*A(I-1,4)
110     CONTINUE
C   NOW WE BACK SUBSTITUTE
            A(NM2,4) = A(NM2,4)/A(NM2,2)
            DO 120 I = 2,NM2
            J = NM1 - I
            A(J,4) = (A(J,4) - A(J,3)*A(J+1,4))/A(J,2)
120     CONTINUE
C   NOW PUT THE VALUES INTO THE S VECTOR
            DO 130 I = 1,NM2
            S(I+1) = A(I,4)
130     CONTINUE
C   GET S(1) AND S(N) ACCORDING TO END CONDITIONS
            GO TO (150,160,170), IEND
C   FOR LINEAR ENDS, S(1) = 0, S(N) = 0.
150     S(1) = 0.
            S(N) = 0.
            RETURN
C   FOR PARABOLIC ENDS, S(1) = S(2), S(N) = S(N-1).
160     S(1) = S(2)
            S(N) = S(N-1)
            RETURN
C   FOR CUBIC ENDS, EXTRAPOLATE TO GET S(1) , S(N).
170     S(1) = ((DX1 + DX2)*S(2) + DX1*S(3))/DX2
            S(N) = ((DXN2 + DXN1)*S(NM1) - DXN1*S(NM2))/DXN2
            RETURN
            END
```

Figure 10.11 (continued)

EXERCISES

Section 10.1

1. Find the individual deviations of the data in Fig. 10.1 from those computed from the least-squares line, $R = 702.2 + 3.395T$. Compare these deviations with those from the line drawn by eye, $R = 700 + 3.500T$. Find the sum of squares of deviations in each case and compare. Note that even though the maximum deviations from each of the two lines are not too different, the sums of squares differ significantly.

▶2. Show that the point whose x-coordinate is the mean of all the x-values and whose y-coordinate is the mean of all the y-values satisfies the least-squares line. Often a change of variable is made to relocate the origin at this point, with a corresponding reduction in the magnitude of the numbers worked with, making them more readily handled by hand or on some desk calculators.

▶3. Find the least-squares line that fits to the following data, assuming the x-values are free of error:

x	y	x	y
1	2.04	4	7.18
2	4.12	5	9.20
3	5.64	6	12.04

(The data are tabulated from $y = 2x$, with perturbations from a table of random numbers.)

4. In the data for Exercise 3, consider that the y-values are free of error and that all the errors are in the x-values. By suitable modifications of the normal equations, now determine the least-squares line for $x = ay + b$. Note that this is not the same line as determined in Exercise 3.

5. In multivariate analysis, the least-squares technique is used to determine the hyperplane from which the sum of the squares of the deviations is a minimum. For z, a function of two independent variables, x and y, find the normal equations to determine the parameters a, b, and c in:

$$z = ax + by + c.$$

Use this relation to find the least-squares values for the constants if the data below is available.

x	0	1.2	2.1	3.4	4.0	4.2	5.6	5.8	6.9
y	0	0.5	6.0	0.5	5.1	3.2	1.3	7.4	10.2
z	1.2	3.4	−4.6	9.9	2.4	7.2	14.3	3.5	1.3

Section 10.2

6. Observe that the data in Table 10.12 seem to be fit by a curve $y = ae^{bx}$ by plotting on semi-log paper and noting that the points then fall near a straight line. The data are for the solubilities of n-butane in anhydrous hydrofluoric acid at high pressures, and were needed in the design of petroleum refineries.

Table 10.12

Temperature, °F	Solubility, weight percent
77	2.4
100	3.4
185	7.0
239	11.1
285	19.6

By plotting on rectilinear graph paper, observe that the relationship is nonlinear.

▶7. Determine the constants for $y = ae^{bx}$ for the data in Exercise 6 by the least-squares method by fitting to the relation $\ln y = \ln a + bx$.

8. It is suspected (from theoretical considerations) that the rate of flow from a fire hose is proportional to some power of the pressure at the nozzle. Determine whether the speculation seems to be true, and what the exponent is from the data in Table 10.13. (Assume that the pressure data are more accurate.)

Table 10.13

Flow, gallons per minute	Pressure, psi
94	10
118	16
147	25
180	40
230	60

9. Since the data of Exercise 8, when plotted on log-log paper (pressure as a function of flow), seem to have a slope of nearly 2, we should expect that fitting a quadratic to the data would be successful. Do this and compare the deviations with those from the power-function relation of Exercise 8.

▶10. The data given in Exercise 3, while perturbations from a linear relation, seem to plot better along a curve because of the accidental occurrence of three negative deviations in succession. Fit a quadratic to the data.

11. To compare the results of an exact polynomial fit according to Chapter 3 with the least-squares procedure, find y-values at $x = 1.5, 2.5, 3.5, 4.5$, and 5.5 for the data of Exercise 3, utilizing a fifth-degree interpolating polynomial. Sketch the interpolating polynomial and compare to the least-squares line of Exercise 3, and the least-squares quadratic of Exercise 10. The slope of the "true" function is 2.0. How do the maximum and minimum values of the slope as determined from the three approximations (use a graphical procedure) compare to the "true" value?

▶12. The data in Table 10.14 seem to fit a cubic equation, but determine by least squares the optimum degree of polynomial.

Table 10.14

x	$f(x)$	x	$f(x)$	x	$f(x)$
0.1	1.9	6.6	36.1	13.2	−48.6
1.1	7.9	7.1	23.7	14.1	−40.2
1.6	24.9	8.2	13.0	15.6	−51.6
2.4	24.9	9.1	20.5	16.1	−30.5
2.5	34.9	9.4	−3.1	17.6	−34.6
4.1	42.7	11.1	−13.0	17.9	−16.4
5.2	29.7	11.4	−28.7	19.1	−13.8
6.1	49.8	12.2	−39.5	20.0	−1.1

13. Repeat Exercise 12, but now using every other point ($x = 0.1, x = 1.6, x = 2.5$, etc.). Do you get the same result for the optimum degree of polynomial? Are the coefficients the same? Repeat again, with the other half of the points.

14. The data of Exercise 12 also suggest a function of the form $y = A + B \sin Cx$. How could the least-squares method be used to evaluate A, B, and C? In solving the normal equations, are any difficulties encountered? What if theory suggested that $C = \pi/10$? Would this make any significant difference in solving the normal equations?

Section 10.3

15. Consider the function $f(x) = 20/(1 + 5x^2)$ over the interval $[-2, 2]$. Compute and graph polynomials of degree 1, 2, 3, 4, and 5 that agree with $f(x)$ at uniformly spaced points on the interval.

16. Confirm the statement that, for a set of data exactly fitted by a cubic, the values of S at the two ends will be linearly related to the adjacent S-values if the spline curve and the cubic polynomial are the same function. If one changes one end value for S, say S_1, does this change the portions of the spline curve in intervals other than the first?

▶17. Find the coefficient matrix and the righthand-side vector for fitting a cubic spline to the following data. Use linearity condition on the terminal S-values:

x	y	x	y
0.15	0.3945	1.07	0.2251
0.76	0.2989	1.73	0.0893
0.89	0.2685	2.11	0.0431

18. Solve the set of equations of Exercise 17 (you may wish to utilize a computer program for this), and then determine the constants of the various cubics. The data are the ordinates of the normal probability function. Compare a few interpolated values with tabulated values of the function, say at $x = 0.30, 0.80, 1.50, 2.00$.

▶19. Fit a cubic spline to the exponential function, utilizing values of e^x for $x = 0.0\,(0.5)$ 2.0, and compare the interpolated with the true values for $x = 0.25, 1.25, 1.75$. Which end condition gives the best fit?

20. Fit a natural cubic spline to $f(x) = 20/(1 + 5x^2)$ on the interval $[-2, 2]$. Use five equispaced points on the function $[x = -2(1)2]$. Graph the spline and compare to the graphs of Exercise 15.

21. Repeat Exercise 20, but use end conditions two and three. Compare results to Exercise 21.

Section 10.4

▶22. Consider this table of data (which are obviously for $f(x) = 1/x$).

x	1.0	1.5	2.0	2.5	3.0
$f(x)$	1.000	0.667	0.500	0.400	0.333

Find values for $f'(x)$ and $f''(x)$ at $x = 1.5, 2.0$, and 2.5 from the cubic spline functions that approximate $f(x)$. Compare to analytical values for $f'(x)$ and $f''(x)$ to determine their errors. Also compare to derivative values computed from central-difference formulas.

23. The comparison in Exercise 22 between the spline and central-difference values may favor the spline method because it is based on approximation by a cubic while the central-difference formulas approximate $f(x)$ with a quadratic. Suppose we fit the function by polynomials of degree 3 and 4. How do estimates of $f'(x)$ and $f''(x)$ from these approximations compare to those from the cubic splines?

24. The natural spline in Exercise 22 is handicapped by setting the second derivatives to zero at the endpoints, while the correct values are 2 and 2/27. Repeat Exercise 23 for a spline which has the correct values for its second derivative at $x = 1$ and $x = 3$.

▶25. Find $\int_0^2 \text{sech } x \, dx$ by integrating the natural cubic spline that fits at five equispaced points on $[0, 2]$. Compare to the analytical value; also compare to the Simpson's $\frac{1}{3}$ rule value.

26. In Exercise 25, the natural spline assumes that the second derivatives at the extremes of the interval are zero. How does the value of the integral change when conditions 2 and 3 are employed?

27. The best values to use in a cubic spline for the values of its second derivatives at the endpoints should be the correct values of $f''(x)$. Repeat Exercise 25, using the correct values for $f''(x)$ where $f(x) = \text{sech } x$.

28. Use the method of splines to solve the differential equation

$$\frac{dy}{dt} = ty + t^2, \qquad y(0) = -1.$$

Get $y(0.2)$, $y(0.4)$, and $y(0.6)$ from the spline function that has

$$S(0) = -1, \qquad S'(0) = 0, \qquad S''(0) = \left. \frac{d^2y}{dt^2} \right|_{t=0}.$$

Compare with a Runge–Kutta solution to find the errors.

▶29. Use the method of splines to solve

$$\frac{dy}{dt} = t \sin \frac{\pi}{2} y, \qquad y(0) = 1.$$

Get $y(0.2)$, $y(0.4)$, and $y(0.6)$ from the spline that has

$$S(0) = 1, \qquad S'(0) = 0, \qquad S''(0) = \left. \frac{d^2y}{dt^2} \right|_{t=0}.$$

Compare with the analytical solution to find the errors.

30. Use the method of splines as described in Section 10.4 to solve the second-order equation

$$\frac{d^2y}{dt^2} = 3y - 2\left(\frac{dy}{dt}\right) + t, \qquad y(0) = 0, \qquad y'(0) = 1.$$

Compare with the analytical solution at $t = 0.25$, $t = 0.5$.

31. The errors of the spline functions that approximate y in Exercises 28 and 29 increase as h increases. A more accurate method would be to first determine y(0.2), and then recalculate the spline approximation, using that point as the initial condition. After getting y(0.4) from this second spline, the process would be repeated to get y(0.6). Solve Exercises 28 and 29, using this more accurate procedure. Compare the errors with those you previously obtained.

Section 10.5

32. Compute $T_{11}(x)$ and $T_{12}(x)$.

▶33. Show that Eq. (10.22) is satisfied for these values of (m, n): (0, 1), (1, 1), (0, 0), (1, 2).

34. a) Graph $T_5(x)$ on $[-1, 1]$.
    b) Extend the graphs of $T_1(x)$, $T_2(x)$, $T_3(x)$ to the interval $[-2, 2]$. Observe that the maximum magnitude of the Chebyshev polynomials is not equal to one outside of $[-1, 1]$.

Section 10.6

35. Reduce Eq. (10.24) to a fourth-degree economized polynomial and show that its maximum error is 0.000791.

36. The error curve for a truncated Maclaurin-series approximation increases monotonically as x varies from 0 to the ends of the interval. This is not true for an economized power series. Exhibit the form of the error curve by plotting the error of Eq. (10.24) on the interval $[0, 1]$.

▶37. Given

$$\arctan x = x - \frac{x^3}{3} + \frac{x^5}{5} - \frac{x^7}{7} + \frac{x^9}{9} - \dots$$

Plot the error over the interval $[-1, 1]$ when the above series is truncated after the term in x^7. Economize the ninth-degree truncated power-series three times (giving a third-degree expression), and plot its error over the interval $[-1, 1]$.

38. Find the first few terms of the Chebyshev series for sin x by rewriting the Maclaurin series in terms of $T_i(x)$ and collecting terms. Express this as a power series. Compare the errors when both series are truncated to third-degree polynomials.

39. A power-series expansion of $(1 + x/5)^{1/2}$ is

$$1 + \frac{x}{10} - \frac{x^2}{200} + \frac{x^3}{2000} - \frac{x^5}{16,000} + \frac{x^7}{800,000} - \dots$$

Convert this to a Chebyshev series including terms to $T_2(x)$. What is the maximum error of the truncated Chebyshev series? Compare this to the error of the power series truncated after the x^2 term.

40. The Chebyshev series, and economized polynomials as well, require us to approximate the function on the interval $[-1, 1]$ only. Show that an appropriate change of variable will change $f(x)$ on $[a, b]$ to $f(y)$ on $[-1, 1]$. Find the linear relation between x and y that makes this transformation.

Section 10.7

41. Find Padé approximations to the following functions, with numerators and denominators each of degree three: (a) $\sin x$; (b) $\cos x$; ▶ (c) e^x.

42. Compare the errors on the interval $[-1, 1]$ for the Padé approximations of Exercise 41 with the errors of the corresponding Maclaurin series.

43. Express these rational functions in continued-fraction form:

   a) $\dfrac{x^2 - 2x + 2}{x^2 + 2x - 2}$
   b) $\dfrac{2x^3 + x^2 + x + 3}{x^2 - x - 4}$
   c) $\dfrac{2x^4 + 45x^3 + 381x^2 + 1353x + 1511}{x^3 + 21x^2 + 157x + 409}$.

   In each case, compare the number of multiplication and division operations in continued fraction form with initial evaluation of the polynomials by nested multiplication.

44. Express the Padé approximations of Exercise 41 as continued fractions.

45. Estimate the errors of the Padé approximations of Exercise 41 by computing the coefficient of the next nonzero term in the numerator. Compare these estimates with the actual errors at $x = 1$, and at $x = -1$.

▶46. The Chebyshev series for $\cos(\pi x/4)$ is

$$0.851632 - 0.146437 T_2 + 0.00192145 T_4 - 9.965 \times 10^{-6} T_6.$$

   Develop a Padé-like rational function from this by the method of Section 10.7, where the function is $R_{4,2}$.

47. Fike (*Computer Evaluation of Mathematical Functions*, Prentice–Hall, 1968) gives this example of a rational fraction approximation to $\Gamma(1 + x)$ on $[0, 1]$:

$$R_{3,4}(x) = \frac{0.999999 + 0.601781x + 0.186145x^2 + 0.0687440x^3}{1 + 1.17899x - 0.122321x^2 - 0.260996x^3 + 0.0609927x^4}.$$

   How could you determine if this is a minimax approximation? Could one find bounds to the error of the equivalent minimax $R_{3,4}(x)$ if it is not?

APPLIED PROBLEMS AND PROJECTS

48. In many situations, experimental data fit to an exponential relation: $f(x) = ae^{bx}$. To determine the constants through least squares, one frequently fits the function $\ell n\, f = \ell n\, a + bx$ to the data by least squares. In this manner one can find the proper values for the constants b and $\ell n\, a$ quite readily, for the log relation is linear in these terms. However, we might wish to determine the constants a and b of the original function, without changing it to a logarithm. Develop a procedure for doing this, which will require solving a set of nonlinear normal equations. Use the data in Exercise 6 to test your procedure; then compare your values for a and b with those obtained in Exercise 7.

49. E. B. Penrod and K. V. Prasanna (in *Solar Energy*, **6**, 1, 1962), measured the total sun energy reaching their test site in Lexington, Ky., during the winter months. The insolation was recorded both for a vertical receiver and for one tilted at an angle of $25°45'$. Fit the data by least-squares polynomials.

Month	Sept.	Oct.	Nov.	Dec.	Jan.	Feb.	Mar.	Apr.
I_{Tilt}	2177	1952	1545	1215	1140	1523	1615	1713
I_{Vert}	1484	1523	1290	1041	963	1208	1143	1043

Do your equations suggest that one set of data could be predicted from the other?

50. Star S in the Big Dipper (Ursa Major) has a regular variation in its apparent magnitude. Leon Campbell and Laizi Jacchia give data for the mean light curve of this star in their book *The Story of Variable Stars* (Blakeston, 1941). A portion of these data is given below.

Phase	−110	−80	−40	−10	30	80	110
Magnitude	7.98	8.95	10.71	11.70	10.01	8.23	7.86

The data is periodic in that the magnitude for Phase $= -120$ is the same as for Phase $= +120$. The spline functions discussed in Section 10.3 do not allow for periodic behavior. For a periodic function, the slope and second derivatives are the same at the two endpoints. Taking this into account, develop a spline that interpolates the above data.

Other data given by Campbell and Jacchia for the same star are:

Phase	−100	−60	−20	20	60	100
Magnitude	8.37	9.40	11.39	10.84	8.53	7.89

How well do interpolants based on your spline function agree with this second set of observations?

51. The data in Exercise 22 exhibit round-off in the second and in the last entries. By recalculating the spline function, with more and less accurate values of $f(1.5)$ and $f(3.0)$, determine how the values of S are affected by the precision of the data. Also calculate how the precision affects the estimates of $f'(x)$ and $f''(x)$. Compare the effects of the precision of the original data, using a cubic spline, with the effects of precision on the values of $f'(x)$ and $f''(x)$ when central-difference formulas are used.

52. The standard technique to solve a second-order differential equation in Chapter 5 was to rewrite it as two simultaneous first-order equations and then apply a first-order procedure to both equations together. Develop a method of solving two simultaneous first-order equations with cubic splines, then use it to solve the second-order example of Section 10.4. Which approach in using splines on second-order problems is the more accurate?

53. Look up the algorithms used on your computer to compute the functions built into FORTRAN. Classify them into Taylor-series formulas, Chebyshev-polynomial formulas, rational functions, and other types. Which of them are minimax? If more than one computer system is accessible to you, compare the different systems. How does BASIC compute these functions?

Bibliography and References

Acton, F. S. (1970): *Numerical Methods That Work*, Harper and Row, New York.

Ahlberg, J. H., E. N. Nilson, and J. L. Walsh (1967): *The Theory of Splines and their Applications*, Academic Press, New York.

Allen, D. N. (1954): *Relaxation Methods*, McGraw-Hill, New York.

Carnahan, Brice, *et al.* (1969): *Applied Numerical Methods*, Wiley, New York.

Carslaw, H. S. and J. C. Jaeger (1959): *Conduction of Heat in Solids*, 2nd ed., Oxford, London.

Conte, S. D. and Carl de Boor (1972): *Elementary Numerical Analysis*, 2nd ed., McGraw-Hill, New York.

Davis, Philip J., and Philip Rabinowitz (1967): *Numerical Integration*, Blaisdell, Waltham, Mass.

Douglas, J. (1962): Alternating Direction Methods for Three Space Variables, *Numerical Mathematics*, 4:41–63.

Fike, C. T. (1968): *Computer Evaluation of Mathematical Functions*, Prentice-Hall, Englewood Cliffs, N.J.

Forsythe, G. E. and W. R. Wasow (1960): *Finite Difference Methods for Partial Differential Equations*, Wiley, New York.

Forsythe, G. E. and C. B. Moler (1967): *Computer Solution of Linear Algebraic Systems*, Prentice-Hall, Englewood Cliffs, N. J.

Fox, L. (1962): *Numerical Solution of Ordinary and Partial Differential Equations*, Addison-Wesley, Reading, Mass.

Gear, C. W. (1967): The Numerical Integration of Ordinary Differential Equations, *Mathematics of Computation*, 21:146–156.

Hamming, R. W. (1962): *Numerical Methods for Scientists and Engineers*, McGraw-Hill, New York.

Hamming, R. W. (1971): *Introduction to Applied Numerical Analysis*, McGraw-Hill, New York.

Hastings, C. (1955): *Approximations for Digital Computers*, Princeton University Press, Princeton, N.J.

Henrici, P. H. (1964): *Elements of Numerical Analysis*, Wiley, New York.

Hornbeck, R. W. (1975): *Numerical Methods*, Quantum, New York.

International Business Machines Corporation (1968): *System/360 Scientific Subroutine Package*, Version III, Programmer's Manual.

Kopal, Zdenek (1955): *Numerical Analysis*, Wiley, New York.

Kunz, Kaiser S. (1957): *Numerical Analysis*, McGraw-Hill, New York.

Love, Carl H. (1966): *Abscissas and Weights for Gaussian Quadrature*, National Bureau of Standards, Monograph 98.

Muller, D. E. (1956): A Method of Solving Algebraic Equations Using an Automatic Computer, *Math Tables and other Aids to Comp.*, 10:208–215.

Peaceman, D. W., and H. H. Rachford (1955): The Numerical Solution of Parabolic and Elliptic Differential Equations, *J. Soc. Ind. Appl. Math.*, 3: 28–41.

Pipes, L. A. (1963): *Matrix Methods for Engineering*, Prentice-Hall, Englewood Cliffs, N.J.

Ralston, Anthony (1965): *A First Course in Numerical Analysis*, McGraw-Hill, New York.

Ralston, Anthony and H. S. Wilf (1967): *Mathematical Methods for Digital Computers*, Wiley, New York.

Richtmyer, R. D. (1957): *Difference Methods for Initial Value Problems*, Interscience, New York.

Scarborough, J. B. (1950): *Numerical Mathematical Analysis*, Johns Hopkins Press, Baltimore.

Smith, G. D. (1965): *Numerical Solution of Partial Differential Equations*, Oxford, London.

Stewart, G. W. (1973): *Introduction to Matrix Computations*, Academic Press, New York.

Traub, J. F. (1964): *Iterative Methods for the Solution of Equations*, Prentice-Hall, Englewood Cliffs, N.J.

Varga, Richard (1959): *p*-Cyclic Matrices: A Generalization of the Young-Frankel Successive Overrelaxation Scheme, *Pacific J. Math*, 9:617–628.

Varga, R. S. (1962): *Matrix Iterative Analysis*, Prentice-Hall, Englewood Cliffs, N.J.

Wilkinson, J. H. (1963): *Rounding Errors in Algebraic Processes*, Prentice-Hall, Englewood Cliffs, N.J.

Wilkinson, J. H. (1965): *The Algebraic Eigenvalue Problem*, Oxford, London.

Appendixes

Appendix A

Some Basic Information from Calculus

Since a number of results and theorems from the calculus are frequently used in the text, we collect here a number of these items for ready reference, and to refresh the student's memory.

Open and Closed Intervals
We use for the open interval $a < x < b$, the notation (a, b), and for the closed interval $a \leq x \leq b$, the notation $[a, b]$.

Continuous Functions
If a real-valued function is defined on the interval (a, b), it is said to be continuous at a point x_0 in that interval if for every $\varepsilon > 0$ there exists a positive nonzero number δ such that $|f(x) - f(x_0)| < \varepsilon$ whenever $|x - x_0| < \delta$ and $a < x < b$. In simple terms, we can meet any criterion of matching the value of $f(x_0)$ (the criterion is the quantity ε) by choosing x near enough to x_0, without having to make x equal to x_0, when the function is continuous.

If a function is continuous for all x-values in an interval, it is said to be continuous on the interval. A function that is continuous on a closed interval $[a, b]$ will assume a maximum value and a minimum value at points in the interval (perhaps the endpoints). It will also assume any value between the maximum and the minimum at some point in the interval.

Similar statements can be made about a function of two or more variables. We then refer to a domain in the space of the several variables instead of an interval.

Sums of Values of Continuous Functions
When x is in $[a, b]$, the value of a continuous function $f(x)$ must be no greater than the maximum and no less than the minimum value of $f(x)$ on $[a, b]$. The sum of n such values must be bounded by $(n)(m)$ and $(n)(M)$, where m and M are the

minimum and maximum values. Consequently the sum is n times some intermediate value of the function. Hence

$$\sum_{i=1}^{n} f(\xi_i) = nf(\xi) \quad \text{if } a \le \xi_i \le b, \quad i = 1, 2, \ldots, n, \quad a \le \xi \le b.$$

Similarly, it is obvious that

$$c_1 f(\xi_1) + c_2 f(\xi_2) = (c_1 + c_2)f(\xi), \quad \xi_1, \xi_2, \xi \text{ in } [a, b],$$

for the continuous function f when c_1 and c_2 are both equal to or greater than one. If the coefficients are positive fractions, dividing by the smaller gives

$$c_1 f(\xi_1) + c_2 f(\xi_2) = c_1\left[f(\xi_1) + \frac{c_2}{c_1}f(\xi_2)\right] = c_1\left(1 + \frac{c_2}{c_1}\right)f(\xi) = (c_1 + c_2)f(\xi),$$

so the rule holds for fractions as well. If c_1 and c_2 are of unlike sign, this rule does not hold unless the values of $f(\xi_1)$ and $f(\xi_2)$ are narrowly restricted.

Mean-Value Theorem for Derivatives

When $f(x)$ is continuous on the closed interval $[a, b]$, then at some point ξ in the interior of the interval

$$f'(\xi) = \frac{f(b) - f(a)}{b - a}, \quad a < \xi < b,$$

provided, of course, that $f'(x)$ exists at all interior points. Geometrically this means that the curve has at one or more interior points a tangent parallel to the secant line connecting the ends of the curve (Fig. A.1).

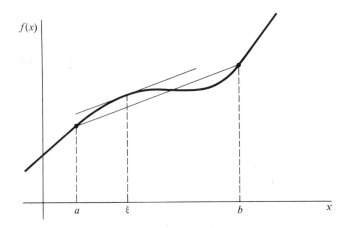

Figure A.1

Mean-Value Theorems for Integrals

If $f(x)$ is continuous and integrable on $[a, b]$, then

$$\int_a^b f(x)\, dx = (b - a)f(\xi), \qquad a < \xi < b.$$

This says, in effect, that the value of the integral is an average value of the function times the length of the interval. Since the average value lies between the maximum and minimum values, there is some point ξ at which $f(x)$ assumes this average value.

If $f(x)$ and $g(x)$ are continuous and integrable on $[a, b]$, and if $g(x)$ does not change sign on $[a, b]$, then

$$\int_a^b f(x)g(x)\, dx = f(\xi)\int_a^b g(x)\, dx, \qquad a < \xi < b.$$

Note that the previous statement is a special case $[g(x) = 1]$ of this last theorem which is called the *second theorem of the mean for integrals*.

Taylor Series

If a function $f(x)$ can be represented by a power series on the interval $(-a, a)$, then the function has derivatives of all orders on that interval and the power series is

$$f(x) = f(0) + f'(0)x + \frac{f''(0)}{2!} x^2 + \frac{f'''(0)}{3!} x^3 + \dots .$$

The above power-series expansion of $f(x)$ about the origin is called a *Maclaurin* series. Note that if the series exists, it is unique and any method of developing the coefficients gives this same series.

If the expansion is about the point $x = a$, we have the Taylor series

$$f(x) = f(a) + f'(a)(x - a) + \frac{f''(a)}{2!} (x - a)^2 + \frac{f'''(a)}{3!} (x - a)^3 + \dots .$$

We frequently represent a function by a polynomial approximation, which we can regard as a truncated Taylor series. Usually we cannot represent a function exactly by this means, so we are interested in the error. Taylor's formula with a remainder gives us the error term. The remainder term is usually derived in elementary calculus texts in the form of an integral:

$$f(x) = f(a) + f'(a)(x - a) + \frac{f''(a)}{2!} (x - a)^2 + \dots + \frac{f^{(n)}(a)}{n!} (x - a)^n$$

$$+ \int_a^x \frac{(x - t)^n}{n!} f^{(n+1)}(t)\, dt.$$

Since $(x - t)$ does not change sign as t varies from a to x, the second theorem of the mean allows us to write the remainder term as

$$\text{Remainder of Taylor series} = \frac{(x - a)^{n+1}}{(n + 1)!} f^{(n+1)}(\xi), \qquad \xi \text{ in } [a, x].$$

The derivative form is the more useful for our purposes. It is occasionally useful to express a Taylor series in a notation that shows how the function behaves at a distance h from a fixed point a. If we call $x = a + h$ in the above, so $x - a = h$, we get

$$f(a + h) = f(a) + f'(a)h + \frac{f''(a)}{2!} h^2 + \ldots + \frac{f^{(n)}(a)}{n!} h^n + \frac{f^{(n+1)}(\xi)}{(n + 1)!} h^{n+1}.$$

Taylor Series for Functions of Two Variables

For a function of two variables, $f(x, y)$, the rate of change of the function can be due to changes in either x or y. The derivatives of f can be expressed in terms of the partial derivatives. For the expansion in the neighborhood of the point (a, b),

$$f(x, y) = f(a, b) + f_x(a, b)(x - a) + f_y(a, b)(y - b)$$

$$+ \frac{1}{2!}[f_{xx}(a, b)(x - a)^2 + 2f_{xy}(a, b)(x - a)(y - b) + f_{yy}(a, b)(y - b)^2]$$

$$+ \ldots.$$

Appendix B

Deriving Formulas by the Method of Undetermined Coefficients

In Chapter 3 we developed formulas for integration and differentiation of functions by replacing the actual function with a polynomial that agrees at a number of points (a so-called *interpolating polynomial*), and then integrating or differentiating the polynomial. These formulas are valuable if one wishes to write computer programs for integration and differentiation, but the most important use is for solving differential equations numerically. Because computers calculate their functional values rather than interpolating in a table, there is less interest today in interpolating polynomials than in earlier times. Hence, there is reason to present an alternative method of deriving the formulas for derivatives and integrals which are needed to solve differential equations.

We shall call the method that we employ in this appendix the *method of undetermined coefficients*. Basically, we impose certain conditions on a formula of desired form and use these conditions to determine values of the unknown coefficients in the formula. Hamming (1962) presents the method in considerable detail.

B.1 DERIVATIVE FORMULAS BY THE METHOD OF UNDETERMINED COEFFICIENTS

Since the derivative of a function is the rate of change of the function relative to changes in the independent variable, we should expect that formulas for the derivative would involve differences between function values in the neighborhood of the point where we wish to evaluate the derivative. It is, in fact, possible to approximate the derivative as a linear combination of such function values. While one can argue that formulas of greatest accuracy will use function values very near to the point in question, the formulas important in practice impose the restriction that only function values at equally spaced x-values are to be used.

For example, we can write a formula for the first derivative in terms of $n + 1$ equispaced points:

$$f'(x_0) = c_0 f(x_0) + c_1 f(x_1) + c_2 f(x_2) + \ldots + c_n f(x_n),$$
$$x_{i+1} - x_i = h = \text{constant.}$$

(B.1)

The more terms we employ, the greater accuracy we shall expect, since more information about the function is being fed in. We shall evaluate the coefficients in the equation by requiring that the formula be exact whenever the function is a polynomial of degree n or less.* (We shall find, throughout this chapter, that the method of undetermined coefficients uses this criterion to develop formulas. It has validity because any function that is continuous on an interval can be approximated to any specified precision by a polynomial of sufficiently high degree. Using polynomials to replace the function also greatly simplifies the work, in contrast to replacing with other functions.)

Let us illustrate the method of undetermined coefficients with a simple case. We shall simplify the notation by defining

$$f_i = f(x_i).$$

If we write the derivative in terms of only two function values, we would have

$$f'_0 = c_0 f_0 + c_1 f_1.$$

(B.2)

We require that the formula be exact if $f(x)$ is a polynomial of degree 1 or less. (The maximum degree is always one less than the number of undetermined constants.) Hence, since the formula is to be exact if $f(x)$ is any first-degree polynomial, it must be exact if either $f(x) = x$ or $f(x) = 1$, for these definitions of $f(x)$ are just two special cases of the general first-degree function $ax + b$. We write Eq. (B.2) for both these cases:

$$f(x) = x, \quad f'(x) = 1, \quad 1 = c_0(x_0) + c_1(x_0 + h),$$
$$f(x) = 1, \quad f'(x) = 0, \quad 0 = c_0(1) + c_1(1).$$

(B.3)

We solve Eqs. (B.3) simultaneously to get $c_0 = -1/h$, $c_1 = 1/h$. Consequently,

$$f'(x_0) \doteq \frac{f_1 - f_0}{h}.$$

(B.4)

The dot over the equal sign in Eq. (B.4) is to remind us that the formula is only an approximation. It is exact if $f(x)$ is a polynomial of degree 1, but not exact if a polynomial of higher degree, or some transcendental function.

---

* Intuitively, it seems reasonable that we can satisfy this criterion, for a polynomial of degree n is determined uniquely by its $n + 1$ coefficients and our formula contains $n + 1$ constants.

Similarly, a three-term formula can be derived by replacing the function with x^2 and x and 1 in

$$f_0' = c_0 f_0 + c_1 f_1 + c_2 f_2.$$

The set of equations to solve is

$$2x_0 = c_0(x_0)^2 + c_1(x_0 + h)^2 + c_2(x_0 + 2h)^2,$$
$$1 = c_0(x_0) + c_1(x_0 + h) + c_2(x_0 + 2h), \qquad \text{(B.5)}$$
$$0 = c_0(1) + c_1(1) + c_2(1).$$

The arithmetic is simplified by letting $x_0 = 0$. That this is valid is readily seen. Imagine the graph of $f(x)$ versus x. The derivative we desire is the slope of the curve at the point where $x = x_0$, which obviously is unchanged by a translation of the axes. Taking $x_0 = 0$ is the equivalent of a translation of axes so that the origin corresponds to x_0. Equations (B.5) become, with this change,

$$0 = c_0(0) + c_1(h)^2 + c_2(2h)^2,$$
$$1 = c_0(0) + c_1(h) + c_2(2h),$$
$$0 = c_0(1) + c_1(1) + c_2(1).$$

Solving, we get $c_0 = -3/2h$, $c_1 = 2/h$, $c_2 = -1/2h$; so a three-term formula for the derivative is

$$f_0' \doteq \frac{-3f_0 + 4f_1 - f_2}{2h}. \qquad \text{(B.6)}$$

We could extend these formulas to include more and more terms, but after a little reflection we can conclude that the original form, Eq. (B.1), is not the best to use. It utilizes only functional values to one side of the point in question, while it would be better to utilize information from both sides of x. After all, the limit in the definition of the derivative is two-sided, and further, we utilize closer and hence more pertinent information by going to both the left and the right.

We expect an improvement over Eq. (B.6) if we begin with

$$f_0' = c_{-1} f_{-1} + c_0 f_0 + c_1 f_1, \qquad \text{(B.7)}$$

where $f_{-1} = f(x_0 - h)$. Adopting the simplification of letting $x_0 = 0$ as before, and taking x^2, x, and 1 for $f(x)$, we have to solve

$$0 = c_{-1}(-h)^2 + c_0(0) + c_1(h)^2,$$
$$1 = c_{-1}(-h) + c_0(0) + c_1(h), \qquad \text{(B.8)}$$
$$0 = c_{-1}(1) + c_0(1) + c_1(1).$$

Completing the algebra, we get $c_{-1} = -1/2h$, $c_0 = 0$, $c_1 = 1/2h$.
The following equation is particularly important:

$$f_0' \doteq \frac{f_1 - f_{-1}}{2h}. \qquad \text{(B.9)}$$

In the next section we will compare its accuracy with Eqs. (B.4) and (B.6). Because the point where the derivative is evaluated is centered among the function values whose differences appear in the formula, it is called a *central difference approximation*. Higher-order central difference approximations to the first derivative, utilizing five or seven or more points, can be derived by this same procedure.

We now apply the method of undetermined coefficients to higher derivatives. We shall discuss only the central difference approximations since they are the more widely used. In terms of values both to the right and left of x_0,

$$f_0'' = c_{-1}f_{-1} + c_0f_0 + c_1f_1.$$

Taking $f(x) = x^2$, x, and 1, we get the relations ($x_0 = 0$),

$$2 = c_{-1}(-h)^2 + c_0(0) + c_1(h)^2,$$
$$0 = c_{-1}(-h) + c_0(0) + c_1(h),$$
$$0 = c_{-1}(1) + c_0(1) + c_1(1).$$

The resulting formula is

$$f_0'' = \frac{f_{-1} - 2f_0 + f_1}{h^2}. \tag{B.10}$$

Equation (B.10), like Eq. (B.9), is particularly useful.

We are not confined to using only functional values in the method of undetermined coefficients.* Suppose we have values of the first derivative as well as functional values. A formula for the second derivative might then be written as

$$f_0'' = c_0f_0 + c_1f_1 + c_3f_0'.$$

As before, we take x^2, x, and 1 for $f(x)$, with $x_0 = 0$:

$$2 = c_0(0)^2 + c_1(h)^2 + c_3(0),$$
$$0 = c_0(0) + c_1(h) + c_3(1),$$
$$0 = c_0(1) + c_1(1) + c_3(0).$$

Solving, we obtain

$$f_0'' = 2\frac{f_1 - f_0 - hf_0'}{h^2}.$$

In the same way, we can derive formulas for the third and fourth derivatives. Derivatives beyond these do not often appear in applied problems. We must use a minimum of four and five terms, however, since only polynomials of degree 3 and 4 have nonzero third or fourth derivatives. For the third-degree formula, complete symmetry with four points is impossible; five-term formulas for both

---

* We must be sure that the set of values is sufficient to determine a polynomial uniquely, however. For example, f_{-1}, f_1, and f_0' will not give a formula for f_0''.

derivatives are therefore given. We present the results only, leaving the derivations as an exercise:

$$f_0''' = \frac{-f_{-2} + 2f_{-1} - 2f_1 + f_2}{2h^3},$$ (B.11)

$$f_0^{iv} = \frac{f_{-2} - 4f_{-1} + 6f_0 - 4f_1 + f_2}{h^4}.$$ (B.12)

B.2   ERROR TERMS FOR DERIVATIVE FORMULAS

In the previous section, we used the method of undetermined coefficients to derive several formulas for the first derivative of a function, utilizing function values at equispaced x-values. These were, using the notation that $f_i = f(x_i)$,

$$f_0' \doteq \frac{f_1 - f_0}{h},$$ (B.13)

$$f_0' \doteq \frac{-3f_0 + 4f_1 - f_2}{2h},$$ (B.14)

$$f_0' \doteq \frac{f_1 - f_{-1}}{2h}.$$ (B.15)

While each of these formulas is not exact, as suggested by the $\doteq$ symbol, we argued heuristically that the error should decrease in each succeeding one. We now wish to develop expressions for the errors.

     We begin with the Taylor-series expansion of $f(x_1) = f(x_0 + h)$ in terms of $x_1 - x_0 = h$:

$$f(x_1) = f(x_0) + hf'(x_0) + \tfrac{1}{2}h^2 f''(x_0) + \tfrac{1}{6}h^3 f'''(x_0) + \ldots.$$

Changing to our subscript notation, and truncating after the term in h, with the usual error term, we have

$$f_1 = f_0 + hf_0' + \tfrac{1}{2}h^2 f''(\xi), \qquad x_0 < \xi < x_0 + h.$$ (B.16)

Solving for f_0', we have

$$f_0' = \frac{f_1 - f_0}{h} - \frac{1}{2}hf''(\xi), \qquad x_0 < \xi < x_0 + h,$$ (B.17)

so that the error term is $-\tfrac{1}{2}hf''(\xi)$ for the derivative formula, Eq. (B.13).

     In Eq. (B.17), the error term involves the second derivative of the function evaluated at a place which is known only within a certain interval. In the majority of applications for numerical differentiation, not only is the point of evaluation uncertain, but the function $f(x)$ is also unknown. If we do not know $f(x)$, we can hardly expect to know its derivatives. We do know, however, that the error involves the first power of $h = x_{i+1} - x_i$, and, in fact, the only way we can change the error is to change h. Making h smaller will decrease h, and in the limit as $h \to 0$ the error will go to zero. Further, as $h \to 0$, $x_1 \to x_o$, and the value of ξ is squeezed into a smaller and smaller interval. In other words, $f''(\xi)$ approaches a fixed value, specifically $f''(x_0)$, as h goes to zero.

The special importance of h in the error term is denoted by a special notation in numerical analysis, the *order relation*. We say the error of Eq. (B.17) is "of order h" and write

$$\text{error} = O(h), \qquad \text{when} \qquad \lim_{h \to 0} (\text{error}) = ch,$$

where c is a fixed value.

To develop the error term for Eq. (B.14), we proceed similarly, except that we write expansions for both f_1 and f_2. It is also necessary to carry terms through h^3 because the second derivatives cancel, resulting in an error term involving the third derivative:

$$f_1 = f_0 + hf_0' + \tfrac{1}{2}h^2 f_0'' + \tfrac{1}{6}h^3 f'''(\xi_1), \qquad x_0 < \xi_1 < x_0 + h;$$
$$f_2 = f_0 + 2hf_0' + \tfrac{1}{2}(2h)^2 f_0'' + \tfrac{1}{6}(2h)^3 f'''(\xi_2), \qquad x_0 < \xi_2 < x_0 + 2h.$$

Note that the values of ξ_1 and ξ_2 may not be identical. If we multiply the first equation by 4, the second by -1, and add $-3f_0$ to their sum, we get

$$-3f_0 + 4f_1 - f_2 = (-3 + 4 - 1)f_0 + 2hf_0' + (2 - 2)f_0'' + \frac{4h^3}{6}[f'''(\xi_1) - 2f'''(\xi_2)].$$

Solving for f_0', we get

$$f_0' = \frac{-3f_0 + 4f_1 - f_2}{2h} + \frac{1}{3}h^2[2f'''(\xi_2) - f'''(\xi_1)], \qquad x_0 < \xi_1 < x_0 + h,$$
$$x_0 < \xi_2 < x_0 + 2h. \qquad \text{(B.18)}$$

The last term of Eq. (B.18) is the error term. As $h \to 0$, the two values of ξ approach the same values. Consequently the error term approaches $\tfrac{1}{3}h^2 f'''(x_0)$. We then conclude that the error of Eq. (B.14) is $O(h^2)$.

For the error of Eq. (B.15), we proceeded similarly. We leave the development as an exercise to show that

$$f_0' = \frac{f_1 - f_{-1}}{2h} - \frac{1}{6}h^2 f'''(\xi), \qquad x_0 - h < \xi < x_0 + h, \qquad \text{(B.19)}$$

$$= \frac{f_1 - f_{-1}}{2h} + O(h^2).$$

The error terms of both Eqs. (B.18) and (B.19) are $O(h^2)$, but the coefficient in Eq. (B.19) is only half the magnitude of the coefficient in Eq. (B.18). The progressive increase in accuracy we anticipated is confirmed.

By similar arguments, we can show that the formulas for third and fourth derivatives, Eqs. (B.11) and (B.12), both have errors $O(h^2)$.

B.3 INTEGRATION FORMULAS BY THE METHOD OF UNDETER-MINED COEFFICIENTS

A numerical integration formula will estimate the value of the integral of the function by a formula involving the values of the function at a number of points in or near the interval of integration. Figure B.1 illustrates the general situation. If we desire to evaluate $\int_a^b f(x)\,dx$, it is obvious that if we could find some average value of the function on the interval $[a, b]$, the integral would be:

$$\int_a^b f(x)\,dx = (b - a)f_{av}.$$

It seems reasonable to assume that f_{av} could be approximated by a linear combination of function values within or near the interval.* We therefore write, similarly to our procedure for derivatives,

$$\int_a^b f(x)\,dx = c_0 f(x_0) + c_1 f(x_1) + \ldots + c_n f(x_n),$$

where the coefficients c_i are to be determined. The points x_i at which the function is to be evaluated can also be left undetermined but, for the formulas which we need to derive, we shall again impose the restriction that they are equally spaced with the value of $x_{i+1} - x_i = \Delta x = h = $ constant. We shall impose the further restriction that the boundaries of the interval of integration, a and b, coincide with two of the x_i-values.

We start with a simple case. Suppose we wished to express the integral in terms of just two functional values, specifically $f(a)$ and $f(b)$. Our formula takes the form

$$\int_a^b f(x)\,dx = c_0 f(a) + c_1 f(b). \qquad (B.20)$$

Obviously, the values of c_0 and c_1 depend on $f(x)$, and probably on the values of a and b as well. The method of undetermined coefficients assumes that $f(x)$ can be approximated by a polynomial over the interval $[a, b]$ and determines c_0 and c_1 so that Eq. (B.20) is exact for all polynomials of a certain maximum degree or less.

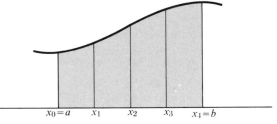

Figure B.1 $x_0 = a$ x_1 x_2 x_3 $x_1 = b$

---

* One could generalize the concept by extending this to include derivatives of the function as well, but we stay with the simpler case.

Equation (B.20) has only two arbitrary constants, so we would not expect that it could be made exact for polynomials of degree higher than the first, since more than two parameters appear in second- or higher-degree polynomials. We therefore force Eq. (B.20) to be exact when $f(x)$ is replaced by either a first-degree polynomial or one of zero degree (a constant function). If it is to be exact for all first-degree polynomials, it must be exact for the very simple function $f(x) = x$. If it is to be exact when the function is any constant, it must hold if $f(x) = 1$. We express these two relations:

$$\int_a^b x \, dx = c_0(a) + c_1(b),$$

$$\int_a^b (1) \, dx = c_1(1) + c_1(1).$$

(B.21)

The integrals of Eqs. (B.21) are easily evaluated, so we have, as conditions that the constants must satisfy,

$$\int_a^b x \, dx = \frac{x^2}{2}\bigg|_a^b = \frac{b^2}{2} - \frac{a^2}{2} = c_0 a + c_1 b,$$

$$\int_a^b dx = x\bigg|_a^b = b - a = c_0 + c_1.$$

Solving these equations gives $c_0 = (b - a)/2$, $c_1 = (b - a)/2$.
    Consequently,

$$\int_a^b f(x) \doteq \frac{b - a}{2}[f(a) + f(b)].$$

(B.22)

This formula is the *trapezoidal rule*, and was derived in Chapter 3 as Eq. (3.10) through an entirely different approach. We have put a dot over the equality sign in Eq. (B.22) to remind ourselves that the relation is only approximately true, for the function $f(x)$ cannot ordinarily be replaced without error by a first-degree polynomial. It is intuitively obvious that unless the interval $[a, b]$ is very small, the error will be considerable.
    We can reduce the error in the above formula by using a higher-degree polynomial to replace $f(x)$, but we would then need to take a linear combination of more than two function values. In fact, we shall have to preserve a balance between the number of undetermined coefficients in the formula and the number of parameters in the polynomial. The number of coefficients hence must be one greater than the degree of the polynomial.
    We now look at the next case, a three-term formula corresponding to replacing $f(x)$ by a quadratic. For three terms, using $x_0 = a$ and $x_2 = b$, with x_1 at the midpoint,

$$\int_a^b f(x) \, dx = c_0 f(a) + c_1 f\left(\frac{a + b}{2}\right) + c_2 f(b).$$

The formula must be exact if $f(x) = x^2$, or $f(x) = x$, or $f(x) = 1$, so

$$\int_a^b x^2 \, dx = \frac{b^3}{3} - \frac{a^3}{3} = c_0 a^2 + c_1 \left(\frac{a+b}{2}\right)^2 + c_2 b^2,$$

$$\int_a^b x \, dx = \frac{b^2}{2} - \frac{a^2}{2} = c_0 a + c_1 \left(\frac{a+b}{2}\right) + c_2 b,$$

$$\int_a^b dx = b - a = c_0 + c_1 + c_2.$$

The solution is $c_0 = c_2 = \frac{1}{3}(b - a)$, $c_1 = \frac{4}{3}(b - a)$. Hence

$$\int_a^b f(x) \, dx \doteq \frac{b-a}{6}\left(f(a) + 4f\left(\frac{a+b}{2}\right) + f(b)\right). \tag{B.23}$$

A more common form of Eq. (B.23) is found by writing $b - a = 2h$ (the interval from a to b is subdivided into two panels) and substituting x_0 for a, x_2 for b, and x_1 for the midpoint. We then have

$$\int_{x_0}^{x_0+2h} f(x) \, dx \doteq \frac{h}{3}(f_0 + 4f_1 + f_2). \tag{B.24}$$

Formula (B.24) is Simpson's $\frac{1}{3}$ rule, a particularly popular formula. Again, it is not exact—as indicated by the dot over the equality sign.

The application of Eqs. (B.22) and (B.24) over an extended interval is straightforward. It is intuitively apparent that the error in these formulas will be large if the interval is not small. (We discuss these errors quantitatively in Section B.5.) To apply these to a large interval of integration and still maintain control over the error, we break the interval into a large number of small subintervals and add together the fromulas applied to the subintervals. When this is done we get the *extended trapezoidal rule:*

$$\int_a^b f(x) \, dx \doteq \frac{b-a}{2n}[f(x_0) + 2f(x_1) + 2f(x_2) + \ldots + 2f(x_{n-1}) + f(x_n)], \tag{B.25}$$

and the extended Simpson's $\frac{1}{3}$ rule:

$$\int_a^b f(x) \, dx \doteq \frac{b-a}{3(2n)}[f(x_0) + 4f(x_1) + 2f(x_2) + 4f(x_3) + \ldots \tag{B.26}$$

$$+ 2f(x_{2x-2}) + 4f(x_{2n-1}) + f(x_{2n})].$$

These forms of the trapezoid rule (Eq. (B.25)) and of Simpson's $\frac{1}{3}$ rule (Eq. (B.26)) are widely used in computer programs for integration. By applying them with n increasing, the error can be made arbitrarily small.* For Simpson's rule, observe that the number of subintervals must be even.

---

* Except for round-off error effects, which eventually will dominate since they are not decreased by small subdivision of the interval and, in fact, may increase as the number of computations increases.

B.4 INTEGRATION FORMULAS USING POINTS OUTSIDE THE INTERVAL

In studying numerical methods to solve differential equations we shall have need for some specific integral formulas that involve function values computed at points outside the interval of integration. Figures B.2, B.3, and B.4 sketch three special cases of importance. In Fig. B.2 the curve that passes through the four points whose abscissas are x_{n-3}, x_{n-2}, x_{n-1}, and x_n is extrapolated to x_{n+1}, and we desire the integral only over the extrapolated interval. Figure B.3 presents the case where the curve passes through four points, and the integral is taken only over the last panel. In Figure B.4 we consider a case where a curve that fits at three points is extrapolated in both directions and integration is taken over four panels.

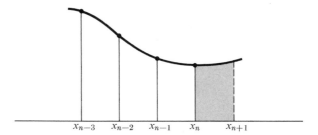

Figure B.2 x_{n-3} x_{n-2} x_{n-1} x_n x_{n+1}

In the derivations of this section it will be convenient to adopt the notation that $f_i = f(x_i)$, so that subscripts on the function indicate the x-value at which it is evaluated.

For the first case (Fig. B.2) we desire a formula of the form

$$\int_{x_n}^{x_n+1} f(x)\, dx = c_0 f_{n-3} + c_1 f_{n-2} + c_2 f_{n-1} + c_3 f_n.$$

With four constants, we can make the formula exact when $f(x)$ is any polynomial of degree 3 or less. Accordingly, we replace $f(x)$ successively by x^3, x^2, x, and 1 to evaluate the coefficients.

It is apparent that the formula must be independent of the actual x-values. To simplify the equations, let us shift the origin to the point $x = x_n$; our integral is then taken over the interval from 0 to h, where $h = x_{n+1} - x_n$:

$$\int_0^h f(x)\, dx = c_0 f(-3h) + c_1 f(-2h) + c_2 f(-h) + c_3 f(0).$$

Figure B.3 x_{n-2} x_{n-1} x_n x_{n+1}

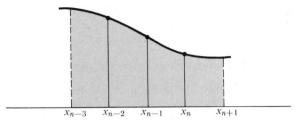

Figure B.4 x_{n-3} x_{n-2} x_{n-1} x_n x_{n+1}

Carrying out the computations by replacing $f(x)$ with the particular polynomials, we have:

$$\frac{h^4}{4} = c_0(-3h)^3 + c_1(-2h)^3 + c_2(-h)^3 + c_3(0),$$

$$\frac{h^3}{3} = c_0(-3h)^2 + c_1(-2h)^2 + c_2(-h)^2 + c_3(0),$$

$$\frac{h^2}{2} = c_0(-3h) + c_1(-2h) + c_2(-h) + c_3(0),$$

$$h = c_0(1) + c_1(1) + c_2(1) + c_3(1).$$

(B.27)

After completion of the algebra we get

$$\int_{x_n}^{x_{n+1}} f(x)\, dx \doteq \frac{h}{24} (-9f_{n-3} + 37f_{n-2} - 59f_{n-1} + 55f_n).$$

(B.28)

For the second case, illustrated by Fig. B.3, we again translate the origin to x_n to simplify. The set of equations analogous to Eqs. (B.27) is:

$$\frac{h^4}{4} = c_0(-2h)^3 + c_1(-h)^3 + c_2(0) + c_3(h)^3,$$

$$\frac{h^3}{3} = c_0(-2h)^2 + c_1(-h)^2 + c_2(0) + c_3(h)^2,$$

$$\frac{h^2}{2} = c_0(-2h) + c_1(-h) + c_2(0) + c_3(h),$$

$$h = c_0(1) + c_1(1) + c_2(1) + c_3(1).$$

These give values of the constants in the integration formula:

$$\int_{x_n}^{x_{n+1}} f(x)\, dx \doteq \frac{h}{24} (f_{n-2} - 5f_{n-1} + 19f_n + 9f_{n+1}).$$

(B.29)

The third case, Fig. B.4, where extrapolation of a quadratic in both directions is involved, leads to the equation:

$$\int_{x_{n-3}}^{x_{n+1}} f(x)\, dx \doteq \frac{4h}{3} (2f_{n-2} - f_{n-1} + 2f_n).$$

(B.30)

Setting up the equations for this case is left as an exercise for the student.

The particular methods for solving differential equations that use the formulas derived in this section are known as *Adams' method* and *Milne's method.*

B.5 ERROR OF INTEGRATION FORMULAS

In each of the formulas that we derived above, we used the symbol $\doteq$ to remind us that the formulas are not exact unless $f(x)$ is in fact a polynomial of a certain degree or less. It is important to derive expressions for the error.

We first determine the error term for the trapezoidal rule over one panel, Eq. (B.22). Begin with a Taylor expansion:

$$F(x_1) = F(x_0) + F'(x_0)h + \tfrac{1}{2}F''(x_0)h^2 + \tfrac{1}{6}F'''(\xi_1)h^3,$$
$$x_0 < \xi_1 < x_1 = x_0 + h. \tag{B.31}$$

Let us define $F(x) = \int_a^x f(t)\, dt$ so that $F'(x) = f(x)$, $F''(x) = f'(x)$, $F'''(x) = f''(x)$. Equation (B.31) becomes, with this definition of $F(x)$,

$$\int_a^{x_1} f(x)\, dx = \int_a^{x_0} f(x)\, dx + f(x_0)h + \tfrac{1}{2}f'(x_0)h^2 + \tfrac{1}{6}f''(\xi_1)h^3.$$

On rearranging, and using subscript notation, we get:

$$\int_a^{x_1} f(x)\, dx - \int_a^{x_0} f(x)\, dx = \int_{x_0}^{x_1} f(x)\, dx = f_0 h + \tfrac{1}{2}f_0'h^2 + \tfrac{1}{6}f''(\xi_1)h^3. \tag{B.32}$$

We now replace the term f_0' by the forward-difference approximation with error term from Section B.2:

$$f_0' = \frac{f_1 - f_0}{h} - \frac{1}{2}hf''(\xi_2), \qquad x_0 < \xi_2 < x_1.$$

Equation (B.32) becomes:

$$\int_{x_0}^{x_1} f(x)\, dx = f_0 h + \tfrac{1}{2}h(f_1 - f_0) - \tfrac{1}{4}h^3 f''(\xi_2) + \tfrac{1}{6}h^3 f''(\xi_1).$$

We can combine the last two terms into $-\tfrac{1}{12}h^3 f''(\xi)$, where ξ also lies in $[x_0, x_1]$. Hence,

$$\int_{x_0}^{x_1} = \frac{h}{2}(f_0 + f_1) - \frac{1}{12}h^3 f''(\xi).$$

The error of the trapezoidal rule over one panel is then $O(h^3)$. This is called the *local error.* Since we normally use a succession of applications of this rule over the interval $[a, b]$ to evaluate $\int_a^b f(x)\, dx$, by subdividing into $n = (b - a)/h$ panels, we need to determine the so-called global error for such intervals of integration.

The global error will be the sum of the local errors in each of the n panels:

$$\text{Global error} = -\tfrac{1}{12}h^3 f''(\xi_1) - \tfrac{1}{12}h^3 f''(\xi_2) - \ldots - \tfrac{1}{12}h^3 f''(\xi_n)$$
$$= -\tfrac{1}{12}h^3 [f''(\xi_1) + f''(\xi_2) + \ldots + f''(\xi_n)].$$

Here the subscripts on ξ indicate the panel in whose interval the value lies. If $f''(x)$ is continuous throughout the interval of integration,

$$\sum_{i=1}^{n} f''(\xi_i) = nf''(\xi), \qquad a < \xi < b.$$

Using $n = (b - a)/h$, we get

$$\text{Global error} = -\frac{1}{12}h^3\left(\frac{b - a}{h}\right) f''(\xi) = -\frac{b - a}{12} h^2 f''(\xi) = O(h^2).$$

Therefore the global error is $O(h^2)$, even though the local error is $O(h^3)$.

Similar treatment shows that the local error for Simpson's $\tfrac{1}{3}$ rule is

$$-\tfrac{1}{90}h^5 f^{iv}(\xi) = O(h^5)$$

and the global error is

$$-\tfrac{1}{180}(b - a)h^4 f^{iv}(\xi) = O(h^4).$$

One can also show that the local error of each of the integration formulas in Section B.4 is $O(h^5)$.

Answers to Selected Exercises

Chapter 1

1. Successive iterates are: 1.5, 1.25, 1.375, 1.4375, 1.40625, 1.42188. For five accurate digits, the error must be ≤ 0.00005, which requires 15 iterations. The error bound is then $\pm 0.5^{15} = \pm 0.00003$.

4. b) 0.4450,   c) 0.9210.

7. 0.55496, 2.2470, −0.80194.

10. −0.458962.

15. b) Roots are 0.618034, 1.73205, −1.61803, −1.73205.

18. 10 iterations. Using Aitken acceleration, 4 iterations are required (two extrapolations).

20. Suitable rearrangements are:

$$x = \sqrt[3]{\frac{-4x^2 + 2x + 5}{2}}, \qquad x = \sqrt{\frac{-2x^3 + 2x + 5}{4}}, \qquad x = \sqrt{\frac{2x + 5}{2x + 4}}.$$

Some of these may converge slowly.

24. Iterates: 1.0, 1.5, 1.75, 1.875, 1.9375, ..., 2.0. Errors: 1.0, 0.5, 0.25, 0.125, 0.0625, ..., 0.0. This is linearly convergent; Aitken acceleration applies.

27. Roots found are 0.790886 and 1.61803, leaving a quadratic whose roots are −0.619223 and −3.78962.

29. After three iterations $r = 4.1$, $s = -5.2$; factor is $x^2 - 4.1x + 5.2$. From the reduced polynomial, the other factor is $x^2 + x + 1$. Zeros of the polynomial are at $2.05 \pm 0.9987i$, $0.5 \pm 8660i$.

33. a) 1.8019, −1.2470, 0.44504.
    d) Two e-values oscillate, indicating two pairs of complex roots. Factors are $x^2 - 4.1x + 5.2$ and $x^2 + x + 1$.

39. Roots are $0 \pm i$, $-2 \pm 4i$. For the particular computer programs used, the execution times were 0.13 sec for the Bairstow method and 0.49 sec for the Muller method.

41. b) Chopped: 0.176×10^6, error 0.6×10^3.
    Rounded: 0.177×10^6, error -0.4×10^3.

43. c) 0.135×10^{-4}. Absolute error -0.24×10^{-7}, relative error -0.18×10^{-2}.

44. b) $(1.32 * 1.98) * 1.01 \neq 1.32 * (1.98 * 1.01)$ with rounding.

47. The results depend on the computer system. With IBM/360, single precision:

    a) 0.9999991   b) 0.9999878   c) 0.9995956

    d) 14.017 after about 1,050,000 terms. It converges because the computer evaluates $1/n$ as 0, for large enough n, in comparison to the current sum.

    e) The value is greater than the result in (d) because less precision is lost due to exponent alignment.

Chapter 2

1. b)
$$AB = \begin{bmatrix} 7 & 0 & -2 \\ 4 & 8 & -9 \\ 2 & -2 & 5 \end{bmatrix}, \qquad BA \text{ is undefined,}$$

$$AC \text{ is undefined,} \qquad Bx = \begin{bmatrix} 3.7 \\ 8.2 \\ -10.9 \\ -11.8 \end{bmatrix}, \qquad Cx \text{ is undefined.}$$

2. a)
$$M^2 = \begin{bmatrix} -1 & 2 & 1 \\ 0 & 0 & 0 \\ 1 & 2 & -1 \end{bmatrix}, \qquad M^3 = \begin{bmatrix} -2 & 0 & 2 \\ 2 & 0 & -2 \\ -2 & 0 & 2 \end{bmatrix}.$$

2. e) $\det(M) = 0$.

6. $x^T = (23/9, 31/18, -19/18) = (2.5555, 1.7222, -1.0555)$.

12. a) $x^T = (1.30, -1.35, -0.275)$.
    b) $x^T = (1.45, -1.59, -0.276)$.
    c) Calculated righthand sides are:

    $(0.02, 1.02, -0.21)$ and $(0.04, 1.03, -0.54)$.

14. $x_1^T = (0.5834, 0.6370, 0.4019, -0.6439)$,
    $x_2^T = (-0.8333, 0.3636, -0.6515, 0.8939)$,
    $x_3^T = (2.419, -0.4401, 2.699, 0.007576)$.

20. b) $x^T = (27.051, 8.2051, 5.7692, 14.872, 53.718)$.

23. b) $7c_1 - 12c_2 - 3c_3 = \bar{0}$.

25. a) $\det(H)$ is very small (about 1.65×10^{-5}); one also cannot avoid a very small divisor in solving the system.
    b) $x^T = (1.11, 0.228, 1.95, 0.797)$.
    c) $x^T = (0.988, 1.42, -0.428, 2.10)$.
    d) $x^T = (1.0000, 0.9995, 1.0017, 0.9990)$.

27. $\det = 51$.

32.
$$H^{-1} = \begin{bmatrix} 16.00 & -120.0 & 239.9 & -139.9 \\ -120.0 & 1199. & -2699. & 1679. \\ 239.9 & -2699. & 6477. & -4198. \\ -139.9 & 1679. & -4198. & 2799. \end{bmatrix}.$$

35. 10.337.

36. a) $x^T = (1592.71, -631.956, -493.653)$.

38. $x^T = (119.5, -47.14, -36.84)$. This is further evidence of ill-condition, in that small changes in the coefficients make large changes in the solution vector.

41. $\dfrac{1}{\text{C.N.}} \dfrac{\|r\|_2}{\|b\|_2} = 1.82 \times 10^{-5}$, $\text{C.N.} \dfrac{\|r\|_2}{\|b\|_2} = 55642$.

$\dfrac{\|e\|_2}{\|\bar{x}\|_2} = 25.5$, $\dfrac{\|e\|_2}{\|x\|_2} = 1.04$.

The relation is verified no matter whether $\|\bar{x}\|$ or $\|x\|$ is used.

45. Using 3-digit arithmetic to compute $\bar{e}$:

$\bar{e}^T = (0.00488, -0.00173, -0.00924)$. $x^T = (0.146, 0.147, -0.165)$.

$(x + e)^T + (0.15088, 0.14527, -0.16592)$, to be compared with $x^T = (0.15094, 0.14525, -0.16592)$. The improvement is remarkable even in one iteration.

48. After six iterations, the Jacobi vector has diverged to $x^T = (3.16, 3.16)$ while with Gauss–Seidel, $x^T = (-154.5, 467.6)$. Gauss–Seidel is diverging much more rapidly.

50. $x^T = (1, -1, 2)$.

54. $(0.72595, 0.50295), (-1.6701, 0.34513)$.

57. $(1.64304, -2.34978), (-2.07929, -3.16173)$.

Chapter 3

2. A sixth-degree polynomial is required to fit exactly. Since the third differences are almost constant [in fact, their variation could be due to round-off errors in $f(x)$], a third-degree polynomial will almost fit all seven points.

4.

x	$\sinh x$	Δf	$\Delta^2 f$	$\Delta^3 f$	$\Delta^4 f$
1.0	1.1752				
		0.1604			
1.1	1.3356		0.0135		
		0.1739		0.0015	
1.2	1.5095		0.0150		0.0005
		0.1889		0.0020	
1.3	1.6984		0.0170		0.0001
		0.2059		0.0021	
1.4	1.9043		0.0191		0.0001
		0.2250		0.0022	
1.5	2.1293		0.0213		0.0002
		0.2463		0.0024	

x	$\sinh x$	Δf	$\Delta^2 f$	$\Delta^3 f$	$\Delta^4 f$
1.6	2.3756		0.0237		0.0005
		0.2700		0.0029	
1.7	2.6456		0.0266		−0.0001
		0.2966		0.0028	
1.8	2.9422		0.0294		0.0005
		0.3260		0.0033	
1.9	3.2682		0.0327		
		0.3587			
2.0	3.6269				

If $\Delta^4 f = 0.0003$,
$$f(0.9) = 1.0271 \text{ vs. } \sinh 0.9 = 1.0265.$$
$$f(2.1) = 4.0219 \text{ vs. } \sinh 2.1 = 4.0219.$$

If $f(1.1)$ were entered as 1.3357, the first two entries in each difference column would change with the first of these by increasing amounts.

8. $f(0.158) = 0.78801$, $f(0.636) = 0.65178$.

11. $0.0625x^3 + 0.14375x^2 + 0.259375x - 0.0254375$.

15. All results are identical $[f(0.385) = 0.74091]$ because all polynomials are the same though of different forms.

18.

Formula	y_0	s	$y(0.92)$
NGF	0.148	2.1	0.3836
NGB	0.518	−0.9	0.3836
GF	0.248	1.1	0.3836
GB	0.370	0.1	0.3836
Bessel	0.309	1.1	0.3836

(The precise value of each $y(0.92)$ was 0.383564).

23. $\binom{s}{2}h^2 f''(\xi)$, $x_0 \leqslant \xi \leqslant x_1$ (based on NGF).
Set the derivative of $[s(s-1)]/2$ equal to zero; this gives a maximum at $s = 0.5$.

28. a) $\Delta(f_0 g_0) = (f_1 g_1) - (f_0 g_0) = f_1 g_1 - f_0 g_1 + f_0 g_1 - f_0 g_0$
$= f_0(g_1 - g_0) + g_1(f_1 - f_0) = f_0 \Delta g_0 + g_1 \Delta f_0.$
   b) $\Delta^n x^n = n!$ by the argument in Section 3.1.
   $\nabla^n E^n x^n = (1 - E^{-1})^n E^n x^n = (E - 1)^n x^n = \Delta^n x^n = n!$
   $\nabla^n x^n = (E^{-1}\Delta)^n x^n = \Delta^n E^{-n} x^n = \Delta^n (E^{-1}x)^n = \Delta^n (x - 1)^n = n!$
   because $(x - 1)^n$ is an nth-degree polynomial whose leading term is x^n.

30. $\Delta^n y_s = (\Delta^n E^{-n}) E^n y_s = \nabla^n y_{s+n}$ so $r = s + n$.

34. $\dfrac{(x - 1)(x + 2)(x - 2)}{-12}(15) + \dfrac{(x - 1)(x + 2)(x + 1)}{12}(9) = -0.5x^3 + 4x^2 + 0.5x - 4.$

36. 1.2218. Actual error $= -0.0004$, estimates $= \begin{cases} -0.00033 & \text{min,} \\ -0.00045 & \text{max.} \end{cases}$

39. $x(25) = 0.6$. Estimated error $= 0.1613$ (min) to 504 (max).

43. $f(1.6, 0.33) = 1.8330$. Compare to the true value of 1.8350.

46. $f(1.62, 0.31) = 1.7524$. (Analytical value $= 1.7515$.)

Chapter 4

1. $f'(0.15)$: a) 2.715, error 0.180   b) 2.862, error 0.033
   c) 2.878, error 0.018.

   $f'(0.19)$: a) 2.170, error 0.116   b) 2.268, error 0.018
   c) 2.278, error 0.008.

   $f'(0.23)$: a) 1.810, error 0.078   b) 1.878, error 0.011
   c) 1.881, error 0.007.

2. $x = 0.15$: a) 0.150 to 0.193   b) 0.017 to 0.034
   c) 0.006 to 0.010.

   $x = 0.19$: a) 0.098 to 0.120   b) 0.010 to 0.017
   c) 0.001 to 0.004.

   $x = 0.23$: a) 0.069 to 0.082   b) 0.006 to 0.010
   c) 7×10^{-4} to 0.002.

   (Round-off errors cause the actual error to fall outside these bounds in some cases.)

9. Use $h^{-n}\Delta^n f$ as an approximation of $f^{(n)}(\xi)$. The best average value for the derivative uses a difference in the center of the range of interpolation. Estimates of errors are:

   $x = 0.15$: a) 0.148   b) 0.015   c) -0.002;
   $x = 0.19$: a) 0.111   b) 0.009   c) -0.002;
   $x = 0.23$: a) 0.075   b) 0.006   c) 0.

   Round-off may invalidate such estimates.

14. Employing central-difference approximations where possible:

x	$f'(x)$	$f''(x)$	x	$f'(x)$	$f''(x)$
0	0.500	-0.03?	0.6	0.434	-0.21
0.1	0.475	-0.03	0.7	0.410	-0.25
0.2	0.492	-0.08	0.8	0.384	-0.27
0.3	0.482	-0.11	0.9	0.356	-0.29
0.4	0.470	-0.19	1.0	0.328	-0.29?
0.5	0.454	-0.19			

17. Successive computations and extrapolations are:

   $\Delta x = 0.08$:   1.9706
                                   1.8865
   $\Delta x = 0.04$:   1.9075   1.8874
                                   1.8875
   $\Delta x = 0.02$:   1.8925

21. a)

h:	0.0001	0.001	0.01	0.1
f':	∞	3.50	3.45	3.52
Error:	∞	-0.074	-0.024	-0.094

   b)

h:	0.0001	0.001	0.01	0.1
f':	5.00	3.50	3.40	3.52
Error:	-1.574	-0.074	0.026	-0.094

26. The error terms are of one order of h greater than expected because

$$\int_0^{n-1} \binom{s}{n} ds = 0 \qquad \text{when } n \text{ is odd.}$$

This can be shown by a change of variable so that the range of integration is symmetrical about the origin; the integrand changes to an odd function for which the integral is zero. For example, with $n = 3$, take $s = t + 1$, giving

$$\int_0^2 \binom{s}{3} ds = \int_{-1}^1 \frac{(t + 1)(t)(t - 1)}{6} dt = \int_{-1}^1 \frac{t^3 - t}{6} dt = 0.$$

28. a) 1.7683   b) 1.7728   c) 1.7904.

31. $h \leq 1.1 \times 10^{-3}$, using the upper bound of the error estimate.

34.

h	Integral	Extrapolations
1.0	0.32045	
		0.34156
0.5	0.33628	0.34154
		0.34136
0.25	0.34009	

38. Using the upper bound to the error, h must be ≤ 0.135. With $h = 0.1$, the integral is 1.718283 (exact is 1.718282).

42. -0.516845.

44. $\int_{x_0}^{x_3} f(x) \, dx = h(3f_0 + \frac{9}{2}\Delta f_0 + \frac{9}{4}\Delta^2 f_0 + \frac{9}{12}\Delta^3 f_0 - \frac{3}{80}\Delta^4 f_0 + \ldots)$.

49. 0.94608.

54. a) 2.8285.

56. $0.785353 + 0.000045 = 0.785398$.

59. b)
$$\begin{pmatrix} 1 & 4 & 2 & 4 & 1 \\ 4 & 16 & 8 & 16 & 4 \\ 2 & 8 & 4 & 8 & 2 \\ 4 & 16 & 8 & 16 & 4 \\ 1 & 4 & 2 & 4 & 1 \end{pmatrix}$$

62. a) 0.140586   b) 0.140587.

65. a) 0.1046.

68. Integrating with x constant, $\Delta x = 0.2$, and taking four intervals in the y-direction: Integral $= 0.64307$.

Chapter 5

1. $y(x) = 1 + x + \frac{3}{2}x^2 + \frac{5}{6}x^3 + \frac{7}{12}x^4 + \frac{17}{60}x^5$;
   $y(0.1) = 1.11589$, $y(0.5) = 2.0245$.

3. $y(0.2) = 1.2015$, $y(0.4) = 1.4130$, $y(0.6) = 1.6487$.

5. With $h = 0.01$, $y(0.1) = 1.11418$, error $= 0.00171$. To reduce the error to 0.00005 (34-fold), we must reduce h 34-fold, or to about 0.00029.

7. $y(0.1) = 1.11587$ (four decimals are correct). The simple Euler method will require about 340 steps and 340 function evaluations for similar accuracy, compared to four steps and eight function evaluations with the modified Euler method.

9.
x:	0.1	0.2	0.3	0.4	0.5
y:	2.2150	2.4630	2.7473	3.0715	3.4394

13.
t:	0.2	0.4	0.6
y:	2.0933	2.1755	2.2493

16. a) $y(0.5) = -0.28326$, error $= -0.00077$.
    b) $y(0.5) = -0.28387$, error $= -0.00016$.
    c) $y(0.5) = -0.28396$, error $= -0.00007$.

21. $y(4)$ predicted $= 4.2229$, $y(4)$ corrected $= 4.1149$. From these, the error in y_c is estimated as 0.0037; equivalently y_c has three correct digits. (The actual error is 0.0082.) Obviously the starting values must have at least three digits of accuracy also.

23.
x:	0.8	1.0	1.2	1.6	2.0
y:	2.3163	2.3780	2.4350	2.5380	2.6294
Error estimate	0.0003	<0.00005	≈ 0	-0.00005	-0.00002

After $x = 1.2$, h was increased to 0.4.

29. By Runge–Kutta:

x:	0	0.2	0.4	0.6
y:	0	0.0004	0.0064	0.0325

By Adams–Moulton:

x:	0.8	0.9	1.0	1.1	1.2	1.25	· 1.3	1.35	1.4
y:	0.1035	0.1669	0.2574	0.3836	0.5581	0.6688	0.7994	0.9541	1.1387

The step size was halved after $x = 0.8$ and $x = 1.2$, to provide accuracy to four decimals.

31. a) $\partial f/\partial y = \sin x$ so $h_{max} = (24/9)/1 = 2.67$.
    b) With $h = 0.267$, D cannot exceed 10×10^{-N} for N-decimal-place accuracy.
    c) If $D = 14.2 \times 10^{-N}$, h cannot exceed $(1/14.2)h_{max}$.

37.

x	y	f	$1 + hf_y$	$\frac{1}{2}h^2 y''$	Est'd. error	Actual error
1.0	1.000	1.000	1.200	0.015	0	0
1.1	1.100	1.331	1.242	0.022	0.019	0.017
1.2	1.233	1.825	1.296	0.035	0.052	0.049
1.3	1.416	2.605	1.368	0.058	0.115	0.111
1.4	1.676	3.933	1.437	0.087	0.234	0.247
1.5	2.069	6.424	1.560	0.173	0.481	0.597
1.6	2.712	11.766	1.782	0.403	1.091	1.834

In the above calculations, average values were used to propagate the errors.

39. $y' = z$, $y(0) = 0$;

$$z' = \frac{M(x)}{EI}(1 - z^2)^{3/2}, \qquad z(0) = 0.$$

41.

t:	0	0.2	0.4	0.6
y:	1	0.982	0.933	0.864
x:	0	0.022	0.093	0.220

44.

x	y	y'
0	1	-1
0.1	0.8950	-1.0995
0.2	0.7802	-1.1956
0.3	0.6561	-1.2847
0.4	0.5236	-1.3629
0.5	0.3840	-1.4263
0.6	0.2389	-1.4715

47.

t	x	Analytical
0	0	0
0.1	0.0715	0.0707
0.2	0.1983	0.1999
0.3	0.2004	0.2026
0.4	-0.0260	-0.0233
0.5	-0.3814	-0.3784
0.6	-0.5934	-0.5977
0.7	-0.4320	-0.4419
0.8	0.1047	0.0924

Chapter 6

1. With $y'(0) = 1.0$, $y(1) = 2.08536$,
   With $y'(0) = 0.8$, $y(1) = 1.81534$.

   Interpolating gives $y'(0) = 0.86271$ for $y(1) = 1.9$;
   Computations with $y'(0) = 0.86271$:

t:	0	0.25	0.50	0.75	1.00
y:	0	0.2388	0.5724	1.0948	1.9000
Anal.:	0	0.2406	0.5750	1.0969	1.9000

5. Using the modified Euler method with $h = 0.2$, initial slope is -2.4270.

t:	0	0.2	0.4	0.6	0.8	1.0
y:	1	0.4494	-0.0119	-0.4449	-0.7863	-1.0000

8. c)

θ:	0	$\pi/8$	$\pi/4$	$3\pi/8$	$\pi/2$
y:	0	0.3851	0.7108	0.9269	1.0

13. c)

θ:	0	$\pi/8$	$\pi/4$	$3\pi/8$	$\pi/2$
y:	-1.0199	-0.9413	-0.7175	-0.3830	0.0105

17. Using $h = 0.2$:

x:	0.2	0.4	0.6	0.8
y:	0.2607	0.7844	1.3592	1.8109

21. a) ± 2.82843 b) ± 3.08459 c) ± 3.17664,
    d) 3.29271, 3.29499.

24. Analytical: $y = Ae^{-x} \sin 2\pi x$

From finite differences, using $h = \frac{1}{8}$ and $k = \pm 6.3623$:

x:	0.25	0.50	0.75
y:	0.7788	−0.0607	−0.4664
Anal.:	0.7788	0	−0.4724

26. $\lambda = 3.49086$, $x^T = (1, 0.9362, 0.7733)$.

29. $\lambda = 3.49086$, -0.83424, 0.34338.

For the inverse, $\lambda = 0.286462$, -1.19869, 2.9122, which are reciprocals of the above values.

Chapter 7

3. $\dfrac{\partial u}{\partial x} = \dfrac{u_{i+1,j} - u_{i-1,j}}{2\Delta x}$;

$$\frac{\partial}{\partial y}\left(\frac{\partial u}{\partial x}\right) = \frac{\dfrac{u_{i+1,j+1} - u_{i-1,j+1}}{2\Delta x} - \dfrac{u_{i+1,j-1} - u_{i-1,j-1}}{2\Delta x}}{2\Delta y}$$

$$= \frac{1}{4h^2}\begin{Bmatrix} -1 & 1 \\ 1 & -1 \end{Bmatrix} u_{i,j}, \text{ when } \Delta x = \Delta y = h.$$

7. Temperatures are:

0.69 2.08 5.56 14.58 38.19
0.69 2.08 5.56 14.58 38.19

10. For points 1 to 8:

74.46, 27.69, 47.86, 65.38, 78.85, 12.90, 23.89, 34.82.

12. Same answers as Exercise 7. Beginning with initial values of 0° at all interior points, 18 iterations are required to achieve a maximum change at any point less than 0.0001°.

15. Same answers as Exercise 7. Beginning with all interior points at 0 and a tolerance for maximum change at any point of 0.0001, the number of iterations required is:

Value of ω:	1.05	1.10	1.15	1.20	1.25	1.30
Number of iterations:	16	14	12	11	12	13

18. a) Each interior point = 0.444.
    b) Values at interior points:

0.211 0.312 0.342 0.312 0.211
0.312 0.472 0.521 0.472 0.312
0.342 0.521 0.577 0.521 0.342
0.312 0.472 0.521 0.472 0.311
0.211 0.312 0.342 0.312 0.211

22. Same values as in Exercise 18(b). The number of iterations to achieve a change in any value less than 0.0001 are:

ω	1.0	1.1	1.2	1.3	1.35	1.4	1.5
No. of iterations	13	11	10	9	10	11	13

The calculated value for ω optimum is 1.33. The smallest number of iterations occurs with that value.

26. There is fourfold symmetry. Values in the upper left quadrant are:

    20.00 20.00 20.00 20.00 20.00
    50.30 59.21 63.16 64.83 65.30
    61.53 72.13 77.33 79.63 80.28

30. Interior points:

    41.08
    19.06 35.17

35. Utilizing symmetry, nine points are sufficient.

    At $r = 1$, clockwise from the center: 10.45, 10.41, 10.27.
    At $r = 2$:   10.79, 10.78, 10.70.
    At $r = 3$:   7.01, 7.06, 7.41.

36. There is sixfold symmetry, so each node is the same, 33.33°.

39. Same answers as Exercise 9. The number of iterations (one vertical and one horizontal traverse) is 8. Compare this to 13 iterations by S.O.R. using optimum ω. S.O.R. with $\omega = 1$ requires 26 iterations.

Chapter 8

2. Using $r = \frac{1}{2}$, at $t = 2.0625$ (10 steps):

x:	0.25	0.50	0.75	1.00	1.25
u:	17.34	32.04	41.86	45.31	41.86
Anal.:	17.70	32.71	42.74	46.26	42.74
Error:	0.36	0.67	0.88	0.95	0.88

6. Values at $t = 268.8$ sec:

	Relative effort	$u(4)$	Error	$u(12)$	Error
$\Delta x = 4,\ r = \frac{1}{2}$:	1	1.25	−0.142	4.75	−0.479
$\Delta x = 4,\ r = \frac{1}{4}$:	2	1.115	−0.007	4.372	−0.101
$\Delta x = 4,\ r = \frac{1}{8}$:	4	1.132	−0.024	4.317	−0.046
$\Delta x = 2,\ r = \frac{1}{2}$:	4	1.115	−0.007	4.372	−0.101
$\Delta x = 1,\ r = \frac{1}{2}$:	16	1.109	−0.001	4.297	−0.026

    One can conclude that decrease of r is more effective than decrease of Δx in increasing accuracy. (It can be shown that $r = \frac{1}{6}$ is particularly favorable, however.)

9.

x:	0.2	0.4	0.6	0.8
At $t = 0.0468$:	−0.00390	−0.00762	−0.01056	−0.01064
At $t = 0.0936$:	−0.00738	−0.01392	−0.01782	−0.01510
At $t = 0.2340$:	−0.01326	−0.02357	−0.02757	−0.02126

12. Values at $t = 15.12$:

x:	0	1	2	3	4
u:	620.1	778.3	833.6	778.3	620.1

14. Computing with $\Delta x = 2''$, $r = \frac{1}{4}$ in the explicit method, the midpoint reaches 200°F in 876.8 sec.

15. After 4, 8, and 20 time steps, values are:

t_4:	0	-0.224	0.504	-0.216	0
t_8:	0	-0.336	0.509	-0.336	0
t_{20}:	0	-0.615	0.870	-0.615	0

    The errors are growing, so the method is unstable.

19. a) $\lambda_{max} = 0.9595$; the formula gives the same value.
    b) $\lambda_{max} = -1.3509$; the formula gives -1.3514. The power method has great difficulties with this matrix.

22. The equations are of the form:

$$u_{i,j,k}^{n+1} = r\left\{\begin{array}{ccc} 1 & 1 & \\ 1 & & 1 \\ 1 & 1 & \end{array}\right\}u^n + (1 - 6r)u_{i,j,k}^n$$

    where $r = k\,\Delta t/c\rho(\Delta x)^2$ and must be less than $\frac{1}{6}$. There are 125 equations to be solved at each time step (assuming no symmetry), and if $r = \frac{1}{8}$, 16 time steps are needed to reach $t = 15.12$.

24. There are three sets of 125 equations, but each set can be decomposed into 25 subsets of 5 equations, each subset being tridiagonal, so at most 1125 coefficients need to be stored; r can be chosen conveniently at unity, so only two time steps will suffice.

Chapter 9

1. The calculated frequency is 409.9 sec^{-1}. If $\Delta x = 70/4$, $\Delta t = 3.050 \times 10^{-4}$ sec. Eight time steps are required for a full cycle. This corresponds to a frequency of 409.84 sec^{-1}.

3. Using $\Delta x = 0.1$, $\Delta t = 0.1$ (for $r = 1$), representative values are:

x:	0.1	0.3	0.5
$t = 0.1$:	0.2939	0.7694	0.9511
$t = 0.7$:	-0.1816	-0.4755	-0.5878
$t = 1.6$:	0.0955	0.2500	0.3090

    These are the same as the analytical results.

6. Representative values [compare these to Exercise 2(c)]:

x:	$L/8$	$L/4$	$3L/8$	$L/2$
$t = \Delta t$:	-0.00651	-0.01140	-0.01434	-0.01531
$t = 2\Delta t$:	-0.01140	-0.02085	-0.02670	-0.02868
$t = 11\Delta t$:	0.01434	0.02670	0.03518	0.03810

Compare these to Exercise 2(d):

$t = \Delta t$:	0.02344	0.4688	0.7031	0.6927
$t = 2\Delta t$:	0.2188	0.4375	0.4115	0.4062
$t = 11\Delta t$:	−0.2031	−0.1615	−0.1406	−0.1302

9. The errors expand (or contract) exactly as in the example in Section 3 or in Exercise 8. Whenever the boundary conditions are specified, the error is zero at that boundary.

12. Some representative values:

(0.25, 0.1021):   0.0186;   (0.50, 0.0884): 0.0216;
(0.75, 0.0791):   0.0144;
(0.25, 0.2042):   0.0333;   (0.50, 0.1768): 0.0408;
(0.75, 0.1581):   0.0269;
(0.25, 0.3062):   0.0402;   (0.50, 0.2652): 0.0546;
(0.75, 0.2372):   0.0360.

Comparison is awkward and would require two-way interpolation, but it is clear that the values do not agree. The values by finite differences are 20 to 50% larger.

14. Beginning with points at $t = 0$ where $x = 0.2$, 0.4, 0.6, and 0.8, intersections of characteristics are:

(0.2993, 0.1006),   (0.4989, 0.1067),   (0.6982, 0.1005),
(0.3915, 0.1946),   (0.5950, 0.2003).

Computing along the characteristics gives $u = 0.5986$, 0.9978, 1.3964, 0.7830, and 1.1900. These are identical to the analytical solution, $u = 2x$.

17. Representative values at points (x, y), with $\Delta t = 0.0018$:

Point:	(0.5, 0.5)	(0.5, 1.0)	(1.5, 3.5)
$t = 0.0036$	0.0625	0.1250	0.1458
$t = 0.0072$	−0.0729	−0.1458	−0.4062
$t = 0.0144$	−0.0911	−0.2135	−0.0078

Chapter 10

2. From the normal equations:

$$a = \frac{N \sum xy - \sum x \sum y}{N \sum x^2 - (\sum x)^2}, \qquad b = \frac{\sum x^2 \sum y - \sum x \sum xy}{N \sum x^2 - (\sum x)^2}.$$

If we substitute $y = \sum Y/N$ and $x = \sum x/N$ into $ax + b = y$, we find that

$$a \frac{\sum x}{N} + b = \frac{\sum Y}{N}.$$

3. $y = 1.908x + 0.025$.

7. $\ell n\, S = 0.18395 + 9.6027 \times 10^{-3} T$,

$S = 1.2019 e^{9.6027 \times 10^{-3} T}$.

10. $y = 0.104x^2 + 1.183x + 0.992$.

12. $f = -6.366 + 24.086x - 3.611x^2 + 0.122x^3$.

When the degree is increased to four, β does not decrease:

Degree: 1   2   3   4
β: 511 532 85   87

17. $$\begin{bmatrix} 4.9523 & -2.7323 & & \\ 0.13 & 1.1568 & 0.18 & \\ & 0.18 & 1.6800 & 0.66 \\ & & 0.4412 & 2.6788 \end{bmatrix} \begin{bmatrix} S_2 \\ S_3 \\ S_4 \\ S_5 \end{bmatrix} = \begin{bmatrix} -0.46275 \\ -0.04359 \\ 0.21212 \\ 0.50507 \end{bmatrix},$$

$S_1 = 5.6923S_2 - 4.6923S_3,$
$S_6 = 1.5758S_5 - 0.57576S_4.$

19.

x:	0.25	1.25	1.75
y (cond 1)	1.2930	3.4684	5.4397
y (cond 2)	1.2771	3.4824	5.7797
y (cond 3)	1.2861	3.4879	5.7615
y (actual)	1.2840	3.4903	5.7646

Condition 3 gives the most accurate interpolated values.

22.

x:	1.5	2.0	2.5	
y':	−0.505	−0.234	−0.162	Spline,
y'':	0.966	0.118	0.168	condition 1.
y':	−0.476	−0.240	−0.164	Spline,
y'':	0.761	0.181	0.122	condition 2.
y':	−0.462	−0.245	−0.161	Spline,
y'':	0.664	0.203	0.132	condition 3.
y':	−0.500	−0.267	−0.167	Central
y'':	0.644	0.268	0.132	differences.
y':	−0.4444	−0.250	−0.1600	Analytical
y'':	0.5926	0.2500	0.1280	values.

The natural spline gives poor results, in comparison to the other techniques.

25. Integral = 1.29919. Analytical value = 1.30176. Simpson's $\frac{1}{3}$ rule = 1.30160. In this example, Simpson's rule has a considerably smaller error.

29.

t:	0.2	0.4	0.6
y:	1.01999	1.07958	1.17524
Anal.:	1.02000	1.07979	1.17765
Value of A:	−0.004931	−0.039018	−0.125819

The values of A were obtained from the relation:

$$A = \frac{2}{h}\left\{\sin\left[\frac{\pi}{2}\left(1 + \frac{h^2}{2} + \frac{Ah^3}{6}\right)\right] - 1\right\}.$$

33. For $(0, 1)$:   integral $= -\sqrt{1 - x^2}\,]^1_{-1} = 0.$

For $(1, 1)$:   integral $= \left(-\dfrac{x}{2}\sqrt{1 - x^2} + \dfrac{1}{2}\sin^{-1} x\right)\Big]_{-1}^{1} = \dfrac{\pi}{2}$.

For $(0, 0)$:   integral $= \sin^{-1} x]_{-1}^{1} = \pi$.

For $(1, 2)$:   (Use integration by parts)

$$\text{integral} = -2x^2\sqrt{1 - x^2} + 4\int x\sqrt{1 - x^2}\,dx$$

$$= (-2x^2\sqrt{1 - x^2} - \tfrac{4}{3}\sqrt{(1 - x^2)^3})]_{-1}^{1} = 0.$$

37. The error of the ninth-degree MacLaurin series is very small near the origin but increases rapidly to 0.04952 at $x = \pm 1$.

The error of P_3 has four maxima (counting the endpoints) and has a maximum error (near $x = \pm 0.4$) of about 0.01975. This is much smaller than the maximum error of the ninth-degree expression.

41. c) $R_{3,3} = \dfrac{1 + (x/2) + (x^2/10) + (x^3/120)}{1 - (x/2) + (x^2/10) - (x^3/120)}$.

Maximum error is -0.00003 at $x = 1$.

46. $R_{4,2} = \dfrac{0.851632 - 0.14202T_2 + 1.1620 \times 10^{-3}T_4}{1 + 5.18619 \times 10^{-3}T_2}$.

Index

absolute error, 7, 41
accuracy of solutions, 106–113
  effect of pivoting, 106
  iterative improvement, 113
  residual, 110
Adams' method, 260–263
Adams–Moulton method, 266–270, 280,
  A.16
  convergence criterion, 273
  criterion of accuracy, 268, 273
  stability, 270
A.D.I. method, 373–376, 412–414
  convergence and stability, 414
  programs, 380, 424–427
Aitken acceleration, 24
algorithm, 5
  Bairstow's method, 44
  Gauss–Seidel iteration, 116
  Gaussian elimination, 82
  halving the interval (bisection), 5
  Jacobi iteration, 115
  linear interpolation (*regula falsi*), 9
  LU decomposition, 92
  modified linear interpolation, 10, 42
  Newton's method, 16
  synthetic division, 19
  $x = g(x)$ form (iteration), 23
alternating-direction implicit method, 373–
  376, 412–414
  program, 380

approximation of functions, 491–503
  Chebyshev expansion, 494
  economized power series, 491
  Padé approximation, 497
  rational function, 495
auxiliary function, 167, 175

backward difference, 154
Bairstow's method, 27–30
  program, 50–51
Bessel polynomial, 161
bisection method, 3
Bolzano method, 3
boundary-value problems, 277, 304–321,
  340
  characteristic value, 318–321
  derivative conditions, 316–318
  finite differences, 310–315
  linear, 306
  nonlinear, 308, 314
  shooting method, 304–310

central difference, 194, 200, A.8
characteristic polynomial, 321
characteristic value, 318
characteristics, 443
Chebyshev polynomials, 488–491, 492
  bounds, 490
  products, 501
Chebyshev series, 493